Tuberculosis Drug Discovery and Development 2019

Tuberculosis Drug Discovery and Development 2019

Editors

Giovanna Riccardi
Claudia Sala

MDPI • Basel • Beijing • Wuhan • Barcelona • Belgrade • Manchester • Tokyo • Cluj • Tianjin

Editors
Giovanna Riccardi
Università degli Studi di Pavia
Italy

Claudia Sala
Fondazione Toscana Life Sciences
Italy

Editorial Office
MDPI
St. Alban-Anlage 66
4052 Basel, Switzerland

This is a reprint of articles from the Special Issue published online in the open access journal *Applied Sciences* (ISSN 2076-3417) (available at: https://www.mdpi.com/journal/applsci/special_issues/Tuberculosis_Drug_Discovery).

For citation purposes, cite each article independently as indicated on the article page online and as indicated below:

LastName, A.A.; LastName, B.B.; LastName, C.C. Article Title. *Journal Name* **Year**, *Article Number*, Page Range.

ISBN 978-3-03943-236-3 (Hbk)
ISBN 978-3-03943-237-0 (PDF)

© 2020 by the authors. Articles in this book are Open Access and distributed under the Creative Commons Attribution (CC BY) license, which allows users to download, copy and build upon published articles, as long as the author and publisher are properly credited, which ensures maximum dissemination and a wider impact of our publications.

The book as a whole is distributed by MDPI under the terms and conditions of the Creative Commons license CC BY-NC-ND.

Contents

About the Editors vii

Claudia Sala, Laurent Roberto Chiarelli and Giovanna Riccardi
Editorial on Special Issue "Tuberculosis Drug Discovery and Development 2019"
Reprinted from: *Appl. Sci.* 2020, 10, 6069, doi:10.3390/app10176069 1

Balachandra Bandodkar, Radha Krishan Shandil, Jagadeesh Bhat and Tanjore S. Balganesh
Two Decades of TB Drug Discovery Efforts—What Have We Learned?
Reprinted from: *Appl. Sci.* , 10, 5704, doi:10.3390/app10165704 5

Christian Lienhardt and Mario C. Raviglione
TB Elimination Requires Discovery and Development of Transformational Agents
Reprinted from: *Appl. Sci.* 2020, 10, 2605, doi:10.3390/app10072605 25

Angelo Iacobino, Lanfranco Fattorini and Federico Giannoni
Drug-Resistant Tuberculosis 2020: Where We Stand
Reprinted from: *Appl. Sci.* 2020, 10, 2153, doi:10.3390/app10062153 31

Paolo Mazzarello
A Physical Cure for Tuberculosis: Carlo Forlanini and the Invention of Therapeutic Pneumothorax
Reprinted from: *Appl. Sci.* 2020, 10, 3138, doi:10.3390/app10093138 49

Catherine Vilchèze
Mycobacterial Cell Wall: A Source of Successful Targets for Old and New Drugs
Reprinted from: *Appl. Sci.* 2020, 10, 2278, doi:10.3390/app10072278 59

Giulia Degiacomi, Juan Manuel Belardinelli, Maria Rosalia Pasca, Edda De Rossi, Giovanna Riccardi and Laurent Roberto Chiarelli
Promiscuous Targets for Antitubercular Drug Discovery: The Paradigm of DprE1 and MmpL3
Reprinted from: *Appl. Sci.* 2020, 10, 623, doi:10.3390/app10020623 95

Vadim Makarov and Katarína Mikušová
Development of Macozinone for TB treatment: An Update
Reprinted from: *Appl. Sci.* 2020, 10, 2269, doi:10.3390/app10072269 115

Caroline Shi-Yan Foo, Kevin Pethe and Andréanne Lupien
Oxidative Phosphorylation—an Update on a New, Essential Target Space for Drug Discovery in *Mycobacterium tuberculosis*
Reprinted from: *Appl. Sci.* 2020, 10, 2339, doi:10.3390/app10072339 127

Young Lag Cho and Jichan Jang
Development of Delpazolid for the Treatment of Tuberculosis
Reprinted from: *Appl. Sci.* 2020, 10, 2211, doi:10.3390/app10072211 161

Raphael Gries, Claudia Sala and Jan Rybniker
Host-Directed Therapies and Anti-Virulence Compounds to Address Anti-Microbial Resistant Tuberculosis Infection
Reprinted from: *Appl. Sci.* 2020, 10, 2688, doi:10.3390/app10082688 173

Rob C. van Wijk, Rami Ayoun Alsoud, Hans Lennernäs and Ulrika S. H. Simonsson
Model-Informed Drug Discovery and Development Strategy for the Rapid Development of Anti-Tuberculosis Drug Combinations
Reprinted from: *Appl. Sci.* **2020**, *10*, 2376, doi:10.3390/app10072376 . . . . **191**

Eduardo M. Bruch, Stéphanie Petrella and Marco Bellinzoni
Structure-Based Drug Design for Tuberculosis: Challenges Still Ahead
Reprinted from: *Appl. Sci.* **2020**, *10*, 4248, doi:10.3390/app10124248 . . . . **211**

Aaron Goff, Daire Cantillon, Leticia Muraro Wildner and Simon J Waddell
Multi-Omics Technologies Applied to Tuberculosis Drug Discovery
Reprinted from: *Appl. Sci.* **2020**, *10*, 4629, doi:10.3390/app10134629 . . . . **231**

Dina Visca, Simon Tiberi, Rosella Centis, Lia D'Ambrosio, Emanuele Pontali, Alessandro Wasum Mariani, Elisabetta Zampogna, Martin van den Boom, Antonio Spanevello and Giovanni Battista Migliori
Post-Tuberculosis (TB) Treatment: The Role of Surgery and Rehabilitation
Reprinted from: *Appl. Sci.* **2020**, *10*, 2734, doi:10.3390/app10082734 . . . . **251**

Rino Rappuoli
Drugs and Vaccines Will Be Necessary to Control Tuberculosis
Reprinted from: *Appl. Sci.* **2020**, *10*, 4026, doi:10.3390/app10114026 . . . . **269**

Carlos Martin, Nacho Aguilo, Dessislava Marinova and Jesus Gonzalo-Asensio
Update on TB Vaccine Pipeline
Reprinted from: *Appl. Sci.* **2020**, *10*, 2632, doi:10.3390/app10072632 . . . . **271**

Dedication

"La tisi non le accorda che poche ore" From "La Traviata" by G. Verdi

Dear readers I spent most of my life in the battle against *Mycobacterium tuberculosis*. My retirement is approaching and this special issue will conclude perfectly my TB research.
I would like to dedicate it to Vita Quinci, one of the most positive student I had in my Lab, affected from Progressive Ossifying Fibrodysplasia, who suddenly died on May 1st. Her last message to me was: "Life is beautiful".

Giovanna Riccardi

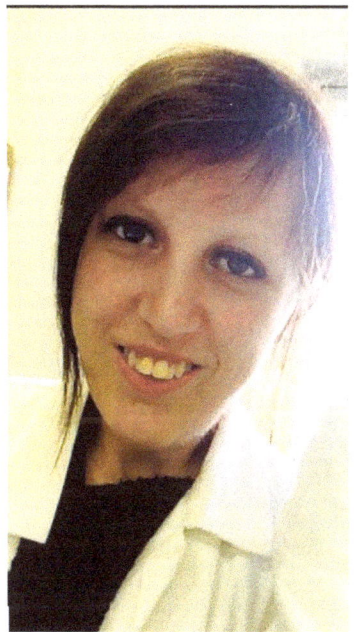

Vita Quinci: "La vita è bella" (life is beautiful)

About the Editors

Giovanna Riccardi 1976—Master Degree cum laude in Biology. 1977–1984—Fellowships aimed to work in Microbiology fields, at the Institute of Microbiology and Plant Physiology, University of Pavia; 1979—EMBO Short-Term fellowship, University of Liverpool (UK); 1984–1998—Researcher at the Department of Genetics and Microbiology, University of Pavia; 1999–2002—Associate Professor of Microbiology at the Department of Experimental, Environmental and Applied Biology, University of Genoa; Since October 2002—Full Professor of Microbiology, at the Department of Genetics and Microbiology (now Department of Biology and Biotechnology), University of Pavia; From January 2010 to December 2012—President of the SIMGBM (Italian Society of General Microbiology and Microbial Biotechnologies); Since August 2011—Member of the European Academy of Microbiology; The two main lines of research she is currently pursuing are:

(1) RESISTANCE MECHANISMS AND TARGET IDENTIFICATION OF NEW DRUGS FOR Mycobacterium tuberculosis;

(2) IDENTIFICATION OF NEW DRUGS AND NEW TARGETS FOR Burkholderia cenocepacia.

FUNDING—Prof. G. Riccardi has obtained several grants from different sources: WHO; CNR-Bilateral Project; CNR-RAISA; MURST 40%; MURST-PRIN-1998, 2001, 2003, 2008; 2017; EC-V, VI and -VII frameworks; Istituto Superiore di Sanità, Fondazione Fibrosi Cistica 2004, 2006, 2009, 2012, 2015. Cystic Fibrosis Foundation-USA 2017. Regarding the EC grants, she was always part of the steering committee. She is the author of several peer-reviewed articles, four book chapters, two International Patent Applications and several national and international communications. H-index Google Scholar: 42. Citations Google Scholar: 6099.

Claudia Sala, PhD. Claudia Sala trained as a molecular microbiologist in the laboratory of Prof. Daniela Ghisotti, at the University of Milan (Italy), where she obtained her degree in Biological Sciences in 2000 and a PhD degree in Genetics and Molecular Biology in 2003. Her PhD thesis dealt with the transcriptional regulation of the furA and katG genes in mycobacteria in response to oxidative stress. In the framework of the EU FP6 "New Medicines for Tuberculosis" (NM4TB) and FP7 "More Medicines for Tuberculosis" (MM4TB), she worked as a post-doctoral fellow in the laboratories of Prof. Stewart Cole, first at the Pasteur Institute in Paris, and then at the Ecole Polytechnique Fédérale de Lausanne, where she was subsequently promoted to senior scientist. She took active part in several research projects, including functional genomics and investigations on the M. tuberculosis Type VII Secretion System, and established the ChIP-Seq and RNA-Seq technologies in M. tuberculosis. She obtained the certificate of Biosafety Level 3 (BSL3) Safety Officer from the Swiss Confederation as well as the FESALA Category B and Category C licenses for performing and directing experiments involving animals. She was the recipient of the Swiss TB Award in 2010. She has recently moved to the Fondazione Toscana Life Sciences (TLS) in Siena, in the group led by Prof. Rino Rappuoli, and performs research on monoclonal antibodies and vaccine development. Her main interests include drug discovery against infectious diseases, vaccinology, host–pathogen interaction and biosafety.

Editorial

Editorial on Special Issue "Tuberculosis Drug Discovery and Development 2019"

Claudia Sala [1], Laurent Roberto Chiarelli [2] and Giovanna Riccardi [2,*]

1. Fondazione Toscana Life Sciences, 53100 Siena, Italy; c.sala@toscanalifesciences.org
2. Department of Biology and Biotechnology "Lazzaro Spallanzani", University of Pavia, 27100 Pavia, Italy; laurent.chiarelli@unipv.it
* Correspondence: giovanna.riccardi@unipv.it; Tel.: +39-0382-985574

Received: 31 August 2020; Accepted: 1 September 2020; Published: 2 September 2020

1. Introduction

Mycobacterium tuberculosis, the etiological agent of human tuberculosis (TB), represents a global challenge to human health since it is the main cause of death by an infectious disease worldwide. Estimations by the World Health Organization (WHO) reported that the tubercle bacillus latently infects approximately one fourth of the world's population, and it is responsible for more than one million deaths every year [1]. Additional factors such as immunodeficiencies [2] and diabetes [3] increase the risk of developing active TB.

The currently available anti-TB therapy is composed of four antibiotics (rifampicin, isoniazid, pyrazinamide and ethambutol) that must be administered for at least 6 months to patients affected by drug-sensitive pulmonary TB [4]. However, the increasing number of multi- and extensively drug-resistant TB cases [5] requires the use of second- or even third-line anti-TB medications, which are characterized by frequent severe side-effects that reduce patients' compliance [6].

Feeding the drug discovery pipeline with the identification of novel chemical entities and promoting the development of those candidate drugs that are presently in clinical trials are therefore of outmost importance in order to shorten anti-TB treatment.

In this Special Issue of Applied Sciences dedicated to "Tuberculosis Drug Discovery and Development", we review the most recent achievements in drug and target identification and present an update on the clinical development of two candidate compounds (macozinone and delpazolid). An overview of technical advancements is included, together with a summary of the anti-TB vaccines which are either in the discovery or clinical phases.

2. The Present Special Issue on "Tuberculosis Drug Discovery and Development 2019"

This Special Issue of Applied Sciences dedicated to "Tuberculosis Drug Discovery and Development" starts with a review article by Bandodkar and colleagues [7] where several drug discovery approaches, which led to the identification of the TB drug candidates currently in the pipeline, are presented. In addition, the authors describe validated and promiscuous drug targets in the context of their experience at AstraZeneca R&D, Bangalore, India. In their article, Lienhardt and Raviglione discuss the ambitious aim of the WHO to reduce TB incidence by 90% by the year 2030 [8], whereas Iacobino and co-authors review the increasing global challenge represented by drug-resistant TB [9]. An interesting paper by Mazzarello closes the initial section by presenting a historical perspective focused on Carlo Forlanini, who invented pneumothorax for TB treatment in 1882, in the same year when Robert Koch identified *M. tuberculosis* as the causative agent of human TB [10].

The Special Issue then features a series of articles dedicated to the most relevant and frequently explored drug targets: the cell wall of *M. tuberculosis* is reviewed by Vilchèze [11], DprE1 and MmpL3 are described by Degiacomi and co-workers [12], and the oxidative phosphorylation pathways are presented by Foo and colleagues [13]. In addition, Gries et al. report on the most recent advances in host-directed therapies and anti-virulence compounds, which could represent a helpful complement to current anti-TB approaches [14]. In the context of additional approaches to standard antibiotic treatment, an article by Visca et al. reviews the importance of post-TB treatment with the roles of surgery and rehabilitation [15]. Two candidate compounds which are in the advanced stages of development complete the section dedicated to novel medications: macozinone [16] and delpazolid [17].

Three papers describe state-of-the-art approaches to TB drug discovery. The first one by van Wijk and co-authors deals with quantitative pharmacology models including machine learning and artificial intelligence [18]; the second one by Bruch and colleagues discusses structure- and target-based approaches to TB drug design [19]; the last one explores the –omics technologies and how they have been exploited so far in TB drug discovery [20].

The Special Issue closes with an Editorial by Rappuoli who highlights the need for new drugs and vaccines to eradicate TB [21] and introduces the final article by Martin and colleagues [22] who wrote an update on the TB vaccine pipeline.

Overall, this Special Issue has gathered together most of the globally known TB professionals, including clinicians, academic staff as well as researchers from the private sector, and provides an extensive overview of the currently available tools and compounds that can help in the fight against TB.

3. Conclusions

The research work described in these sixteen reviews that constitute the Applied Sciences Special Issue provides an extremely useful example of the achieved results in the field of tuberculosis drug development. Moreover, readers can find information regarding the new approaches that are in progress to identify new antitubercular drugs, as well as novel drug targets.

We are extremely grateful to all of the authors for their excellent contribution to this Special Issue dedicated to Tuberculosis. We would also like to thank the reviewers who carefully evaluated the submitted manuscripts. Finally, special thanks to Ms. Marin Ma for her technical support.

Funding: This research received no external funding.

Conflicts of Interest: The authors declare no conflict of interest.

References

1. WHO. *Global Tuberculosis Report 2019*; WHO: Geneva, Switzerland, 2019.
2. du Bruyn, E.; Peton, N.; Esmail, H.; Howlett, P.J.; Coussens, A.K.; Wilkinson, R.J. Recent progress in understanding immune activation in the pathogenesis in HIV-tuberculosis co-infection. *Curr. Opin. HIV AIDS* **2018**, *13*, 455–461. [CrossRef] [PubMed]
3. Ferlita, S.; Yegiazaryan, A.; Noori, N.; Lal, G.; Nguyen, T.; To, K.; Venketaraman, V. Type 2 Diabetes Mellitus and Altered Immune System Leading to Susceptibility to Pathogens, Especially *Mycobacterium tuberculosis*. *J. Clin. Med.* **2019**, *8*, 2219. [CrossRef] [PubMed]
4. WHO. *The End-TB Strategy*; WHO: Geneva, Switzerland, 2014.
5. Mabhula, A.; Singh, V. Drug-resistance in *Mycobacterium tuberculosis*: Where we stand. *MedChemComm* **2019**, *10*, 1342–1360. [CrossRef] [PubMed]
6. Pontali, E.; Raviglione, M.C.; Migliori, G.B. Regimens to treat multidrug-resistant tuberculosis: Past, present and future perspectives. *Eur. Respir. Rev. Off. J. Eur. Respir. Soc.* **2019**, *28*. [CrossRef] [PubMed]
7. Bandodkar, B.; Shandil, R.; Bhat, J.; Balganesh, T. Two Decades of TB Drug Discovery Efforts—What Have We Learned? *Appl. Sci.* **2020**, *10*, 5704. [CrossRef]
8. Lienhardt, C.; Raviglione, M. TB Elimination Requires Discovery and Development of Transformational Agents. *Appl. Sci.* **2020**, *10*, 2605. [CrossRef]
9. Iacobino, A.; Fattorini, L.; Giannoni, F. Drug-Resistant Tuberculosis 2020: Where We Stand. *Appl. Sci.* **2020**, *10*, 2153. [CrossRef]

10. Mazzarello, P. A Physical Cure for Tuberculosis: Carlo Forlanini and the Invention of Therapeutic Pneumothorax. *Appl. Sci.* **2020**, *10*, 3138. [CrossRef]
11. Vilchèze, C. Mycobacterial Cell Wall: A Source of Successful Targets for Old and New Drugs. *Appl. Sci.* **2020**, *10*, 2278. [CrossRef]
12. Degiacomi, G.; Belardinelli, J.; Pasca, M.; De Rossi, E.; Riccardi, G.; Chiarelli, L. Promiscuous Targets for Antitubercular Drug Discovery: The Paradigm of DprE1 and MmpL3. *Appl. Sci.* **2020**, *10*, 623. [CrossRef]
13. Foo, C.; Pethe, K.; Lupien, A. Oxidative Phosphorylation—An Update on a New, Essential Target Space for Drug Discovery in *Mycobacterium tuberculosis*. *Appl. Sci.* **2020**, *10*, 2339. [CrossRef]
14. Gries, R.; Sala, C.; Rybniker, J. Host-Directed Therapies and Anti-Virulence Compounds to Address Anti-Microbial Resistant Tuberculosis Infection. *Appl. Sci.* **2020**, *10*, 2688. [CrossRef]
15. Visca, D.; Tiberi, S.; Centis, R.; D'Ambrosio, L.; Pontali, E.; Mariani, A.; Zampogna, E.; van den Boom, M.; Spanevello, A.; Migliori, G. Post-Tuberculosis (TB) Treatment: The Role of Surgery and Rehabilitation. *Appl. Sci.* **2020**, *10*, 2734. [CrossRef]
16. Makarov, V.; Mikušová, K. Development of Macozinone for TB treatment: An Update. *Appl. Sci.* **2020**, *10*, 2269. [CrossRef]
17. Cho, Y.; Jang, J. Development of Delpazolid for the Treatment of Tuberculosis. *Appl. Sci.* **2020**, *10*, 2211. [CrossRef]
18. van Wijk, R.; Ayoun Alsoud, R.; Lennernäs, H.; Simonsson, U. Model-Informed Drug Discovery and Development Strategy for the Rapid Development of Anti-Tuberculosis Drug Combinations. *Appl. Sci.* **2020**, *10*, 2376. [CrossRef]
19. Bruch, E.; Petrella, S.; Bellinzoni, M. Structure-Based Drug Design for Tuberculosis: Challenges Still Ahead. *Appl. Sci.* **2020**, *10*, 4248. [CrossRef]
20. Goff, A.; Cantillon, D.; Muraro Wildner, L.; Waddell, S. Multi-Omics Technologies Applied to Tuberculosis Drug Discovery. *Appl. Sci.* **2020**, *10*, 4629. [CrossRef]
21. Rappuoli, R. Drugs and Vaccines Will Be Necessary to Control Tuberculosis. *Appl. Sci.* **2020**, *10*, 4026. [CrossRef]
22. Martin, C.; Aguilo, N.; Marinova, D.; Gonzalo-Asensio, J. Update on TB Vaccine Pipeline. *Appl. Sci.* **2020**, *10*, 2632. [CrossRef]

© 2020 by the authors. Licensee MDPI, Basel, Switzerland. This article is an open access article distributed under the terms and conditions of the Creative Commons Attribution (CC BY) license (http://creativecommons.org/licenses/by/4.0/).

Review

Two Decades of TB Drug Discovery Efforts—What Have We Learned?

Balachandra Bandodkar [1], Radha Krishan Shandil [2], Jagadeesh Bhat [3] and Tanjore S. Balganesh [3,*]

[1] Pharmaron Beijing Co., Ltd., 6, Taihe Road, BDA, Beijing 100176, China; balachandra.bandodkar@pharmaron.com
[2] Foundation for Neglected Disease Research [FNDR], Plot 20A, KIADB Industrial Area, Doddaballapur, Bengaluru 561203, India; rk.shandil@fndr.in
[3] Gangagen Biotechnologies P Ltd., #12, 5th Cross, Raghavendra Layout, Tumkur Road, Yeshwantpur, Bengaluru 560022, India; bjagadeesh@gangagen.com
* Correspondence: balganesh@gangagen.com

Received: 3 May 2020; Accepted: 11 August 2020; Published: 17 August 2020

Abstract: After several years of limited success, an effective regimen for the treatment of both drug-sensitive and multiple-drug-resistant tuberculosis is in place. However, this success is still incomplete, as we need several more novel combinations to treat extensively drug-resistant tuberculosis, as well newer emerging resistance. Additionally, the goal of a shortened therapy continues to evade us. A systematic analysis of the tuberculosis drug discovery approaches employed over the last two decades shows that the lead identification path has been largely influenced by the improved understanding of the biology of the pathogen *Mycobacterium tuberculosis*. Interestingly, the drug discovery efforts can be grouped into a few defined approaches that predominated over a period of time. This review delineates the key drivers during each of these periods. While doing so, the author's experiences at AstraZeneca R&D, Bangalore, India, on the discovery of new antimycobacterial candidate drugs are used to exemplify the concept. Finally, the review also discusses the value of validated targets, promiscuous targets, the current anti-TB pipeline, the gaps in it, and the possible way forward.

Keywords: tuberculosis; *Mycobacterium tuberculosis*; drug discovery; drug development; target-based screening; phenotypic screening; antituberculosis agents; antimycobacterial; anti-TB drug pipeline; privileged targets; promiscuous targets; lead generation

---

Chemotherapy for the treatment of tuberculosis has evolved over a period of several decades, starting from the 1950s. The discovery of drugs with superior effectiveness that are part of the current regimen has been facilitated through the success of several novel approaches and technologies. While the medical need has at times hastened newer approaches to be adopted, the main driver for the improvement has been the increased understanding of the biology of the pathogen, as well as its interaction with the human host. In this review, we try to discuss the discovery of new drugs in groups, with the groups sharing a common key driver that precipitated the changes of the treatment regimen. As 'newer aspects of the biology' of the pathogen became known, they provided opportunities to build novel drug discovery approaches. The paper also uses as examples the approaches, the results, and the learning gathered as a part of the antituberculosis (anti-TB) program at Astra Zeneca, R and D, Bangalore, India (AZI).

Several of the drugs that have been discovered are based on the learning that followed the introduction of each of the new compounds into the drug regimen. The data have several pointers; while each of the successful drugs was discovered building on the then state of the knowledge, the newer drugs themselves also helped in the further understanding of the pathogen biology. This is even reflected in the current set of drugs that are in late-stage development or have been recently introduced into the anti-TB regimen.

The anti-TB drugs currently in use, and those in the late stages of clinical development, can be broadly pooled into the following groups:

1. Serendipitous drug discovery—early chemotherapy;
2. Modification of drug scaffolds;
3. Revisiting targets that have clinically validated drugs against them (referred to as established targets);
4. Target-based screening;
5. Phenotypic screening.

1. Serendipitous Drug Discovery: Early Chemotherapy

The first successful chemotherapy and cure of an infectious disease is indeed the discovery and design of the 'first line' therapy for the treatment of tuberculosis—the design and development of which was completed in the 1960s. Even today, this regimen is the therapy of choice for treating drug-sensitive tuberculosis (DSTB). The drugs in the first-line treatment for DSTB, isoniazid, rifampicin, pyrazinamide, and ethambutol became anti-TB drugs based on their activity on *Mycobacterium tuberculosis* (MTB) cells in vitro, followed by testing in animal models and their rapid introduction into humans [1]. This progression was driven by the medical need, as no chemotherapy existed before these drugs were discovered.

It is interesting to note that two biological observations and a hypothesis on potential 'chemical structures' that may interfere with the observed biological process were the first starting points of anti-TB drug discovery. Aspirin was shown to be a potent stimulator of the TB bacilli's 'oxygen consumption'; analogs of aspirin were then postulated to be inhibitors of this process. This led to the synthesis of a number of aspirin-like structures, of which para-aminosalicylic acid (PAS) became a successful anti-TB drug [1–3]. The second observation was that niacin helped in the recovery of guinea pigs infected with MTB, as well as the observation that niacin helped in faster recovery of TB patients, raising the possibility that niacin was acting as a 'vitamin' [4,5]. Chemical synthesis focused on making derivatives of niacin led to the design of three anti-TB drugs, namely isoniazid (Inh), pyrazinamide (Pza) and ethionamide (Eth) [6]. Two of these are even today the most potent drugs for the treatment of TB. This early period of chemotherapy also included extensive search for natural products with antibacterial activity; Rifamycin and Streptomycin were natural products that showed potent activity against MTB cells and were also introduced into the treatment of TB.

In 1979, Mitchison observed that 10 to 12 drugs were available for the treatment of tuberculosis, which could be classified in terms of their effectiveness [7]. The choice of combinations was dictated by animal toxicity of the individual drugs and in human trials, which assessed time taken to the sputum negative state, cure as reflected by relapse rates, the emergence of resistant strains, and compliance. Once an effective combination had been proven, it became to be referred to as the 'Short Course Chemotherapy' and was adopted systematically all over the world [8]. This was the first successful conquering of an infectious disease. In about 20 years, TB patients went from complete helplessness to an effective cure achieved with drug treatment.

The key learning from this pioneering era was as follows:

1. In vitro MIC (Minimum Inhibitory Concentration) was insufficient to predict efficacy in humans. Requirement for a combination therapy of drugs.
2. The best regimen required six months of treatment to achieve cure.
3. Each of the drugs in the combination had a unique role to play in leading to the cure.

The cure was defined as the lack of relapse.

The last two points remain, to date, the biggest challenge in finding and developing novel combinations. The era of 1950 to 1980 had 10 drugs, of which the most efficacious combination of four drugs was identified through rapid testing in humans. In spite of more than two decades of sustained research into the biology of MTB, the traits of a new drug that could contribute to both the 'cure' and the shortening of therapy are still unknown.

2. Modification of Drug Scaffolds: Analogs of Known Drugs or the Literature Compounds

The initiation points for this approach were natural product scaffolds or scaffolds known to have activity against MTB cells in vitro but did not possess drug-like properties. Efforts to address this were mainly medicinal-chemistry driven, focused on understanding structure–activity relationship (SAR) on potency and animal toxicity. The advances in chemistry in terms of novel reactions and the use of combinatorial chemistry resulted in rapid diversification of key scaffolds to yield potent analogs. Some of the examples of scaffolds were the nitroimidazoles and several newer rifampicins, isoniazid, and ethambutol analogs [9–12].

2.1. Nitroimidazole as a Starting Point

Among the diverse scaffolds tested in this approach, the 'nitroimidazole' starting point has been the most successful. The antibiotic 5-nitroimidazole was used for treating bacterial infections of the gut and was also shown to be active on anaerobic bacteria [13]. Metronidazole, a drug still in use for the treatment of amoebic infections, was one of the first successful derivatives of 5-nitroimidazole. CGI-17341, a bicyclic imidazofuran, was one of the derivatives and was found to be a potent anti-Tb molecule [14]. Continued chemistry on this molecule led to several analogs, among which PA-824 (Pretomanid) [15], OPC-67683 (Delamanid) [16] have recently been registered as anti-TB drugs and are constituents of the current Multi drug resistant (MDR) regimen [15,16]. This progression is shown in Figure 1.

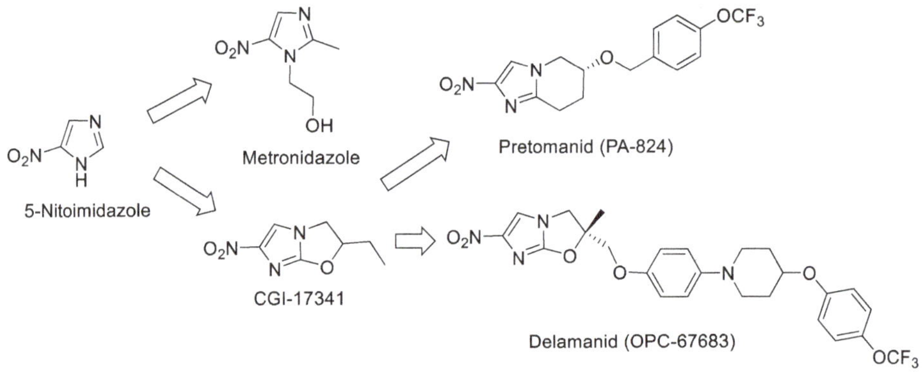

Figure 1. The progression of 5-Nitoimidazole derivatives.

2.2. Rifampicin Analogs

Rifampicin was obtained through the chemical modification of the natural product rifamycin [17]. Derivatives of rifampicin, like rifabutin, were synthesized and found to have favorable properties in terms of compatibility with anti-HIV drugs but could not overcome the cross-resistance with rifampicin [18].

2.3. Ethambutol Derivatives

Increased throughput in chemistry also contributed to finding new leads. A combinatorial library created around ethambutol led to the discovery of a clinical candidate SQ-109 [19,20]. SQ-109 was found to be active even on ethambutol-resistant strains of MTB. SQ109 has been shown to target the Mycobacterial Membrane Protein Large 3 (Mmpl3) [21]. It is interesting to note that ethambutol does not inhibit Mmpl3.

2.4. Isoniazid Analogs

Among the many Inh analogs synthesized, Sudoterb (LL 3858) was successfully progressed to Phase 1 [22,23]. Inh, because of its simplicity of structure, remains an attractive starting point for analog-based discovery.

The approach involving modifications of existing drug scaffolds has two important aims: firstly, to discover novel analogs that are active on MTB strains resistant to the parent drug, and secondly, to design 'drug-friendly' molecules. This approach has yielded two, drugs and a third is in development: Pretomanid (Pre), Delamanid (Del), and SQ-109, respectively.

The key learning from these examples is that sustained effort can lead to useful drugs, even though the starting scaffolds have issues.

3. Revisiting Established Targets: Revisiting Targets Proven as Druggable by Using Broad-Spectrum Compounds

TB was declared a global emergency by WHO in 1996 [24]. The emergence of drug-resistant TB and the complete lack of drugs capable of treating patients with MDR TB led the drug development community to investigate alternate approaches to rapidly induct novel drugs into the regimen. This prompted investigations into the feasibility of introducing the existing broad-spectrum antibacterial drugs into the TB treatment regimen. Several antibacterial classes that have been shown to be active on MTB in vitro were investigated in clinical trials.

This approach, which is now classified as 'repurposing', has successfully delivered new options for TB treatment. The key classes that have added to the anti-TB portfolio are the following:

3.1. Protein-Synthesis Inhibitors

Streptomycin, a protein-synthesis inhibitor and a well-established anti-TB drug, has been shown to be effective in treating MTB patients, but its use is limited because of it not being an oral drug and the toxicities associated with it for prolonged use [25]. Several novel protein-synthesis inhibitors that have been approved as antibacterials were also tested for their antimycobacterial activity. The oxazolidinone [26] class of compounds, despite its limitations of myelotoxicity, hold a significant position in the treatment of MDR and Extensively drug-resistant tuberculosis (XDR)TB in the current anti-TB treatment regimen. Linezolid [27,28] is currently a part of the drug regimen for the treatment of MDR TB, XDR and non-responding TB (NRTB), while newer oxazolidinones like Posizolid [29–31] and Sutezolid [32,33] have been tested in advanced clinical trials. Newer oxazolidinones like the Delpazolid [34] and Contezolid [35] are also in clinical trials.

3.2. Beta Lactams as Antimycobacterials

Several broad-spectrum antibacterials like meropenem, a beta-lactam, have also been shown to have activity against MTB in in vitro models, as well as in studies measuring Early Bactericidal Activity (EBA) in humans [36].

The key 'unknowns' in the development of broad-spectrum antibiotics as anti-TB treatment are twofold:

- Priming of resistance against the antibiotic among normal gut bacteria due to the long-term treatment required for TB. This priming could lead to the selection of resistant mutants and subsequently spread of resistance to other pathogens in the gut.
- Effect of the antibiotic on 'latent MTB bacteria': Latent bacteria could also be primed, leading to probability of a drug-resistant infection on reactivation.

Despite these concerns, even after several years of the use of rifampicin for the treatment of Drug sensitive (DS) TB and moxifloxacin for the treatment of MDR TB, the extent of 'priming' caused is not clear, and some of the fears could well be unfounded. This could also be a reflection of the use of these drugs only in combinations or our inability to monitor the impact systematically.

The key learning from this approach of including broad-spectrum antibiotics into the combination regimen to treat MTB patients has been as follows:

- Drugs with several new targets, like the ones discussed above, can be introduced into novel combinations, thus enabling the treatment of drug resistant (DR) MTB patients.
- The effectiveness of drugs like moxifloxacin or linezolid establish the vulnerability of the target, thus promoting the search for new compounds that can inhibit the same target. This approach of revisiting 'established/vulnerable targets' continues to be explored by using several newer assets, like new libraries, which are novel screening formats, including those enabled by the availability of the molecular structures. Two novel compounds that have entered clinical development, GSK070 shown in Figure 2a [37] and SPR20 shown in Figure 2b [38], are examples of this approach. GSK 3036656 (GSK-070) belongs to the oxaborole class of compounds and has been shown to be a Leucine tRNA synthase inhibitor. The compound is currently in Phase 2 clinical trials [37]. SPR 720 is a GyrB ATPase inhibitor that belongs to the benzimidazole class. The molecule is also in Phase 2 clinical trials [38,39]. A very recent report shows that SPR 720 obtained an orphan disease status from FDA to treat non-tuberculosis mycobacteria (NTM) [40].

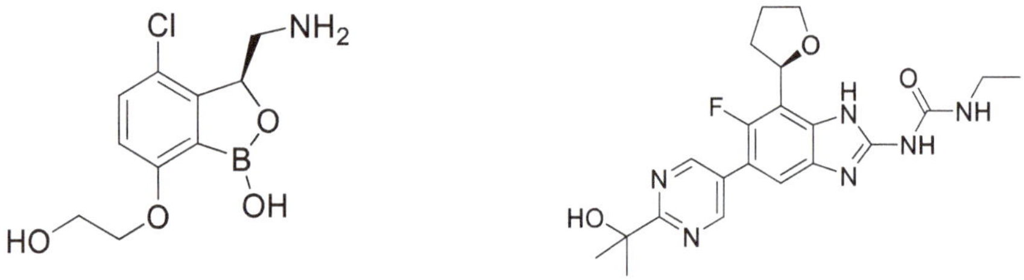

Figure 2. Two novel compounds that have entered clinical development (a) GSK070 and (b) SPR20.

3.3. Gyrase Inhibitors

The fluoroquinolone class (Moxifloxacin, Levofloxacin, Ofloxacin, and Gatifloxacin) of compounds are potent inhibitors of the DNA gyrase enzyme and are proven antibacterials. Several of these were shown to be active on the MTB bacilli in vitro. Researchers at the National Tuberculosis Institute, India, tested the usefulness of ofloxacin as a part of the anti-TB regimen and showed it to be effective in the clinical trial [41]. Multiple members of this class of compounds have undergone clinical trials as part of an anti-TB regimen; moxifloxacin [42] is now a part of the standard regimen to treat drug-resistant TB infections. Section 3.4 covers the target-based TB drug discovery efforts at AstraZeneca, with major emphasis on gyrase inhibitors.

3.4. The AstraZeneca India (AZI) Effort

Gyrase as a target: One of the favorite targets for anti-TB drug discovery is the 'gyrase enzyme'. This is because of the multiple steps involved in the mechanistic of the 'negative supercoiling' enzyme reaction, several steps of which have been shown to be inhibitable [43]. Additionally, the availability of the several crystal structures of the enzyme has also helped in developing diverse screening approaches, as well as in building SAR of the identified inhibitors.

AZI employed multiple 'hit' generation approaches like high-throughput screening (HTS) of the AZ library, fragment library screening, targeted library screening, and pharmacophore-based screening, as well as virtual screening, in the quest for robust novel inhibitors. Shirude and Hameed [44] reviewed the features of the diverse set of inhibitors identified by the different groups. Several novel chemical entities were identified and are being investigated further (Table 1).

Table 1. Gyrase inhibitors identified at AstraZeneca (AZ).

Approach	Target	Inhibitor Series	Mechanism of Inhibition	Reference
Following known series	GyrB	Pyrrolamides	ATPase inhibitor	[45]
Pharmacophore based library	GyrB	Thiazolopyridine ureas	ATPase inhibitor	[46]
Pharmacophore based library, scaffold morphing	GyrB	Thiazolopyridone ureas	ATPase inhibitor	[47]
High throughput screening	GyrB	Aminopyrazinamides	ATPase, MTB gyrase specific. Novel binding mode.	[48]
Focused library screening	GyrB	Aminopiperidine	Non-ATP site binders, different from fluoroquinolones	[49]
Scaffold hopping	GyrB	Benzimidazoles	Non-ATP site binders, different from FQs	[50]

The key learning from these extensive efforts on 'revisiting gyrase as an established target' are as follows:

- A variety of novel chemical structures could be identified as potent starting points (Table 1).
- Several of these enzyme inhibitors showed an IC_{50}-MIC correlation.
- The inhibitors worked through different mechanisms; hence, they have a potential to avoid cross-resistance.
- These inhibitors were shown to have a higher potency against the MTB enzyme target, as compared to other bacterial gyrases, that translated into selectivity in their antimicrobial activity.

4. Target-Based Screening

The target-based lead identification approach was given a major impetus because of the following developments:

- The availability of the MTB genome sequence in 1998 [51] promised a new era both for studying the 'biology' of the pathogen and investigating novel pathways suitable for drug development. Several publications appeared, proving the essentiality of biochemical targets based on gene knockout studies in vitro, as well as investigations on the survival of the gene knockouts of MTB in the mouse model, confirming the essentiality of a variety of metabolic targets in vivo [52].
- Chris Lipinski et al. published the 'Lipinski rule of 5' for oral drugs [53]. The poor physiochemical properties of lead compounds vis a vis the 'Lipinski rule of 5' was shown to be the leading reason for the failure of potent compounds in the clinical trials, which was the direct result of poor pharmacokinetics. This led to the understanding of the concept of 'lead-like compounds' [54]. These rules served as guidelines for the selection/prioritization of 'hits' with a higher probability of being converted to drugs with favorable pharmacokinetics.

- This period also saw an enhanced efficiency in solving crystal structures; the macromolecular structures of several mycobacterial enzymes were solved and became key tools for the rationale design of novel inhibitors. A consortium of structural genomics groups was formed for facilitating the determination and analysis of structures of proteins from MTB [55].
- In addition to these facilitators of drug development, several public–private partnership consortia were also formed, which propelled interactions between the pharmaceutical industry and academic laboratories to engage in the TB drug discovery process. Examples include 'The Action TB program' (1996); the 'Global Alliance for TB' (GATB), which was launched in 2000; and the EU FW6/7 program entitled 'New Medicines for Tuberculosis' (NM4TB), which was launched in 2005, followed by the 'More Medicines for Tuberculosis' (MM4TB) program. Several pharmaceutical companies turned to target-based screening to find novel 'lead compounds' with a potential to inhibit the growth of MTB cells [56].
- Examples of efforts at AZI toward finding potent inhibitors by using target-based screening are summarized in Table 2. Though there are several publications on these efforts, it is interesting to view them in the context of the learning and the impact. A few of these programs were also in collaboration between AZI and partners of the NM4TB and MM4TB, the EU FW6, and seven programs, as well as with the Global Alliance for TB.

Table 2. Target-based efforts at AstraZeneca.

Approach	Target	Pathway	Inhibitor Class	Status Reached	Remarks	Reference
Known inhibitors	Acetolactate synthase	Branched chain amino acid biosynthesis	Triazolopyrimidine	IC_{50} ~30 nM MIC 0.5 µg/mL	Possible auxotrophy	[57,58]
Virtual screening	MtSK	Aromatic amino acid biosynthesis	Pyrazolones	IC_{50} ~0.6 µM MIC 0.5 µg/mL	Possible promiscuity	[59,60]
Targeted Library	Pkn B	Cell division	Bis anilinopyrimidines (BAP)	IC_{50} ~40 pM MIC 8 µg/mL	Possible redundancy	[61]
HTS	PanK	Coenzyme A (CoA) biosynthesis pathway	Triazoles (ATP competitive)	Nanomolar IC_{50}, not translating into microbial inhibition.	Possible Target vulnerability	[62–64]
HTS	PanK	Coenzyme A (CoA) biosynthesis pathway	Quinolones (ATP competitive)	Nanomolar IC_{50}, not translating into microbial inhibition.	Possible Target vulnerability	[62–64]
HTS	PanK	Coenzyme A (CoA) biosynthesis pathway	Biarylacetic acids (ATP non-competitive)	Nanomolar IC_{50}, not translating into microbial inhibition.	Possible Target vulnerability	[62–64]
HTS	TMK	DNA synthesis	3-Cyanopyridone	Nanomolar IC_{50}, not translating into microbial inhibition.	Possible redundancy	[65]
Fragment screening	TMK	DNA synthesis	1,6-Naphthyridinone	Nanomolar IC_{50}, not translating into microbial inhibition.	Possible redundancy	[65]
HTS	NdH2	Energy Metabolism	Quinolyl pyrimidines	Ndh2 IC_{50}: 96 nM MIC: <0.5 µg/mL	In progress	[66]

The key learning derived from these extensive studies are as follows:

a. The patterns obtained throw important light on both the characteristics of the target and the characteristics of the inhibitor.
b. The question of target vulnerability: Gene knockout studies for PanK demonstrated the essentiality of the gene product. However, elegant studies with the PanK target showed that modulation of enzyme levels (inferred based on regulating gene expression) altered the sensitivity of the MTB cell to inhibitors of the enzyme [62]. This change in vulnerability of the target was the reason attributed to the failure to obtaining bactericidal effects with the potent inhibitors.

c. Type of inhibition: Most of the inhibitors described in Table 2 were 'competitive' in nature, competing with the ATP site for binding. High-throughput screening mostly identifies 'competitive inhibitors' because of the design of the assay. The inhibition brought about by this type of inhibition is influenced by the 'substrate concentration', resulting in the modulation of target vulnerability [65].
d. Promiscuous targets: These types of targets are discussed later, in the lead generation approaches.
e. Established target: As described under Section 3.4, several inhibitors that have progressed to anti-TB drugs establish the 'vulnerability' of a target; ATP synthase (ATPS) is one such target. Bedaquiline, a non-competitive inhibitor of ATPS, is a front-line treatment for MDR TB [67,68].

5. Phenotypic Screening

This is a biology-driven effort where compound libraries are screened for antimycobacterial activity against mycobacterial cells in culture, and the 'hits' are progressed based on the potency both in vitro and in vivo [69]. This cell-based approach has been successful from the early days of TB discovery, and most of the frontline drugs were progressed based on the structure–activity relationship that was evaluated directly on the whole cells. Current-day phenotypic-screening approaches include the ability to identify inhibitors that are active against MTB cells growing in a variety of microenvironments that represent the replicating, the non-replicating, or bacilli in different physiological states [70]. Genetic probes have also facilitated the unravelling of the biological targets of the inhibitors.

Some of the examples of phenotypic-screening-derived compounds that have become drugs, or those in late clinical development, are discussed here.

a. **Bedaquiline:** Approved in 2012 for the treatment of MDR TB, Bedaquiline (TMC-207, R207910, shown in Figure 3) has the reputation of being the first FDA approved TB drug in more than four decades. It was discovered by Johnson & Johnson by screening more than 70,000 compounds against *Mycobacterium. smegmatis* (M.sm). The compound was first described in 2004, at the Interscience Conference on Antimicrobial Agents and Chemotherapy (ICAAC). The target of TMC-207 was later conclusively proven to be the MTB ATP synthase [68]. Subsequently, in 2012, Bedaquiline became an essential part of the MDR regimen [71,72]. Several combinations of Bedaquiline are also being investigated for their potential to reduce the duration of therapy [73]. Discovery of Bedaquiline and the target link has made MTB ATP synthase an attractive validated target.

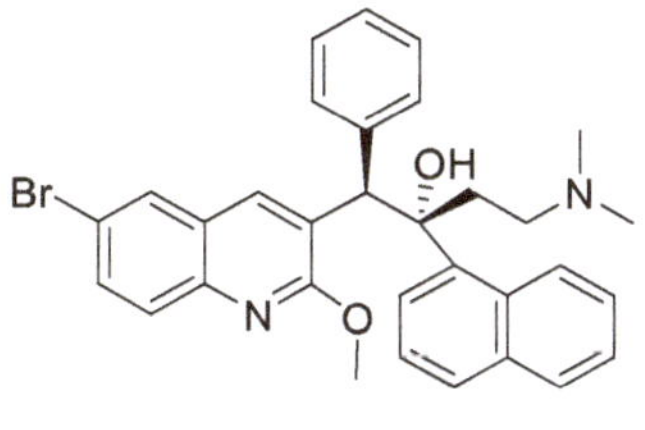

Bedaquiline (TMC-207, R207910)

Figure 3. Structure of Bedaquiline (TMC-207, R207910).

b. **Benzothiazinone (BTZ) and Macozinone (PBTZ):** BTZ was discovered at the Russian academy of science and developed with NM4TB, and it is presently being clinically evaluated under the Innovative Medicines for Tuberculosis (iM4TB). The target for BTZ has been shown as decaprenylphosphoryl-β-D-ribose-2′-oxidase (DprE1) by genetic and biochemical studies [74,75]. Based on SAR explorations, a piperazine derivative, PBTZ-169 (Macozinone), which has superior pharmacokinetics (PK), as compared to that of BTZ [76], was discovered and is currently in Phase 2 clinical trials, in addition to BTZ 043. Structures of BTZ and PBTZ are shown in Figure 4.

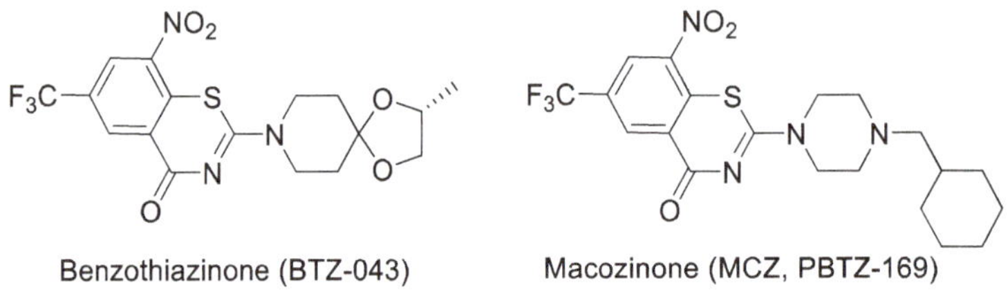

Figure 4. Structures of BTZ and PBTZ.

c. **Azaindoles:** These compounds emerged from a literature search for compounds similar to Q-203 [77,78]. The compound had an MIC but was not bactericidal. Scaffold morphing efforts on this compound at AZ Bangalore in a collaboration with TB alliance led to the identification of TBA 7371 [79–81], a potent DprE1 inhibitor with non-covalent binding. TBA 7371, jointly owned by Global Alliance for TB and the Foundation for Neglected Disease Research (FNDR), was developed by GATB through Phase 1 safety trials in humans. Recently, GATB has licensed TBA 7371 to Gates Medical Research Institute (GMRI); the news was disclosed in Union World conference of lung health at Hyderabad, India, in November 2019. GMRI is currently testing TBA7371 in Phase 2A clinical trials in TB patients [82]. The progression is depicted in Figure 5.

Figure 5. Evolution of TBA7371 which is in Phase 2A clinical trials in TB patients.

d. **Q203:** Telacebec (Q203) is a compound that was discovered by the researchers at Institute Pasteur, Korea, and is structurally very similar to the one discussed above [77,78]. This compound inhibits the cytochrome bc1 complex of MTB that is critical for the electron transport chain. The compound is in Phase 2 clinical trials [83]. Structure of Telacebec is shown in Figure 6.

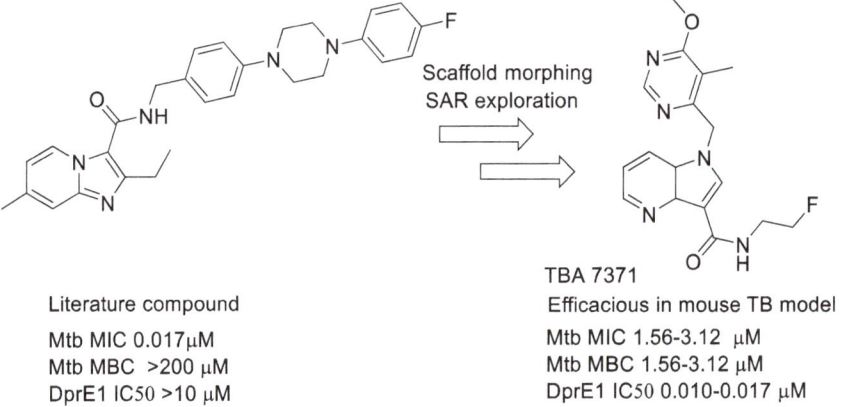

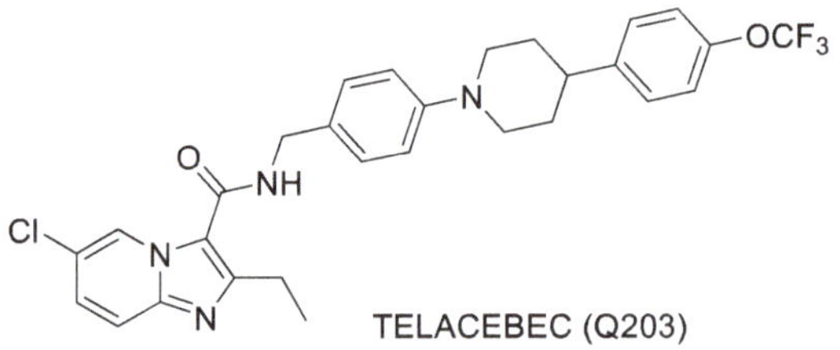

Figure 6. Structure of Telacebec.

e. **OPC-167832:** First discussed in 2016 at the 47th Union World Conference on Lung Health in Liverpool, UK, this is a carbostyril derivative that inhibits DPRE1. The structure shown in Figure 7 represents a balance of hydrophobic and hydrophilic residues. The compound is currently in Phase 2 clinical trials [84]. Structure of OPC-167832 is shown in Figure 7.

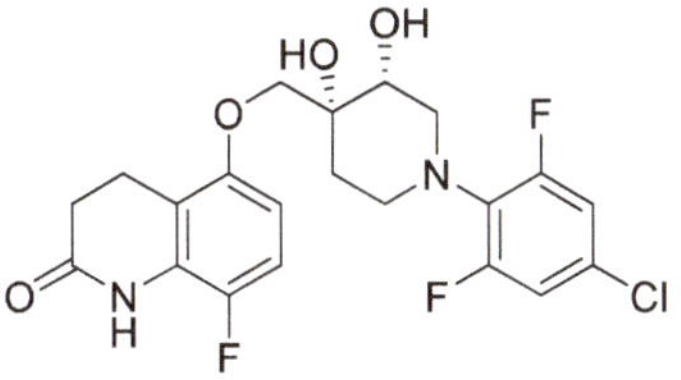

Figure 7. Structure of OPC-167832.

Phenotypic-Screening Efforts at AstraZeneca India

The continued success of phenotypic screening as a viable alternative to discover novel inhibitors of MTB was based on its ability to circumvent the failure to convert enzyme inhibition into antibacterial activity. The advent of 'rapid sequencing technology' provided an impetus to this approach because of the ability to rapidly sequence the genome to elucidate mechanism of action. This led AZ India to focus on phenotypic screening for lead generation. A diversity library of 100,000 compounds was assembled and used for screening directly against both *M. smegmatis* or *Mycobacterium bovis* BCG cells as surrogates for MTB. Several novel scaffolds and their cognate targets were identified, confirming the validity of this approach; a short list of the results obtained is shown in Table 3. Many of the novel scaffolds identified were found to target DprE1 or the ATP synthase in target enzyme assays.

Table 3. Phenotypic screening-based efforts at AstraZeneca India (AZI).

Screening Mode	Target	Compound	Comments	Reference
HTS	DprE1	Benzothiazole	Promiscuous target	[85,86]
HTS	DprE1	4-Aminoquinoline piperidine amides	Promiscuous target	[87]
HTS	DprE1	Pyrazolopryridones	Promiscuous target	[88]
HTS	DprE1	Azaindoles	Promiscuous target	[79,80]
Scaffold hopping	DprE1	Benzimidazole	Promiscuous target	[81]
HTS	ATPS	Squaramide	Established/Validated Target	[89]
HTS	ATPS	Imidazopyridine	Established/Validated Target	[89]
Scaffold morphing	ATPS	Diaminoquinazoline	Established/Validated Target	[90]

6. The Anti-TB Pipeline—The Learning and the Gaps

6.1. The 'Evolution' of the Anti-TB Drug Pipeline

The anti-TB drug portfolio went from a 'no treatment' option to a successful therapy in ~20 years. The main properties of drugs in the regimen that was labeled 'Short Course Chemotherapy' were potency against the microbe, the ability to prevent relapse, and compatibility in terms of safety. The next phase, ~1980s, was the search for a compound that could shorten the course of the treatment. In the absence of 'physiologically' relevant in vitro and in vivo models to progress compounds for this property, it was indeed a 'black box' experimentation. However, with the spread of MDR TB, this focus turned to finding compounds 'active on MDR TB' and compatibility with the 'patient comorbidities', like HIV infection [91]. The concept of introducing 'broad-spectrum antibiotics' for the treatment of TB was a direct consequence of this shift; even the original trial that included 'ofloxacin' into the regimen was for reducing the duration of therapy [25]. Studies in animal models with a combination containing moxifloxacin showed that the combination had a potential to reduce the duration of therapy [92], and this was one of the end points in the trials with moxifloxacin in humans. However, the final results showed that, while the drug was effective, the combination did not reduce the duration of therapy, although it was efficacious on MDR, as well as on DS TB patients. The drug was introduced into the regimen to treat MDR TB [93].

The need for efficacious drugs for treating MDR TB became the 'new end point' for a drug. Both Bedaquiline and Delamanid fulfilled this criterion and are now parts of the new regimen [67,94].

There has been a concerted effort to understand the 'persister' population that is hypothesized to be difficult to eradicate and hence the need for prolonged duration of therapy [95,96]. The non-replicating persister (NRP) state has been the subject of intense investigations and several models representing this state have been developed. Both Bedaquiline and Delamanid are active against MTB bacilli growing under such conditions [16,67]. Further trials to investigate if this property of the new drugs will contribute to the reduction in therapy duration needs to be performed. In parallel, the portfolio is sure to see 'adjunct therapy' compounds that modulate the host response [92–95]. Repurposed candidate drugs offer faster development options and are also expected to enrich the TB portfolio [97–103]. These could also be weapons against the 'persister' population.

6.2. Lead Generation Approaches

The evolution of the anti-TB drug regimen, including the current portfolio, has followed a traceable pattern of periods. These periods are recognizable by the 'classes' of compounds that were introduced into the regimen. The different lead generation approaches and their accelerators are shown in Figure 8. Accelerators are defined as new knowledge or global exigencies. Each of these accelerators influenced lead generation methodology, to follow a certain approach during a certain period.

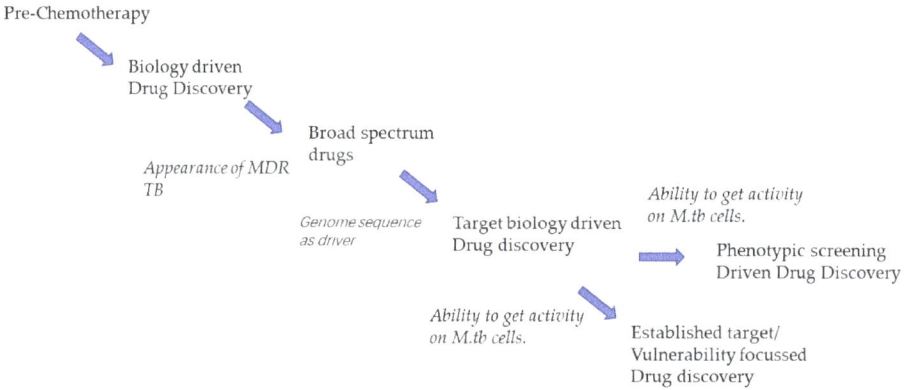

Figure 8. Evolution of lead generation approaches.

What is the current trend of preclinical 'hit' molecules being identified by using phenotypic screening? Mdluli et al. [104] have discussed this trend which indicates that the majority of these 'hits' act on the targets, namely DprE1, Mmpl3, and Cytochrome bcc complex. This has led to the coining of the term 'promiscuous targets'. The AstraZeneca 'hit' compound list also shows the same trend (Table 3). Why are these targets prone to get hit in phenotypic screens? The ultimate consequence of the so called 'promiscuity' on drug discovery will be fully answered by the clinical trial outcome of the compounds TBA737, BTZ043, PBTZ-169, OPC 167832 (DprE1 and Mmpl3), and Q203 (bcc1 complex) hitting such targets [105].

Another interesting trend is the increased number of compounds in later stages of development that are inhibitors of 'established targets'. This again adds to the concept of 'privileged targets' of MTB (privileged because the inhibition of these targets converts to an antimicrobial effect on MTB). Perhaps the so-called 'promiscuous targets' can also be classified as 'privileged targets'.

6.3. Has the Pipeline Evolved Based on the Significant Increase in the Knowledge of the 'Biology' of MTB?

In 1996, Mitchison in his Garrod lecture postulated the presence of different populations of MTB in the lungs of the TB patient [106]. Mitchison postulated the presence of the rapidly multiplying and the slow-growing populations of MTB, as well as the intracellular and extracellular niches of the organism. The last two decades have uncovered several shifts in the physiological state of the pathogen that are responses to both the immune mechanism of the host or the 'niche' in which the pathogen is proliferating [107]. The newer drugs like Bedaquiline, Pretomanid, and Delamanid have been shown to be active against MTB cells in the NRP state in vitro, as well as against the MTB bacilli multiplying within the macrophage [108,109]. However, these drugs were initially identified as inhibitors of the rapidly multiplying pathogen under in vitro culture conditions. It is generally accepted that the MTB population in the human host is a heterogeneous collection of the multiple 'physiological states' [69] that could indeed be a dynamic interchanging state. This raises two questions:

1. What is the value of having novel drugs that are active predominantly on one of the physiological states like the hypoxic or the intracellular state? Metronidazole is a known antiparasitic drug that is also active on MTB growing under oxygen-limiting conditions [110]. The addition of metronidazole to a cocktail of drugs for the treatment of MDR TB did not show clinical benefit [111]. On the other hand, several inhibitors of the intracellular survival of the pathogen in the macrophage have been identified, have been shown to be active in animal models of infection [34,67,77,83,112], and are undergoing clinical evaluation. These inhibitors potentially will facilitate the 'killing' of the pathogen within the macrophage, which in turn may reflect in 'faster' cure.
2. How do we develop molecules with activity against only specific subpopulations? The accepted path through the regulatory system is by showing superiority of the new combination over the existing standard of care (SOC) combination—a steep bar to cross. Until the beginning of this decade, very few efficacious drugs were available for the treatment of MDR TB patients. This provided the opportunity for newer efficacious drugs to show non-inferiority against the poorly active regimen, a bar, which was reasonable to go across. However, the current combination regimen for MDR TB treatment is efficacious, and newer compounds will have to demonstrate a clear advantage over the current regimen. Identification of novel biomarkers that can be evaluated during Phase 2a (Early Bactericidal Activity trials) could help in positioning these compounds that are active on specific populations to the best advantage.

Undoubtedly, there is a rapid expansion in the knowledge of the pathogen biology and its interaction with the host. Additionally, several novel inhibitors for pathways of the host that influence the survival of the bacilli have been identified. However, translational research must be supported to evaluate the potential of these compounds to become anti-TB drugs.

6.4. Combination Therapy: Finding Novel Combinations

What constitutes an 'ideal' combination? Faster time for sputum conversion, faster cure, tolerability, lower rates of drug resistance, lower relapse rates, combination suitable for treating pulmonary and extrapulmonary TB, and other properties, like compatible with comorbidities, etc. The concept of searching for novel combinations is recent; it was first advocated by the GATB [113] by studying a variety of combinations in humans using EBA. The two recently approved drugs, Bedaquiline and Delamanid, were progressed as add-ons to a cocktail of second-line drugs. Pretomanid, in combination with Bedaquiline and linezolid, has been approved for the treatment of XDR TB patients for whom there is very limited choice [73].

Another question is, can compounds active on the same target be compatible in the same combination? It is possible if the two inhibitors bind to different sites of the same targets, like in the examples of BTZ, PBTZ, and TBA7371. If the resistant mutants for the individual compounds do not confer cross-resistance, can such an approach bring additional advantage to the treatment? This can be relevant in the context of the so-called 'promiscuous targets' in TB and the diverse chemical inhibitors of these targets reported in recent years, e.g., benzothiazinones, azaindoles, and Q 203.

7. Going Forward

It is interesting to view the current anti-TB portfolio as a glass 'half full'. Novel efficacious compounds are in the late stages of clinical development [114–116]. This will offer the opportunity of designing new combinations. Several novel molecules are in the late discovery phase that will diversify the repertoire of potential anti-TB molecules. The challenge is to find simpler paths through the regulatory system that can demonstrate advantages with the molecule, as well as ensure safety. This will need concerted discussion with multiple stakeholders, which is already happening [117,118].

The secret to 'shortening therapy' needs to be unraveled, whether this will continue to be a hit-and-miss experimentation or further knowledge of the biology of the pathogen and will allow rational experimentation is not clear. Can adjunct therapy with immuno-modulators help, or will compounds acting on subpopulations that represent the 'difficult to treat' cells facilitate faster cure? Such questions need to be evaluated. This requires an urgent need to find novel translational approaches compatible with the regulatory framework to achieve this shift that, in turn, can create a paradigm shift in our modus operandi of how we treat this affliction.

8. Conclusions

Finding new drugs for the treatment of tuberculosis has been and continues to be a challenge despite the increased efforts over the last two decades. Lead generation approaches for MTB drug discovery have undergone several changes, mostly driven by an increased understanding of the biology of the pathogen, as well as the rapidly expanding knowledge on the biology of the interaction of the pathogen with the human host. This review chronicles the advances in a systematic study, starting from 'serendipitous discovery to phenotypic screening'. The technological advance mirrors the increased understanding of the biology of the pathogen. Interestingly, much of this increased understanding is also a consequence of the introduction of newer drugs for the treatment of tuberculosis.

Author Contributions: The authors contributed to the review in the following manner: Original idea of writing review (T.S.B., B.B. and R.K.S.), Conceptualization and frame work of the article (T.S.B., R.K.S., B.B. and J.B.), Data analysis (B.B., R.K.S. and J.B.) writing of review (T.S.B., B.B. and R.K.S.), data curation and editing (B.B., R.K.S. and J.B.) and writing of the article (T.S.B., B.B., R.K.S. and J.B.). All authors have read and agreed to the published version of the manuscript.

Funding: This research received no external funding.

Conflicts of Interest: The authors declare no conflict of interest. All AstraZeneca results discussed in this paper are from published sources and available in the public domain.

References

1. TB Facts.org, Information about Tuberculosis. Available online: https://tbfacts.org/history-of-tb-drugs/ (accessed on 13 April 2020).
2. Lehmann, J. Para-aminosalicylic acid in the treatment of Tuberculosis. *Lancet* **1946**, *247*, 15–16. [CrossRef]
3. Lehmann, J. Twenty years afterwards, Historical notes on the Discovery of the Antituberculosis Effect of Para-Aminosalicylic Acid (PAS) and the First Clinical Trials. *Am. Rev. Respir. Dis.* **1964**, *90*, 953–956.
4. Lehmann, E. Nicotinic acid in therapy of pulmonary tuberculosis; preliminary therapeutic report. *Dtsch. Med. Wochenschr.* **1952**, *77*, 1480–1481. [CrossRef]

5. Murray, F.M. Nicotinamide: An Oral Antimicrobial Agent with Activity against Both Mycobacterium tuberculosis and Human Immunodeficiency Virus. *Clin. Infect. Dis.* **2003**, *36*, 453–460. [CrossRef]
6. Thayer, J.D.; Seligman, R.B. The anti-tuberculous activity of some derivatives of p-aminosalicylic acid, nicotinic acid, and isonicotinic acid. *Antibiot. Chemother. (Northfield)* **1955**, *5*, 129–131.
7. Mitchison, D.A. Treatment of tuberculosis. The Mitchell lecture 1979. *J. R. Coll. Physicians Lond.* **1980**, *14*, 91.
8. Aquinas, M. Short-course therapy for tuberculosis. *Drugs* **1982**, *24*, 118–132. [CrossRef]
9. Mohammad, A. Rifampin and Their Analogs: A Development of Antitubercular Drugs. *World J. Org. Chem.* **2013**, *1*, 14–19.
10. Ramani, A.V.; Monika, A.; Indira, V.L.; Karyavardhi, G.; Venkatesh, J.; Jeankumar, V.U.; Manjashetty, T.H.; Yogeeswari, P.; Sriram, D. Synthesis of highly potent novel anti-tubercular isoniazid analogues with preliminary pharmacokinetic evaluation. *Bioorg. Med. Chem. Lett.* **2012**, *22*, 2764–2767. [CrossRef]
11. Yamamoto, S.; Toida, I.; Watanabe, N.; Ura, T. In vitro Antimycobacterial Activities of Pyrazinamide Analogs. *Antimicrob. Agents Chemother.* **1995**, *39*, 2088–2091. [CrossRef]
12. Häusler, H.; Kawakami, R.P.; Mlaker, E.; Severn, W.B.; Stütz, A.E. Ethambutol Analogues as Potential Antimycobacterial Agents. *Bioorg. Med. Chem. Lett.* **2001**, *11*, 1679–1681. [CrossRef]
13. Goldman, P. The development of 5-nitroimidazole for the treatment and prophylaxis of anaerobic bacterial infections. *J. Antimicrob. Chemother.* **1982**, *10*, 23–33. [CrossRef]
14. Ashtekar, D.R.; Costa-Perira, R.; Nagrajan, K.; Vishvanathan, N.; Bhatt, A.D.; Rittel, W. In vitro and in vivo activities of the nitroimidazole CGI 17341 against Mycobacterium tuberculosis. *Antimicrob. Agents Chemother.* **1993**, *37*, 183–186. [CrossRef]
15. Stover, C.K.; Warrener, P.; Van Devanter, D.R.; Sherman, D.R.; Arain, T.M.; Langhorne, M.H.; Anderson, S.W.; Towell, J.A.; Yuan, Y.; McMurray, D.N.; et al. A small-molecule nitroimidazopyran drug candidate for the treatment of tuberculosis. *Nature* **2000**, *405*, 962–966. [CrossRef]
16. Liu, Y.; Matsumoto, M.; Ishida, H.; Ohguro, K.; Yoshitake, M.; Gupta, R.; Geiter, L.; Hafkin, J. Delamanid: From discovery to its use for pulmonary multidrug-resistant tuberculosis (MDR-TB). *Tuberculosis* **2018**, *111*, 20–30. [CrossRef]
17. Sensi, P. History of the development of rifampin. *Rev. Infect. Dis.* **1983**, *5*, S402–S466. [CrossRef]
18. Janin, Y.L. Antituberculosis drugs: Ten years of research. *Bioorg. Med. Chem.* **2007**, *15*, 2479–2513. [CrossRef]
19. Bogatcheva, E.; Hanrahan, C.F.; Nikonenko, B.; Samala, R.; Chen, P.; Gearhart, J.; Barbosa, F.; Einck, L.; Nacy, A.C.A.; Protopopova, M. Identification of new diamine scaffolds with activity against Mycobacterium tuberculosis. *J. Med. Chem.* **2006**, *49*, 3045–3048. [CrossRef]
20. Yendapally, R.; Lee, R.E. Design, synthesis, and evaluation of novel ethambutol analogues. *Bioorg. Med. Chem. Lett.* **2008**, *18*, 1607–1611. [CrossRef]
21. Kapil, T.; Regina, W.; David, B.K.; Kriti, A.; Vinod, N.; Elizabeth, F.; Barnes, S.W.; John, R.W.; David, A.; Clifton, E.B., III; et al. SQ109 targets MmpL3, a membrane transporter of trehalose monomycolate involved in mycolic acid donation to the cell wall core of Mycobacterium tuberculosis. *Antimicrob. Agents Chemother.* **2012**, *56*, 1797–1809.
22. Sinha, N.; Jain, S.; Tilekar, A.; Upadhayaya, R.S.; Kishore, N.; Jana, G.H.; Arora, S.K. Synthesis of isonicotinic acid N'-arylidene-N-2-oxo-2-(4-aryl-piperazin-1-yl) ethyl-hydrazides as antituberculosis agents. *Bioorg. Med. Chem. Lett.* **2005**, *15*, 1573–1576. [CrossRef]
23. Sinha, R.K.; Arora, S.K.; Sinha, N.; Modak, V.M. In vivo activity of LL4858 against Mycobacterium tuberculosis. In Proceedings of the 44th Annual Interscience Conference on Antimicrobial Agents and Chemotherapy (ICAAC-2004), Washington, DC, USA, 30 October–2 November 2004.
24. World Health Organization. *Tuberculosis—A Global Emergency Case Notification Update: February 1996*; WHO Reference Number: WHO/TB/96.197; WHO: Geneva, Switzerland, 1996.
25. Hinshaw, H.; Pyle, M.M.; Feldman, W.H. Streptomycin in tuberculosis. *Am. J. Med.* **1947**, *2*, 429–435. [CrossRef]
26. Schecter, G.F.; Scott, C.; True, L.; Raftery, A.; Flood, J.; Mase, S. Linezolid in the Treatment of Multidrug-Resistant Tuberculosis. *Clin. Infect. Dis.* **2010**, *50*, 49–55. [CrossRef]
27. Maartens, G.; Benson, C.A. Linezolid for Treating Tuberculosis: A Delicate Balancing Act. *BioMedicine* **2015**, *2*, 1568–1569. [CrossRef]
28. Lee, M.; Song, T.; Kim, Y.; Jeong, I.; Cho, S.N.; Barry, C.E., III. Linezolid for XDR-TB—Final Study Outcomes. *N. Engl. J. Med.* **2015**, *373*, 290–291. [CrossRef]

29. Balasubramanian, V.; Solapure, S.; Shandil, R.; Gaonkar, S.; Mahesh, K.N.; Reddy, J.; Deshpande, A.; Bharath, S.; Kumar, N.; Wright, L.; et al. Pharmacokinetic and pharmacodynamic evaluation of AZD5847 in a mouse model of tuberculosis. *Antimicrob. Agents Chemother.* **2014**, *58*, 4185–4190. [CrossRef]
30. Balasubramanian, V.; Solapure, S.; Iyer, H.; Ghosh, A.; Sharma, S.; Kaur, P.; Deepthi, R.; Subbulakshmi, V.; Ramya, V.; Ramachandran, V.; et al. Bactericidal activity and mechanism of action of AZD5847, a novel oxazolidinone for treatment of tuberculosis. *Antimicrob. Agents Chemother.* **2013**, *58*, 495–502. [CrossRef]
31. Werngren, J.; Wijkander, M.; Perskvist, N.; Balasubramanian, V.; Sambandamurthy, V.K.; Rodrigues, C.; Hoffner, S. In vitro activity of AZD5847 against geographically diverse clinical isolates of Mycobacterium tuberculosis. *Antimicrob. Agents Chemother.* **2014**, *58*, 4222–4223. [CrossRef]
32. Williams, K.N.; Brickner, S.J.; Stover, C.K.; Zhu, T.; Ogden, A.; Tasneen, R.; Tyagi, S.; Grosset, J.H.; Nuermberger, E.L. Addition of PNU-100480 to First-Line Drugs Shortens the Time Needed to Cure Murine Tuberculosis. *Am. J. Respir. Crit. Care Med.* **2009**, *180*, 371–376. [CrossRef]
33. Lanoix, J.P.; Nuermberger, E. Sutezolid: Oxazolidinone antibacterial treatment of tuberculosis. *Drugs Future* **2013**, *38*, 387–394. [CrossRef]
34. Choi, Y.; Lee, S.W.; Kim, A.; Jang, K.; Nam, H.; Cho, Y.L.; Yu, K.-S.; Chung, J.-Y. Safety, tolerability and pharmacokinetics of 21 day multiple oral administration of a new oxazolidinone antibiotic, LCB01-0371, in healthy male subjects. *Antimicrob. Chemother.* **2018**, *73*, 183–190. [CrossRef]
35. Shoen, C.; DeStefano, M.; Hafkin, B.; Cynamon, M. In vitro and in vivo activity of contezolid (MRX-I) against M. tuberculosis. *Antimicrob. Agents Chemother.* **2018**, *62*, e00493-18. [CrossRef]
36. ClinicalTrails.gov. Phase 2 Trial to Evaluate the Early Bactericidal Activity, Safety and Tolerability of Meropenem Plus Amoxycillin/CA and Faropenem Plus Amoxycillin/CA in Adult Patients with Newly Diagnosed Pulmonary Tuberculosis. Available online: https://clinicaltrials.gov/ct2/show/NCT02349841 (accessed on 15 March 2020).
37. Palencia, A.; Li, X.; Bu, W.; Choi, W.; Ding, C.Z.; Easom, E.E.; Feng, L.; Hernandez, V.; Houston, P.; Liu, L.; et al. Discovery of Novel Oral Protein Synthesis Inhibitors of Mycobacterium tuberculosis That Target Leucyl-tRNA Synthetase. *Antimicrob. Agents Chemother.* **2016**, *60*, 6271–6280. [CrossRef]
38. Shoen, C.; DeStefano, M.; Pucci, M.; Cynamon, M. Evaluating the Sterilizing Activity of SPR720 in Combination Therapy against Mycobacterium Tuberculosis Infection in Mice, ASM Microbe 2019. Session P439 Poster AAR-749. Available online: https://www.newtbdrugs.org/pipeline/compound/spr720 (accessed on 19 March 2020).
39. Shoen, C.; Pucci, M.; DeStefano, M.; Cynamon, M. Efficacy of SPR720 and SPR750 Gyrase Inhibitors in a Mouse Mycobacterium tuberculosis Infection Model, ASM Microbe 2017. Session 336. Poster 43. Available online: https://www.abstractsonline.com/pp8/#!/4358/presentation/6167 (accessed on 25 March 2020).
40. Spero Therapeutics Receives FDA Orphan Drug Designation for SPR720 for the Treatment of on tuberculous Mycobacterial (NTM) Infection. Available online: https://www.globenewswire.com/news-release/2020/03/11/1998722 (accessed on 6 April 2020).
41. Narayanan, P. Shortening short course chemotherapy: A randomised clinical trial for treatment of smear positive pulmonary tuberculosis with regimens using ofloxacin in the intensive phase. *Indian J. Tuberc.* **2002**, *49*, 27–38.
42. Stephen, H.G. The role of moxifloxacin in tuberculosis therapy. *Eur. Respir. Rev.* **2016**, *25*, 19–28.
43. Nagaraja, V.; Godbole, A.A.; Henderson, S.R.; Maxwell, A. DNA topoisomerase I and DNA gyrase as targets for TB therapy. *Drug Discov. Today* **2017**, *22*, 510–518. [CrossRef]
44. Shirude, P.S.; Hameed, S. Nonfluoroquinolone-Based Inhibitors of Mycobacterial Type II Topoisomerase as Potential Therapeutic Agents for TB. *Annu. Rep. Med. Chem.* **2012**, *47*, 319–330.
45. Solapure, S.; Mukherjee, K.; Nandi, V.; Waterson, D.; Shandil, R.; Balganesh, M.; Sambandamurthy, V.K.; Raichurkar, A.K.; Deshpande, A.; Ghosh, A.; et al. Optimization of Pyrrolamides as Mycobacterial GyrB ATPase Inhibitors: Structure-Activity Relationship and In Vivo Efficacy in a Mouse Model of Tuberculosis. *Antimicrob. Agents Chemother.* **2014**, *58*, 61–70.
46. Kale, M.G.; Raichurkar, A.; Waterson, D.; McKinney, D.; Manjunatha, M.R.; Kranthi, U.; Koushik, K.; Jena, L.K.; Shinde, V.; Rudrapatna, S.; et al. Thiazolopyridine Ureas as Novel Antitubercular Agents Acting through Inhibition of DNA Gyrase B. *J. Med. Chem.* **2013**, *56*, 8834–8848.

47. Kale, R.R.; Kale, M.G.; Waterson, D.; Raichurkar, A.; Hameed, S.P.; Manjunatha, M.R.; Reddy, B.K.; Malolanarasimhan, K.; Shinde, V.; Koushik, K.; et al. Thiazolopyridoneureas as DNA gyrase B inhibitors: Optimization of antitubercular activity and efficacy. *Bioorg. Med. Chem. Lett.* **2014**, *24*, 870–879. [CrossRef]
48. Shirude, P.S.; Madhavapeddi, P.; Tucker, J.A.; Murugan, K.; Patil, V.; Basavarajappa, H.D.; Raichurkar, A.V.; Humnabadkar, V.; Hussein, S.; Sharma, S.; et al. Aminopyrazinamides: Novel and Specific GyrB Inhibitors that Kill Replicating and Nonreplicating Mycobacterium tuberculosis. *ACS Chem. Biol.* **2013**, *8*, 519–523. [CrossRef]
49. Hameed, P.S.; Patil, V.; Solapure, S.; Sharma, U.; Madhavapeddi, P.; Raichurkar, A.; Chinnapattu, M.; Manjrekar, P.; Shanbhag, G.; Puttur, J.; et al. Novel N-Linked Aminopiperidine-Based Gyrase Inhibitors with Improved hERG and in Vivo Efficacy against Mycobacterium tuberculosis. *J. Med. Chem.* **2014**, *57*, 4889–4905.
50. Hameed, P.S.; Raichurkar, A.; Madhavapeddi, P.; Menasinakai, S.; Sharma, S.; Kaur, P.; Nandishaiah, R.; Panduga, V.; Reddy, J.; Sambandamurthy, V.K.; et al. Benzimidazoles: Novel Mycobacterial Gyrase Inhibitors from Scaffold Morphing. *ACS Med. Chem. Lett.* **2014**, *5*, 820–825. [CrossRef]
51. Cole, S.T.; Brosch, R.; Parkhill, J.; Garnier, T.; Churcher, C.; Harris, D.; Gordon, S.; Eiglmeier, K.; Gas, S.; Barry, C.E.; et al. Deciphering the biology of Mycobacterium tuberculosis from the complete genome sequence. *Nature* **1998**, *393*, 537–544. [CrossRef]
52. Vashisht, R.; Bhat, A.G.; Kushwaha, S.; Bhardwaj, A.; OSDD Consortium; Brahmachari, S.K. Systems level mapping of metabolic complexity in Mycobacterium tuberculosis to identify high-value drug target. *J. Transl. Med.* **2014**, *12*, 263. [CrossRef]
53. Lipinski, C.A.; Lombardo, F.; Dominy, B.W.; Feeney, P.J. Experimental and computational approaches to estimate solubility and permeability in drug discovery and development settings. *Adv. Drug Deliv. Rev.* **1997**, *23*, 3–25. [CrossRef]
54. Teague, S.J.; Davis, A.M.; Leeson, P.D.; Oprea, T. The Design of Lead like Combinatorial Libraries. *Angew. Chem. Int. Ed. Eng.* **1999**, *28*, 3743–3748. [CrossRef]
55. Terwilliger, T.C.; Park, M.; Waldo, G.; Berendzen, J.; Hung, L.-W.; Kim, C.-Y.; Smith, C.; Sacchettini, J.; Bellinzoni, M.; Bossi, R.; et al. The TB structural genomics consortium: A resource for Mycobacterium tuberculosis biology. *Tuberculosis* **2003**, *83*, 223–249. [CrossRef]
56. Yuan, T.; Sampson, N.S. Hit Generation in TB Drug Discovery: From Genome to Granuloma. *Chem. Rev.* **2018**, *118*, 1887–1916. [CrossRef]
57. Patil, V.; Kale, M.; Raichurkar, A.; Bhaskar, B.; Prahlad, D.; Balganesh, M.; Nandan, S.; Shahul Hameed, P. Design and synthesis of triazolopyrimidineacylsulfonamides as novel anti-mycobacterial leads acting through inhibition of acetohydroxyacid synthase. *Bioorg. Med. Chem. Lett.* **2014**, *24*, 2222–2225. [CrossRef]
58. Balganesh, M.; Nandan, S. Combination Chemotherapy for Tuberculosis by Synergistic Action of Rifampicin as the RNA Polymerase Inhibitors with Acetolactate Synthase Inhibitors. PCT International Application WO 2007132189 A1 20071122, 22 November 2007.
59. Bandodkar, B.S.; Naik, M.; Ghorpade, S.; Kale, M.; Shanbhag, G.; Patil, V.; Solapure, S.; Balganesh, M.; Shandil, R.K.; Balasubramanian, B.; et al. Lead Generation via Virtual Screening: Discovery of Pyrazolones as Potent Antimycobacterial Leads through structure based virtual screening of shikimate kinase. In Proceedings of the ICAAC 2009, San Francisco, CA, USA, 12–15 September 2009.
60. Bandodkar, B.S.; Schmitt, S. Pyrazolone Derivatives for the Treatment of Tuberculosis. PCT International Application No. WO/2007/020426 A1, 22 February 2007.
61. Bandodkar, B.S. Oral presentation, "A decade of learning". In Proceedings of the CSIR-NM4TB Symposium, Bangalore, India, 14 December 2009.
62. Venkatraman, J.; Bhat, J.; Solapure, S.M.; Sandesh, J.; Sarkar, D.; Aishwarya, S.; Mukherjee, K.; Datta, S.; Malolanarasimhan, K.; Bandodkar, B.; et al. Screening, Identification, and Characterization of Mechanistically Diverse Inhibitors of the Mycobacterium Tuberculosis Enzyme, Pantothenate Kinase (CoaA). *J. Biomol. Screen.* **2012**, *17*, 293. [CrossRef] [PubMed]
63. Reddy, B.K.K.; Landge, S.; Ravishankar, S.; Patil, V.; Shinde, V.; Tantry, S.; Kale, M.; Raichurkar, A.; Menasinakai, S.; Mudugal, N.V.; et al. Assessment of Mycobacterium tuberculosis Pantothenate Kinase Vulnerability through Target Knockdown and Mechanistically Diverse Inhibitors. *Antimicrob. Agents Chemother.* **2014**, *8*, 3312–3326. [CrossRef]

64. Björkelid, C.; Bergfors, T.; Raichurkar, A.K.V.; Mukherjee, K.; Malolanarasimhan, K.; Bandodkar, B.; Jones, T.A. Structural and Biochemical Characterization of Compounds Inhibiting Mycobacterium tuberculosis Pantothenate Kinase. *J. Biol. Chem.* **2013**, *288*, 18260–18270. [CrossRef]
65. Naik, M.; Raichurkar, A.; Bandodkar, B.S.; Varun, B.V.; Bhat, S.; Kalkhambkar, R.; Murugan, K.; Menon, R.; Bhat, J.; Paul, B.; et al. Structure Guided Lead Generation for M. tuberculosis Thymidylate Kinase (Mtb TMK): Discovery of 3-Cyanopyridone and 1,6-Naphthyridin-2-one as Potent Inhibitors. *J. Med. Chem.* **2015**, *58*, 755–766.
66. Shirude, P.S.; Paul, B.; Choudhury, N.R.; Kedari, C.; Bandodkar, B.; Ugarkar, B.G. Quinolinyl Pyrimidines: Potent Inhibitors of NDH-2 as a Novel Class of Anti-TB Agents. *ACS Med. Chem. Lett.* **2012**, *3*, 736–740. [CrossRef]
67. Andries, K. A diarylquinoline drug active on the ATP Synthase of Mycobacterium tuberculosis. *Science* **2006**, *307*, 223–227. [CrossRef]
68. Koul, A.; Dendouga, N.; Vergauwen, K.; Molenberghs, B.; Vranckx, L.; Willebrords, R.; Ristic, Z.; Lill, H.; Dorange, I.; Guillemont, J.; et al. Diarylquinolines target subunit c of mycobacterial ATP synthase. *Nat. Chem. Biol.* **2007**, *3*, 323–324. [CrossRef]
69. Grzelak, E.M.; Choules, M.P.; Gao, W.; Cai, G.; Wan, B.; Wang, Y.; McAlpine, J.B.; Cheng, J.; Jin, Y.; Lee, H.; et al. Strategies in anti-Mycobacterium tuberculosis drug discovery based on phenotypic screening. *J. Antibiot.* **2019**, *72*, 719–728. [CrossRef]
70. Manjunatha, U.H.; Smith, P.W. Perspective: Challenges and opportunities in TB drug discovery from phenotypic screening. *Bioorg. Med. Chem.* **2015**, *23*, 5087–5097. [CrossRef]
71. Matteelli, A.; Carvalho, A.C.; E Dooley, K.; Kritski, A. TMC207: The first compound of a new class of potent anti-tuberculosis drugs. *Future Microbiol.* **2010**, *5*, 849–858. [CrossRef]
72. Deoghare, S. Bedaquiline: A new drug approved for treatment of multidrug-resistant tuberculosis. *Indian J. Pharmacol.* **2013**, *45*, 536–537. [CrossRef] [PubMed]
73. Conradie, F.; Diacon, A.H.; Ngubane, N.; Howell, P.; Everitt, D.; Crook, A.M.; Mendel, C.M.; Egizi, E.; Moreira, J.; Timm, J.; et al. Bedaquiline, pretomanid and linezolid for treatment of extensively drug resistant, intolerant or non-responsive multidrug resistant pulmonary tuberculosis. *N. Engl. J. Med.* **2020**, *382*, 893–902. [CrossRef] [PubMed]
74. Makarov, V.; Manina, G.; Mikusova, K.; Möllmann, U.; Ryabova, O.; Saint-Joanis, B.; Dhar, N.; Pasca, M.R.; Buroni, S.; Lucarelli, A.P.; et al. Benzothiazinones Kill Mycobacterium tuberculosis by Blocking Arabinan Synthesis. *Science* **2009**, *324*, 801–804. [CrossRef] [PubMed]
75. Trefzer, C.; Rengifo-Gonzalez, M.; Hinner, M.J.; Schneider, P.; Makarov, V.; Cole, S.T.; Johnsson, K. Benzothiazinones: Prodrugs That Covalently Modify the Decaprenylphosphoryl-D-ribose 2′-epimerase DprE1 of Mycobacterium tuberculosis. *J. Am. Chem. Soc.* **2010**, *132*, 13663–13665. [CrossRef] [PubMed]
76. Makarov, V.; Lechartier, B.; Zhang, M.; Neres, J.; Sar, A.M.; A Raadsen, S.; Hartkoorn, R.C.; Ryabova, O.B.; Vocat, A.; Decosterd, L.A.; et al. Towards a new combination therapy for tuberculosis with next generation benzothiazinones. *EMBO Mol. Med.* **2014**, *6*, 372–383. [CrossRef] [PubMed]
77. No, Z.; Kim, J.; Brodin, P.B.; Seo, M.J.; Kim, Y.M.; Cechetto, J.; Jeon, H.; Genovesio, A.; Lee, S.; Kang, S.; et al. Anti-Infective Compounds. PCT Publication No. WO 2011/113606 A1, 22 September 2011.
78. No, Z.; Kim, J.; Brodin, P.; Seo, M.J.; Park, E.; Cechetto, J.; Jeon, H.; Genovesio, A.; Lee, S.; Kang, S.; et al. Anti-Infective pyrido (1,2-a) pyrimidines. PCT Publication No. WO 2011/085990 A1, 21 July 2011.
79. Shirude, P.S.; Shandil, R.; Sadler, C.; Naik, M.; Hosagrahara, V.; Hameed, S.; Shinde, V.; Bathula, C.; Humnabadkar, V.; Kumar, N.; et al. Azaindoles: Noncovalent DprE1 Inhibitors from Scaffold Morphing Efforts, Kill Mycobacterium tuberculosis and Are Efficacious In Vivo. *J. Med. Chem.* **2013**, *56*, 9701–9708. [PubMed]
80. Chatterji, M.; Shandil, R.; Manjunatha, M.R.; Solapure, S.; Ramachandran, V.; Kumar, N.; Saralaya, R.; Panduga, V.; Reddy, J.; Kr, P.; et al. 1, 4-Azaindole, a Potential Drug Candidate for Treatment of Tuberculosis. *Antimicrob. Agents Chemother.* **2014**, *58*, 5325–5331. [CrossRef] [PubMed]
81. Manjunatha, M.R.; Shandil, R.K.; Panda, M.; Sadler, C.; Ambady, A.; Panduga, V.; Kumar, N.; Mahadevaswamy, J.; Sreenivasaiah, M.; Narayan, A.; et al. Scaffold Morphing to Identify Novel DprE1 Inhibitors with Antimycobacterial Activity. *ACS Med. Chem. Lett.* **2019**, *10*, 1480–1485.
82. Early Bactericidal Activity of TBA-7371 in Pulmonary Tuberculosis, Identifier: NCT04176250. Available online: https://clinicaltrials.gov (accessed on 2 April 2020).

83. Pethe, K.; Bifani, P.J.; Jang, J.; Kang, S.; Park, S.; Ahn, S.; Jiricek, J.; Jung, J.; Jeon, H.K.; Cechetto, J.; et al. Discovery of Q203, a potent clinical candidate for the treatment of tuberculosis. *Nat. Med.* **2013**, *19*, 1157–1160. [CrossRef]
84. Otsuka Awarded Grant to Advance Development of Novel Anti-Tuberculosis Compound OPC-167832 with Delamanid. Available online: https://www.businesswire.com/news/home/20180129005073/en/ (accessed on 11 April 2020).
85. Landge, S.; Mullick, A.B.; Nagalapur, K.; Neres, J.; Subbulakshmi, V.; Murugan, K.; Ghosh, A.; Sadler, C.; Fellows, M.D.; Humnabadkar, V.; et al. Discovery of benzothiazoles as antimycobacterial agents: Synthesis, structure–activity relationships and binding studies with Mycobacterium tuberculosis decaprenylphosphoryl-b-D-ribose 20-oxidase. *Bioorg. Med. Chem.* **2015**, *23*, 7694–7710. [CrossRef]
86. Landge, S.; Ramachandran, V.; Kumar, A.; Neres, J.; Murugan, K.; Sadler, C.; Fellows, M.D.; Humnabadkar, V.; Vachaspati, P.; Raichurkar, A.; et al. Nitroarenes as Antitubercular Agents: Stereoelectronic Modulation to Mitigate Mutagenicity. *Chemmedchem* **2016**, *11*, 331–339. [CrossRef] [PubMed]
87. Naik, M.; Humnabadkar, V.; Tantry, S.J.; Panda, M.; Narayan, A.; Guptha, S.; Panduga, V.; Manjrekar, P.; Jena, L.K.; Koushik, K.; et al. 4-Aminoquinolone Piperidine Amides: Noncovalent Inhibitors of DprE1 with Long Residence Time and Potent Antimycobacterial Activity. *J. Med. Chem.* **2014**, *57*, 5419–5434.
88. Panda, M.; Ramachandran, S.; Ramachandran, V.; Shirude, P.S.; Humnabadkar, V.; Nagalapur, K.; Sharma, S.; Kaur, P.; Guptha, S.; Narayan, A.; et al. Discovery of Pyrazolopyridones as a Novel Class of Noncovalent DprE1 Inhibitor with Potent Anti-Mycobacterial Activity. *J. Med. Chem.* **2014**, *57*, 4761–4771. [PubMed]
89. Tantry, S.J.; Markad, S.D.; Shinde, V.; Bhat, J.; Balakrishnan, G.; Gupta, A.K.; Ambady, A.; Raichurkar, A.; Kedari, C.; Sharma, S.; et al. Discovery of Imidazo[1,2-a]pyridine Ethers and Squaramides as Selective and Potent Inhibitors of Mycobacterial Adenosine Triphosphate (ATP) Synthesis. *J. Med. Chem.* **2017**, *60*, 1379–1399. [PubMed]
90. Tantry, S.J.; Shinde, V.; Balakrishnan, G.; Markad, S.D.; Gupta, A.K.; Bhat, J.; Narayan, A.; Raichurkar, A.; Jena, L.K.; Sharma, S.; et al. Scaffold morphing leading to evolution of 2,4-diaminoquinolines and aminopyrazolopyrimidines as inhibitors of the ATP synthesis pathway. *MedChemComm* **2016**, *7*, 1022–1032. [CrossRef]
91. Gandhi, N.R.; Nunn, P.; Dheda, K.; Schaaf, H.S.; Zignol, M.; Van Soolingen, D.; Jensen, P.; Bayona, J. Multidrug-resistant and extensively drug-resistant tuberculosis: A threat to global control of tuberculosis. *Lancet* **2010**, *375*, 1830–1843. [CrossRef]
92. Nuermberger, E.; Yoshimatsu, T.; Tyagi, S.; O'Brien, R.J.; Vernon, A.N.; Chaisson, R.E.; Bishai, W.R.; Grosset, J. 2004. Moxifloxacin-containing Regimen Greatly Reduces Time to Culture Conversion in Murine Tuberculosis. *Am. J. Respir. Crit. Care Med.* **2003**, *169*, 421–426. [CrossRef]
93. Gillespie, S.H.; Crook, A.M.; McHugh, T.D.; Mendel, C.M.; Meredith, S.K.; Murray, S.R.; Pappas, F.; Phillips, P.P.; Nunn, A.J. Four-Month Moxifloxacin-Based Regimens for Drug-Sensitive Tuberculosis. *N. Eng. J. Med.* **2014**, *351*, 1577–1587. [CrossRef]
94. Mabhula, A.; Singh, V. Drug-resistance in Mycobacterium tuberculosis: Where we stand. *MedChemComm* **2019**, *10*, 1342–1360. [CrossRef]
95. Muñoz-Elías, E.J.; Timm, J.; Botha, T.; Chan, W.T.; Gomez, J.E.; McKinney, J.D. Replication Dynamics of *Mycobacterium tuberculosis* in Chronically Infected Mice. *Infect. Immun.* **2005**, *73*, 546–551. [CrossRef]
96. Mitchison, D.A. Role of individual drugs in the chemotherapy of tuberculosis. *Int. J. Tuberc. Lung Dis.* **2004**, *4*, 796–806.
97. Rayasam, G.V.; Balganesh, T.S. Exploring the potential of adjunct therapy in tuberculosis. *Trends Pharmacol. Sci.* **2015**, *36*, 506–513. [CrossRef] [PubMed]
98. Zumla, A.; Chakaya, J.; Hoelscher, M.; Ntoumi, F.; Rustomjee, R.; Vilaplana, C.; Yeboah-Manu, D.; Rasolofo, V.; Munderi, P.; Singh, N.; et al. Towards host-directed therapies for tuberculosis. *Nat. Rev. Drug Discov.* **2015**, *14*, 511–512. [CrossRef] [PubMed]
99. Zumla, A.; Rao, M.; Wallis, R.S.; Kaufmann, S.H.; Rustomjee, R.; Mwaba, P.; Vilaplana, C.; Yeboah-Manu, D.; Chakaya, J.; Ippolito, G.; et al. Host-directed therapies for infectious diseases: Current status, recent progress and future prospects. *Lancet Infect. Dis.* **2016**, *16*, e47–e63. [CrossRef]
100. Kumar, D.; Nath, L.; Kamal, A.; Varshney, A.; Jain, A.; Singh, S.; Rao, K.V.; Singh, S. Genome-wide analysis of the host intracellular network that regulates survival of Mycobacterium tuberculosis. *Cell* **2010**, *140*, 731–743. [CrossRef]

101. Singhal, A.; Jie, L.; Kumar, P.; Hong, G.S.; Leow, M.K.-S.; Paleja, B.; Tsenova, L.; Kurepina, N.; Chen, J.; Zolezzi, F.; et al. Metformin as adjunct antituberculosis therapy. *Sci. Transl. Med.* **2014**, *263*, 263ra159. [CrossRef]
102. Mishra, R.; Kohli, S.; Malhotra, N.; Bandyopadhyay, P.; Mehta, M.; Munshi, M.; Adiga, V.; Ahuja, V.K.; Shandil, R.K.; Rajmani, R.S.; et al. Targeting redox heterogeneity to counteract drug tolerance in replicating Mycobacterium tuberculosis. *Sci. Transl. Med.* **2019**, *11*, eaaw6635. [CrossRef]
103. Padmapriyadarsini, C.; Bhavani, P.K.; Natrajan, M.; Ponnuraja, C.; Kumar, H.; Gomathy, S.N.; Guleria, R.; Jawahar, S.M.; Singh, M.; Balganesh, T.; et al. Evaluation of metformin in combination with rifampicin containing antituberculosis therapy in patients with new, smear-positive pulmonary tuberculosis (METRIF): Study protocol for a randomised clinical trial. *BMJ Open* **2019**, *9*, e024363. [CrossRef]
104. Mdluli, K.; Kaneko, T.; Upton, A. The Tuberculosis Drug Discovery andDevelopment Pipeline and Emerging Drug Targets. *Cold Spring Harb. Perspect. Med.* **2015**, *5*, a021154. [CrossRef]
105. Roy, K.K.; Wani, M.A. Emerging opportunities of exploiting electron transport chain pathway for drug resistant tuberculosis drug discovery. *Expert Opin. Drug Discov.* **2020**, *15*, 231–241. [CrossRef]
106. Mitchison, D.A. The action of antituberculosis drugs in short-course chemotherapy. *Tubercle* **1985**, *66*, 219–225. [CrossRef]
107. Jindani, A.; Doré, C.J.; Mitchison, D.A. The bactericidal and sterilising activities of antituberculosis drugs during the first 14 days. *Am. J. Respir. Crit. Care Med.* **2003**, *167*, 1348–1354. [CrossRef]
108. Hernandez-Pando, R. Persistence of DNA from Mycobacterium tuberculosis in superficially normal lung tissue during latent infection. *Lancet* **2004**, *356*, 2133–2137. [CrossRef]
109. Mitchison, D.A. The search for new sterilizing anti-tuberculosis drugs. *Front. Biosci.* **2004**, *9*, 1059–1072. [CrossRef] [PubMed]
110. Waynel, L.G.; Hilda, A. Metronidazole is bactericidal to dormant cells of Mycobacterium tuberculosis. *Antimicrob. Agents Chemother.* **1994**, *38*, 2054–2058. [CrossRef] [PubMed]
111. Carroll, M.W. Efficacy and safety of metronidazole for pulmonary multidrug resistant tuberculosis. *Antimicrob. Agents Chemother.* **2013**, *57*, 3903–3909. [CrossRef]
112. Tyagi, S.; Nuermberger, E.; Yoshimatsu, T.; Williams, K.; Rosenthal, I.; Lounis, N.; Bishai, W.; Grosset, J. Bactericidal activity of the nitroimidazopyranPA-824 in a murine model of tuberculosis. *Antimicrob. Agents Chemother.* **2005**, *49*, 2289–2293. [CrossRef]
113. Dooley, K.E.; Hanna, D.; Mave, V.; Eisenach, K.; Savic, R.M. Advancing the development of new tuberculosis treatment regimens: The essential role of translational and clinical pharmacology and microbiology. *PLoS Med.* **2019**, *16*, e1002842. [CrossRef]
114. Nimmo, C.; Naidoo, K.; O'Donnell, M.; Bolhuis, M.S.; Van Der Werf, T.S.; Akkerman, O.W.; Conradie, F.; Everitt, D.; Crook, A.M. Treatment of Highly Drug-Resistant Pulmonary Tuberculosis. *N. Engl. J. Med.* **2020**, *382*, 893–902.
115. Diacon, A.H.; Dawson, R.; Von Groote-Bidlingmaier, F.; Symons, G.; Venter, A.; Donald, P.R.; Van Niekerk, C.; Everitt, D.; Winter, H.; Becker, P.; et al. 14-Day bactericidal activity of PA-824, bedaquiline, pyrazinamide, and moxifloxacin combinations: A randomised trial. *Lancet* **2012**, *380*, 986–993. [CrossRef]
116. McHugh, T.D.; Honeyborne, I.; Lipman, M.; Zumla, A. Revolutionary new regimens for multidrug resistant tuberculosis. *Lancet* **2018**, *19*, 233–234. [CrossRef]
117. Working Group on New TB Drugs. Available online: https://www.newtbdrugs.org/ (accessed on 13 April 2020).
118. TB Facts.org, New-Tb-Drugs. Available online: https://tbfacts.org/new-tb-drugs/ (accessed on 13 April 2020).

© 2020 by the authors. Licensee MDPI, Basel, Switzerland. This article is an open access article distributed under the terms and conditions of the Creative Commons Attribution (CC BY) license (http://creativecommons.org/licenses/by/4.0/).

Commentary

TB Elimination Requires Discovery and Development of Transformational Agents

Christian Lienhardt [1] and Mario C. Raviglione [2,*]

[1] Unité Mixte Internationale TransVIHMI (UMI 233 IRD–U1175 INSERM, Université de Montpellier)-Institut de Recherche pour le Développement (IRD), 34000 Montpellier, France; christian.lienhardt@ird.fr
[2] Centre for Multidisciplinary Research in Health Science (MACH), Università di Milano, Milan, Italy and Global Studies Institute, Université de Genève, 1211 Genève, Switzerland
* Correspondence: mario.raviglione@unimi.it

Received: 24 March 2020; Accepted: 7 April 2020; Published: 10 April 2020

Abstract: The World Health Organization (WHO) End Tuberculosis (TB) Strategy has set ambitious targets to reduce 2015 TB incidence and deaths by 80% and 90%, respectively, by the year 2030. Given the current rate of TB incidence decline (about 2% per year annually), reaching these targets will require new transformational tools and innovative ways to deliver them. In addition to improved tests for early and rapid detection of TB and universal drug-susceptibility testing, as well as novel vaccines for improved prevention, better, safer, shorter and more efficacious treatments for all forms of TB are needed. Only a handful of new drugs are currently in phase II or III clinical trials, and a few combination regimens are being tested, mainly for drug-resistant TB. In this article, capitalising on an increasingly rich medicine pipeline and taking advantage of new methodological designs with great potential, the main areas where progress is needed for a transformational improvement of treatment of all forms of TB are described.

Keywords: tuberculosis treatment; biomarkers; drug combination; clinical trial

On 14 May 2014, the 67th World Health Assembly (WHA) endorsed a resolution detailing the global strategy to control and eliminate tuberculosis (TB) in the 2015–2030 Sustainable Development Goal (SDG) era [1]. The 3-pillar strategy was branded as the "End TB Strategy" [2]. Additional to two pillars devoted to patient-centred care and to health system policies, a third pillar is fully devoted to research and the need of innovations. This pillar and its components are fundamental in reaching ambitious international targets set as part of the new strategy to "End TB": to reduce 2015 TB incidence and deaths by 80% and 90%, respectively, by the year 2030. The simple projection model underpinning such figures, presented at the 67th WHA, is based on previous empirical experiences showing the plausibility of declining trends conducive to those targets. The large-scale interventions promoted by the End TB Strategy would be able to reduce TB incidence at a much higher annual rate than the current 1.5–2% per year [3]. For instance, it is known that in the Netherlands, United Kingdom, Germany and other western European countries, TB incidence was declining at 8%–10% per year during the late 1950s and 1960s thanks to wide access to diagnosis and effective chemotherapy, screening of people at risk, affordable care and social protection mechanisms [4]. It is also known that additional intensive interventions, including large-scale preventive therapy, were associated with even faster declines reaching 17% per year among the small Inuit populations of Alaska and North-Western Territory of Canada [5]. However, while these declines could in theory be achieved with optimal implementation of existing diagnostic and treatment tools—which are better than those available 70 years ago—the needed acceleration towards 15%–20% incidence decline per year to reach the 2030 targets will require new transformational tools and innovative ways to deliver them. These tools need to cover all aspects of TB care and its cascade, as well as prevention.

1. Which New Tools Are Necessary to End the Tuberculosis (TB) Epidemic?

First of all, achieving the End TB targets will require improved tests for early and rapid detection of TB and for universal drug-susceptibility testing (DST) to reach more patients when they first seek care so that one can cut transmission early and accelerate the decline in TB incidence and mortality. Diagnostics must be rapid, precise, connectable, available at the point-of-care and effective in detecting both active disease and latent infection [6]. Molecular diagnostics and sequencing technology available today are powerful tools, but still need relatively sophisticated laboratory capacity, and are not offering a point of care solution thus limiting their efficacy in the field.

Second, we need better, safer, shorter and more efficacious treatment for all forms of TB. The current research and development pipeline—although the most populated seen in the past few decades—still shows only less than 10 new agents in phase 1 trials, a handful of new drugs in phase 2 or 3 trials, and a few combination regimens being tested mainly for drug-resistant TB [7]. Completely new classes of anti-TB compounds are scarce, and potential synergistic combinations still unknown. Furthermore, targeting better treatment of active TB towards elimination goals is clearly not enough, as the reservoir of latent TB infection needs to be tackled as well. Mathematical models show that elimination may not be possible without targeting simultaneously latent infection and active disease [8]. Thus, addressing the vast pool of at-risk individuals among the estimated 1.7 billion people who may be latently infected is paramount, and solutions must be found that are simple and can be implemented safely among those at the highest risk of disease. Currently, the pipeline for treatment of latent infection includes a few trials using essentially known drugs such as isoniazid, rifampicin and rifapentine in different combinations, dosage and duration.

Lastly, a vaccine would be the ultimate solution if found to be highly effective, safe, able to prevent pre-exposure infection as well as reactivation. However, at the moment, the most advanced and promising vaccine candidate among the 14 in the current pipeline offers at best a 50% protection among those with latent infection, and this still needs confirmation [9]. A fully effective potent vaccine is not envisaged for several years, and will certainly not be available on time to allow reaching the 2030 targets.

Notwithstanding the importance of new, rapid diagnostics and efficacious vaccines, efforts to populate the pipeline and accelerate anti-TB drug research are a top priority for investors and researchers alike. Lately, after several years of stagnation, financial investments have been growing slowly passing 800 million US$ in 2019 [10]. However, this is far from sufficient to truly accelerate research toward new agents and regimens. The modest target of 2 billion US$/year, promoted within an existing international plan [11], is not at reach at the moment.

At the same time, efforts in drug research begun in the early 2000s resulted in some successes with the discovery and development of agents such as bedaquiline and delamanid that are recommended for the treatment of drug-resistant TB. Recently, the combination of two new drugs—pretomanid and bedaquiline—with a re-purposed agent, linezolid, has resulted in unprecedented high rates of cure among advanced forms of multidrug-resistant TB (MDR-TB) and of extensively drug-resistant TB (XDR-TB) (Nix-TB trial) [12]. The demonstrated 90% cure rates observed in a single-arm, open-label trial in South Africa has prompted approval by the US Food and Drug Administration (FDA), despite the high frequency of adverse events, including bone marrow suppression and peripheral neuropathy, due to linezolid. These advances show that with wise investments and a well-thought strategy that pursues development of a full regimen composed of new and existing or re-purposed drugs rather than of individual new drugs-as promoted by the WHO Target Regimen Profiles [13] -may be conducive to success despite the obvious challenges that TB research poses. This now needs to be consolidated and accelerated, so that we can hope to obtain a shorter and safer novel regimen that can treat TB irrespective of pre-existing drug resistance (and thus with reduced need for drug-resistance testing).

2. Which New Drugs and Regimens?

First, further progress would require improving molecules of known classes. The Nix-TB trial has raised much hope for the development of a short fully oral regimen for the treatment of severe drug-resistant TB. The use of the tested regimen may, however, be limited in clinical practice due to severe side-effects caused by linezolid. While a trial is on-going to test lower dose of linezolid, [14] new oxazolidinones are being developed to try and reduce toxic effects. Four oxazolidinones are now in early phases of the clinical development pipeline, i.e., contezolid, delpazolid, sutezolid and TBI-223 [7]. Similarly, the use of clofazimine, another important component of the therapeutic armamentarium against drug-resistant TB, is hampered by potential undesirable skin pigmentation. Novel riminophenazine derivatives are being developed with the goal of maintaining potent anti-tuberculosis activity while lowering side effects–such as TBI-166, a compound currently in clinical development in China [15].

The second requirement is that of developing new agents or, better classes of anti-TB drugs.

While improving existing classes has the advantage of initiating the risky discovery process with a well-characterized, validated compound, some level fast adaptation of bacterial populations can be expected. Ideally, therefore, research and development should produce entirely new classes, targets and modes of action to avoid cross-resistance to existing antibiotics. Thus, in parallel with improvements in existing drug and regimen models, innovative approaches, including discovery of novel chemical scaffolds and identification of new targets, are urgently needed.

The discovery of bedaquiline opened the way to investigating the possibility to alter the energy metabolism in mycobacteria, in particular the oxidative phosphorylation pathway, as a novel target pathway in drug discovery, leading to the depletion of ATP synthesis of *M. tuberculosis*. New classes of antibacterial drugs interfering with elements of this pathway have been shown to be highly active in combating latent mycobacterial infections. The discovery of Q203, a candidate drug targeting the cytochrome bc1 complex in the respiratory chain, has highlighted the importance of this new target pathway. Inhibiting the bacterial energy metabolism might be a key feature of novel and sterilizing drug combinations for the treatment of TB [16]. Furthermore, Q203 was shown to have good synergy with bedaquiline in the murine chronic infection model, indicating promising potential for new treatment regimens [17]. Thus, the combination of drugs targeting various elements of the oxidative phosphorylation pathway could lead to a completely new regimen for drug-susceptible and multi-drug resistant tuberculosis.

A series of new compounds are currently developed that focus on newly identified targets. Four of these inhibit DprE1 (decaprenylphosphoryl-β-D-ribose 2-epimerase), a flavoenzyme that catalyses a key step in the synthesis of the complex cell wall of *M. tuberculosis*: macozinone, BTZ-043, OPC-167832 and TBA-7371 [7]. Another compound targets the leucyl-tRNA synthetase (LeuRS), which is essential for protein synthesis (GSK3036656).

The third approach is to identify efficient and seamless development processes to accelerate testing of novel treatment regimens. The current TB treatment development pathway is complex, lengthy and costly, partly due to the fact that some drugs are still being developed individually, and partly due to the lack of reliable surrogate markers of treatment outcomes and the lack of predictive quantitative relationships between Phase II and Phase III outcomes [18]. Currently, Phase II TB drug development includes 14-day early bactericidal activity monotherapy studies to identify the maximally efficacious dose for a new chemical entity (Phase IIA), followed by 2-month serial sputum colony count studies (Phase IIB), in which the efficacy of treatment combinations is usually studied with time-to-sputum-culture-conversion as the primary endpoint. In addition to long duration, this approach suffers multiple weaknesses, including an inadequate exploration of dose-response, the lack of means to determine early the optimal combination of drugs to be tested and the optimal duration of therapy, as well as the inability to study multiple regimens in parallel [19].

Research is needed to identify biomarkers that could predict relapse and guide selection of suitable drug combination(s) and treatment duration(s) so as to accelerate drug development in TB. Novel approaches are being explored to identify early and streamline suitable drug combinations to advance from early to late phases of development taking into account new developments in pharmacokinetic and pharmacodynamic methodology and modelling [20], as well as novel adaptive designs [21]. The "multi-arm multi-stage" (MAMS) design, initially used in oncology, allows testing a broad range of combinations and dose levels without requiring a large sample size, dropping early arms that do not meet pre-specified efficacy threshold [22]. Recently, a newer approach has been proposed that combines Phase II and Phase III features. In this design, named "selection trial with extended post-treatment follow-up" (STEP), limited long-term follow-up data on relapse are collected, together with data on culture conversion, permitting estimation of a Bayesian prediction interval for the likely results of a future Phase III trial [23]. Such Phase IIB/C studies, with arms testing different doses and durations, coupled with the use of novel biomarkers for sterilising cures—these being either RNA expression, cytokine, bacterial or radiological markers—would strengthen and accelerate the process for identifying candidate regimens likely to succeed in Phase III [24], as well as prospectively validating novel biomarkers against the relapse endpoint.

3. Conclusions

We are now at a crossroads in new anti-tuberculosis drug and regimen development. New initiatives, such as the the European Union (EU) Innovative Medicines Initiative IMI2 call and the Gates Medical Research Institute alliance, have the potential to accelerate research and synergize with existing efforts carried out by academia, public-private partnerships, such as the TB Alliance, and industry. If researchers collaborate and join forces towards the common aim of innovative regimens, there is a possibility of pursuing effectively a transformational improvement of treatment of all forms of TB [25]. If, however, such initiatives compete for limited resources and resist cooperation, the targets expressed at the time of the Moscow Conference in late 2017 and subsequently at the United Nations General Assembly in 2018 may remain a dream.

Author Contributions: The two authors contributed equally to the writing of the article. All authors have read and agreed to the published version of the manuscript.

Funding: This research received no external funding.

Conflicts of Interest: The authors declare no conflict of interest.

References

1. World Health Organization. *Sixty-Seventh World Health Assembly WHA67.1*; World Health Organization: Geneva, Switzerland, 2014.
2. World Health Organization. *Global Strategy and Targets for Tuberculosis Prevention, Care and Control after 2015*; Agenda item 12.1, 21 May 2014; World Health Organization: Geneva, Switzerland, 2014.
3. Uplekar, M.; Weil, D.; Lönnroth, K.; Jaramillo, E.; Lienhardt, C.; Dias, H.M.; Falzon, D.; Floyd, K.; Gargioni, G.; Getahun, H.; et al. WHO's End TB Strategy. *Lancet* **2015**, *385*, 1799–1801. [CrossRef]
4. World Health Organization. *Global Tuberculosis Report 2019*; World Health Organization: Geneva, Switzerland, 2019.
5. Styblo, K. *Epidemiology of Tuberculosis*; The Royal Netherlands Tuberculosis Association (KNCV): The Hague, The Netherlands, 1991.
6. Grzybowski, S.; Styblo, K.; Dorken, E. Tuberculosis in Eskimos. *Tubercle* **1976**, *57*, 1–58. [CrossRef]
7. Walzl, G.; McNerney, R.; du Plessis, N.; Bates, M.; McHugh, T.D.; Chegou, N.N.; Zumla, A. Tuberculosis: Advances and challenges in development of new diagnostics and biomarkers. *Lancet Infect Dis.* **2018**, *18*, e199–e210. [CrossRef]
8. Stop TB Partnership's Working Group on New TB Drugs Clinical Pipeline. Available online: https://www.newtbdrugs.org/pipeline/clinical (accessed on 30 March 2020).
9. Dye, C.; Williams, B.G. Eliminating human tuberculosis in the 21st century. *J. R. Soc. Interface* **2008**, *5*, 653–662. [CrossRef] [PubMed]

10. Tait, D.R.; Hatherill, M.; Van Der Meeren, O.; Ginsberg, A.M.; Van Brakel, E.; Salaun, B.; Scriba, T.J.; Akite, E.J.; Ayles, H.M.; Bollaerts, A.; et al. Final Analysis of a Trial of M72/AS01$_E$ Vaccine to Prevent Tuberculosis. *N. Engl. J. Med.* **2019**, *381*, 2429–2439. [CrossRef]
11. Treatment Action Group. *Tuberculosis Research Funding Trends 2005–2018*; Treatment Action Group: New York, NY, USA, 2019; ISBN 978-0-9983966-8-2.
12. The Lancet Commission on TB. Building a Tuberculosis-Free World: The Lancet Commission on Tuberculosis. *Lancet* **2019**. [CrossRef]
13. Conradie, F.; Diacon, A.H.; Ngubane, N.; Howell, P.; Everitt, D.; Crook, A.M.; Mendel, C.M.; Egizi, E.; Moreira, J.; Timm, J.; et al. Treatment of Highly Drug-Resistant Pulmonary Tuberculosis. *N. Engl. J. Med.* **2020**, *382*, 893–902. [CrossRef]
14. World Health Organization. *Target Regimen Profiles for TB Treatment: Candidates: rifampicin-Susceptible, Rifampicinresistant and pan-TB Treatment Regimens*; World Health Organization: Geneva, Switzerland, 2016.
15. Zhang, D.; Liu, Y.; Zhang, C.; Zhang, H.; Wang, B.; Xu, J.; Fu, L.; Yin, D.; Cooper, C.B.; Ma, Z.; et al. Synthesis and biological evaluation of novel 2-methoxypyridylamino-substituted riminophenazine derivatives as antituberculosis agents. *Molecules* **2014**, *19*, 4380–4394. [CrossRef]
16. Bald, D.; Villellas, C.; Lu, P.; Koul, A. Targeting energy metabolism in *Mycobacterium tuberculosis*, a new paradigm in antimycobacterial drug discovery. *mBio* **2017**, *8*, e00272-17. [CrossRef]
17. Shi-Yan Foo, C.; Pethe, K.; Lupien, A. Oxidative Phosphorylation—an Update on a New, Essential Target Space for Drug Discovery in Mycobacterium tuberculosis. *Appl. Sci.* **2020**, *10*, 2339. [CrossRef]
18. Ginsberg, A.M.; Spigelman, M. Challenges in tuberculosis drug research and development. *Nat Med.* **2007**, *13*, 290–294. [CrossRef]
19. Lienhardt, C.; Nahid, P. Advances in clinical trial design for development of new TB treatments: A call for innovation. *PLoS Med.* **2019**, *16*, e1002769. [CrossRef]
20. Dooley, K.E.; Hanna, D.; Mave, V.; Eisenach, K.; Savic, R.M. Advancing the development of new tuberculosis treatment regimens: The essential role of translational and clinical pharmacology and microbiology. *PLoS Med.* **2019**, *16*, e1002842. [CrossRef] [PubMed]
21. Davies, G.R.; Phillips, P.P.J.; Jaki, T. Adaptive clinical trials in tuberculosis: Applications, challenges and solutions. *Int. J. Tuberc. Lung Dis.* **2015**, *19*, 626–634. [CrossRef]
22. Boeree, M.J.; Heinrich, N.; Aarnoutse, R.; Diacon, A.H.; Dawson, R.; Rehal, S.; Kibiki, G.S.; Churchyard, G.; Sanne, I.; Ntinginya, N.E.; et al. High-dose rifampicin, moxifloxacin, and SQ109 for treating tuberculosis: A multi-arm, multi-stage randomised controlled trial. *Lancet Infect Dis.* **2017**, *17*, 39–49. [CrossRef]
23. Phillips, P.P.J.; Dooley, K.E.; Gillespie, S.H.; Heinrich, N.; Stout, J.E.; Nahid, P.; Diacon, A.H.; Aarnoutse, R.E.; Kibiki, G.S.; Boeree, M.J.; et al. A new trial design to accelerate tuberculosis drug development: The Phase IIC Selection Trial with Extended Post treatment follow-up (STEP). *BMC Med.* **2016**, *14*, 51. [CrossRef] [PubMed]
24. Davies, G.; Hoelscher, M.; Boeree, M.; Hermann, D. Accelerating the transition of new tuberculosis drug combinations from Phase II to Phase III trials: New technologies and innovative designs. *PLoS Med.* **2019**, *16*, e1002851. [CrossRef]
25. Lienhardt, C.; Nunn, A.; Chaisson, R.; Vernon, A.A.; Zignol, M.; Nahid, P.; Delaporte, E.; Kasaeva, T. Advances in clinical trial design: Weaving tomorrow's TB treatments. *PLoS Med.* **2020**, *17*, e1003059. [CrossRef]

© 2020 by the authors. Licensee MDPI, Basel, Switzerland. This article is an open access article distributed under the terms and conditions of the Creative Commons Attribution (CC BY) license (http://creativecommons.org/licenses/by/4.0/).

Review

Drug-Resistant Tuberculosis 2020: Where We Stand

Angelo Iacobino, Lanfranco Fattorini * and Federico Giannoni

Istituto Superiore di Sanità, Department of Infectious Diseases, Via Regina Elena 299, 00161 Rome, Italy; angelo.iacobino@iss.it (A.I.); federico.giannoni@iss.it (F.G.)
* Correspondence: lanfranco.fattorini@iss.it

Received: 3 March 2020; Accepted: 15 March 2020; Published: 22 March 2020

Featured Application: This comprehensive overview of drug-resistant tuberculosis will be useful for researchers to expand their knowledge beyond mechanisms other than chromosomal mutations, and for the development of novel drugs/drug combinations, hoping to shorten the therapy of the disease.

Abstract: The control of tuberculosis (TB) is hampered by the emergence of multidrug-resistant (MDR) *Mycobacterium tuberculosis* (Mtb) strains, defined as resistant to at least isoniazid and rifampin, the two bactericidal drugs essential for the treatment of the disease. Due to the worldwide estimate of almost half a million incident cases of MDR/rifampin-resistant TB, it is important to continuously update the knowledge on the mechanisms involved in the development of this phenomenon. Clinical, biological and microbiological reasons account for the generation of resistance, including: (i) nonadherence of patients to their therapy, and/or errors of physicians in therapy management, (ii) complexity and poor vascularization of granulomatous lesions, which obstruct drug distribution to some sites, resulting in resistance development, (iii) intrinsic drug resistance of tubercle bacilli, (iv) formation of non-replicating, drug-tolerant bacilli inside the granulomas, (v) development of mutations in Mtb genes, which are the most important molecular mechanisms of resistance. This review provides a comprehensive overview of these issues, and releases up-dated information on the therapeutic strategies recently endorsed and recommended by the World Health Organization to facilitate the clinical and microbiological management of drug-resistant TB at the global level, with attention also to the most recent diagnostic methods.

Keywords: tuberculosis; *Mycobacterium tuberculosis*; rifampin; isoniazid; mechanisms of resistance; mutations; granulomas; caseum; cell envelope; dormancy

1. Introduction

Mycobacterium tuberculosis (Mtb) is the etiologic agent of tuberculosis (TB), the leading cause of death from a single infectious disease agent worldwide [1]. In 2018, the World Health Organization (WHO) estimates of the global burden of TB were 10 million cases and 1.45 million deaths. Furthermore, about 1.7 billion people are known to be latently infected with Mtb, with about 10% of them reactivating to active TB in their lifetime. The current antibiotic treatment of active, drug-susceptible TB, requires administration of a combination therapy for 6 months, including the first-line drugs rifampin (RIF), isoniazid (INH), pyrazinamide (PZA) and ethambutol (EMB) for 2 months, followed by RIF and INH for 4 months. To prevent reactivation of latent TB, a long treatment is also used, consisting of at least 6 months of INH, or 3 to 4 months of RIF plus INH [1,2].

Poor regimen selection, inadequate drug supply and poor adherence of patients to the 6-months therapy may lead to development of drug-resistant Mtb strains, including multidrug-resistant (MDR: resistant at least to INH and RIF) and extensively-drug-resistant (XDR) strains [MDR resistant to a fluoroquinolone (FQ) and a second-line injectable drug [kanamycin (KM), amikacin (AM), capreomycin

(CM)] [3]. Shortening the duration of therapy could increase adherence to treatment and reduce development of MDR and XDR TB.

The goal of this Review is to give a comprehensive overview of the interplay of clinical, biological and microbiological factors involved in the development of drug-resistant TB.

2. Epidemiology of Drug Resistant TB

WHO reported that in 2018 there were an estimated 484,000 incident cases of MDR/rifampin-resistant (RR) TB cases, including about 378,000 MDR-TB cases and 214,000 deaths [1]. The average proportion of MDR-TB cases with XDR-TB was 6.2%. The countries accounting for 50% of the global burden of MDR/RR-TB were India (27%), China (14%) and the Russian Federation (9%). Among 24 countries with a high TB or MDR-TB burden and representative data to second-line drugs, the proportion of MDR/RR TB cases with resistance to any FQ including ofloxacin (OFL), levofloxacin (LFX) and moxifloxacin (MFX) was 20.8%. At the global level, 3.4% of new cases (patients never treated with anti-TB medicines, or treated for < 1 month) and 18% of previously treated cases (patients treated for ≥ 1 month in the past) had MDR/RR-TB, with the highest proportion occurring in the former Soviet Union (FSU) countries. In the low incidence countries of the European Economic Area, the MDR-TB was more prevalent among migrants (particularly from the FSU) than the native population [4,5].

3. Recent WHO Recommendations for the Treatment of MDR/RR TB

In the last decades, WHO made great efforts to facilitate and improve the treatment of patients with MDR-TB in high burden countries using various actions including the Directly Observed Treatment Strategy (DOTS)-Plus, to stress the use of second-line drugs in low- and middle-income settings, but the cure rate was lower than the WHO 2015 target of at least 75% to 90% [6]. For instance, treatment success for MDR/RR-TB cases started on treatment in 2016 in India, China and Russian Federation was 48%, 52% and 54%, respectively [1].

In 2018, the results from an individual patient data meta-analysis involving 12,030 patients from 25 countries showed that treatment success and death of pulmonary MDR-TB were significantly reduced after the administration of the newer or repurposed drugs linezolid (LZD), later generation FQs, bedaquiline (BDQ), clofazimine (CFZ) and carbapenems [7]. On the basis of this and other studies, in March 2019, WHO released a new drug classification and new recommendations for the treatment of MDR-TB [8–10]. The second-line drugs were reorganized into three groups, including priority drugs [Group A: LFX or MFX, BDQ, LZD], preferentially used drugs [Group B: CFZ, cycloserine (CS) or terizidone (TRD)], and other drugs [Group C: EMB, delamanid (DLM), PZA, imipenem-cilastatin (IPM-CLN) or meropenem (MPM) (administered with clavulanic acid, CLV), AM or streptomycin (SM), ethionamide (ETO) or protionamide (PTO), para-aminosalicylic acid (PAS)].

In summary, WHO recommended an injection-free therapy (groups A and B drugs) at the initiation of MDR-TB treatment. Group C agents (oral and parenteral) should be administered when groups A and B drugs cannot be used. The commonly used second-line injectable drugs KM and CM were associated with worse outcomes [7], and were no longer recommended for the treatment of MDR-TB; AK and SM may be administered only if drug susceptibility testing (DST) confirms susceptibility.

To further reduce the burden of drug-resistant TB in the near future, in December 2019, WHO also released a Rapid Communication to inform countries and stakeholders that a regimen containing BDQ, pretomanid (PRT, formerly PA-824) and LNZ (BPaL regimen) may be used under operational research conditions conforming to WHO standards for the treatment of XDR-TB patients [11]. This communication was released after a previous announcement of the Global TB Alliance in the second half of 2019, following the decision of the United States Food and Drug Administration (FDA) to administer BPaL (Nix-TB trial by the Global TB Alliance) to adults with pulmonary XDR-TB or intolerant/not responsive MDR-TB [12].

4. Drug Resistance Mechanisms in TB

If the 6-months combination therapy for the treatment of drug-susceptible TB is adequately taken, patients achieve cure rates of > 95%, and the development of resistance by simultaneous mutations to various drugs is very rare [13]. The resistance developed by Mtb to any antimicrobial agent is not due to a single mechanism, but to the interplay of biological, clinical and microbiological reasons, including:

1. Nonadherence of patients to their 6-months therapy and/or errors of physicians in the therapy management (human errors), that increases the risk of developing genetically drug resistant bacilli [14,15];
2. Complexity and poor vascularization of granulomatous lesions, which obstruct drug distribution to some sites, further leading to suboptimal drug concentration and the development of phenotypic and genetic resistance [16,17];
3. Naturally occurring high levels of antibiotic resistance in tubercle bacilli (intrinsic resistance) [18–20];
4. Formation of non-replicating (NR) drug-tolerant bacilli inside the granulomas (phenotypic resistance) [21,22];
5. Development of genetically resistant bacilli by chromosomal mutations (acquired resistance) [23–25].

4.1. Human Errors and Advances in MDR-TB Management

Human errors may contribute to the development of drug-resistance because of the improper use of anti-TB drugs. Two pathways lead to the development of genetic resistance: (i) primary resistance, when a person is infected with a drug-resistant strain, and (ii) acquired resistance, when a person infected with a drug susceptible strain is inadequately treated with drugs, allowing the selection of resistant mutants [13]. The first case mostly occurs in highly crowded communities (e.g., prisons), or in countries with high MDR-TB prevalence, where it is essential to rapidly diagnose and treat patients, so as to reduce transmission [15]. In the second case, it is essential to follow the WHO recommendations on how to adequately treat the TB patients whose disease is caused by a drug-susceptible strain. The clinicians need also to ensure that infection control measures are established, particularly when MDR-patients are hospitalized [26].

To implement the Stop TB Strategy (developed from the DOTS framework), WHO identified a number of factors contributing to poor treatment outcomes, including the acquisition of acquired drug-resistant TB [13]. They were: (i) Inappropriate treatment by health care providers (inappropriate or absent guidelines, poor training of physicians and nurses, sub-optimal education of patients, poor management of adverse drug reactions, no monitoring of treatment, poorly organized or funded TB control programs); (ii) Inadequate drug supply (poor quality medicines, stock-outs, poor storage conditions, wrong dose or combination); (iii) Inadequate drug intake or treatment response by patients (lack of information on treatment adherence, adverse effects, malabsorption).

Common clinical errors in MDR-TB management, particularly in developing countries, include the addition of a single drug to a failing regimen, failure to recognize existing drug resistance, failure to provide directly observed therapy and to manage nonadherence, suboptimal dosages of second-line drugs to decrease side effects, drug treatment based on clinical facts while waiting for DST results [13,14]. In any case, it is important to know that only drug combinations decrease the risk of selection of resistant strains.

Since the treatment of MDR/XDR-TB cases is difficult, WHO recommended that their management be performed by a multidisciplinary team (TB Consilium) at local, regional and/or national levels, including experts (e.g., clinicians, microbiologists, public health officers) with different professional backgrounds [13,15]. In medium and high incidence countries, TB Consilia are important for accessing second-line drugs and/or new drugs BDQ and DLM.

A comparative analysis of the TB Consilia for management of difficult-to-treat MDR/XDR TB cases in Europe and Latin America has been reported [27]. In October 2018, a Global TB Consilium was launched by the Global TB Network, with the goal to provide to a clinician a response on difficult TB cases within 48 h [15,28].

4.2. Complexity of TB Granulomas

Long lasting therapies are also attributable to the complex pathology of TB. In the lungs of patients with active and latent TB, a spectrum of heterogeneous granulomatous lesions coexist, ranging from well-vascularized cellular granulomas, in which a rim of lymphocytes surrounds macrophages and neutrophils, to avascular caseous granulomas, characterized by a necrotic center with a cheese-like aspect (caseum) formed by the lysis of host cells and bacteria [29,30]. In these lesions, tubercle bacilli range from actively replicating (AR) stages, particularly in cellular granulomas, to dormant, slowly-replicating or NR stages, typical of hypoxic caseous granulomas [31]. In Mtb-infected rabbits, the fraction unbound of a drug penetrates the caseum via passive diffusion, and caseum binding of a drug is proportional to hydrophobicity (cLogP) and aromatic ring count [32].

The current 4-drugs therapeutic regimen (RIF-INH-PZA-EMB) is effective against AR intracellular bacilli in cellular granulomas, while NR extracellular bacilli localized in pH-neutral, caseous granulomas are refractory to drug action [17,33–36]. The necrotic center of caseous granulomas contains NR bacilli phenotypically resistant to several drugs (drug-tolerant persisters), with the exception of rifamycins, which are known to sterilize caseum in ex-vivo assays [35,36]. Spatial and temporal differences in drug distribution and the kinetics of accumulation of drugs in specific lesion compartments may create local windows of monotherapy that increase the risk of the emergence of drug-resistance [17,37]. This is in keeping with the knowledge that genetically resistant mutants of Mtb may emerge from the persistence phase of some TB drugs, due to hydroxyl radical-mediated genome-wide random mutagenesis [38–40]. In this view, drug combinations should contain complementary drugs preferentially distributing in lesions in which their most vulnerable target population resides [17].

In the event of caseous granulomas expansion, the necrotic centers fuse with the airway structures of bronchi to form pulmonary cavities in which are found both extracellular bacilli from liquefied caseum and intracellular bacilli derived from the lysis of infected macrophages of the cavity walls. In contact with the atmospheric oxygen, these bacilli rapidly proliferate in the lumen of cavities, and later appear in the sputum of TB patients [17]. Due to high bacterial load in pulmonary cavities, genetically resistant bacilli with chromosomal mutations may be generated, playing an important role in the development of resistance [16]. Noticeably, in comparison with paired sputum isolates, additional resistances were found in Mtb isolates recovered from surgically resected cavities of the same patient [41]. A single founder Mtb strain underwent genetic mutations during treatment, leading to the acquisition of additional drug resistance in different sections of the lung of the same patient, preferentially in the cavity wall [42]. In keeping with this observation, drug-specific gradients in the walls of human pulmonary cavities were reported to be associated with the development of acquired resistance in patients with MDR-TB, due to the low level of some drugs in the cavities centers, where there is a high number of replicating bacilli [43]. In the latter study, spatial heterogeneity of drug concentrations across the pulmonary cavity resulted in the development of mutations in the Mtb genes *gyrA* (FQ resistance) and *gydB* (aminoglycoside resistance), consistent with evolution from MDR- to XDR-TB after about five months of therapy [43]. Overall, these observations indicate that acquired Mtb resistance may be related to the formation of drug-penetration gradients in TB lesions generating suboptimal drug concentrations in non-vascularized caseous granulomas and in liquefied caseum in the cavity centers [16,17,43].

4.3. Intrinsic Drug-Resistance of Mtb

During the evolution, Mtb developed mechanisms of intrinsic resistance to antibiotics involving cell envelope, efflux systems and other mechanisms (drug degradation and modification, target modification), which allowed the organism to reach high drug resistance levels. Some examples of these mechanisms are provided in the following sections.

4.3.1. Cell Envelope

The constituents of the mycobacterial cell envelope are: the cytoplasmic membrane, the periplasmic space (PS), a network of peptidoglycan (PG), the arabinogalactan (AG), the long-chain mycolic acids (MA) and the capsule, made of a loose matrix of glucans and secreted proteins [44]. As to the first-line TB drugs, the bactericidal agent INH inhibits MA synthesis, while the bacteriostatic EMB inhibits AG synthesis and may sensitize Mtb to other drugs [44].

It is assumed that the innermost hydrophilic layers of PG and AG hinder the penetration of hydrophobic molecules. Instead, in the external part of the envelope, the PG and AG layers are linked to the hydrophobic MA layer, formed by long-chain fatty acids that restrict the penetration of hydrophilic drugs [18,20]. In principle, more lipophilic drugs, such as rifamycins, macrolides, and some FQs, diffuse by passive transport into and through the lipid-rich cell wall [45,46]. In early studies, mutants defective in the biosynthesis of cell wall components were very useful to demonstrate the role of the cell wall in the intrinsic resistance to drugs. For instance, a mycolate defective *Mycobacterium smegmatis* mutant showed increased susceptibility to RIF, chloramphenicol (CF), novobiocin and erythromycin [47,48]. Also, insertions in genes involved in the mycolate biosynthesis of Mtb (*mymA* operon) showed enhanced chemical penetration and sensitivity to RIF, INH, PZA and ciprofloxacin [49].

Small hydrophilic drugs traverse the cell wall of bacteria via water-filled porins, without energy consumption. *M. tuberculosis* encodes at least two porin-like proteins (OmpA, encoded by *Rv0899* and *Rv1698*), but the role of porins in Mtb drug uptake and susceptibility needs to be further investigated [18,20,50,51]. Penetration of hydrophilic β-lactam antibiotics through the mycobacterial cell was about 100 times lower than in the *Escherichia coli* cell wall [20]. The β-lactamases, probably in conjunction with slow drug penetration, were shown to be major determinants of Mtb resistance to β-lactams [52,53]. In Mtb, the PG is remodeled by nonclassical L,D-transpeptidases (LDT). The structural basis and the inactivation mechanism of LDT and the active role of carbapenems were investigated, providing a basis for their potential use in inhibiting Mtb [54]. Indeed, the carbapenems IPM-CLN and MPM (both to be used with CLV, available only in formulations combined with amoxicillin) have been listed as add-on drugs in the recent WHO treatment guidelines of MDR/XDR TB [8].

Overall, it is thought that anti-TB drugs have the peculiarity of being more lipophilic than many other antimicrobial agents, likely due to improved penetration through the waxy mycobacterial cell wall [45,46]. However, the issue is perhaps more complex, since some studies showed that lipophilicity is an important but not exclusive factor of compound permeability [50,55].

4.3.2. Drug Efflux

Efflux pumps (EPs) are transmembrane proteins that provide resistance by expelling the drugs from the interior to the exterior of the cell. Five EP families are known, organized on the basis of energetic and structural characteristics: the ATP-binding cassette (ABC) superfamily, the major facilitator superfamily (MFS), the multidrug and toxic compound extrusion (MATE) family, the small multidrug resistance (SMR) family and the resistance nodulation division (RND) superfamily [18,19,46,56,57]. The ATP-energized ABC members are primary transporters, while the others are secondary transporters energized by proton gradients (MFS, SMR, RND) or sodium gradients (MATE). The EP of Mtb belongs to the ABC (representing 2.5% of the entire Mtb genome), MFS and RND superfamilies, and to the SMR family.

Following exposure of Mtb to sub-inhibitory concentration of INH and EMB, EP genes are overexpressed resulting in the development of low-level resistance for a prolonged period of time. After several weeks, a high level of acquired resistance develops, caused by chromosomal mutations in the genes encoding the target proteins [58,59]. These observations indicate that inappropriate TB treatment may generate pressure by sub-inhibitory drug concentrations that increase drug efflux, allowing a subsequent selection of mutants with high-level resistance [46,57].

Several EPs are known to be associated with resistance. For instance, Mtb exposure to INH induces the overexpression of *MmpL7* and *mmR* EP genes [60,61]. Furthermore, several EPs are involved in resistance to several drugs. Thus, the EP Tap mediates low-level resistance to tetracycline (TC) and aminoglycosides, whereas EPs encoded by the *Rv0194* gene is associated with resistance to β-lactams, SM, TC, CF and vancomycin. Mutations in the *Rv0678* gene caused an up-regulation of the transport protein, MmpL5, which caused EP-mediated cross-resistance to both BDQ and CFZ [60,62,63]. This is a potentially dangerous evolution of Mtb against antibiotics particularly in recent times, since BDQ and CFZ have just been included in the new WHO treatment guidelines of MDR/XDR-TB [8,11].

A strategy used to inhibit efflux-mediated drug resistance is efflux inhibition by non-antibiotic molecules that block the EP or inhibit the EP energy sources [57,64]. The most studied inhibitors are verapamil (VP), thioridazine (TZ), reserpine, piperine, protonophores [57,64], to be used in combination with anti-TB drugs in order to decrease or abolish the drug resistance caused by EP activity. Verapamil, an FDA-approved calcium channel blocker, decreased the MICs of BDQ, CFZ and other drugs [57]. This synergism was confirmed in various studies, but it was found that the effect of VP was not due to intra-mycobacterial drug accumulation, but on the disruption of membrane functions [65]. In Mtb-infected mice, VP increased the bioavailability and efficacy of BQ but not CFZ [66]. EP inhibitors are not presently used for the treatment of human TB, with the exception of TZ, which was administered in compassionate therapy for some XDR-TB cases [67].

4.3.3. Other Mechanisms

The most important mechanisms of the intrinsic drug resistance of Mtb are considered to be the lipid-rich cell wall and the EP, but other systems are known to neutralize toxic chemicals and antibiotics, including drug inactivation or modification, and target modification.

Among drug inactivating enzymes, Mtb β-lactamases are less effective than those of other bacteria to hydrolyse β-lactams, but their activity, together with slow penetration across the cell wall and low affinity for penicillin-binding proteins, is good enough to render Mtb intrinsically resistant to most β-lactams [18,20]. The most important Mtb β-lactamase (BlaC) is thought to localize in the PS, and shows broad substrate specificity, including carbapenems, which are usually resistant to the β-lactamases of other bacteria. BlaC is inhibited by CLV that, as mentioned above, must be added to IPM- or MPM-containing prescriptions, as salvage WHO regimens for treatment of MDR/XDR-TB [8].

As to aminoglycosides (KM, AK) and cyclic peptides (CP), Mtb is able to inactivate them by acetylation performed by the enhanced intracellular survival protein encoded by *eis*, whose expression is upregulated by the MDR transcription regulator WhiB7 [20]. Promoter mutations lead to an overexpression of *eis*, resulting in low-level resistance to KM, but not AK [68].

M. tuberculosis is naturally resistant to macrolides (e.g., clarithromycin and azithromycin) because of the inducible erm(37), a ribosomal RNA methyltransferase which alters ribosomes by methylating the 23S rRNA [69,70]. Other *erm* genes conferring inducible resistance to FQs were found in non-tuberculous mycobacteria [71]. Intrinsic resistance to FQ is also attributed to a pentapeptide repeat protein called MfpA, which mimic the size, shape and surface charge of duplex DNA by resembling the 3D structure of a DNA double helix [20,72].

As to resistance to the broad-spectrum agent fosfomycin, Mtb is intrinsically resistant to this agent, which inhibits the MurA enzyme, involved in the biosynthesis of PG, because a cysteine residue in the active site of Mtb MurA is changed into aspartic acid [73].

5. Phenotypic Drug-Resistance of Mtb

Caseous granulomas and the cavities of the lungs of TB patients harbor subpopulations of NR bacilli which are phenotypically drug-resistant but genetically susceptible, commonly referred to as persisters. Characterized by a transient, non-heritable drug tolerance, persisters are capable of withstanding bactericidal drug concentrations, and once the antibiotic is removed, to resume growth with genetic features identical to the original strain.

The level of resistance to different antimicrobial agents varies with the in vitro stress model used [31–34,74–77], including hypoxia (Wayne dormancy model) [78,79], nutrient starvation [80], acids and/or nitric oxide [81,82], stationary phase [83], antibiotic-starved strains [84] and others, or their variants. In the Wayne model, in which dormant bacilli are obtained by a gradual adaptation to anaerobiosis through the self-generated formation of an oxygen gradient, nonreplicating persistence (NRP) stages 1 and 2 were observed [78]. NRP-2 cells developed a thickened outer layer that helped in restricting RIF entry [85]. Our group used the Wayne model at different pHs: pH 6.6, the pH or culture media [86], pH 5.8, to mimic the environment of cellular granulomas [87], pH 7.3, to mimic the environment of caseous granulomas [88]. We found that at pH 5.8, several drugs killed NR bacilli, with the best being the rifamycins RIF and rifapentine (RFP), while at pH 7.3, only RIF and RFP killed dormant bacilli out of 12 drugs tested [88]. Since the rifamycins were the only agents sterilizing caseum obtained from rabbits [35,36], our model could mimic caseum to measure drug activity against NR Mtb in this environment. In hypoxia at pH 7.3, we found that RIF plus nitazoxanide (a nitro-compound for anaerobic infections) killed NR Mtb cells, while the combination currently used for human TB therapy (RIF-INH-PZA-EMB) did not [89].

Two kinds of persisters are known [74]: (i) Class I, rare, generated in a replicating population, formed continuously and in a purely stochastic manner. They are bacilli phenotypically tolerant to different antibiotics by different mechanisms, and it is likely that the overall population can be killed by drug combinations; (ii) Class II, abundant, involving almost all of the cells in a population, e.g., in the stationary phase, hypoxic conditions, nutrient starvation. Growth arrest is associated with resistance to a large number of drugs, and it is likely that new kinds of antibiotics are necessary to overcome these cells [74].

Dormancy is not necessary or sufficient for Mtb persistence, indicating that persistence is a phenomenon more complex than dormancy, and that additional characteristics are needed to define the persister phenotypes, which depends on the NR model used [90]. A poor correlation was found between the transcriptomes of class I persisters enriched by cycloserine [91] and class II persisters obtained under hypoxia, the stationary phase or nutrient starvation [74]. On the other hand, persister diversity is expected also from the different host environments in which these specialized cells live, ranging from the intracellular location in the phagosomes to extracellular life in the caseum. In BDQ-treated guinea pigs, persisting bacilli where located in the acellular rim of necrotic lesions, morphologically similar to human TB lung lesions [92].

The state of non-replication is associated with phenotypic drug-tolerance, but different stresses may induce phenotypically different bacilli. Few compounds were dual active molecules with bactericidal activity against both replicating and NR Mtb. They included RIF, BDQ, PRT and MFX, which target RNA polymerase, ATP synthase, cell wall synthesis/cell respiration, and DNA gyrase, respectively [24,31,74]. In BALB/c mice, persisters were eradicated by regimens containing high-dose RIF and BDQ [93,94]. In BALB/c mice, C3HeB/FeJ caseum-forming mice and athymic nude mice, PRT contributed significantly to the efficacy of BDQ-containing regimens, with either LZN (BPaL regimen) or MFX and PZA (BPaMZ regimen) [95].

Interestingly, RIF-resistant or MXF resistant mutants carrying mutations in *rpoB* or *gyrA* genes emerged at high frequency from the persistent phase of Mtb cells exposed to RIF for prolonged periods. These cells carried elevated levels of the hydroxyl radical, which inflicted genome-wide mutations facilitating resistance to the same, or another, antibiotic [38,39]. In consideration of the long TB therapy, these observations may have clinical significance in the emergence of drug-resistant mutants if local monotherapy occurs in patients who do not correctly take multi-drug TB therapy.

In this view, it was postulated that persisters behave as an evolutionary reservoir from which drug-resistant mutants can emerge [22].

Thus, targeting NR persisters could reduce the duration of antibiotic treatment and rate of post-treatment relapse [74,96]. Researches aimed at better understanding the relationship between persistence and resistance, and at finding novel drug combinations for killing both AR and NR bacilli, will provide new strategies to shorten TB therapy.

6. Acquired Drug-Resistance of Mtb

A cocktail of different drugs is used to treat TB. Each molecule binds to one or more target, thus inhibiting their functions. The continuous drug exposure during lengthy treatments and the noncompliance of patients to drug regimens, pushes Mtb to select for mutations in genes encoding drug targets, responsible for development of the majority of resistances in clinical strains [20]. A list of the major target genes that, in the case of mutation, confer resistance to the drugs of the WHO groups A, B and C, is shown in Table 1 [97–125]. Many excellent reviews report the genetic mechanisms involved in this resistance to RIF, INH, KM, CP and other drugs [18,23,24,98,118,126].

Table 1. Drugs of the World Health Organization (WHO) groups A, B and C, and list of the most common drug resistance-related target genes.

Group	Drug	Target Gene/s	Gene Product (Function Affected)	References
A	LFX or MFX	gyrA	DNA gyrase, subunit A (DNA replication)	[97,98]
		gyrB	DNA gyrase, subunit B (DNA replication)	[98,99]
	BDQ	atpE	ATP synthase, subunit F0 (ATP synthesis)	[100,101]
		rv0678	Transcriptional regulator (drug efflux)	[100,101]
	LNZ	rplC	50S ribosomal protein L3 (protein synthesis)	[100,102]
		rrl	23S RNA (protein synthesis)	[100,102]
B	CFZ	rv0678	Transcriptional regulator (drug efflux)	[100,103]
		rv1979c	(Possible permease involved in aminoacid transport)	[100,103]
		rv2535c	(PepQ putative aminopeptidase)	[103,104]
	CS or TRD	alr	Alanine racemase (peptidoglycan synthesis)	[105]
C	EMB	embCAB	Arabinofuranosyltransferases (arabinogalactan synthesis)	[106,107]
		ubiA	Phosphoribosyltransferase (cell wall synthesis)	[108,109]
	DLM	ddn	Deazaflavin (F_{420})-dependent nitroreductase (mycolic acid synthesis)	[110]
		fgd-1	Glucose-6-phosphate dehydrogenase (F_{420} synthesis)	[110]
		fbiA	Protein FbiA (F_{420} synthesis)	[110]
		fbiB	Protein FbiB (F_{420} synthesis)	[110]
		fbiC	Protein FbiC (F_{420} synthesis)	[110]
	PZA	pncA	Pyrazinamidase (conversion of PZA into pyrazinoic acid, resulting in dysfunctions of membrane potential)	[98,111]

Table 1. *Cont.*

Group	Drug	Target Gene/s	Gene Product (Function Affected)	References
		rpsA	30S ribosomal protein S1 (m-RNA trans-translation)	[23,111]
		panD	Aspartate decarboxylate (panthotenate synthesis)	[111,112]
		clpc1	ATP-dependent ATP-ase (protein degradation)	[23,113]
	IPM-CLN or MPM	rv2518c	LdtB, nonclassical, L,D-transpeptidase (peptidoglycan synthesis)	[19,54,114,115]
		rv3682	PonA2, penicillin-binding protein (peptidoglycan synthesis)	[19,114]
		Rv2068c	blaC (β-lactamase)	[116]
	AM	rrs	16S ribosomal RNA (protein synthesis)	[98,117]
	SM	rpsl	ribosomal protein S12 (protein synthesis)	[98,118]
		rrs	16S ribosomal RNA (protein synthesis)	[98,117]
		gidB	(putative 16S rRNA methyltransferase)	[119,120]
	ETO or PTO	rv0565c	Monoxygenase (activation of pro-drugs ETO and PTO)	[121]
		ethA	Monooxygenase (activation of ETO and PTO)	[19,122]
		mymA	Monooxygenase (activation of ETO and PTO)	[121,123]
		katG	Catalase-peroxidase (activation of ETO, PTO, INH)	[122]
		inhA	Enoyl-ACP reductase (mycolic acid synthesis)	[98,122]
	PAS	thyA	Thymidylate synthase	[23,124]
		folC	Dihydropholate synthase	[23,125]
		dfrA	Dihydropholate reductase	[23,125]

Phenotypic testing is still considered a gold standard for Mtb DST, which is accurate, but takes at least two weeks for results [98]. However, a pivotal role has been recently played by the more and more rapid molecular methods to diagnose drug-resistant TB by the identification of chromosomal mutations, including line probe assays, the Xpert MTB/RIF system (Cepheid, Sunnyvale, CA, USA), target gene sequencing, whole genome sequencing (WGS), point-of-care nucleic acid amplification devices [9,127,128].

The Treatment Action Group (TAG) recently released the pipeline report 2019 on TB diagnostics [129]. The TAG-stratified DST tests for decentralized and centralized laboratories. Useful information was provided on what it is currently in TB diagnostics, including tests already recommended by the WHO, and on which tests are expected to be available soon. As to the decentralized tests, the Xpert MTB/RIF assay (sensitivity and specificity for RIF resistance of 96% and 98%, respectively) was recommended by the WHO in 2010, and entered in the market in the same year. The sensitivity of this assay increased with the 2017 rollout of the Xpert MTB/RIF Ultra cartridge. In 2020, there is expected the WHO evaluation and market entry of Xpert XDR, which will detect resistance to INH,

MFX, OFL, AK and KM [129]. In 2013, another company (Molbio Diagnostics, Goa, India) released its systems Truenat MTB and Truenat MTB-RIF Dx onto the Indian market [129]. In January 2020, a rapid WHO Communication reported that the Truenat systems MTB, MTB Plus and MTB RIF Dx assays showed comparable accuracy with Xpert MTB/RIF and Xpert Ultra for Mtb detection (Truenat MTB and Truenat MTB Plus), and for sequential RIF resistance detection (Truenat MTB RIF Dx) [130]. Furthermore, the data for Truenat MTB RIF Dx showed similar accuracy to the WHO approved commercial line probe assays indicated by the TAG for centralized DST [GenoType MTBDR*plus* Version 2.0 (Hain Lifescience, Nehren, Germany) and Nipro NTM+MDRTB detection kit2 (Nipro, Osaka, Japan)] [129,130]. Other systems marketed in 2015–2019 and on the pathway to the WHO evaluation for the centralized determination of molecular resistance to INH and RIF are: Cobas MTB-RIF/INH (Roche, Basel, Switzerland), BD MAX MDR-TB (Becton Dickinson, Sparks, MA, USA), real-time MTB-RIF/INH Resistance assay (Abbott, Abbott Park, IL, USA) and FluoroType MTBDR version 2.0 (RIF, INH) (Hain Lifescience) [129,131–133].

Finally, the WGS technology is capable of identifying the complete drug-resistance profile of an Mtb strain, ideally enabling clinicians to obtain the best anti-TB treatment [98,128,130,134]. However, more data are still needed to correlate genetic mutations with phenotypic resistance, in order to definitely guide the clinical care.

In this view, the initiative of the Comprehensive Resistance Prediction for Tuberculosis: an International Consortium (CRyPTIC) project aims at understanding the relationship between genotypes and resistance by sequencing 100,000 whole TB genomes from various countries, in parallel with comprehensive DST assays. Overall, at this stage, the WGS still needs more studies, but it is commonly believed that this technology will be the future of rapid, centralized DST [129,135].

7. Conclusions

Drug-resistant TB is a significant challenge for the successful control of the disease worldwide. A comprehensive review of clinical, biological and microbiological issues favoring resistance development has been provided, helping in the development of new tools for the rapid diagnosis and treatment of drug-resistant TB. The review was based on the most recent updates on drug resistance mechanisms reported in the literature, and on the international recommendations of WHO to facilitate the clinical and microbiological management of MDR/XDR TB at global level.

Author Contributions: Conceptualization, writing, reviewing and editing: A.I., L.F., F.G. All authors have read and agree to the published version of the manuscript.

Funding: This research received no external funding.

Conflicts of Interest: The authors declare no conflict of interest.

References

1. World Health Organization. *Global Tuberculosis Report 2019*; World Health Organization: Geneva, Switzerland, 2019.
2. Zenner, D.; Beer, N.; Harris, R.J.; Lipman, M.C.; Stagg, H.R.; van der Werf, M.J. Treatment of Latent Tuberculosis Infection: An Updated Network Meta-analysis. *Ann. Intern. Med.* **2017**, *67*, 248–255. [CrossRef]
3. Nahid, P.; Mase, S.R.; Migliori, G.B.; Sotgiu, G.; Bothamley, G.H.; Brozek, J.L.; Cattamanchi, A.; Cegielski, J.P.; Chen, L.; Daley, C.L.; et al. Treatment of Drug-Resistant Tuberculosis. An Official ATS/CDC/ERS/IDSA Clinical Practice Guideline. *Am. J. Respir. Crit. Care Med.* **2019**, *200*, e93–e142. [CrossRef] [PubMed]
4. Mustazzolu, A.; Borroni, E.; Cirillo, D.M.; Giannoni, F.; Iacobino, A.; Italian Multicentre Study on Resistance to Antituberculosis Drugs (SMIRA); Fattorini, L. Trend in rifampicin-, multidrug- and extensively drug-resistant tuberculosis in Italy, 2009–2016. *Eur. Respir. J.* **2018**, *52*. [CrossRef] [PubMed]
5. Bernard, C.; Brossier, F.; Sougakoff, W.; Veziris, N.; Frechet-Jachym, M.; Metivier, N.; Renvoisé, A.; Robert, J.; Jarlier, V. MDR-TB Management group of the NRC. A surge of MDR and XDR tuberculosis in France among patients born in the Former Soviet Union. *Euro Surveill.* **2013**, *18*. [CrossRef]

6. Kibret, K.T.; Moges, Y.; Memiah, P.; Biadgilign, S. Treatment outcomes for multidrug-resistant tuberculosis under DOTS-Plus: A systematic review and meta-analysis of published studies. *Infect. Dis. Poverty.* **2017**, *6*. [CrossRef] [PubMed]
7. Ahmad, N.; Ahuja, S.D.; Akkerman, O.W.; Alffenaar, J.C.; Anderson, L.F.; Baghaei, P.; Bang, D.; Barry, P.M.; Bastos, M.L.; Behera, D.; et al. Collaborative Group for the Meta-Analysis of Individual Patient Data in MDR-TB treatment–2017. *Lancet.* **2018**, *392*, 821–834. [CrossRef]
8. World Health Organization. *WHO Consolidated Guidelines on Drug-Resistant Tuberculosis Treatment*; World Health Organization: Geneva, Switzerland, 2019.
9. Dheda, K.; Gumbo, T.; Maartens, G.; Dooley, K.E.; Murray, M.; Furin, J.; Nardell, E.A.; Warren, R.M. On behalf of the Lancet Respiratory Medicine drug-resistant tuberculosis Commission group. The Lancet Respiratory Medicine Commission: 2019 update: Epidemiology, pathogenesis, transmission, diagnosis, and management of multidrug-resistant and incurable tuberculosis. *Lancet Respir. Med.* **2019**, *7*, 820–826. [CrossRef]
10. Lange, C.; Chesov, D.; Furin, J.; Udwadia, Z.; Dheda, K. Revising the definition of extensively drug-resistant tuberculosis. *Lancet Respir. Med.* **2018**, *6*, 893–895. [CrossRef]
11. World Health Organization. *Rapid Communication: Key Changes to the Treatment of Drug-Resistant Tuberculosis*; World Health Organization: Geneva, Switzerland, 2019.
12. TB Alliance. Available online: https://www.tballiance.org/news/fda-advisory-committee-votes-favorably-question-effectiveness-and-safety-pretomanid-combination (accessed on 24 February 2020).
13. World Health Organization. *Companion Handbook to the WHO Guidelines for the Programmatic Management of Drug-Resistant Tuberculosis*; World Health Organization: Geneva, Switzerland, 2014.
14. Monedero, I.; Caminero, J.A. Common errors in multidrug-resistant tuberculosis management. *Expert Rev. Respir. Med.* **2014**, *8*, 15–23. [CrossRef]
15. Tiberi, S.; Pontali, E.; Tadolini, M.; D'Ambrosio, L.; Migliori, G.B. Challenging MDR-TB clinical problems—The case for a new Global TB Consilium supporting the compassionate use of new anti-TB drugs. *Int. J. Infect. Dis.* **2019**, *80*, S68–S72. [CrossRef]
16. Strydom, N.; Gupta, S.V.; Fox, W.S.; Via, L.E.; Bang, H.; Lee, M.; Eum, S.; Shim, T.; Barry, C.E.; Zimmerman, M.; et al. Tuberculosis drugs' distribution and emergence of resistance in patient's lung lesions: A mechanistic model and tool for regimen and dose optimization. *PLoS Med.* **2019**, *16*. [CrossRef] [PubMed]
17. Dartois, V. The path of anti-tuberculosis drugs: From blood to lesions to mycobacterial cells. *Nat. Rev. Microbiol.* **2014**, *12*, 159–167. [CrossRef] [PubMed]
18. Singh, R.; Dwivedi, S.P.; Gaharwar, U.S.; Meena, R.; Rajamani, P.; Prasad, T. Recent updates on drug resistance in *Mycobacterium tuberculosis*. *J. Appl. Microbiol.* **2019**. [CrossRef]
19. Nasiri, M.J.; Haeili, M.; Ghazi, M.; Goudarzi, H.; Pormohammad, A.; Imani Fooladi, A.A.; Feizabadi, M.M. New Insights in to the Intrinsic and Acquired Drug Resistance Mechanisms in Mycobacteria. *Front. Microbiol.* **2017**, *8*. [CrossRef] [PubMed]
20. Nguyen, L. Antibiotic resistance mechanisms in *M. tuberculosis*: An update. *Arch. Toxicol.* **2016**, *90*, 1585–1604. [CrossRef] [PubMed]
21. Brauner, A.; Fridman, O.; Gefen, O.; Balaban, N.Q. Distinguishing between resistance, tolerance and persistence to antibiotic treatment. *Nat. Rev. Microbiol.* **2016**, *14*, 320–330. [CrossRef] [PubMed]
22. Cohen, N.R.; Lobritz, M.A.; Collins, J.J. Microbial persistence and the road to drug resistance. *Cell Host Microbe.* **2013**, *13*, 632–642. [CrossRef]
23. Hameed, H.M.A.; Islam, M.M.; Chhotaray, C.; Wang, C.; Liu, Y.; Tan, Y.; Li, X.; Tan, S.; Delorme, V.; Yew, W.W.; et al. Molecular Targets Related Drug Resistance Mechanisms in MDR- XDR- and TDR-*Mycobacterium tuberculosis* Strains. *Front. Cell Infect. Microbiol.* **2018**, *8*. [CrossRef]
24. Lohrasbi, V.; Talebi, M.; Bialvaei, A.Z.; Fattorini, L.; Drancourt, M.; Heidary, M.; Darban-Sarokhalil, D. Trends in the discovery of new drugs for *Mycobacterium tuberculosis* therapy with a glance at resistance. *Tuberculosis* **2018**, *109*, 17–27. [CrossRef]
25. Islam, M.M.; Hameed, H.M.A.; Mugweru, J.; Chhotaray, C.; Wang, C.; Tan, Y.; Liu, J.; Li, X.; Tan, S.; Ojima, I.; et al. Drug resistance mechanisms and novel drug targets for tuberculosis therapy. *J. Genet. Genom.* **2017**, *44*, 21–37. [CrossRef]
26. World Health Organization Regional Office for Europe. *Guiding Principles to Reduce Tuberculosis Transmission in the WHO European Region*; World Health Organization Regional Office for Europe: Copenhagen, Denmark, 2018.

27. D'Ambrosio, L.; Bothamley, G.; Caminero Luna, J.A.; Duarte, R.; Guglielmetti, L.; Muñoz Torrico, M.; Payen, M.C.; Saavedra Herrera, N.; Salazar Lezama, M.A.; Skrahina, A.; et al. Team approach to manage difficult-to-treat TB cases: Experiences in Europe and beyond. *Pulmonology.* **2018**, *24*, 132–141. [CrossRef] [PubMed]
28. World Association for Infectious Diseases and Immunological Disorders. Available online: http://www.waidid.org/site/workinggroups (accessed on 24 February 2020).
29. Barry, C.E.; Boshoff, H.I.; Dartois, V.; Dick, T.; Ehrt, S.; Flynn, J.; Schnappinger, D.; Wilkinson, R.J.; Young, D. The spectrum of latent tuberculosis: Rethinking the biology and intervention strategies. *Nat. Rev. Microbiol.* **2009**, *7*, 845–855. [CrossRef] [PubMed]
30. Lenaerts, A.; Barry, C.E.; Dartois, V. Heterogeneity in tuberculosis pathology, microenvironments and therapeutic responses. *Immunol. Rev.* **2015**, *264*, 288–307. [CrossRef] [PubMed]
31. Iacobino, A.; Piccaro, G.; Giannoni, F.; Mustazzolu, A.; Fattorini, L. Fighting tuberculosis by drugs targeting nonreplicating *Mycobacterium tuberculosis* bacilli. *Int. J. Mycobacteriol.* **2017**, *6*, 213–221. [CrossRef] [PubMed]
32. Sarathy, J.P.; Zuccotto, F.; Hsinpin, H.; Sandberg, L.; Via, L.E.; Marriner, G.A.; Masquelin, T.; Wyatt, P.; Ray, P.; Dartois, V. Prediction of Drug Penetration in Tuberculosis Lesions. *ACS Infect. Dis.* **2016**, *2*, 552–563. [CrossRef] [PubMed]
33. Lanoix, J.P.; Lenaerts, A.J.; Nuermberger, E.L. Heterogeneous disease progression and treatment response in a C3HeB/FeJ mouse model of tuberculosis. *Dis. Model. Mech.* **2015**, *8*, 603–610. [CrossRef]
34. Iacobino, A.; Piccaro, G.; Giannoni, F.; Mustazzolu, A.; Fattorini, L. Activity of drugs against dormant *Mycobacterium tuberculosis. Int. J. Mycobacteriol.* **2016**, *5* (Suppl. 1), S94–S95. [CrossRef]
35. Sarathy, J.P.; Liang, H.H.; Weiner, D.; Gonzales, J.; Via, L.E.; Dartois, V. An In Vitro Caseum Binding Assay that Predicts Drug Penetration in Tuberculosis Lesions. *J. Vis. Exp.* **2017**, *123*. [CrossRef]
36. Sarathy, J.P.; Via, L.E.; Weiner, D.; Blanc, L.; Boshoff, H.; Eugenin, E.A.; Barry, C.E.; Dartois, V.A. Extreme Drug Tolerance of *Mycobacterium tuberculosis* in Caseum. *Antimicrob. Agents Chemother.* **2018**, *62*. [CrossRef]
37. Prideaux, B.; Via, L.E.; Zimmerman, M.D.; Eum, S.; Sarathy, J.; O'Brien, P.; Chen, C.; Kaya, F.; Weiner, D.M.; Chen, P.Y.; et al. The association between sterilizing activity and drug distribution into tuberculosis lesions. *Nat. Med.* **2015**, *21*, 1223–1227. [CrossRef]
38. Sebastian, J.; Swaminath, S.; Nair, R.R.; Jakkala, K.; Pradhan, A.; Ajitkumar, P. De Novo Emergence of Genetically Resistant Mutants of *Mycobacterium tuberculosis* from the Persistence Phase Cells Formed against Antituberculosis Drugs In Vitro. *Antimicrob. Agents Chemother.* **2017**, *61*. [CrossRef] [PubMed]
39. Nair, R.R.; Sharan, D.; Sebastian, J.; Swaminath, S.; Ajitkumar, P. Heterogeneity of ROS levels in antibiotic-exposed mycobacterial subpopulations confers differential susceptibility. *Microbiology.* **2019**, *165*, 668–682. [CrossRef] [PubMed]
40. Piccaro, G.; Pietraforte, D.; Giannoni, F.; Mustazzolu, A.; Fattorini, L. Rifampin induces hydroxyl radical formation in *Mycobacterium tuberculosis. Antimicrob. Agents Chemother.* **2014**, *58*, 7527–7533. [CrossRef] [PubMed]
41. Kempker, R.R.; Rabin, A.S.; Nikolaishvili, K.; Kalandadze, I.; Gogishvili, S.; Blumberg, H.M.; Vashakidze, S. Additional drug resistance in *Mycobacterium tuberculosis* isolates from resected cavities among patients with multidrug-resistant or extensively drug-resistant pulmonary tuberculosis. *Clin. Infect. Dis.* **2012**, *54*, e51–e54. [CrossRef] [PubMed]
42. Kaplan, G.; Post, F.A.; Moreira, A.L.; Wainwright, H.; Kreiswirth, B.N.; Tanverdi, M.; Mathema, B.; Ramaswamy, S.V.; Walther, G.; Steyn, L.M.; et al. *Mycobacterium tuberculosis* growth at the cavity surface: A microenvironment with failed immunity. *Infect. Immun.* **2003**, *71*, 7099–7108. [CrossRef]
43. Dheda, K.; Lenders, L.; Magombedze, G.; Srivastava, S.; Raj, P.; Arning, E.; Ashcraft, P.; Bottiglieri, T.; Wainwright, H.; Pennel, T.; et al. Drug-Penetration Gradients Associated with Acquired Drug Resistance in Patients with Tuberculosis. *Am. J. Respir. Crit. Care Med.* **2018**, *198*, 1208–1219. [CrossRef]
44. Dulberger, C.L.; Rubin, E.J.; Boutte, C.C. The mycobacterial cell envelope—A moving target. *Nat. Rev. Microbiol.* **2020**, *18*, 47–59. [CrossRef]
45. Sarathy, J.P.; Dartois, V.; Lee, E.J. The role of transport mechanisms in *Mycobacterium tuberculosis* drug resistance and tolerance. *Pharmaceuticals* **2012**, *5*, 1210–1235. [CrossRef]
46. Machado, D.; Girardini, M.; Viveiros, M.; Pieroni, M. Challenging the Drug-Likeness Dogma for New Drug Discovery in Tuberculosis. *Front. Microbiol.* **2018**, *9*. [CrossRef]

47. Nguyen, L.; Chinnapapagari, S.; Thompson, C.J. FbpA-Dependent biosynthesis of trehalose dimycolate is required for the intrinsic multidrug resistance, cell wall structure, and colonial morphology of *Mycobacterium smegmatis*. *J. Bacteriol.* **2005**, *187*, 6603–6611. [CrossRef]
48. Liu, J.; Nikaido, H. A mutant of *Mycobacterium smegmatis* defective in the biosynthesis of mycolic acids accumulates meromycolates. *Proc. Natl. Acad. Sci. USA.* **1999**, *96*, 4011–4016. [CrossRef] [PubMed]
49. Singh, A.; Gupta, R.; Vishwakarma, R.A.; Narayanan, P.R.; Paramasivan, C.N.; Ramanathan, V.D.; Tyagi, A.K. Requirement of the *mymA* operon for appropriate cell wall ultrastructure and persistence of *Mycobacterium tuberculosis* in the spleens of guinea pigs. *J. Bacteriol.* **2005**, *187*, 4173–4186. [CrossRef] [PubMed]
50. Gygli, S.M.; Borrell, S.; Trauner, A.; Gagneux, S. Antimicrobial resistance in *Mycobacterium tuberculosis*: Mechanistic and evolutionary perspectives. *FEMS Microbiol. Rev.* **2017**, *41*, 354–373. [CrossRef] [PubMed]
51. Siroy, A.; Mailaender, C.; Harder, D.; Koerber, S.; Wolschendorf, F.; Danilchanka, O.; Wang, Y.; Heinz, C.; Niederweis, M. Rv1698 of *Mycobacterium tuberculosis* represents a new class of channel-forming outer membrane proteins. *J. Biol. Chem.* **2008**, *283*, 17827–17837. [CrossRef] [PubMed]
52. Chambers, H.F.; Moreau, D.; Yajko, D.; Miick, C.; Wagner, C.; Hackbarth, C.; Kocagöz, S.; Rosenberg, E.; Hadley, W.K.; Nikaido, H. Can penicillins and other beta-lactam antibiotics be used to treat tuberculosis? *Antimicrob. Agents Chemother.* **1995**, *39*, 2620–2624. [CrossRef] [PubMed]
53. Quinting, B.; Reyrat, J.M.; Monnaie, D.; Amicosante, G.; Pelicic, V.; Gicquel, B.; Frère, J.M.; Galleni, M. Contribution of beta-lactamase production to the resistance of mycobacteria to beta-lactam antibiotics. *FEBS Lett.* **1997**, *406*, 275–278. [CrossRef]
54. Van Rijn, S.P.; Zuur, M.A.; Anthony, R.; Wilffert, B.; van Altena, R.; Akkerman, O.W.; de Lange, W.C.M.; van der Werf, T.S.; Kosterink, J.G.W.; Alffenaar, J.C. Evaluation of Carbapenems for Treatment of Multi- and Extensively Drug-Resistant *Mycobacterium tuberculosis*. *Antimicrob. Agents Chemother.* **2019**, *63*. [CrossRef]
55. Janardhan, S.; Ram Vivek, M.; Narahari Sastry, G. Modeling the permeability of drug-like molecules through the cell wall of *Mycobacterium tuberculosis*: An analogue based approach. *Mol. Biosyst.* **2016**, *12*, 3377–3384. [CrossRef]
56. De Rossi, E.; Aínsa, J.A.; Riccardi, G. Role of mycobacterial efflux transporters in drug resistance: An unresolved question. *FEMS Microbiol. Rev.* **2006**, *30*, 36–52. [CrossRef]
57. Rodrigues, L.; Parish, T.; Balganesh, M.; Ainsa, J.A. Antituberculosis drugs: Reducing efflux=increasing activity. *Drug Discov. Today.* **2017**, *22*, 592–599. [CrossRef]
58. Srivastava, S.; Musuka, S.; Sherman, C.; Meek, C.; Leff, R.; Gumbo, T. Efflux-pump-derived multiple drug resistance to ethambutol monotherapy in *Mycobacterium tuberculosis* and the pharmacokinetics and pharmacodynamics of ethambutol. *J. Infect. Dis.* **2010**, *201*, 1225–1231. [CrossRef] [PubMed]
59. Machado, D.; Couto, I.; Perdigão, J.; Rodrigues, L.; Portugal, I.; Baptista, P.; Veigas, B.; Amaral, L.; Viveiros, M. Contribution of efflux to the emergence of isoniazid and multidrug resistance in *Mycobacterium tuberculosis*. *PLoS ONE* **2012**, *7*, e34538. [CrossRef] [PubMed]
60. Rodrigues, L.; Villellas, C.; Bailo, R.; Viveiros, M.; Aínsa, J.A. Role of the Mmr efflux pump in drug resistance in *Mycobacterium tuberculosis*. *Antimicrob. Agents Chemother.* **2013**, *57*, 751–757. [CrossRef] [PubMed]
61. Pasca, M.R.; Guglierame, P.; De Rossi, E.; Zara, F.; Riccardi, G. mmpL7 gene of *Mycobacterium tuberculosis* is responsible for isoniazid efflux in *Mycobacterium smegmatis*. *Antimicrob. Agents Chemother.* **2005**, *49*, 4775–4777. [CrossRef]
62. Hartkoorn, R.C.; Uplekar, S.; Cole, S.T. Cross-resistance between clofazimine and bedaquiline through upregulation of MmpL5 in *Mycobacterium tuberculosis*. *Antimicrob. Agents Chemother.* **2014**, *58*, 2979–2981. [CrossRef]
63. Ismail, N.; Peters, R.P.H.; Ismail, N.A.; Omar, S.V. Clofazimine Exposure In Vitro Selects Efflux Pump Mutants and Bedaquiline Resistance. *Antimicrob. Agents Chemother.* **2019**, *63*. [CrossRef]
64. Pule, C.M.; Sampson, S.L.; Warren, R.M.; Black, P.A.; van Helden, P.D.; Victor, T.C.; Louw, G.E. Efflux pump inhibitors: Targeting mycobacterial efflux systems to enhance TB therapy. *J. Antimicrob. Chemother.* **2016**, *71*, 17–26. [CrossRef]
65. Chen, C.; Gardete, S.; Jansen, R.S.; Shetty, A.; Dick, T.; Rhee, K.Y.; Dartois, V. Verapamil Targets Membrane Energetics in *Mycobacterium tuberculosis*. *Antimicrob. Agents Chemother.* **2018**, *62*. [CrossRef]
66. Xu, J.; Tasneen, R.; Peloquin, C.A.; Almeida, D.V.; Li, S.Y.; Barnes-Boyle, K.; Lu, Y.; Nuermberger, E. Verapamil Increases the Bioavailability and Efficacy of Bedaquiline but Not Clofazimine in a Murine Model of Tuberculosis. *Antimicrob. Agents Chemother.* **2017**, *62*. [CrossRef]

67. Abbate, E.; Vescovo, M.; Natiello, M.; Cufré, M.; García, A.; Gonzalez Montaner, P.; Ambroggi, M.; Ritacco, V.; van Soolingen, D. Successful alternative treatment of extensively drug-resistant tuberculosis in Argentina with a combination of linezolid, moxifloxacin and thioridazine. *J. Antimicrob. Chemother.* **2012**, *67*, 473–477. [CrossRef]
68. Georghiou, S.B.; Magana, M.; Garfein, R.S.; Catanzaro, D.G.; Catanzaro, A.; Rodwell, T.C. Evaluation of genetic mutations associated with *Mycobacterium tuberculosis* resistance to amikacin, kanamycin and capreomycin: A systematic review. *PLoS ONE* **2012**, *7*, e33275. [CrossRef] [PubMed]
69. Buriánková, K.; Doucet-Populaire, F.; Dorson, O.; Gondran, A.; Ghnassia, J.C.; Weiser, J.; Pernodet, J.L. Molecular basis of intrinsic macrolide resistance in the *Mycobacterium tuberculosis* complex. *Antimicrob. Agents Chemother.* **2004**, *48*, 143–150. [CrossRef] [PubMed]
70. Madsen, C.T.; Jakobsen, L.; Buriánková, K.; Doucet-Populaire, F.; Pernodet, J.L.; Douthwaite, S. Methyltransferase Erm (37) slips on rRNA to confer atypical resistance in *Mycobacterium tuberculosis*. *J. Biol. Chem.* **2005**, *280*, 38942–38947. [CrossRef] [PubMed]
71. Nash, K.A.; Brown-Elliott, B.A.; Wallace, R.J., Jr. A novel gene, erm(41), confers inducible macrolide resistance to clinical isolates of *Mycobacterium abscessus* but is absent from *Mycobacterium chelonae*. *Antimicrob. Agents Chemother.* **2009**, *53*, 1367–1376. [CrossRef] [PubMed]
72. Khrapunov, S.; Brenowitz, M. Stability, denaturation and refolding of *Mycobacterium tuberculosis* MfpA, a DNA mimicking protein that confers antibiotic resistance. *Biophys. Chem.* **2011**, *159*, 33–40. [CrossRef]
73. Alderwick, L.J.; Harrison, J.; Lloyd, G.S.; Birch, H.L. The Mycobacterial Cell Wall–Peptidoglycan and Arabinogalactan. *Cold Spring Harb. Perspect. Med.* **2015**, *5*, a021113. [CrossRef]
74. Gold, B.; Nathan, C. Targeting Phenotypically Tolerant *Mycobacterium tuberculosis*. *Microbiol. Spectr.* **2017**, *5*. [CrossRef]
75. Lipworth, S.; Hammond, R.J.; Baron, V.O.; Hu, Y.; Coates, A.; Gillespie, S.H. Defining dormancy in mycobacterial disease. *Tuberculosis* **2016**, *99*, 131–142. [CrossRef]
76. Hammond, R.J.H.; Baron, V.O.; Lipworth, S.; Gillespie, S.H. Enhanced Methodologies for Detecting Phenotypic Resistance in Mycobacteria. *Methods Mol. Biol.* **2018**, *1736*, 85–94. [CrossRef]
77. Batyrshina, Y.R.; Schwartz, Y.S. Modeling of *Mycobacterium tuberculosis* dormancy in bacterial cultures. *Tuberculosis* **2019**, *117*, 7–17. [CrossRef]
78. Wayne, L.G.; Hayes, L.G. An in vitro model for sequential study of shiftdown of *Mycobacterium tuberculosis* through two stages of nonreplicating persistence. *Infect. Immun.* **1996**, *64*, 2062–2069. [CrossRef] [PubMed]
79. Sohaskey, C.D.; Voskuil, M.I. In vitro models that utilize hypoxia to induce non-replicating persistence in Mycobacteria. *Methods Mol. Biol.* **2015**, *1285*, 201–213. [CrossRef] [PubMed]
80. Betts, J.C.; Lukey, P.T.; Robb, L.C.; McAdam, R.A.; Duncan, K. Evaluation of a nutrient starvation model of *Mycobacterium tuberculosis* persistence by gene and protein expression profiling. *Mol. Microbiol.* **2002**, *43*, 717–731. [CrossRef] [PubMed]
81. Voskuil, M.I.; Schnappinger, D.; Visconti, K.C.; Harrell, M.I.; Dolganov, G.M.; Sherman, D.R.; Schoolnik, G.K. Inhibition of respiration by nitric oxide induces a *Mycobacterium tuberculosis* dormancy program. *J. Exp. Med.* **2003**, *198*, 705–713. [CrossRef]
82. de Carvalho, L.P.; Lin, G.; Jiang, X.; Nathan, C. Nitazoxanide kills replicating and nonreplicating *Mycobacterium tuberculosis* and evades resistance. *J. Med. Chem.* **2009**, *52*, 5789–5792. [CrossRef]
83. Hu, Y.; Mangan, J.A.; Dhillon, J.; Sole, K.M.; Mitchison, D.A.; Butcher, P.D.; Coates, A.R. Detection of mRNA transcripts and active transcription in persistent *Mycobacterium tuberculosis* induced by exposure to rifampin or pyrazinamide. *J. Bacteriol.* **2000**, *182*, 6358–6365. [CrossRef]
84. Sala, C.; Dhar, N.; Hartkoorn, R.C.; Zhang, M.; Ha, Y.H.; Schneider, P.; Cole, S.T. Simple model for testing drugs against nonreplicating *Mycobacterium tuberculosis*. *Antimicrob. Agents Chemother.* **2010**, *54*, 4150–4158. [CrossRef]
85. Jakkala, K.; Ajitkumar, P. Hypoxic Non-replicating Persistent *Mycobacterium tuberculosis* Develops Thickened Outer Layer That Helps in Restricting Rifampicin Entry. *Front. Microbiol.* **2019**, *10*. [CrossRef]
86. Filippini, P.; Iona, E.; Piccaro, G.; Peyron, P.; Neyrolles, O.; Fattorini, L. Activity of drug combinations against dormant *Mycobacterium tuberculosis*. *Antimicrob. Agents Chemother.* **2010**, *54*, 2712–2715. [CrossRef]
87. Piccaro, G.; Giannoni, F.; Filippini, P.; Mustazzolu, A.; Fattorini, L. Activities of drug combinations against *Mycobacterium tuberculosis* grown in aerobic and hypoxic acidic conditions. *Antimicrob. Agents Chemother.* **2013**, *57*, 1428–1433. [CrossRef]

88. Iacobino, A.; Piccaro, G.; Giannoni, F.; Mustazzolu, A.; Fattorini, L. *Mycobacterium tuberculosis* Is Selectively Killed by Rifampin and Rifapentine in Hypoxia at Neutral pH. *Antimicrob. Agents Chemother.* **2017**, *61*. [CrossRef] [PubMed]
89. Iacobino, A.; Giannoni, F.; Pardini, M.; Piccaro, G.; Fattorini, L. The Combination Rifampin-Nitazoxanide, but Not Rifampin-Isoniazid-Pyrazinamide-Ethambutol, Kills Dormant *Mycobacterium tuberculosis* in Hypoxia at Neutral pH. *Antimicrob. Agents Chemother.* **2019**, *63*. [CrossRef] [PubMed]
90. Orman, M.A.; Brynildsen, M.P. Dormancy is not necessary or sufficient for bacterial persistence. *Antimicrob. Agents Chemother.* **2013**, *57*, 3230–3239. [CrossRef] [PubMed]
91. Keren, I.; Minami, S.; Rubin, E.; Lewis, K. Characterization and transcriptome analysis of *Mycobacterium tuberculosis* persisters. *MBio* **2011**, *2*, e00100-11. [CrossRef]
92. Lenaerts, A.J.; Hoff, D.; Aly, S.; Ehlers, S.; Andries, K.; Cantarero, L.; Orme, I.M.; Basaraba, R.J. Location of persisting mycobacteria in a Guinea pig model of tuberculosis revealed by r207910. *Antimicrob. Agents Chemother.* **2007**, *51*, 3338–3345. [CrossRef]
93. Liu, Y.; Pertinez, H.; Ortega-Muro, F.; Alameda-Martin, L.; Harrison, T.; Davies, G.; Coates, A.; Hu, Y. Optimal doses of rifampicin in the standard drug regimen to shorten tuberculosis treatment duration and reduce relapse by eradicating persistent bacteria. *J. Antimicrob. Chemother.* **2018**, *73*, 724–731. [CrossRef]
94. Hu, Y.; Pertinez, H.; Liu, Y.; Davies, G.; Coates, A. Bedaquiline kills persistent *Mycobacterium tuberculosis* with no disease relapse: An in vivo model of a potential cure. *J. Antimicrob. Chemother.* **2019**, *74*, 1627–1633. [CrossRef]
95. Xu, J.; Li, S.Y.; Almeida, D.V.; Tasneen, R.; Barnes-Boyle, K.; Converse, P.J.; Upton, A.M.; Mdluli, K.; Fotouhi, N.; Nuermberger, E.L. Contribution of Pretomanid to Novel Regimens Containing Bedaquiline with either Linezolid or Moxifloxacin and Pyrazinamide in Murine Models of Tuberculosis. *Antimicrob. Agents Chemother.* **2019**, *63*. [CrossRef]
96. Van den Bergh, B.; Fauvart, M.; Michiels, J. Formation, physiology, ecology, evolution and clinical importance of bacterial persisters. *FEMS Microbiol. Rev.* **2017**, *41*, 219–251. [CrossRef]
97. Pang, Y.; Zong, Z.; Huo, F.; Jing, W.; Ma, Y.; Dong, L.; Li, Y.; Zhao, L.; Fu, Y.; Huang, H. In Vitro Drug Susceptibility of Bedaquiline, Delamanid, Linezolid, Clofazimine, Moxifloxacin, and Gatifloxacin against Extensively Drug-Resistant Tuberculosis in Beijing, China. *Antimicrob. Agents Chemother.* **2017**, *61*. [CrossRef]
98. Miotto, P.; Tessema, B.; Tagliani, E.; Chindelevitch, L.; Starks, A.M.; Emerson, C.; Hanna, D.; Kim, P.S.; Liwski, R.; Zignol, M.; et al. A standardised method for interpreting the association between mutations and phenotypic drug resistance in *Mycobacterium tuberculosis*. *Eur. Respir. J.* **2017**, *50*. [CrossRef]
99. Mahmood, N.; Abbas, S.N.; Faraz, N.; Shahid, S. Mutational analysis of *gyrB* at amino acids: G481A & D505A in multidrug resistant (MDR) tuberculosis patients. *J. Infect. Public Health.* **2019**, *12*, 496–501. [CrossRef] [PubMed]
100. Ismail, N.; Omar, S.V.; Ismail, N.A.; Peters, R.P.H. Collated data of mutation frequencies and associated genetic variants of bedaquiline, clofazimine and linezolid resistance in *Mycobacterium tuberculosis*. *Data Brief.* **2018**, *20*, 1975–1983. [CrossRef] [PubMed]
101. Ismail, N.; Ismail, N.A.; Omar, S.V.; Peters, R.P.H. In Vitro Study of Stepwise Acquisition of *rv0678* and *atpE* Mutations Conferring Bedaquiline Resistance. *Antimicrob. Agents Chemother.* **2019**, *63*. [CrossRef] [PubMed]
102. Pi, R.; Liu, Q.; Jiang, Q.; Gao, Q. Characterization of linezolid-resistance-associated mutations in *Mycobacterium tuberculosis* through WGS. *J. Antimicrob. Chemother.* **2019**, *74*, 1795–1798. [CrossRef] [PubMed]
103. Zhang, S.; Chen, J.; Cui, P.; Shi, W.; Zhang, W.; Zhang, Y. Identification of novel mutations associated with clofazimine resistance in *Mycobacterium tuberculosis*. *J. Antimicrob. Chemother.* **2015**, *70*, 2507–2510. [CrossRef]
104. Almeida, D.; Ioerger, T.; Tyagi, S.; Li, S.Y.; Mdluli, K.; Andries, K.; Grosset, J.; Sacchettini, J.; Nuermberger, E. Mutations in *pepQ* Confer Low-Level Resistance to Bedaquiline and Clofazimine in *Mycobacterium tuberculosis*. *Antimicrob. Agents Chemother.* **2016**, *60*, 4590–4599. [CrossRef]
105. Evangelopoulos, D.; Prosser, G.A.; Rodgers, A.; Dagg, B.M.; Khatri, B.; Ho, M.M.; Gutierrez, M.G.; Cortes, T.; de Carvalho, L.P.S. Comparative fitness analysis of D-cycloserine resistant mutants reveals both fitness-neutral and high-fitness cost genotypes. *Nat. Commun.* **2019**, *10*, 4177. [CrossRef]
106. Sun, Q.; Xiao, T.Y.; Liu, H.C.; Zhao, X.Q.; Liu, Z.G.; Li, Y.N.; Zeng, H.; Zhao, L.L.; Wan, K.L. Mutations within *embCAB* Are Associated with Variable Level of Ethambutol Resistance in *Mycobacterium tuberculosis* Isolates from China. *Antimicrob. Agents Chemother.* **2017**, *62*. [CrossRef]

107. Chen, X.; He, G.; Wang, S.; Lin, S.; Chen, J.; Zhang, W. Evaluation of Whole-Genome Sequence Method to Diagnose Resistance of 13 Anti-tuberculosis Drugs and Characterize Resistance Genes in Clinical Multi-Drug Resistance *Mycobacterium tuberculosis* Isolates From China. *Front. Microbiol.* **2019**, *10*. [CrossRef]
108. Lingaraju, S.; Rigouts, L.; Gupta, A.; Lee, J.; Umubyeyi, A.N.; Davidow, A.L.; German, S.; Cho, E.; Lee, J.I.; Cho, S.N.; et al. Geographic Differences in the Contribution of *ubiA* Mutations to High-Level Ethambutol Resistance in *Mycobacterium tuberculosis*. *Antimicrob. Agents Chemother.* **2016**, *60*, 4101–4105. [CrossRef] [PubMed]
109. He, L.; Wang, X.; Cui, P.; Jin, J.; Chen, J.; Zhang, W.; Zhang, Y. *ubiA* (Rv3806c) encoding DPPR synthase involved in cell wall synthesis is associated with ethambutol resistance in *Mycobacterium tuberculosis*. *Tuberculosis* **2015**, *95*, 149–154. [CrossRef] [PubMed]
110. Fujiwara, M.; Kawasaki, M.; Hariguchi, N.; Liu, Y.; Matsumoto, M. Mechanisms of resistance to delamanid, a drug for *Mycobacterium tuberculosis*. *Tuberculosis* **2018**, *108*, 186–194. [CrossRef] [PubMed]
111. Nusrath Unissa, A.; Hanna, L.E. Molecular mechanisms of action, resistance, detection to the first-line anti tuberculosis drugs: Rifampicin and pyrazinamide in the post whole genome sequencing era. *Tuberculosis* **2017**, *105*, 96–107. [CrossRef] [PubMed]
112. Shi, W.; Chen, J.; Zhang, S.; Zhang, W.; Zhang, Y. Identification of Novel Mutations in LprG (*rv1411c*), *rv0521*, *rv3630*, *rv0010c*, *ppsC* and *cyp*128 Associated with Pyrazinoic Acid/Pyrazinamide Resistance in *Mycobacterium tuberculosis*. *Antimicrob. Agents Chemother.* **2018**, *62*. [CrossRef]
113. Zhang, S.; Chen, J.; Shi, W.; Cui, P.; Zhang, J.; Cho, S.; Zhang, W.; Zhang, Y. Mutation in *clpC1* encoding an ATP-dependent ATPase involved in protein degradation is associated with pyrazinamide resistance in *Mycobacterium tuberculosis*. *Emerg. Microbes Infect.* **2017**, *6*. [CrossRef]
114. Lun, S.; Miranda, D.; Kubler, A.; Guo, H.; Maiga, M.C.; Winglee, K.; Pelly, S.; Bishai, W.R. Synthetic lethality reveals mechanisms of *Mycobacterium tuberculosis* resistance to β-lactams. *MBio* **2014**, *5*, e01767-14. [CrossRef]
115. Cohen, K.A.; El-Hay, T.; Wyres, K.L.; Weissbrod, O.; Munsamy, V.; Yanover, C.; Aharonov, R.; Shaham, O.; Conway, T.C.; Goldschmidt, Y.; et al. Paradoxical Hypersusceptibility of Drug-resistant *Mycobacterium tuberculosis* to β-lactam Antibiotics. *EBioMedicine* **2016**, *9*, 170–179. [CrossRef]
116. Iannazzo, L.; Soroka, D.; Triboulet, S.; Fonvielle, M.; Compain, F.; Dubée, V.; Mainardi, J.L.; Hugonnet, J.E.; Braud, E.; Arthur, M.; et al. Routes of Synthesis of Carbapenems for Optimizing Both the Inactivation of L,D-Transpeptidase LdtMt1 of *Mycobacterium tuberculosis* and the Stability toward Hydrolysis by β-Lactamase BlaC. *J. Med. Chem.* **2016**, *59*, 3427–3438. [CrossRef]
117. Ramakrishna, V.; Singh, P.K.; Prakash, S.; Jain, A. Second Line Injectable Drug Resistance and Associated Genetic Mutations in Newly Diagnosed Cases of Multidrug-Resistant Tuberculosis. *Microb. Drug Resist.* **2020**. [CrossRef]
118. Perdigão, J.; Portugal, I. Genetics and roadblocks of drug resistant tuberculosis. *Infect. Genet. Evol.* **2019**, *72*, 113–130. [CrossRef] [PubMed]
119. Verma, J.S.; Gupta, Y.; Nair, D.; Manzoor, N.; Rautela, R.S.; Rai, A.; Katoch, V.M. Evaluation of *gidB* alterations responsible for streptomycin resistance in *Mycobacterium tuberculosis*. *J. Antimicrob. Chemother.* **2014**, *69*, 2935–2941. [CrossRef] [PubMed]
120. Wong, S.Y.; Lee, J.S.; Kwak, H.K.; Via, L.E.; Boshoff, H.I.; Barry, C.E. Mutations in *gidB* confer low-level streptomycin resistance in *Mycobacterium tuberculosis*. *Antimicrob. Agents Chemother.* **2011**, *55*, 2515–2522. [CrossRef]
121. Hicks, N.D.; Carey, A.F.; Yang, J.; Zhao, Y.; Fortune, S.M. Bacterial Genome-Wide Association Identifies Novel Factors That Contribute to Ethionamide and Prothionamide Susceptibility in *Mycobacterium tuberculosis*. *MBio* **2019**, *10*. [CrossRef]
122. Rueda, J.; Realpe, T.; Mejia, G.I.; Zapata, E.; Rozo, J.C.; Ferro, B.E.; Robledo, J. Genotypic Analysis of Genes Associated with Independent Resistance and Cross-Resistance to Isoniazid and Ethionamide in *Mycobacterium tuberculosis* Clinical Isolates. *Antimicrob. Agents Chemother.* **2015**, *59*, 7805–7810. [CrossRef] [PubMed]
123. Grant, S.S.; Wellington, S.; Kawate, T.; Desjardins, C.A.; Silvis, M.R.; Wivagg, C.; Thompson, M.; Gordon, K.; Kazyanskaya, E.; Nietupski, R.; et al. Baeyer-Villiger Monooxygenases EthA and MymA Are Required for Activation of Replicating and Non-replicating *Mycobacterium tuberculosis* Inhibitors. *Cell Chem. Biol.* **2016**, *23*, 666–677. [CrossRef]
124. Pandey, B.; Grover, S.; Kaur, J.; Grover, A. Analysis of mutations leading to para-aminosalicylic acid resistance in *Mycobacterium tuberculosis*. *Sci. Rep.* **2019**, *9*. [CrossRef]

125. Wei, W.; Yan, H.; Zhao, J.; Li, H.; Li, Z.; Guo, H.; Wang, X.; Zhou, Y.; Zhang, X.; Zeng, J.; et al. Multi-omics comparisons of p-aminosalicylic acid (PAS) resistance in *folC* mutated and un-mutated *Mycobacterium tuberculosis* strains. *Emerg. Microbes Infect.* **2019**, *8*, 248–261. [CrossRef]
126. Dookie, N.; Rambaran, S.; Padayatchi, N.; Mahomed, S.; Naidoo, K. Evolution of drug resistance in *Mycobacterium tuberculosis*: A review on the molecular determinants of resistance and implications for personalized care. *J. Antimicrob. Chemother.* **2018**, *73*, 1138–1151. [CrossRef]
127. Meehan, C.J.; Goig, G.A.; Kohl, T.A.; Verboven, L.; Dippenaar, A.; Ezewudo, M.; Farhat, M.R.; Guthrie, J.L.; Laukens, K.; Miotto, P.; et al. Whole genome sequencing of *Mycobacterium tuberculosis*: Current standards and open issues. *Nat. Rev. Microbiol.* **2019**, *17*, 533–545. [CrossRef]
128. World Health Organization. *The Use of Next-Generation Sequencing Technologies for the Detection of Mutations Associated with Drug Resistance in Mycobacterium tuberculosis Complex: Technical Guide*; World Health Organization: Geneva, Switzerland, 2018.
129. Pipeline Report 2019. Diagnostics. Treatment Action Group (TAG). Available online: https://www.treatmentactiongroup.org/wp-content/uploads/2019/12/pipeline_tb_diagnotics_2019_db_final.pdf (accessed on 24 February 2020).
130. World Health Organization. Rapid Communication: Molecular Assays as Initial Test for the Diagnosis of Tuberculosis and Rifampicin Resistance. Available online: https://apps.who.int/iris/bitstream/handle/10665/330395/9789240000339-eng.pdf (accessed on 24 February 2020).
131. Beutler, M.; Plesnik, S.; Mihalic, M.; Olbrich, L.; Heinrich, N.; Schumacher, S.; Lindner, M.; Koch, I.; Grasse, W.; Metzger-Boddie, C.; et al. A Pre-Clinical Validation Plan to Evaluate Analytical Sensitivities of Molecular Diagnostics Such as BD MAX MDR-TB, Xpert MTB/Rif Ultra and FluoroType MTB. *PLoS ONE* **2020**, *15*, e0227215. [CrossRef] [PubMed]
132. Ruiz, P.; Causse, M.; Vaquero, M.; Gutierrez, J.B.; Casal, M. Evaluation of a New Automated Abbott RealTime MTB RIF/INH Assay for Qualitative Detection of Rifampicin/Isoniazid Resistance in Pulmonary and Extra-Pulmonary Clinical Samples of *Mycobacterium tuberculosis*. *Infect. Drug Resist.* **2017**, *10*, 463–467. [CrossRef] [PubMed]
133. de Vos, M.; Derendinger, B.; Dolby, T.; Simpson, J.; van Helden, P.D.; Rice, J.E.; Wangh, L.J.; Theron, G.; Warren, R.M. Diagnostic Accuracy and Utility of FluoroType MTBDR, a New Molecular Assay for Multidrug-Resistant Tuberculosis. *J. Clin. Microbiol.* **2018**, *56*. [CrossRef] [PubMed]
134. Advani, J.; Verma, R.; Chatterjee, O.; Pachouri, P.K.; Upadhyay, P.; Singh, R.; Yadav, J.; Naaz, F.; Ravikumar, R.; Buggi, S.; et al. Whole Genome Sequencing of *Mycobacterium tuberculosis* Clinical Isolates From India Reveals Genetic Heterogeneity and Region-Specific Variations That Might Affect Drug Susceptibility. *Front. Microbiol.* **2019**, *10*. [CrossRef] [PubMed]
135. Allix-Béguec, C.; Arandjelovic, I.; Bi, L.; Beckert, P.; Bonnet, M.; Bradley, P.; Cabibbe, A.M.; Cancino-Muñoz, I.; Caulfield, M.J.; Chaiprasert, A.; et al. Prediction of Susceptibility to First-Line Tuberculosis Drugs by DNA Sequencing. *N. Engl. J. Med.* **2018**, *379*, 1403–1415. [CrossRef] [PubMed]

© 2020 by the authors. Licensee MDPI, Basel, Switzerland. This article is an open access article distributed under the terms and conditions of the Creative Commons Attribution (CC BY) license (http://creativecommons.org/licenses/by/4.0/).

Article

A Physical Cure for Tuberculosis: Carlo Forlanini and the Invention of Therapeutic Pneumothorax

Paolo Mazzarello

Department of Brain and Behavioral Sciences and University Museum System, University of Pavia, 27100 Pavia, Italy; paolo.mazzarello@unipv.it; Tel.: +39-382-984711

Received: 26 February 2020; Accepted: 28 April 2020; Published: 30 April 2020

Abstract: Carlo Forlanini (1847–1918), a medical doctor professor at the universities of Turin and Pavia, was the inventor of artificial pneumothorax, a method that allowed a first significant victory in the long war of medicine against pulmonary tuberculosis. The article outlines a portrait of this important clinician and focuses on the therapeutic innovation he introduced for the treatment of this infectious disease.

Keywords: Carlo Forlanini; tuberculosis; artificial pneumothorax

1. Annus Mirabilis

Sometimes it seems that, at certain moments, the lines of history converge in determining radical changes in scientific thought and practice. The year 1882 was certainly one of these for the history of infectious diseases. On March 24 of that year, Robert Koch announced the isolation and identification of *Mycobacterium tuberculosis* as the causative agent of tuberculosis. The session of the Berlin Physiological Society, during which the German medical doctor gave the report on the discovery of this new microorganism, was one of the most dramatic moments in the history of medicine: finally, the elusive nature of one of the main 19th-century scourges found a unitary etiological explanation. Paul Ehrlich (who would be awarded the Nobel Prize for medicine in 1908), present at the time of the conference, always remembered that evening as "the most important scientific experience of my life" [1]. Against all those who still rejected the contagious nature of tuberculosis, invoking constitutional, degenerative, food or toxic explanations, Koch advanced his overwhelming evidence. When the work was published a few weeks later, it quickly became a source of methodological inspiration for new microbiological investigations.

The year 1882 was also a milestone in the history of tuberculosis for a second important contribution, the publication of a revolutionary therapeutic proposal for the cure of pulmonary phthisis, that would have great importance until the discovery of the treatment of the disease with streptomycin: the "therapeutic" or "artificial pneumothorax" conceived by the Italian medical doctor Carlo Forlanini [2–5] in Figure 1.

Formulated at first only as a theoretical hypothesis (one of the few cases in medicine in which the public hypothetical prediction precedes the practical realization), it would go on to become, in the hands of its proponents during the first half of the twentieth century, an invention generally applied to treat many selected cases of phthisis, the wasting disease due to tuberculosis of the lungs. The method was also destined to be depicted by an important literary figure. In Thomas Mann's "Magic Mountain", set in the years immediately preceding the First World War, the group of patients who underwent this treatment—which was at the time the main effective direct therapy—consider themselves as part of a special club. In Mann's words: "They have formed a group, for of course a thing like the pneumothorax brings people together. They call themselves the Half-Lung Club" [6,7]. The novel by the Scottish doctor-writer Archibald Cronin, "The Citadel", also dedicates, from a descriptive point of view, an important and precise space to the treatment with artificial pneumothorax.

Figure 1. Carlo Forlanini. University Museum System, University of Pavia, Italy.

The cure continued to be the primary one until the discovery of streptomycin, the first effective antibiotic therapy for tuberculosis infection, introduced in 1943 by the microbiologists Selman Abraham Waksman, Albert Schatz and Elizabeth Bugie. However, pneumothorax did not immediately disappear from the therapeutic possibilities and continued to be used until the 1970s, in those cases where the presence of large lung cavities made patients easily susceptible to stagnation of infectious processes.

2. The Path to an Invention

Born in Milan on 11 June 1847, Carlo Forlanini, as the son of a doctor, had breathed medicine since childhood. Tuberculosis was also present early in his life, because his mother was killed by pulmonary phthisis. After high school studies, Forlanini enrolled in the medical faculty of the University of Pavia in 1864 as a pupil of the prestigious Borromeo College. It was a time of great scientific and political enthusiasm because the University of Pavia was at the forefront in the applications of new theories on diseases based on Rudolf Virchow's book "Die Cellularpathologie" and, at the same time, the young students were enthusiastic about the revolutionary movements led by Giuseppe Garibaldi, who in a few years would have brought about the unification of Italy. In 1866 Forlanini was a voluntary fighter in the Garibaldi ranks, in the battles of Monte Suello and Bezzecca. Back in Pavia he published his first scientific work in 1868 and linked himself to a fraternal friendship with Camillo Golgi, who was at the beginning of his stellar scientific career [8]. In 1870 he graduated in medicine under the guidance of the ophthalmologist Antonio Quaglino; the following year, Forlanini was at work in the chronically ill division of the Ospedale Maggiore in Milan. Meanwhile, he began to develop a great interest in the study of lung diseases by starting to conceive devices that could increase lung ventilation. In these endeavors he found help from his younger brother, Enrico, a gifted engineer who was one of the pioneers of air flight (the Milan Linate airport of 1937 is named after him). In 1875, Carlo Forlanini founded the Pneumotherapy Society and the Pneumatic Institute of Milan, and from 1881, he was also chief physician for skin diseases. Meanwhile, he began a series of attempts to treat tuberculosis and lung diseases with forced respiratory movements and with aerotherapy to increase blood flow and air supply to the lung. With the scientific collaboration of his brother Enrico, he would also build some of the first hyperbaric chambers in Figure 2.

Figure 2. Hyperbaric chamber. Museum for the History of the University of Pavia.

He observed some positive results in the treatment of pulmonary emphysema and also some mild improvement in phthisis. Soon, however, Forlanini began to develop opposite therapeutic considerations for pulmonary tuberculosis. It was not the respiratory act of advantage to the sick lung, but rest. In fact, a lesion in the pulmonary tissue, as in the case of tuberculosis, could have worsened with its movements up to the formation and the subsequent development of the cavities, which then became continuously enlarged during each respiratory act. From this consideration, he was convinced that the proper therapy to stop and block the progress of pulmonary disease was to put the lungs to rest through their artificial collapse. In fact, there had already been occasional previous observations that correlated the blockage of the movement of a lung, due to a spontaneous pneumothorax, with the improvement of a tuberculous process there.

In 1822, a British doctor, James Carson, after some experiments on rats, had also made attempts to cure pulmonary tuberculosis by causing pneumothorax on humans, but the attempt had failed [9]. However, other doctors, such as the French physicians Emile Toussaint in 1880 and Hippolyte Hérard in 1881, had reported improvement of the pulmonary tuberculosis process after lung collapse following spontaneous pneumothorax or hydro-pneumothorax [4,10,11]. Forlanini's studies on the mechanism of development of the tuberculosis process in the lung and the careful reading of previous studies, in particular Toussaint's observations, made him think of an articulated therapeutic suggestion: to make a gas penetrate the pleural cavity (interpleural space), causing the collapse of the lung below. In July 1882, Forlanini proposed his method [10], which he applied only in 1888 with his first intervention of artificial pneumothorax. Two years later, in an article by two of his collaborators, we find the first reference to four patients treated with the method [12]. Forlanini's great merit was to propose a technology through which to achieve the purpose of resting a collapsed lung, allowing the drainage and obliteration of tubercular cavity and then the scarring of the underlying lesions. The improvement of the patient became not only local, but also general, because the lung mechanical constriction of artificial pneumothorax reduced the pulmonary inflammatory source of the disease state, and perhaps could improve the body's overall immune reaction to the pathological process. It was also proposed that the method decreased the absorption of toxic factors released by the tuberculosis lesion. Forlanini observed a reduction and disappearance of the cough; the reduction and disappearance of sputum; the fall of fever and sweating; and the disappearance of asthenia. The patient became euphoric, increased his state of nutrition and had the reduction or the disappearance of the toxic-infectious state. Forlanini also noted favorable effects on the other lung, if affected by phthisis, as a consequence of the general improvement [13]. Finally, he argued that there was also an indirect therapeutic effect on initial laryngeal and intestinal tuberculosis.

To develop the appropriate technique for a repetitive injection of a small amount of gas into the pleural cavity (200–250 mL) through a large hypodermic needle, he asked for help and advice from his brother Enrico. Thus, medical expertise combined with the technological skills of the creative engineer, and the result was the new apparatus. At first, Forlanini made several attempts with atmospheric air, but then he realized that the oxygen component of this gas was being reabsorbed too quickly and so he switched to nitrogen [14,15]. In this way, the lung was immobilized for longer. To carry out the pneumothorax, Forlanini made use of two apparatuses, one for the production of nitrogen, the second for administering under pressure (controlled by a pressure gauge) the gas in the interpleural space. The production of nitrogen in the first apparatus took place through a chemical reaction by means of pyrogallic-acid which avidly absorbs the oxygen of the air. The nitrogen was then transferred to the second apparatus connected to a needle, which was introduced very slowly into the previously selected intercostal space in Figure 3. With preliminary experiments on animals and corpses, Forlanini had come to the conclusion that, once the parietal pleura was perforated, as soon as the tip of the needle touched the visceral pleura, the underlying elastic tissue retracted, dragging the visceral pleura with it, while the gas entered the interpleural space. In this way, if the introduction was slow, the visceral pleura and the underlying lung were not pierced. Forlanini usually started the treatment by introducing around 200 cc, and never more than 400 cc (to avoid the risk of subcutaneous emphysema, i.e., the passage of gas into the tissues of the chest wall) or less than 100 cc. [14]. After a few days, nitrogen was administered again to achieve a complete lung collapse. Then, pneumothorax was repeated every 15–20 days and continued for many months, usually up to two years or more.

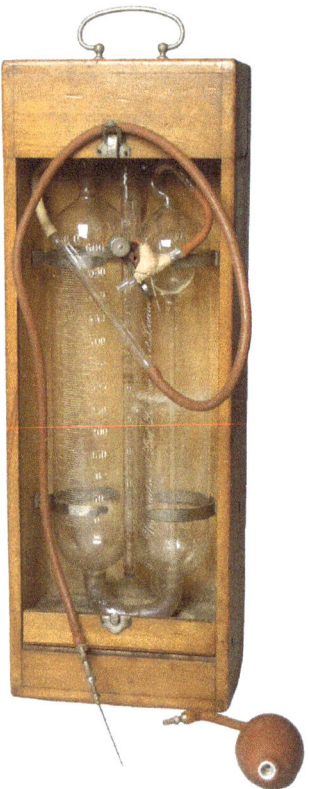

Figure 3. Artificial pneumothorax: Forlanini's apparatus for the injection of nitrogen into the pleural cavity. Museum for the History of the University of Pavia.

3. Success and Difficulty

In 1884, Forlanini began his university career at the University of Turin and obtained the position of "extraordinarius" professor of "Clinica medica preparatoria" (preparatory medical clinic). Soon after, he reached the peak of his university career, becoming full professor of special medical pathology. Among Forlanini's pupils at the University of Turin, the most important was Scipione Riva Rocci, who, under the supervision of the master, developed the pneumatic cuff sphygmomanometer, which soon became a worldwide symbol of practical medicine. In 1898, Forlanini returned to Pavia, first on a provisional teaching of special medical demonstrative pathology and, from 1900, in the prestigious position of full professor of medical clinic. Here a school of talented doctors gathered around him, like Riva Rocci, who moved to Pavia from Turin, Umberto Carpi De Resmini and Eugenio Morelli [5].

Meanwhile, Forlanini made his operation known in Figure 4 with a report held on 2 April 1894 to the eleventh international medical congress of Rome [16] and, the following year, with a lecture to the Italian congress of internal medicine [17]. However, the reaction by the medical community was not very lively and artificial pneumothorax also found oppositions and objections. With the spread of the operation and its application by physicians who did not know the method very well, and who used inadequate criteria for the selection of candidate patients, the number of serious and even fatal complications greatly increased [18]. This is despite the fact that the security of the method steadily improved in the first decades of the new century. Among the complications, those reported and also discussed by Forlanini included empyema, emphysema and cerebral embolism, especially gaseous. A further complication was pleural eclampsia, i.e., a syncope whose symptoms Forlanini described in this way: "the patient falls, suddenly, in abandonment on the bed, unconscious, with his head bent to one side, or back; the eye wide open, the bulbs fixed at the front or rotated up or to one side; the pupil of maximum amplitude, rigid; the face of an extreme pallor, sometimes cyanotic, cold, with bluish streaks in the neck, chest, limbs; clenched teeth; stiff trunk and limbs; breathless and imperceptible pulse rhythm;-in short, an impressive syndrome, for the real impotence of the symptoms, for the suddenness, for the thought that the accident follows an operational act and can be fatal" [14]. Moreover, artificial pneumothorax was sometimes ineffective due to pleural adhesions that prevented lung collapse. To overcome these difficulties, the Swedish physician Hans Christian Jacobaeus introduced the section of endopleural adhesions using a thoracoscope and a cautery [19]. Some alternative methods to artificial pneumothorax for treatment of pulmonary tuberculosis were also developed. An attempt to put the lung at rest to limit diaphragm excursions was made in 1913 by Ernst Sauerbruch, who introduced phrenicotomy [20]. In the years preceding the First World War, the foundations of pulmonary tuberculosis surgery were also laid. Already in 1891, Theodore Tuffier removed the pulmonary apex in a case of cavitated tuberculosis; in 1893 David Lawson introduced pneumonectomy; and two years later, William Macewen removed a left lung for tuberculosis (the patient was still alive in 1940). In 1908 Ludolph Brauer published the first radical thoracoplasty and in the following twenty years, lung surgery for pulmonary tuberculosis continuously developed [21–24]. However, artificial pneumothorax remained the first choice when an intervention to treat the disease had to be considered.

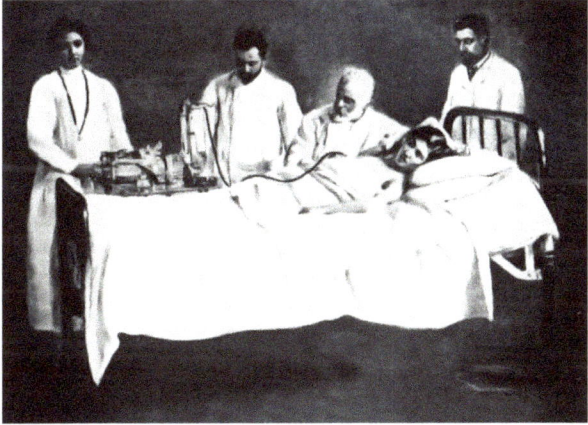

Figure 4. Carlo Forlanini while practicing artificial pneumothorax. Museum for the History of the University of Pavia.

4. Disputes and Diffusion

A bad surprise caught Forlanini when, in 1906, he read an article by Ludolph Brauer of Marburg, published in an authoritative German journal, in which the invention of the method was attributed to the important American surgeon John Benjamin Murphy [25]. The latter had developed a version of the operation, presumably without knowing the work of the Italian, and had described it in 1898 at the meeting of the American Medical Association [26–28]. Murphy was the first to adopt the new x-ray technology (Wilhelm Roentgen had discovered x-ray in 1895) to control pneumothorax therapy results. Forlanini immediately reacted to Brauer's article and published in another authoritative German journal the results obtained by him in 25 cases of pulmonary tuberculosis, also to claim the priority of the invention [29]. Moreover he criticized Murphy's treatment modalities, as they involved an incision of the skin and the chest muscles (and not a single puncture as he did) and again the large amount of nitrogen introduced (1000–3000 mL) by the American surgeon, which led to excessive compression of the lung. Forlanini's article contributed to the diffusion of his invention in the German medical world and therefore in transalpine Europe. From this moment on, Forlanini promoted the method, also with conferences for doctors. In 1906–1907, there was a technological development that made the apparatus more applicable, when Christian Saugman, of the Vejlefjord sanatorium in Denmark, perfected it by inserting in the pneumothorax instrument a water pressure monitor, which allowed a precise control of the gas introduced. Moreover, in 1912, Maurizio Ascoli proposed a partial collapse therapy that allowed a bilateral pneumothorax to treat cases of involvement of the two lungs in the pathological process.

On April 1912, Forlanini held a memorable lecture in Rome entitled "Artificial pneumothorax in the treatment of pulmonary phthisis", at the seventh international congress on tuberculosis [30]. The presentation obtained a great ovation and, after thirty years of studies on pneumothorax, Forlanini at last obtained full international recognition.

In England, the method was not used before 1911; subsequently, it achieved considerable success. In 1917, the doctor Clive Riviere wrote in his textbook "The pneumothorax treatment of pulmonary tuberculosis": "No more hopeful ray of sunshine has ever come to illuminate the dark kingdoms of disease than that introduced into the path of the consumptive through the discovery of artificial pneumothorax" [31]. Riviere confirmed this judgment ten years later in the "British Journal of Tuberculosis" [32]. Other British doctors expressed an overall positive opinion on the treatment [33–47], which was also applied to children [48]. Its level of popularization is also demonstrated by the fact that one of the important episodes in the novel "The Citadel" (1937) by the Scottish writer Archibald Joseph Cronin concerns precisely the description of an artificial pneumothorax operation. However, the spread of the method, after 1912, reached all European countries and, with some delay, also the United States. In Imperial Russia, the therapeutic pneumothorax was already used in 1910 by Arkadij N. Rubel who, two years later, published the first monograph on the topic, contributing to its knowledge in the country [18,49]. In the 1930s, in Stalinist Russia, the application of this method took on political connotations. The treatment was considered an "aristocratic therapy", therefore counter-revolutionary, and in the end the pneumothorax device was attacked by propaganda as a "killing machine". One of the leading Russian physicians advocating pneumothorax therapy, Volf S. Kholtsman, director of the Moscow Oblast Tuberculosis Institute, was arrested in 1939 during the Stalinist purges and then executed two years later.

5. The Battle and the War

In 1912, with the success at the Seventh International Congress of Rome, Forlanini became a famous scientist worldwide. In that same year he was proposed for the first time to the Nobel Prize for Medicine and, over the following years, Forlanini was unsuccessfully nominated for this award at least twenty times. His friend and fellow professor at the University of Pavia, Camillo Golgi (Nobel Prize winner for 1906), proposed him three times with the motivation that the invention of pneumothorax was "of great benefit to humanity", a phrase that underlined the consonance of the candidate's merits with the will of Alfred Nobel. However, the Nobel Prize was not awarded in the years of the First World War between 1915 and 1918, and in this last year Forlanini died. The Italian doctor thus had the possibility of winning the prize only in the three years between 1912 and 1914. There were, however, some reasons analyzed in detail by Hansson and Polianski [18] that probably prevented him from being awarded in this short period of time. In fact, all the nominations came from Italian scientists; the method had not yet reached the safety standards that it would obtain a few years later; and there were other alternative surgical operations, less successful, but used in the treatment of pulmonary tuberculosis.

In 1913, Forlanini was nominated senator of the kingdom and member of the council of public education in Figure 5. During the years of the First World War, his health gradually declined. However, he had the satisfaction of seeing his invention widely used in clinics all over the world. Forlanini died in Nervi, on the Ligurian Riviera, on 25 May 1918. Immediately after his death, Golgi published a very heartfelt obituary [50].

Figure 5. Carlo Forlanini in the last years of his life. Museum for the History of the University of Pavia.

This great physician won his battle against tuberculosis, the first of a series that will follow with other protagonists. His invention has not completely disappeared from medical practice and is still used in special cases, for example in pre-thoracoscopic artificial pneumothorax [51], a fact that would have pleased Forlanini. Unfortunately, however, the war against tuberculosis is still going on and the final victory has not yet been achieved.

Funding: This research received no external funding.

Conflicts of Interest: The author declare no conflict of interest.

References

1. Gradmann, C. *Laboratory Disease. Robert Koch's Medical Bacterology*; The Johns Hopkins University Press: Baltimore, MD, USA, 2009; p. 69.
2. Bottero, A. *Carlo Forlanini Inventore del Pneumotorace Artificiale*; Ulrico Hoepli: Milano, Italy, 1947.
3. Spina, G. Carlo Forlanini: Storia di una invenzione geniale, il pneumotorace terapeutico. *Med. Torac.* **1982**, *4*, 473–479.

4. Sakula, A. Carlo Forlanini, inventor of artificial pneumothorax for treatment of pulmonary tuberculosis. *Thorax* **1983**, *38*, 326–332. [CrossRef] [PubMed]
5. Porro, A.; Franchini, A.F. Forlanini Carlo. In *Dizionario Biografico degli Italiani*; Istituto della Enciclopedia Italiana: Roma, Italy, 1997; Volume 49, pp. 3–7.
6. Mann, T. *The Magic Mountain*; Alfred, A., Ed.; Knopf: New York, NY, USA, 1949; pp. 50–51.
7. Garbarino, M.C.; Cani, V.; Mazzarello, P. A century ago: Carlo Forlanini and the first successful treatment of tuberculosis. *Lancet* **2018**, *392*, 475. [CrossRef]
8. Mazzarello, P. *Golgi. A Biography of the Founder of Modern Neuroscience*; Oxford University Press: New York, NY, USA, 2010; pp. 35–70.
9. Keers, R.Y. Two forgotten pioneers. James Carson and George Bodington. *Thorax* **1980**, *35*, 483–489. [CrossRef] [PubMed]
10. Forlanini, C. A contribuzione della terapia chirurgica della tisi polmonare. Ablazione del polmone? Pneumotorace artificiale? In *Scritti di Carlo Forlanini*; L. Cappelli Editore: Roma, Italy, 1928; pp. 400–432.
11. Rist, É. *25 Portraits de Médecins Français 1900–1950*; Masson: Paris, France, 1955.
12. Cavallero, G.; Riva Rocci, S. La funzione respiratoria negli individui affetti da riduzione di area polmonare respirante. *G. Internazionale Sci. Med.* **1890**, *12*, 361–390.
13. Forlanini, C. Cura della tisi polmonare col pneumotorace prodotto artificialmente. In *Scritti di Carlo Forlanini*; L. Cappelli Editore: Roma, Italy, 1928; pp. 486–505.
14. Forlanini, C. Pneumotorace artificiale (nella tisi). Indicazioni, tecnica dell'atto operative, suoi accidenti. In *Scritti di Carlo Forlanini*; L. Cappelli Editore: Roma, Italy, 1928; pp. 579–619.
15. Forlanini, C. Riassunto, con note originali, dell'articolo "Sulla cura della tubercolosi polmonare mediante il pneumotorace artificiale" del Prof. Chr. Saugmann. In *Scritti di Carlo Forlanini*; L. Cappelli Editore: Roma, Italy, 1928; pp. 539–578.
16. Forlanini, C. Primi tentativi di pneumotorace artificiale nella tisi polmonare. *Gazz. Medica Torino* **1894**, *45*, 381–384, 401–403.
17. Forlanini, C. Primo caso di tisi polmonare monolaterale avanzato curato felicemente col pneumotorace artificiale. In *Scritti di Carlo Forlanini*; L. Cappelli Editore: Roma, Italy, 1928; pp. 469–471.
18. Hansson, N.; Polianski, I.J. Therapeutic pneumothorax and the Nobel Prize. *Ann. Thorac. Surg.* **2015**, *100*, 761–765. [CrossRef]
19. Jacobaeus, H.C. Endopleurale Operationen unter der Leitung des Thorakoskops. *Beitr. Klin. Tuberk.* **1916**, *35*, 1–35. [CrossRef]
20. Sauerbruch, E.F. Die Beeinflussung von Lungenerkrankungen durch Künstliche Lähmung des Zwerchfells (phrenikotomie). *Munch. Med. Wochenschr.* **1913**, *60*, 625–626.
21. Lowson, D. A case of pneumonectomy. *British. Med. J.* **1893**, *1*, 1152–1154. [CrossRef]
22. Tuffier, T. Résection du sommet du poumon droit pour tuberculose au début. Résultat éloigné (18 mois). *Bull. Soc. Chir.* **1892**, *18*, 726–727.
23. Tuffier, T. *Chirurgie du Poumon en Particulier Dans les Cavernes Tuberculeuses et la Gangrène Pulmonaire*; Masson & Cie: Paris, France, 1897.
24. Mann, R. Tuberculosis. *Int. J. STD AIDS* **1991**, *2*, 13–19. [CrossRef] [PubMed]
25. Brauer, L. Die Behandlung der einseitigen Lungenphthisis mit künstlichem Pneumothorax (nach Murphy). *Münch. Med. Woch.* **1906**, *53*, 338–339.
26. Murphy, J.B. Surgery of the lungs. *JAMA* **1898**, *31*, 151–165, 208–216, 281–297, 341–356. [CrossRef]
27. Murphy, J.B. *The Story of Clinical Pulmonary Tuberculosis*; Brown, L., Ed.; Williams & Wilkins: Baltimore, MD, USA, 1941; pp. 261–265.
28. Papagiannis, A.; Lazaridis, G.; Zarogoulidis, K.; Papaiwannou, A.; Karavergou, A.; Lampaki, S.; Baka, S.; Mpoukovinas, I.; Karavasilis, V.; Kioumis, I.; et al. Pneumothorax: An up to date "Introduction". *Ann. Transl. Med.* **2015**, *3*, 53.
29. Forlanini, C. Zur Behandlung der Lungenschwindsucht durch künstlich erzeugten Pneumothorax. *Dtsch. Med. Woch.* **1906**, *32*, 1401–1405. [CrossRef]
30. Forlanini, C. Le pneumothorax artificiel dans le traitement de la phtisie pulmonaire. In *Scritti di Carlo Forlanini*; L. Cappelli Editore: Roma, Italy, 1928; pp. 999–1012.
31. Riviere, C. *The Pneumothorax Treatment of Pulmonary Tuberculosis*; Frowde, H., Ed.; Hodder & Stoughton: London, UK, 1917.

32. Riviere, C. Artificial pneumothorax in the treatment of pulmonary tuberculosis. *Br. J. Tuberc.* **1927**, *21*, 123–124.
33. Horton-Smith Hartley, P. Artificial pneumothorax in the treatment of pulmonary tuberculosis. *Br. J. Tuberc.* **1927**, *21*, 121–123.
34. Pearson, S.V. Artificial pneumothorax in the treatment of pulmonary tuberculosis. *Br. J. Tuberc.* **1927**, *21*, 124–125.
35. Burrell, L.S.T. Artificial pneumothorax in the treatment of pulmonary tuberculosis. *Br. J. Tuberc.* **1927**, *21*, 125–126.
36. Wynn, W.H. Artificial pneumothorax in the treatment of pulmonary tuberculosis. *Br. J. Tuberc.* **1927**, *21*, 126.
37. Morriston Davies, H. Artificial pneumothorax in the treatment of pulmonary tuberculosis. *Br. J. Tuberc.* **1927**, *21*, 126–127.
38. De Carle Woodcock, H. Artificial pneumothorax in the treatment of pulmonary tuberculosis. *Br. J. Tuberc.* **1927**, *21*, 127–128. [CrossRef]
39. Hudson, B. Artificial pneumothorax in the treatment of pulmonary tuberculosis. *Br. J. Tuberc.* **1927**, *21*, 128–129.
40. Haef, F.R.G. Artificial pneumothorax in the treatment of pulmonary tuberculosis. *Br. J. Tuberc.* **1927**, *21*, 129–131. [CrossRef]
41. Morland, A. Artificial pneumothorax in the treatment of pulmonary tuberculosis. *Br. J. Tuberc.* **1927**, *21*, 131.
42. Wingfield, R.C. Artificial pneumothorax in the treatment of pulmonary tuberculosis. *Br. J. Tuberc.* **1927**, *21*, 132.
43. Johnston, J.M. Artificial pneumothorax in the treatment of pulmonary tuberculosis. *Br. J. Tuberc.* **1927**, *21*, 133.
44. Tattersall, N. Artificial pneumothorax in the treatment of pulmonary tuberculosis. *Br. J. Tuberc.* **1927**, *21*, 133–134.
45. Ford, A.P. Artificial pneumothorax in the treatment of pulmonary tuberculosis. *Br. J. Tuberc.* **1927**, *21*, 134–135.
46. Allon Pask, E.H. Artificial pneumothorax in the treatment of pulmonary tuberculosis. *Br. J. Tuberc.* **1927**, *21*, 135–136.
47. Hutchinson, R.C. Artificial pneumothorax in the treatment of pulmonary tuberculosis. *Br. J. Tuberc.* **1927**, *21*, 136.
48. Soltau, E. Artificial pneumothorax in the treatment of pulmonary tuberculosis. *Br. J. Tuberc.* **1927**, *21*, 137.
49. Polianski, I.J. Bolshevik disease and Stalinist terror: On the historical casuistry of artificial pneumothorax. *Med. Hist.* **2015**, *59*, 32–43. [CrossRef] [PubMed]
50. Golgi, C. *Carlo* Forlanini. *Rendiconti R. Istituto Lombardo* **1918**, *51* (Suppl. 2), 654–658.
51. Faurschou, P.; Viskum, K. Artificial pneumothorax by the Veress cannula: Efficacy and safety. *Resp. Med.* **1997**, *91*, 402–405. [CrossRef]

© 2020 by the author. Licensee MDPI, Basel, Switzerland. This article is an open access article distributed under the terms and conditions of the Creative Commons Attribution (CC BY) license (http://creativecommons.org/licenses/by/4.0/).

Review

Mycobacterial Cell Wall: A Source of Successful Targets for Old and New Drugs

Catherine Vilchèze

Einstein College of Medicine, 1301 Morris Park Avenue, the Bronx, New York, NY 10461, USA; catherine.vilcheze@einsteinmed.org

Received: 27 February 2020; Accepted: 19 March 2020; Published: 27 March 2020

Abstract: Eighty years after the introduction of the first antituberculosis (TB) drug, the treatment of drug-susceptible TB remains very cumbersome, requiring the use of four drugs (isoniazid, rifampicin, ethambutol and pyrazinamide) for two months followed by four months on isoniazid and rifampicin. Two of the drugs used in this "short"-course, six-month chemotherapy, isoniazid and ethambutol, target the mycobacterial cell wall. Disruption of the cell wall structure can enhance the entry of other TB drugs, resulting in a more potent chemotherapy. More importantly, inhibition of cell wall components can lead to mycobacterial cell death. The complexity of the mycobacterial cell wall offers numerous opportunities to develop drugs to eradicate *Mycobacterium tuberculosis*, the causative agent of TB. In the past 20 years, researchers from industrial and academic laboratories have tested new molecules to find the best candidates that will change the face of TB treatment: drugs that will shorten TB treatment and be efficacious against active and latent, as well as drug-resistant TB. Two of these new TB drugs block components of the mycobacterial cell wall and have reached phase 3 clinical trial. This article reviews TB drugs targeting the mycobacterial cell wall in use clinically and those in clinical development.

Keywords: tuberculosis; discovery; mode of action; drug resistance; toxicity; target

1. Introduction

In 1882, when Robert Koch made his ground-breaking announcement that he had discovered, isolated and cultured the bacterium responsible for tuberculosis (TB), there were no curative options for people infected with TB. In Roman times, the personal physician of the emperor Marcus Aurelius (161–180) was prescribing sea trips, fresh air and milk to treat TB patients [1]. English, French and German physicians revisited this concept 17 centuries later. In 1840, the English physician George Bodington wrote an essay "On the treatment and cure of pulmonary consumption," where he recommended fresh, cold and open air, exercise, a healthy diet and wine to treat TB patients. He opposed the treatment popular at the time: confinement of TB patients and the use of drugs such as digitalis and antimony potassium tartrate [2]. The journal Lancet published a harsh review of this essay, describing it as "very crude ideas and unsupported assertions" [3], but later admitted that they had been wrong [4]. Other English physicians followed Bodington's ideas of fresh-air treatment for TB patients but were dismissed by the medical intelligentsia [5]. On the other side of the Channel, the French physician Amédée Latour prescribed sunshine, fresh air, exercise and rich food containing 1/8 oz of sea salt every morning to treat TB [6]. The German physician Hermann Brehmer advocated high altitude, fresh air, exercise and a rich diet with some alcohol for the treatment of TB patients [5]. He opened the first sanatorium in 1854 in Görbersdorf, Prussia, to implement his concept. Sanatoria were beneficial to TB patients with early stages of disease. Lower mortality rates were observed in sanatoria compared to TB patients treated at home [1,7]. Sanatoria would close one hundred years later with the introduction of the first multiantimycobacterial drug regimen to treat TB.

The first drug against the etiologic agent of TB, *Mycobacterium tuberculosis,* was streptomycin, isolated from the soil bacterium *Streptomyces griseus* and shown to have activity against *M. tuberculosis* in 1944 [8,9]. Streptomycin was tested on a 21-year-old woman with advanced pulmonary TB and gave "impressive therapeutic effects" [10,11]. Unfortunately, resistance to streptomycin developed quickly [12,13]. In 1946, Jorgen Lehmann published the discovery of the antimycobacterial activity of para-aminosalicylic acid (PAS) in vitro, in guinea pigs and in TB patients [14]. The addition of PAS to streptomycin treatment drastically reduced the emergence of streptomycin-resistant strains in TB patients but did not abolish it [15]. It would take the introduction of a new TB drug, isoniazid (INH), in 1952 to achieve a successful treatment for TB [16].

INH is one of the most effective drugs against *M. tuberculosis*, which is still used to this day to treat active and latent *M. tuberculosis* infections. INH is part of a four-drug regimen (INH, rifampicin (RIF), ethambutol (EMB) and pyrazinamide (PZA)) to treat drug-susceptible *M. tuberculosis* infection. While RIF targets the RNA polymerase RpoB and PZA's mechanism of action is still unclear, INH and EMB target the mycobacterial cell wall.

The *M. tuberculosis* cell wall has an intricate and unique structure composed of a thick peptidoglycan layer and an outer membrane made up mostly of various lipopolysaccharides and fatty acids with imbedded glycolipids and wax esters. This lipid-rich cell wall forms a low-permeability barrier that protects *M. tuberculosis* against most antibiotics. This is one of the many challenges facing TB drug development and the main reason drug target-based screening has been rather unsuccessful [17,18]. Bacteria require an intact cell membrane in order to survive; therefore, the biosynthesis of cell wall components could be considered a weakness to exploit with new and more potent drugs. The ever-expanding threat of drug resistance should motivate greater alacrity in developing these new therapies. Global drug surveillance data from the World Health Organization indicates that in 2018, half a million people were infected with rifampicin- or multidrug-resistant (MDR) TB, and 6% of them had extensively drug-resistant (XDR) TB [19]. While an MDR *M. tuberculosis* strain is "only" resistant to INH and RIF, an XDR strain is resistant to at least five TB drugs (INH, RIF and three second-line TB drugs). Treatment for drug-resistant TB is very long (minimum 20 months), requiring multiple drugs with potential severe adverse reactions. MDR TB treatment's success rate is approximately 55%, whereas XDR TB is barely 39%. New drugs and shorter drug regimens are actively sought to increase these success rates. This article covers TB drugs specifically targeting the mycobacterial cell wall currently used in clinics or in clinical development, focusing on their discovery, activity, toxicity, mode of action and resistance.

2. The Mycobacterial Cell Wall

The biosynthesis of the mycobacterial cell wall components has been extensively described recently [20–22], and only a summary of the most pertinent points for this article is presented here. The proteins implicated or targeted by TB drugs discussed below are underlined.

The three main components of the cell wall are the peptidoglycan (PG), the lipopolysaccharides (arabinogalactan, lipoarabinomannan, and lipomannan) and the outer membrane, which contains mycolic acids, various glycolipids and phthiocerol dimycocerosates (Figure 1).

The *M. tuberculosis* peptidoglycan is located outside of the mycobacterial inner membrane, conferring rigidity, integrity and shape to the cell [22]. The peptidoglycan is a polysaccharide composed of alternating *N*-acetylglucosamine and muramic acid (either *N*-acetylated or *N*-glycolylated) residues linked by β (1→4) bonds [23–26]. Strands of polysaccharides are acylated on the muramic acid residues by the pentapeptide L-alanyl-D-isoglutaminyl-*meso*-diaminopimelyl-D-alanyl-D-alanine synthesized by the Mur ligases (MurC/D/E/F). The acylation reaction is performed by the D-Ala:D-Ala ligase DdlA. The pentapeptides are cross-linked to form the peptidoglycan. Two types of cross-linkages are observed in *M. tuberculosis*: (i) the D,D-transpeptidase activity of the penicillin binding proteins PonA1 and PonA2 cross-links meso-diaminopimelic acid and D-alanine to form a 3→4 linkage; (ii) the

L,D-transpeptidases (LdtMt1 to 5) cross-link two meso-diaminopimelate residues to form a 3→3 linkage [22].

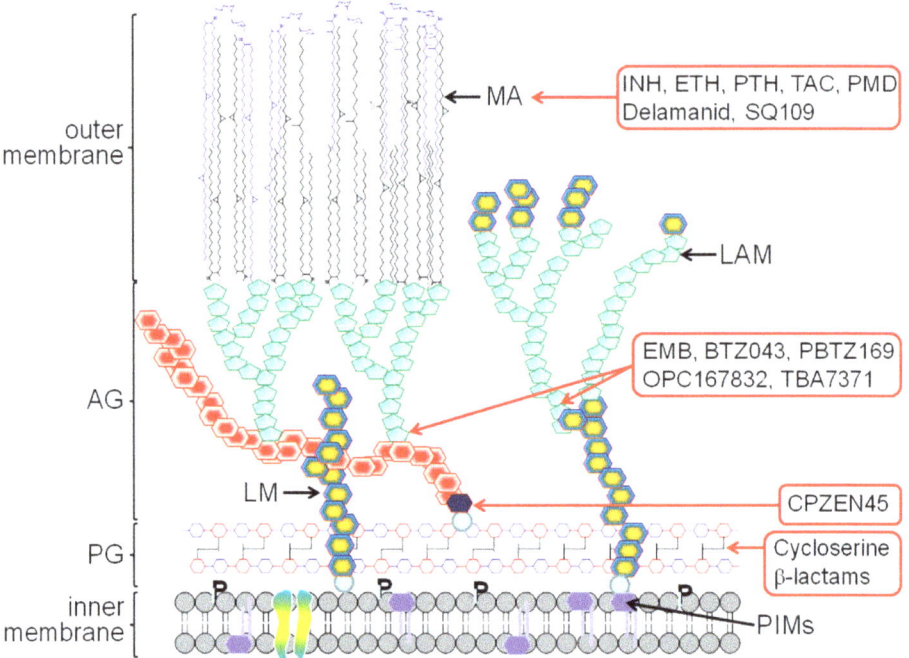

Figure 1. Illustration of the mycobacterial cell wall. The outermost layer of the cell wall, the capsule, is omitted. The cell wall components/sites inhibited by the tuberculosis (TB) drugs discussed herein are indicated. Abbreviations: AG, arabinogalactan; LAM, lipoarabinomannan; LM, lipomannan; MA, mycolic acids; PG, peptidoglycan; P, phospholipid; PIMs, phosphatidyl-*myo*-inositol mannosides; INH, isoniazid; ETH, ethionamide; PTH, prothionamide; TAC, thiacetazone; PMD, pretomanid; EMB, ethambutol.

The arabinogalactan is a branched polysaccharide composed of arabinose (Ara*f*) and galactose (Gal*f*) residues in the furanose configuration [27]. The first step in the arabinogalactan biosynthesis is the formation of the linker that anchors the arabinogalactan complex to the peptidoglycan via the *N*-glycolylated-muramic acid residues [28]. This linker is a decaprenyl-diphospho-*N*-acetylglucosamine-rhamnosyl molecule produced by the successive transfer of *N*-acetylglucosamine-1-phosphate to decaprenyl-phosphate by WecA followed by the transfer of L-rhamnose by WbbL. On this linker, 30 linear Gal*f* residues are added by the galactofuranosyl transferases GlfT1 and GlfT2. An Ara*f* unit is then transferred to the galactan chain using the arabinose donor decaprenylphosphoryl-D-arabinose (DPA). DPA is formed through several steps, starting with phospho-α-D-ribosyl-1-pyrophosphate (pRpp). UbiA adds a decaprenyl group to form decaprenol-1-monophosphate 5-phosphoribose (DPPR), which is dephosphorylated and epimerized by DprE1 and DprE2 to form DPA. The arabinofuranosyltransferase AftA catalyzes the transfer of the first unit of Ara*f* to the galactan chain. The arabinosyltransferases EmbA and EmbB catalyze the further addition of Ara*f* to form the arabinan. The final product, the arabinogalactan, is a linear galactan to which highly branched arabinans are attached. The arabinan anchors the mycolic acids forming the mycolyl-arabinogalactan-peptidoglycan (mAGP) complex.

Mycolic acids are the hallmark of the *M. tuberculosis* cell wall, an essential component regulating the permeability, acid-fast staining, viability and virulence of *M. tuberculosis* [29]. As stated above, mycolic acids are found attached to the arabinose part of the arabinogalactan complex but can also be found as a free form or bound to other saccharides to form trehalose mono/dimycolates (TMM/TDM) and glucose monomycolate. Mycolic acids are long-chain fatty acids composed of a meromycolate chain containing up to 62 carbons with various modifications (*cis/trans* cyclopropane, keto, or methoxy groups) to which a saturated C_{26} alkyl chain is attached at the α position (Figure 2). The biosynthesis of mycolic acids starts with two fatty acid synthases (FAS): the eukaryotic-like type I (FAS-I) and the prokaryotic-like type II (FAS-II) (Figure 2). FAS-I is a multidomain polypeptide [30] that synthesizes the α-C_{26} alkyl chain and also provides a $C_{14/16}$ fatty acyl-CoA to be elongated into the meromycolate chain by the FAS-II system. This elongation reaction starts with the condensation of $C_{14/16}$ fatty acyl-CoA with malonyl-Acyl Carrier Protein (ACP) by the β-ketoacyl-ACP synthase III FabH. The resulting β-ketoacyl-ACP intermediate is delivered to FAS-II, which performs cycles of elongation using independent enzymes: the reductase MabA, the heterodimer dehydratases HadAB and HadBC, the enoyl-reductase InhA and the condensases KasA and KasB (Figure 2). The modifications (*cis/trans* cyclopropane, keto, or methoxy groups) of the meromycolate chains are introduced either during the elongation by FAS-II or when the meromycolate chain is fully formed [31]. The meromycolate chain is activated by the fatty acid AMP synthase FadD32 to a meromycolyl-AMP, which is coupled to the carboxylated α-C_{26} fatty acyl-CoA (from FAS-I) by the polyketide synthase Pks13. Reduction of the resulting mycolic β-ketoester by the mycolyl reductase CmrA yields a mature mycolic acid. The biosynthesis of mycolic acids takes place in the cytoplasm. Transfer of the mycolic acids to the cell envelope occurs via the formation of a trehalose monomycolate (TMM) which is translocated by the efflux pump MmpL3 (Mycobacterial membrane protein Large). The mycolyltransferase Antigen 85 complex (*fbpA*, *fbpB*, *fbpC*) then condenses mycolic acids to the arabinogalactan releasing a molecule of trehalose.

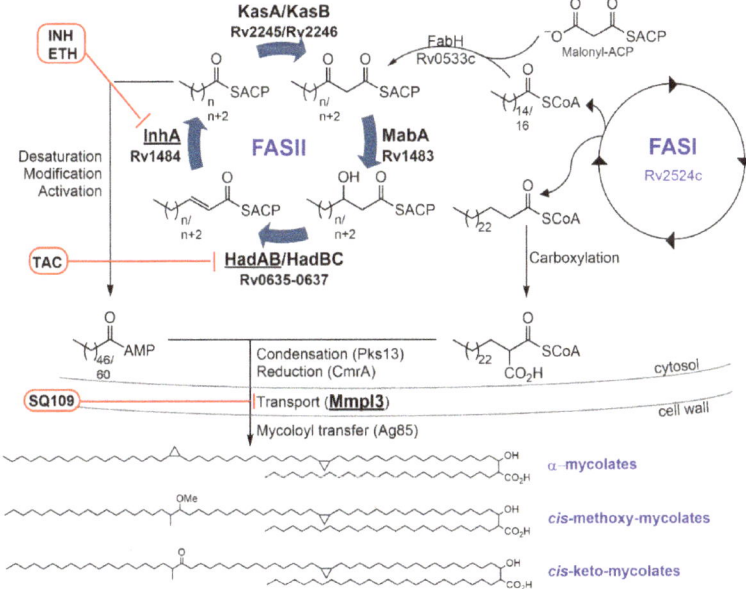

Figure 2. Schematic representation of the biosynthesis of mycolic acids. The three families of mycolic acids in *Mycobacterium tuberculosis* are the α-, methoxy- and keto-mycolic acids. Only the cis configuration of the cyclopropane group of the methoxy- and keto-mycolic acids is shown, but *M. tuberculosis* also produces the trans-methoxy- and trans-keto-mycolic acids. The known targets of the mycolic acid inhibitors in clinical use or in clinical trial are InhA for isoniazid (INH) and ethionamide (ETH); HadAB for thiacetazone (TAC); and Mmpl3 for SQ109.

Besides the mAGP complex, the mycobacterial cell wall also includes lipomannan (LM), lipoarabinomannan (LAM) and phosphatidyl-*myo*-inositol mannosides (PIMs). These lipids are anchored in the inner membrane and play an important role in *M. tuberculosis* growth, survival and virulence [32–34]. In the outer membrane, trehalose-containing glycolipids and phthiocerol dimycoserates are found interlaced with the mycolic acids. These lipids are crucial for their interactions with the host and its immune system [35].

Specific components of the cell wall are essential for *M. tuberculosis* survival in the host. It is, therefore, not surprising that two of the four first-line TB drugs target the cell wall. In the past 20 years, potent new TB drugs have been discovered that target the peptidoglycan, arabinogalactan, LAM, and mycolic acids or their transfer to the cell wall. This article focuses on the cell wall inhibitors included in the TB pharmacopeia since the 1950s/1960s (old TB drugs) and the new ones in clinical development.

3. Old TB Drugs

3.1. Isoniazid—Target: Mycolic Acids

Discovery. In 1949, Colin Hinshaw and Walsh McDermott went to Germany to investigate reports that TB patients were successfully treated with a new synthetic compound, Conteben [36]. Conteben, a thiosemicarbazone also known as thiacetazone and Tibione (see thiacetazone section below), had been discovered by Gerhard Domagk at Bayer in West Germany. Hinshaw and McDermott returned to the USA with a supply of Conteben. After testing of Conteben in US hospitals, they concluded that "a prompt and thorough series of experimental and clinical trials in the United States" was justified along with experimenting with other thiosemicarbazones [37]. Two US pharmaceutical companies, Hoffman-La Roche and E. R. Squibb & Sons, quickly developed a series of thiosemicarbazone analogs but none showed better activity than Conteben [38–40] until the benzene ring in Conteben was replaced with a pyridine ring leading to the simultaneous discovery of Rimifon at Hoffman-La Roche [39] and Nydrazid at Squibb [41]. In parallel, Domagk at Bayer developed the thiosemicarbazone analog Neoteben [42]. Rimifon, Nydrazid and Neoteben had antiTB activity that far exceeded streptomycin, PAS and any other analog synthesized so far and share the same chemical structure: 4-pyridinecarboxylic acid hydrazide. Hoffman-La Roche, Squibb and Bayer had simultaneously and independently discovered the most potent TB drug at the time: isoniazid (INH). Ironically, none of the pharmaceutical companies could patent their discovery, as INH had been synthesized 40 years earlier by two Polish graduate students Hans Meyer and Josef Mally [36]. The results with INH were so striking that chemotherapy became the leading route for TB treatment, and soon after the sanatoria closed. Almost 70 years later, INH remains the cornerstone of TB chemotherapy for the treatment of drug-susceptible and latent *M. tuberculosis* infections.

Activity and toxicity. INH is a first-line TB drug. INH is an oral, highly soluble in water (140 g/L), bactericidal drug with a minimum inhibitory concentration (MIC) ranging from 0.1 to 0.7 μm against *M. tuberculosis*. In vitro, INH rapidly reduces the number of *M. tuberculosis* bacteria by 2- to 3-$\log_{10}$s during the first four days of treatment [43]. This bactericidal activity is only observed in exponentially growing *M. tuberculosis* cultures. INH has no activity in stationary or persistent *M. tuberculosis*. In mice, a similar pattern is observed, where INH is bactericidal only on actively dividing *M. tuberculosis* [44]. INH is readily absorbed and reaches concentrations in tissues and organs above its MIC.

The main adverse effect of INH is hepatotoxicity. TB patients more likely to develop liver damage when taking INH are slow acetylators [45,46]. INH is acetylated into the inactive molecule AcINH by the human *N*-acetyltransferase-2 (NAT2), which is expressed mostly in the liver and gastrointestinal tract. Genetic polymorphisms in NAT2 renders this acetylation reaction either slow or fast, dividing TB patients between slow and fast acetylators. INH metabolism also involves an amidase metabolizing INH into an hydrazine (Hz), which is then acetylated by NAT2 to form a toxic acylhydrazine (AcHz) and a non-toxic diacylhydrazine (DiAcHz) [46]. Hz, AcHz and their metabolites generated by the liver microsomal cytochrome P450 enzymes have been linked to liver toxicity [47,48]. Fast acetylators have a lower risk of liver toxicity by producing more of the DiAcHz metabolite and less of the AcHz metabolite(s) [49].

Mode of action. INH enters *M. tuberculosis* cells by passive diffusion as a prodrug. A prodrug is the ideal drug where the compound has, if possible, no effect on eukaryotic cells, yet, once activated by a pathogen-specific enzyme, leads to the death of the pathogen [50]. Curiously, the TB pharmacopeia is composed of an array of prodrugs approved for clinical therapy as well as in the developmental phase [51]. INH is activated by the mycobacterial catalase peroxidase KatG (Rv1908c) [52] into what is most likely an isonicotinoyl radical that reacts with nicotinamide adenine dinucleotide (NAD^+) to form the INH-NAD adduct [53–56]. This adduct binds to and inhibits InhA [54,57–60], the NADH-dependent enoyl-ACP reductase [61,62] of the FAS-II system [63], leading to the inhibition of mycolic acid biosynthesis and mycobacterial cell death [60,64–66]. Although the mechanism of INH action seems rather straightforward, elucidation of the molecular details took almost fifty years [67].

Resistance. Shortly after the antimycobacterial activity of INH was published in 1952, the first report of INH-resistant clinical isolates appeared less than one year later [68]. In 1954, Gardner Middlebrook demonstrated that INH-resistant mutants isolated in vitro were catalase-negative [69]. It will then take another 40 years to discover the genetic basis of this phenotype and decipher the main mechanism of resistance to INH: mutation in *katG*, the gene encoding the INH activator. Zhang and colleagues showed that 1) complementation of an INH-resistant *Mycobacterium smegmatis* mutant with a single copy of *katG* restored INH susceptibility [52]; 2) two highly INH-resistant clinical isolates had *katG* deletion [52]; and 3) *M. tuberculosis* INH-resistant mutants regained INH susceptibility when transformed with *katG* [70]. Since then, more than 300 *katG* mutations covering 99% of the gene's length (*katG* has 2223 base pairs (bp)) have been identified in INH-resistant laboratory and clinical strains [71]. The most frequent mutation in clinical isolates is the Ser315Thr. Actually, each of the three bases of the serine codon (AGC) can be mutated leading to Asn, Arg, Ile, Gly or Leu amino acid change. Mutations in KatG alter its catalase peroxidase and oxidase activities causing a defect in KatG's ability to activate INH [72–74]. Thus, most *M. tuberculosis* clinical strains carrying *katG* mutations are highly resistant to INH [75]. KatG enzymatic activities can also be disrupted by mutations in *furA* (*Rv1909c*), a gene encoding a ferric uptake regulation protein and a negative regulator of *katG* transcription. Isogenic strains carrying the mutations a-10c and g-7a in the intergenic region between *katG* and *furA* had reduced *katG* expression leading to a decrease in INH oxidase activity and a modest increase in INH resistance [76].

Resistance to a drug can occur through either target mutation (preventing the binding of the drug to its target) or target overexpression (titration of the drug). This is the case for the second most common mutations in INH-resistant clinical isolates: mutations in *inhA* and its promoter region. The c-15t mutation in the *inhA* promoter region increases *inhA* mRNA levels by 20-fold resulting in higher InhA protein levels and an 8-fold MIC increase in *M. tuberculosis* [60]. This mutation is found in about one third of the INH-resistant clinical isolates but more often in XDR TB cases than MDR or INH-monoresistant TB cases, suggesting that the c-15t mutation could be a marker for XDR TB [77]. In contrast to *katG* mutations, clinical strains with *inhA* mutations (either in the gene or in the promoter region) have a low INH resistance phenotype [75]. There are another 20 different mutations identified in the *inhA* promoter region [71]. Mutations in the target of INH *inhA*, an essential gene, are rare, with only 17 identified so far [71]. The first *inhA* mutation (Ser94Ala) was isolated in vitro in *M. smegmatis* during a screening for mutants co-resistant to INH and ethionamide (ETH), a second-line TB drug [57]. This mutant led to the hypothesis that InhA was the primary target of both INH and ETH [57]. Introduction of the Ser94Ala mutation into wild-type *M. tuberculosis* H37Rv was shown to be sufficient to confer INH and ETH resistance [60] and to decrease the binding of the INH-NAD adduct to InhA [60,78]. These observations strongly supported the conclusion that INH and ETH target InhA. The Ser94Ala mutation has been found in INH-resistant *M. tuberculosis* clinical isolates carrying no mutations in *katG* [75,79–81].

Mutations in many other genes have been identified in INH-resistant laboratory and clinical strains such as *kasA*, *mshA*, *ndh*, *nudC*, *ahpC*, and *nat* (the *M. tuberculosis* N-acetyltransferase) to cite a few, but very often these mutations were present in clinical strains already carrying a *katG* or *inhA* mutation or were also found in INH-susceptible strains questioning their roles in INH resistance [71].

Area of investigation. For the past 70 years, INH has had an essential role in TB treatment and control. One of INH downsides is its lack of activity against dormant/persistent *M. tuberculosis*. The reasons for this lack of activity in the dormant form of *M. tuberculosis* are still up for debate; however, it is known that KatG has limited activity in the dormant state [44]. With KatG being a major factor in INH resistance and a potential player in INH shortcoming in dormant *M. tuberculosis*, new InhA inhibitors that would not require activation by KatG have been actively sought [82–89]. Compounds such as GlaxoSmithKline's thiadiazole GSK693 [90] or the diazaborine AN12855 [91,92] are promising leads with good oral bioavailability, low toxicity, activity against *katG*-deficient *M. tuberculosis* and in vivo efficacy similar to INH.

3.2. Ethambutol—Target: Arabinogalactan/LAM

Discovery. Ethambutol ((+)-2,2'-(ethylenediimino)di-1-butanol, EMB) was discovered at the Lederle Laboratories division of the American Cyanamid Company in New York (USA) in 1961. During a random screening of synthetic compounds in mice, N,N'-diisopropylethylenediamine was found to protect mice from an *M. tuberculosis* infection. Following an intensive campaign of structure–activity relationship (SAR), ethambutol, a di-hydroxylated derivative of N,N'-diisopropylethylenediamine, was synthesized [93–95]. Its activity against *M. tuberculosis* in infected mice was four times more potent than streptomycin and protected mice infected with streptomycin- or isoniazid-resistant *M. tuberculosis* strains.

Activity and toxicity. EMB is water soluble (solubility 7.58 g/L) and easily taken up by *M. tuberculosis* [96]. EMB is bacteriostatic, with a MIC ranging from 5 to 34 μm against *M. tuberculosis*. EMB is a first-line TB drug, given for the first two months of TB treatment alongside INH, RIF and PZA. EMB adverse effects include liver and ocular toxicity (decreased vision, color blindness) although these effects are reversible once EMB treatment is stopped [97].

Mode of action. The target of EMB has not been definitively determined. It was initially thought that EMB inhibited the synthesis of metabolites needed for *M. tuberculosis* replication [96] or hampered RNA synthesis [98]. A set of studies in *M. smegmatis* determined that EMB treatment resulted in the inhibition of mycolic acid transfer to the cell wall [99]; the cellular accumulation of TMM, TDM and free mycolic acids [100]; the inhibition of D-arabinose incorporation into the arabinogalactan complex and arabinomannan [101]; and the accumulation of DPA [102], the arabinose donor in the synthesis of arabinan [103]. While the synthesis of arabinan for the arabinogalactan complex was completely inhibited by EMB, EMB only partially inhibited the synthesis of the arabinan of LAM [104], leading to the conclusion that EMB inhibited different arabinosyl transferases responsible for the formation of arabinan in the cell wall [105].

Recently, EMB was postulated to target the glutamate racemase, MurI (Rv1338) [106]. MurI racemizes L-glutamate to D-glutamate, which is required for peptidoglycan biosynthesis. In an enzymatic assay, EMB partially inhibited the MurI racemisation reaction. Docking experiments suggested that EMB could bind to MurI and act as a competitive inhibitor of MurI substrate. Further experiments are required to confirm the role of MurI in *M. tuberculosis* inhibition by EMB.

Resistance. Characterization of EMB-resistant *M. tuberculosis* clinical isolates revealed a connection between EMB resistance and a cluster of genes (*embCAB*, *Rv3793–3795*) [107,108] encoding arabinosyltransferases involved in the biosynthesis of the mycobacterial cell wall arabinan [109]. Initially, the most common mutations found in EMB-resistant but not in EMB-susceptible clinical isolates were at position 306 of *embB* (M306I, M306V and M306L) [107]. When the *embB* mutations M306L, M306V and M306I were introduced into wild-type *M. tuberculosis* strains, the resulting strains were found to be EMB-resistant, suggesting that the *embB* mutations were a molecular marker for EMB resistance [110,111]. However, *embB* mutations were found in EMB-susceptible clinical isolates, leading to questioning the role of *embB* in EMB resistance [112–115]. Additionally, EMB-monoresistant clinical isolates did not harbor mutation in *embB*, suggesting a different target for EMB [113]. EMB-susceptible strains carrying *embB* mutations were resistant to other antituberculosis drugs pointing to the *emb* locus as a marker for MDR or XDR *M. tuberculosis* strains [110,112,115].

Safi and colleagues have postulated that the high-level EMB resistance found in clinical isolates was a result of an accumulation of various mutations, each individually causing a low-level resistance to EMB but together provided a high degree of resistance [110]. The authors demonstrated that, in vitro, a stepwise introduction of specific mutations in *embB*, *Rv3806c* (*ubiA*) and *embC* in *M. tuberculosis* led to a strain 8-fold more resistant to EMB than wild-type *M. tuberculosis*. *ubiA* encodes a DPPR synthase involved in DPA biosynthesis. Mutations in *ubiA* increase intracellular DPA levels, which might bind to Emb, leading to EMB resistance [116]. However, the role of *ubiA* in EMB resistance is questionable since mutations in *ubiA* are found in both EMB-susceptible and EMB-resistant *M. tuberculosis* clinical isolates [117].

Area of investigation. SAR on ethambutol was performed to improve EMB antimycobacterial activity while reducing its toxicity. The outcome was SQ109, a compound now in phase 2 clinical trial, and with a very different mechanism of action than EMB (see below).

3.3. Ethionamide/Prothionamide—Target: Mycolic Acids

Discovery. In 1954, Thomas Gardner and colleagues published the synthesis of an INH analog, where a thioamide group replaced the acyl hydrazide group in INH [118]. This thioisonicotinamide compound was active in a mouse model of *M. tuberculosis* infection but was found to be less potent than INH. This publication was noticed by a French team who found that the thioamide derivative was more active than streptomycin and more importantly, potent against INH-, streptomycin- and PAS-resistant *M. tuberculosis* strains. They synthesized a series of α-alkyl derivatives of the thioisonicotinamide and found that adding an ethyl or propyl group at the α position generated two compounds with activity in vitro and in mice greater than streptomycin but not as potent as INH [119]. They had synthesized ethionamide (ETH) and prothionamide (PTH).

Activity and toxicity. ETH is a bactericidal drug, with a MIC ranging from 6 to 15 μm against drug-susceptible *M. tuberculosis*. ETH and PTH are poorly soluble in water (0.84 and 0.28 g/L, respectively). ETH and PTH are in the group C (other core second-line agents) of second-line TB drugs used interchangeably for the treatment of MDR and XDR TB cases. ETH and PTH are oral drugs with some severe adverse effects (hepatoxicity, gastrointestinal disorders, neurotoxicity) [120].

Mode of action. Winder and colleagues were the first to demonstrate that ETH inhibited mycolic acid biosynthesis [121]. They noticed that ETH affected ***Mycobacterium bovis*** BCG similarly to INH although they observed no cross resistance between INH and ETH leading to the conclusion that the mode of action of INH and ETH were not identical [121]. Actually, cross resistance between INH and ETH had been observed in TB patients a few years earlier. Several studies noted that TB patients treated with INH developed resistance to both INH and ETH although the patients had never received ETH [122–124]. This suggested that ETH and INH shared a common mechanism of action. Winder and colleagues had postulated that INH and ETH "might differ in the means by which they reach the sensitive site". That was prescient since INH and ETH were eventually revealed as prodrugs activated by different enzymes. While INH is activated by a catalase peroxidase, ETH is activated by the NADPH-specific flavin adenine dinucleotide-containing Baeyer–Villiger monooxygenase EthA (Rv3854c also called EtaA) [125,126]. In vitro, activation of ETH by EthA leads first to the formation of ethionamide S-oxide, which is further metabolized by EthA to form either 2-ethyl-4-amidopyridine [126] or (2-ethyl-pyridin-4-yl) methanol [125] via radical intermediate(s). Once activated, the mechanism of action of ETH is very similar to INH. The activated form of ETH reacts with NAD$^+$ to form an ETH*-NAD adduct [127]. The structures of this adduct as well as the PTH*-NAD adduct were determined by X-ray crystallography, revealing an ethyl-isonicotinoyl or a propyl-isonicotinoyl covalently attached to the nicotinamide portion of reduced NAD$^+$ and bound to InhA. Inhibition of InhA by these adducts results in mycolic acid biosynthesis inhibition.

Wang and colleagues postulated that the activated form of ETH was an iminoyl radical but expressed doubt that this activation reaction was caused by EthA alone and hypothesized that additional enzymes might be involved [127]. Another study also concluded that another mechanism of activation might exist for ETH when demonstrating that deletion of *ethA* and its regulator *ethR* (*Rv3855*) in *M. tuberculosis* caused only a modest increase (3-fold) in ETH resistance [128]. EthA has two close homologs: the monooxygenases Rv3083 (MymA) and Rv0565c [125]. Grant and colleagues determined that MymA was indeed an activator of ETH by showing that loss of function of MymA or overexpression of *mymA* conferred resistance or hypersusceptibility to ETH, respectively [129]. In this study, the authors also noted that transposon mutants in *Rv0565c* as well as in two other genes encoding Baeyer–Villiger monooxygenases (*Rv1393c* and *Rv3049c*) did not lead to ETH resistance. However, a recent study determined that overexpression or deletion of *Rv0565c* led to ETH hypersusceptibility or low resistance, respectively, in *M. tuberculosis*, leading the authors to conclude that Rv0565c was an additional activator of ETH in *M. tuberculosis* [130]. In both studies on *mymA* and *Rv0565c*, no biochemical evidence was provided to show that MymA or Rv0565c actually activates ETH and what would be the resulting activated molecule(s).

Resistance. Mutations in ETH-resistant laboratory and clinical *M. tuberculosis* strains are found in the activators of ETH (*ethA*, *mymA*, *Rv0565c*), the negative regulator of *ethA* (*ethR*) and ETH target (*inhA* gene and promoter region) [129–131]. Mutations in the *inhA* promoter region are more frequent than mutations in the activator(s) of ETH [131]. Mutations in *inhA* or in its promoter region are found in up to 68% of ETH-resistant clinical isolates while mutations in ETH activator *ethA* are usually found in no more than 55% of ETH-resistant clinical isolates, and some of these *ethA* mutations can also be found in ETH-susceptible strains [131]. The mutations in *ethA* cover 91% of the gene (from 2 to 1341 bp; *ethA* has 1470 bp) [71]. Mutations in *ethR* have been identified in ETH-resistant clinical strains carrying the c-15t mutation in the *inhA* promoter region [131]. Recently, cyclic dimeric guanosine monophosphate, a bacterial second messenger, was shown to boost the binding of EthR to *ethA* promoter, causing a decrease in *ethA* transcription levels and ETH resistance [132].

Other genes have been implicated in ETH resistance. *M. tuberculosis* strains deleted for *mshA*, a gene encoding the glycosyltransferase involved in the biosynthesis of mycothiol, a major low-molecular-weight thiol, are eight times more resistant to ETH than their parental strains [133]. Mycothiol is thought to play a role in ETH resistance by increasing the rate of ETH activation by EthA [133]. In *M. smegmatis* and *M. bovis* BCG, mutants co-resistant to INH and ETH were isolated carrying mutations in *ndh*, a gene encoding an NADH dehydrogenase whose function is to oxidize NADH into NAD$^+$. In this case, *ndh* mutants were shown to accumulate NADH. Excess NADH could act as a competitive inhibitor for the binding of the ETH-NAD to InhA, triggering ETH resistance [134].

Area of investigation. Regulation of *ethR* expression plays a role in ETH resistance and susceptibility. While overexpression of *ethR* was shown to cause ETH resistance, inhibition of *ethR* triggers higher levels of *ethA* expression and increases ETH susceptibility [135]. Baulard and colleagues have taken advantage of this point and generated EthR inhibitors to boost ETH activity. Screening of chemical libraries and SAR on potential EthR inhibitors led to a first series of EthR inhibitors, which by themselves had no activity on *M. tuberculosis*, yet, when combined with ETH, significantly "boosted" ETH activity in vitro and in *M. tuberculosis*-infected mice [136]. The most active of this first generation of boosters, the oxadiazole BDM41906, decreased the MIC for ETH to the nM range [137]; however, BDM41906 had no "boosting" activity in *M. tuberculosis* strains carrying *ethA* mutations [138]. The second generation of ETH booster, the spiroisoxazoline SMARt (Small Molecule Aborting Resistance)-420, does not inhibit EthR but instead triggers a different activation mechanism for ETH. This new activation system uses the oxidoreductase Rv0077c as the activator, which is negatively regulated by Rv0078. SMARt-420 binds to Rv0078, releasing the activity of Rv0077c. SMARt-420 increases the susceptibility of *M. tuberculosis* to ETH 40-fold. In addition, ETH-resistant *M. tuberculosis* strains carrying an *ethA* mutation regain ETH susceptibility when SMARt-420 is present, both in vitro and in mice [138]. SMARt-420 can, therefore, be used in the context of ETH resistance due to *ethA* mutations but not due to *inhA* mutations.

3.4. Cycloserine—Target: Peptidoglycan

Discovery. In contrast to INH, EMB and ETH, which are synthetic compounds, cycloserine is a natural product produced by *Streptomyces* with broad-spectrum activity. Cycloserine (D-4-amino-3-isoxazolidone, also called oxamycin or seromycin) was isolated from *Streptomyces garyphalus* [139], *Streptomyces lavendulae* [140], *Streptomyces roseochromogenus* [141] and *Streptomyces orchidaceus* [142] in the early 1950s.

Activity and toxicity. Cycloserine is an oral drug, highly soluble in water (877 g/L). Cycloserine is a bacteriostatic second-line TB drug, with a MIC ranging from 50 to 250 µm against drug-susceptible *M. tuberculosis*. An analog of cycloserine, terizidone, a compound made of two molecules of cycloserine, is also used as a second-line TB drug to treat MDR TB. Cycloserine was shown to inhibit the growth of *M. tuberculosis* in vitro [143]. In vivo, cycloserine is rather inactive in a mouse or guinea pig model of *M. tuberculosis* infection [143–145] but effective in *M. tuberculosis*-infected monkeys and humans [146–148]. The differences in cycloserine activity in animal models were correlated with levels of D-alanine in sera [149] and rate of cycloserine excretion [146]. In a cohort study in China, TB patients with MDR TB had a better outcome when cycloserine was added to the treatment but that was not the case for TB patients infected with pre-XDR or XDR *M. tuberculosis* strains [150]. As an inhibitor of peptidoglycan biosynthesis, cycloserine has the distinctive feature of having a unique mechanism of action thus preventing any cross resistance with other first-line and second-line TB drugs. Cycloserine is, therefore, a useful addition to second-line TB drugs, although its use might be limited by its severe toxicity. Serious adverse effects were observed during cycloserine treatment such as neuropathy and behavioral changes [151]. Cycloserine is, therefore, contraindicated in patients suffering from severe depression, suicidal tendencies, kidney failure, epilepsy or seizures.

Mode of action. The mechanism of action of cycloserine was primarily deciphered in *Staphylococcus aureus.* Cycloserine is an analog of D-alanine and works as an antagonist of D-alanine [152]. Cycloserine inhibits the alanine racemase Alr (Rv3423c), which converts L-alanine to D-alanine, and the D-Ala:D-Ala ligase DdlA (Rv2981c) [153] preventing the integration of alanine into the pentapeptide core of the peptidoglycan. In *M. tuberculosis*, DdlA is thought to be the primary target of cycloserine [154–156].

Resistance. D-cycloserine-resistant *M. tuberculosis* mutants were isolated as early as 1957 in TB patients treated with cycloserine [157]. Cycloserine resistance has been associated with mutations in the genes encoding the alanine transporter CycA (Rv1704c), the L-alanine dehydrogenase Ald (Rv2780) and the alanine racemase Alr. A Gly122Ser mutation in *cycA* is present in the naturally cycloserine-resistant *M. bovis* BCG vaccine strain [158]. Complementation of BCG with a cosmid containing *M. tuberculosis cycA* renders BCG more susceptible to cycloserine than the parental strain leading the authors to conclude that *cycA* may be a factor in cycloserine resistance in BCG [158]. Desjardins and colleagues demonstrated that deletion of *ald* increased the resistance to cycloserine 2-fold in *M. tuberculosis*, while complementation of the *ald* mutant with a plasmid expressing *M. tuberculosis ald* only partially restored cycloserine susceptibility. Notably, complementation of BCG with the *ald* plasmid did not alter the strain resistance to cycloserine [159]. Mutations in *alr* (M319T, Y364D, R373L and c-8t in the promoter region) have been identified in XDR TB strains isolated from TB patients treated with cycloserine [160]. Further, the clinical isolate with the R373L mutation in *alr* also contained a deletion in *ald*.

In an in vitro experiment, 18 cycloserine-resistant mutants were obtained by culturing *M. tuberculosis* H37Rv on plates containing increasing concentrations of cycloserine (from 0.2 to

3 mM) and characterized using whole-genome sequencing [161]. Mutants were only obtained on plates containing less than 0.8 mM of cycloserine. A mutation was identified in *alr* (D344N), but no mutations in *ddlA*, *ald* or *cycA* were found. Fifteen novel mutations were identified in genes involved in various pathways however no complementation or allelic exchange experiments were performed to confirm that these novel mutations were indeed involved in cycloserine resistance.

In a recent in vitro study [162], the authors showed that spontaneous cycloserine-resistant mutants emerged at a lower frequency (10^{-10}–10^{-11}) than rifampicin- (10^{-9}) or isoniazid-resistant (10^{-8}) mutants in *M. tuberculosis*. The authors characterized 11 independent cycloserine-resistant *M. tuberculosis* mutants by whole-genome sequencing and found mutations in *alr* (gene (D322N) and promoter region) but again not in *ddlA*, *ald* or *cycA*. None of these 11 cycloserine-resistant mutants showed any cross-resistance to other first-line and second-line TB drugs. The *alr* promoter mutation upregulated *alr* gene transcript, causing an increase in Alr protein level by up to 30-fold. The *alr* D322N mutation present in 8/11 mutants reduced the binding affinity for cycloserine by 240 fold while having limited effect on Alr enzymatic activity. Evangelopoulos and colleagues postulated that the *alr* mutations protected the enzymatic function of both cycloserine targets by decreasing the affinity of cycloserine to Alr and increasing the levels of D-alanine preventing the binding of cycloserine to DdlA. Their main conclusion was that DdlA was the "lethal target" for cycloserine.

3.5. Thiacetazone—Target: Mycolic Acids

Discovery. Gerhard Domagk was awarded the Nobel Prize in 1939 for the discovery of prontosil, a prodrug that releases *p*-aminobenzenesulfonamide, the first sulfonamide drug active against *Staphylococcus* and *Streptococcus* infections, but not against *M. tuberculosis* infections [163]. While prontosil was quickly replaced by penicillin to fight antibacterial infections, Domagk and his team continued working on the synthesis of related compounds, the thiosemicarbazides, and discovered that one in particular, thiacetazone (TAC, also called Tibione, Tb I, Conteben), had impressive activity against *M. tuberculosis* in guinea pigs [164]. In the late 1940s, TAC was tested in Germany on patients with various forms of TB and found to have promising activity although severe adverse effects were recorded [37,165].

Activity and toxicity. TAC is an oral, inexpensive, effective and bacteriostatic drug, with a MIC ranging from 0.3 to 5 µm against drug-susceptible *M. tuberculosis*. TAC is poorly soluble in water (0.09 g/L). TAC is part of the group D3 (add-on agents) of second-line TB drugs. TAC has been given in combination with INH in resource-poor countries [166,167]. Its utilization as a TB drug was mostly discontinued because of its high toxicity (skin disorder, gastrointestinal symptoms, conjunctivitis, vertigo, liver and kidney damage) especially in HIV-positive TB patients [168]. TAC is, therefore, not recommended in HIV-positive TB patients.

Mode of action. TAC is a prodrug that is activated by the monooxygenase EthA [127,169]. Once activated, TAC inhibits the dehydratase HadAB of the FAS-II system by forming a disulfide bound with a cysteine (Cys61) residue of HadA [170,171]. During the elongation of fatty acyl-ACPs by FAS-II, HadA binds the acyl-ACP while HadB performs the dehydratase reaction. When covalently bound to the Cys61 residue of HadA, TAC obstructs the acyl-ACP channel preventing the binding of the fatty acyl-ACP [172]. The covalent bond between TAC and HadA Cys61 blocks the activity of HadAB leading to inhibition of mycolic acid biosynthesis [173].

Resistance. The mode of action of TAC was discovered through analysis of TAC resistance. Since TAC treatment of *M. tuberculosis* results in inhibition of mycolic acids, Belardinelli and Morbidoni overexpressed every gene involved in the FAS-II system. Only overexpression of the dehydratase operon *hadABC* or its dimer *hadBC* resulted in highly TAC-resistant *M. tuberculosis* strains (MIC > 0.2 mm) [170]. Curiously, the level of TAC resistance was much lower when the dimer *hadAB* was overexpressed in *M. tuberculosis* (MIC 10 µm). This was puzzling since HadAB is the target of TAC. Moreover, sequencing of spontaneous *M. tuberculosis* TAC-resistant mutants revealed mutations in HadA (Cys61Ser, Cys61Gly) but also in HadC (Val85Phe, Thr123Ala, Lys157Arg, Ala151Val) [170,174].

Grzegorzewicz and colleagues set up to unmask the role of *hadC* in TAC resistance [175]. The authors found that mutations in *hadC* protected *M. tuberculosis* from TAC and compensated for HadAB inhibition by TAC. The authors proposed that *hadC* mutations prevented TAC from reaching the active site of HadAB [175].

TAC resistance is also mediated by mutations in *mmaA4* (*Rv0642c*), a gene encoding a methoxy mycolic acid synthase required for the synthesis of keto- and methoxy-mycolic acids [170,176]. The methyltransferases involved in the modifications of the meromycolic acids such as MmaA4 interact with the proteins of the FAS-II system including the dehydratase heterodimers HadAB and HadBC [31]. Mutations in *mmaA4* might, therefore, modify the conformation of HadAB preventing the binding of TAC to HadA and causing TAC resistance [175].

Area of investigation. In the late 1940s, the discovery of the antiTB activity of the thiosemicarbazide TAC propelled pharmaceutical companies into an extensive search for TAC analogs with antiTB properties. This led to the discovery of the most important antimycobacterial drug INH (see above). This family of compounds might still reveal interesting molecules with pharmaceutical properties. Recently, new TAC analogs were synthesized with promising activities against *M. tuberculosis* [174].

4. New Generation of Cell Wall Inhibitors

4.1. Target: Mycolic Acids

4.1.1. SQ109

Discovery. A collaboration between a pharmaceutical laboratory (Sequella) and an academic (Laboratory of Host Defenses NIAID/NIH) laboratory led to the synthesis and screening of a chemical library composed of 63,238 molecules based on the ethylenediamine core of EMB [177]. EMB was targeted for its good antimycobacterial activity in TB patients and lack of previous SAR studies. The goal was to create a more potent yet less toxic EMB analog. The screening selected 170 compounds that had an MIC of less than 6 mg/L against *M. tuberculosis* (discarding inactive or compounds that could not cross the cell wall) and had an inhibitory effect on the cell wall (determined by monitoring upregulation of the *iniBAC* operon promoter, a phenotype observed in cell wall inhibitors [178]). Based on the molecular structures of these 170 compounds, a new chemical library composed of 30,000 molecules was synthesized and retested against *M. tuberculosis* (in vitro and in macrophages). Eleven compounds passed those screens and were then assessed in mice [179]. The compound with the best antimycobacterial activity, pharmacological and toxicity data was SQ109.

Activity. SQ109 is a lipophilic, oral drug with poor water solubility (1.7 mg/L). SQ109 is bactericidal against drug-susceptible, MDR and XDR *M. tuberculosis* strains, with a MIC ranging from 0.4 to 3 µm. In vitro, SQ109 synergizes with the first-line TB drugs INH and RIF as well as with second-line TB drugs such as cycloserine, moxifloxacin, amikacin and bedaquiline (BDQ) [180]. In *M. tuberculosis*-infected mice, treatment with SQ109 for 28 days resulted in a dose-dependent reduction in lung and spleen burdens [181]. SQ109 was not as effective as INH in eliminating *M. tuberculosis* with organ burdens between 1 and 2.5 $\log_{10}$ higher in the SQ109 treatment compared to INH treatment [181]. However, SQ109 peak concentration in the lung and spleen was higher than SQ109 MIC and 45-fold higher than in plasma [181].

Mode of action. Surprisingly, this double SAR based on EMB resulted in a drug, SQ109, with a different mode of action than EMB. EMB-resistant *M. tuberculosis* strains were fully susceptible to SQ109 [179]. Transcriptional response of SQ109-treated *M. tuberculosis*, while consistent with other cell wall inhibitors, did not match EMB's response [182]. Lipid analysis of SQ109-treated *M. tuberculosis* revealed an inhibition of mycolic acid attachment to the arabinogalactan, a depletion of TDM and an accumulation of TMM [183]. To decipher the mechanism behind this result, isolation of spontaneous SQ109-resistant mutants was attempted without success (the mutation rate for SQ109 is exceptionally low (10^{-11})) [183]. Instead, an analog of SQ109 was used to isolate spontaneous resistant mutants. These mutants were co-resistant to SQ109 and carried a mutation in the mycolic acid flippase encoded by *mmpl3* (*Rv0206c*), involved in the translocation of TMM across the plasma membrane [184]. Analyses carried out in *M. smegmatis* revealed that SQ109 binds to Mmpl3 at the proton transportation site inhibiting the proton motive force for substrate translocation [185].

Clinical trial. In a 14 day study where SQ109 was given alone or in combination with RIF, SQ109 was shown to be safe in TB patients, with only gastrointestinal issues being disclosed [186]. The SQ109 plasma concentration was lower than the MIC for the drug and, in contrast to the in vitro results, the combination of SQ109 and RIF led to a decrease in SQ109 availability [186]. In mice, SQ109 had been shown to be metabolized by the cytochrome P450 isoenzymes CYP_2D_6 and CYP_2C_{19} [181]. Since RIF induces the expression of CYP_2C enzymes in humans [187], RIF might lower the effective dose of SQ109. On the other hand, the presence of SQ109 had no effect on the plasma concentration of RIF. The result of this 14 day trial was that SQ109 alone had no bactericidal effect, and there was no synergy or additive effect with the combination RIF and SQ109 [186]. In a phase 2 trial (NCT01785186) performed in newly diagnosed pulmonary TB patients, treatment with SQ109 in combination with RIF/INH/PZA failed to improve culture conversion compared to RIF/INH/PZA [188]. A phase 2b clinical trial in Russia on MDR TB patients revealed that SQ109, combined with standard drug therapy, was safe, well tolerated and effective (80% of sputum negative patients after 24 weeks of treatment with SQ109 compared to 61% without SQ109) [189].

Area of investigation. With the success of SQ109 and the essentiality of *mmpl3* in *M. tuberculosis*, novel inhibitors of Mmpl3 have been actively sought. Inhibition of Mmpl3 is not specific to the chemical structure of SQ109 and numerous chemical scaffolds are being identified as Mmpl3 inhibitors [185,190,191].

4.1.2. Pretomanid (PA-824)

Discovery. Another challenge in TB drug discovery is to discover new molecules that will eliminate dormant or persistent *M. tuberculosis*, a subpopulation of bacteria that are genetically drug-susceptible but 'resistant' (or tolerant) to drug treatment. One of the most used methods to study persistent *M. tuberculosis* is the Wayne model where *M. tuberculosis* is subjected to gradual oxygen depletion to allow for a slow entry into anaerobiosis. To find potential inhibitors of dormant *M. tuberculosis*, nitroimidazoles were tested, as they inhibit bacterial anaerobes. Metronidazole, an antibacterial and antiprotozoal drug with activity against Gram-positive and Gram-negative anaerobes [192] showed little, if any, activity against *M. tuberculosis* in mice [193]. However, a bicyclic nitroimidazole (CGI 17341) was reported to have potent in vitro and in vivo antimycobacterial activity. CGI 17341 showed no cross-resistance with other TB drugs but was highly mutagenic [194]. A series of more than 300 analogs was then synthesized to solve the CGI 17341 mutagenicity issue [195]. This led to the discovery of the nitroimidazole PA-824 now called pretomanid (PMD).

Activity. PMD is an oral drug, poorly soluble in water (0.012 g/L). PMD is active against replicating and non-replicating (anaerobic), drug-susceptible and drug-resistant (MIC ranging from 0.08 to 0.7 μm) *M. tuberculosis* strains. In a mouse model of *M. tuberculosis* infection, PMD given orally for 10 days performed as well as INH in lowering organ bacterial burdens. Increasing the concentration of PMD (from 25 mg/kg to 100 mg/kg) led to a significant reduction in organ burden compared to INH [195]. Similar results were obtained in *M. tuberculosis*-infected guinea pigs treated with PMD. The toxic threshold in mice was 1 g/kg for a single dose of PMD and 0.5 g/kg for a daily dose given for 28 days [195]. In a drug combination experiment, the lung burden in mice infected with the drug-susceptible *M. tuberculosis* H37Rv strain was 2.5 $\log_{10}$ lower in mice receiving PMD/BDQ/PZA than in mice treated with standard regimen (INH/RIF/EMB/PZA) after one month of treatment. None of the mice treated for 2 months with PMD/BDQ/PZA relapsed (mice were kept for an additional three months at the end of the treatment to assess relapse). The addition of moxifloxacin (MXF) to this combination did not improve the outcome during the first month of treatment but did decrease the number of mice relapsing after 1.5 month treatment [196]. In a subsequent experiment,

M. tuberculosis-infected mice were treated with BDQ/MXF/PZA +/− PMD [197]. The addition of PMD to the BDQ/MXF/PZA treatment decreased the lung burden an extra 1 $\log_{10}$ after one month of treatment compared to BDQ/MXF/PZA alone. In both groups, mice were relapse free after two months of treatment [197]. In addition, treatment of *M. tuberculosis*-infected mice with BDQ/linezolid (LZD) +/− PMD revealed that the addition of PMD had a significant impact on the lung burden. PMD in combination with BDQ/LZD resulted in a 7.2 $\log_{10}$ CFU *M. tuberculosis* killing in the lungs and relapse-free mice after two and three months of treatment, respectively. In contrast, the BDQ/LZD treatment resulted in a 5.2 log reduction in CFU in the lungs, but more than 90% of the mice relapsed even after 4 months of treatment [197].

Mode of action. Like many other TB drugs, PMD is a prodrug. PMD is activated by a deazaflavin (F420)-dependent nitroreductase (Ddn, Rv3547) to generate a des nitro metabolite releasing nitric oxide (NO) [198]. The presence of the des nitro metabolite and rate of NO release were linked specifically to the anaerobic killing activity of PMD. PMD treatment of *M. tuberculosis* also causes a decrease in ketomycolates production and an accumulation of hydroxymycolates leading to the hypothesis that PMD inhibits an enzyme or affects a cofactor responsible for the oxidation of hydroxymycolic acids to ketomycolic acids [195]. A transcriptional analysis of *M. tuberculosis* treated with PMD under replicating conditions revealed that PMD action had similarity with cell wall inhibitors as well as inhibitors of the respiratory chain [199]. This data confirmed the dual action of PMD as an inhibitor of a subfamily of mycolic acids as well as a NO generator leading to the inhibition of the respiratory chain under anaerobic conditions. This combination of a cell wall inhibitor and NO generator makes PMD a unique antimycobacterial drug, effective against active and dormant TB that can shorten the duration of chemotherapy.

Resistance. The main mechanism of resistance to PMD is through mutations in its activator Ddn. In vitro selection of spontaneous PMD-resistant mutants in *M. tuberculosis* led to the isolation of 183 mutants [200]. The mutation rate was relatively high ranging from 10^{-5} to 10^{-7}. Most of the mutations were in *ddn* (SNPs, base pair deletion or insertion, early stop codon). Mutations in *fbiABC* (*Rv3261*, *Rv3262*, *Rv1173*) encoding F420 biosynthesis proteins and *fgd1* (*Rv0407*) encoding a F420-dependent glucose-6-phosphate dehydrogenase were also identified, but no complementation experiments were performed to confirm the role of these mutations in PMD resistance.

Clinical trial. A phase 2a clinical trial (NCT01215851) evaluated the early bactericidal activity (EBA), safety and tolerability of PMD combined with BDQ, PZA and/or MXF in newly diagnosed drug-susceptible, smear-positive pulmonary TB patients [201]. These combinations were well tolerated and seemed safe. The combination PMD/PZA/MXF was more bactericidal than PMD/BDQ or PMD/PZA and as potent as the standard regimen (INH/RIF/EMB/PZA) [201]. In a phase 2b clinical trial (NCT02193776), the efficacy, safety and tolerability of the combination PMD/BDQ/PZA were assessed in newly diagnosed drug-susceptible, smear-positive pulmonary TB patients treated for 8 weeks [202]. TB patients converted more rapidly to sputum negativity with PMD/BDQ/PZA than the standard regimen. One arm of the study also looked at the combination PMD/BDQ/PZA/MXF for the treatment of newly diagnosed, MXF-sensitive, MDR, pulmonary TB patients. MDR TB patients converted to sputum negativity within 8 weeks of treatment with PMD/BDQ/PZA/MXF. A phase 2c trial (NCT03338621) is currently in progress to evaluate the efficacy, safety and tolerability of a 4 month PMD/BDQ/PZA/MXF treatment given to patients infected with drug-sensitive TB or a 6 month treatment given to drug-resistant TB patients. No results have been posted yet.

The Nix-TB trial (phase 3, NCT02333799) was set to test the efficacy, tolerability, safety and pharmacodynamics of the combination PMD/BDQ/LZD given for six months on 50 patients infected with XDR TB and 24 patients infected with treatment-intolerant or non-responsive MDR TB. 88% of XDR TB cases and 92% of MDR TB cases had favorable outcomes at the end of treatment (no clinical infection, culture negative 6 months post treatment). 23% of the patients had manageable adverse effects [203]. On August 14, 2019, the Food and Drug Administration (FDA) approved the use of PMD in combination with BDQ and LZD for the treatment of pulmonary TB only in the case of non-responsive MDR, XDR and treatment-intolerant *M. tuberculosis* infections. This new regimen is all oral, short (6 months), more efficacious (sputum conversion in less than 6 weeks) and uses fewer drugs (3) compared to the treatment for highly resistant TB, which entails daily injections for 6 months followed by daily treatment with five drugs for 12 to 18 months (https://www.fda.gov/media/128001/download).

4.1.3. Delamanid (OPC-67683)

Discovery. Upon the discovery of the activity of CGI 17341 against *M. tuberculosis*, researchers at Otsuka Pharmaceutical Co., Japan, performed a SAR study on the compound to reduce its mutagenicity and increase its potency against *M. tuberculosis* [204]. The result was the nitroimidazole OPC-67683, now called delamanid.

Activity. Delamanid is an oral, non-mutagenic compound, poorly soluble in water (0.002 g/L). Delamanid is active in vitro (MIC 8–45 nM [205]) against drug-susceptible and MDR *M. tuberculosis* strains and highly efficacious in vivo [206]. Delamanid is also active against non-replicating *M. tuberculosis* [207]. Delamanid is only active against members of the MTB complex as well as some non-tuberculous mycobacteria and shows no activity against bacterial microflora [205]. In a mouse model of acute and chronic *M. tuberculosis* infections, delamanid performed better than PMD. After 3 weeks of treatment, in the acute model of infection, delamanid and PMD reduced the lung bacterial burden by 1 $\log_{10}$ and 0, respectively. In the chronic model of infection, delamanid and PMD reduced the lung bacterial burden by 3 and 2 $\log_{10}$, respectively [207]. Delamanid is included in the Group D2 (add-on agents) of second-line TB drugs to treat RIF-resistant and MDR TB patients.

Mode of action. Similar to PMD, delamanid is a prodrug activated by the deazaflavin-dependent nitroreductase Ddn to form a des nitro metabolite and release NO. Delamanid inhibits the biosynthesis of methoxy- and keto-mycolic acids but not of α-mycolic acids [206]. Deciphering the mechanism of action of delamanid by isolating spontaneous in vitro resistant mutants has led to the identification of mutations in genes involved only in delamanid activation (*ddn, fgd1, fbiABC*), not in its target [208]. Frequency of mutation was relatively high (10^{-5} to 10^{-6}). Complementation restored drug susceptibility except for one *fbiB* mutant that required complementation with a plasmid containing both *fbiB* and *fbiA*. Two delamanid-resistant clinical isolates were also analyzed in that study and shown to have mutations in *ddn* (L107P and a 59–101 bp deletion). The target of delamanid is yet to be discovered, as no mutant in pathways independent to delamanid's mode of activation has been isolated. The question remains as to whether delamanid and PMD target mycolic acid biosynthesis or whether the inhibition of the biosynthesis of these specific mycolic acids observed during delamanid or PMD treatment of *M. tuberculosis* is only a consequence of the inhibition of these compounds' target(s).

Clinical trial. As of 2019, delamanid is one of three new TB drugs in phase 3 clinical development along with PMD and BDQ (www.newtbdrugs.org). In a clinical trial in MDR TB patients, delamanid was added for 2 months to an optimized background treatment regimen (OBR), resulting in 45% sputum culture conversion compared to 30% sputum culture conversion for the patients getting OBR and placebo [209]. In this study, the group of patients receiving higher doses of delamanid (200 mg, twice a day) had a higher incidence of palpitation and prolonged QT intervals than the groups that received a lower concentration (100 mg, twice a day) of delamanid or placebo. In a subsequent six-month delamanid trial, a lower mortality rate was observed in TB patients that received delamanid for 6 months rather than the previous 2 months trial, but no significant difference in successful treatment outcome was recorded between the 2 month and 6 month delamanid trial [210,211]. A phase 3 clinical trial on MDR TB patients tested the addition of delamanid for the first six months of the 24 month OBR regimen. There was no statistical difference in time to sputum culture conversion or rate of adverse events whether the patients received delamanid or not with their OBR treatment [212]. Overall, delamanid is a well-tolerated drug with good safety data. In 2014, delamanid was approved for the treatment of pulmonary MDR TB in adults in Europe, Japan and Korea.

Delamanid, PMD and SQ109 are the only drugs in advanced clinical development so far targeting mycolic acid biosynthesis or mycolic acid incorporation into the cell wall. Academic and pharmaceutical laboratories are developing new inhibitors of enzymes involved in mycolic acid biosynthesis. Inhibitors of KasA [213], InhA [83,90–92], Pks13 [214–217] and Mmpl3 [185,190,191,218,219] are being tested.

4.2. Target: Arabinogalactan/LAM

4.2.1. CPZEN-45

Discovery. Caprazamycin B, a liponucleoside isolated from *Streptomyces*, is a non-toxic antibiotic with good activity in vitro (MIC 3–11 µm) against drug-susceptible and MDR *M. tuberculosis* strains but insoluble in water [220]. To increase its hydrophilicity, SAR was performed and led to the nucleoside caprazene-45 (CPZEN-45) [221].

Activity. CPZEN-45 is a water-soluble drug (solubility ≈ 10 g/L) with an MIC of 2–5 µm and 10 µm against drug-susceptible and MDR *M. tuberculosis*, respectively. Unlike caprazamycin B, which is active against several Gram-positive bacteria, CPZEN-45 has no activity against *S. aureus*, *Streptococcus pneumonia* or *Enterococcus faecalis* [222]. CPZEN-45 is specific to slow-growing pathogens. In mice intravenously infected with the drug-susceptible *M. tuberculosis* H37Rv strain, a 30-day subcutaneous treatment with CPZEN-45 was as effective in reducing lung burden as INH and better than RIF alone [223]. Furthermore, the combination INH/RIF/CPZEN-45 resulted in at least a 1 $\log_{10}$ better killing of *M. tuberculosis* than the combination INH/RIF. In a subsequent experiment, mice were intravenously infected with an XDR *M. tuberculosis* strain and treated with CPZEN-45 at doses ranging from 6.3 to 200 mg/kg. After 30 days, the mice treated with the highest concentration of CPZEN-45 had a 1.5 $\log_{10}$ better reduction in CFUs in the lungs than the mice receiving the lowest concentration [223].

Mode of action. CPZEN-45 is an inhibitor of *M. tuberculosis* WecA (Rv1302) with an IC_{50} of 7 nM [222]. WecA is involved in the first step of the arabinogalactan biosynthesis forming the anchor point between peptidoglycan and arabinogalactan. Transcriptionally silencing *wecA* is bactericidal in *M. tuberculosis* in vitro and bacteriostatic ex vivo, validating WecA as candidate for drug development [224].

Clinical trial. CPZEN-45 is in the early stage of clinical development.

4.2.2. BTZ043

Discovery. A sulfur-based chemical library was tested for antibacterial and antifungal activities. A class of compounds, the nitrobenzothiazinones, was found to have specific activity against mycobacteria. The most promising hit BTZ038 was a racemic molecule and synthesis of its *S* enantiomer gave BTZ043 [225].

Activity. BTZ043 is lipophilic, bactericidal, and is active against drug-susceptible, MDR and XDR *M. tuberculosis* strains, with an MIC ranging from 2 to 70 nM. In contrast to the novel TB drugs targeting the mycolic acids, BTZ043 was less effective in non-replicating conditions, suggesting that BTZ043 would have to be used in combination with other drugs to be effective against TB [225]. BTZ043 was tested with first-, second-line and in-development TB drugs and showed no antagonistic effects. BTZ043 was additive with most of the drugs tested and synergistic with BDQ [226]. The compound is as effective in a mouse model of *M. tuberculosis* infection as INH. BTZ043 was well tolerated in rats, had low interaction with the CYP450 enzymes and showed no mutagenic or genotoxic properties [227].

Mode of action. BTZ043 is a prodrug activated by DprE1 (Rv3790). DprE1 reduces the nitro group in BTZ043, yielding a nitroso metabolite. This metabolite reacts with the thiol group of a cysteine (Cys387) residue in the substrate-binding site of DprE1 to form a covalent bond that irreversibly inhibits DprE1, classifying BTZ043 as a suicide inhibitor [228]. DprE1 catalyzes the first step in the epimerisation of decaprenylphosphoryl ribose (DPR) to decaprenylphosphoryl arabinose (DPA), the arabinose donor for the synthesis of arabinogalactan and LAM. Treatment of *M. tuberculosis* with BTZ043, therefore, results in inhibition of DPA formation and ultimately inhibition of both arabinogalactan and LAM [225].

Resistance. Mutation at the Cys387 position of *dprE1* is the main mechanism of resistance to BTZ043. Frequency of mutation is low ($<10^{-8}$); however, the resulting mutants *dprE1* Cys387Ser and Cys387Gly are highly resistant to BTZ043 (250-10,000-fold) [225].

Clinical trial. As of December 2019, BTZ043 is in Phase 1b/2a clinical trial to assess safety, tolerability, interaction with first-line TB drugs and early bactericidal activity in newly diagnosed patients infected with uncomplicated, smear-positive, drug-susceptible TB.

4.2.3. PBTZ169

Discovery. SAR on BTZ043 to improve its pharmacologic properties yielded PBTZ169 (now called macozinone), a piperazine derivative. Compared to BTZ043, PBTZ169 has the advantage of having no chiral center which facilitates its chemical synthesis.

Activity. PBTZ169 is a very potent bactericidal benzothiazinone with an MIC of 0.6 nM against *M. tuberculosis*. PBTZ169 has better potency, pharmacodynamics and is 10 times less cytotoxic than BTZ043. PBTZ169 has poor solubility in water (0.9 g/L). PBTZ169 is synergistic with BDQ and clofazimine (CFZ) but not with other new TB drugs such as delamanid, linezolid, meropenem or sutezolid [229]. Interestingly, this synergistic phenotype was also observed in non-replicating conditions where PBTZ169 has no activity [230]. In a mouse model of chronic *M. tuberculosis* infection, the reduction in lung and spleen burden was similar between INH and PBTZ169 after 4 weeks of treatment [231]. Furthermore, in a mouse model of chronic *M. tuberculosis* infection, the lung burden was reduced by 2, 4 and 4.6 $\log_{10}$ after 28 days of treatment with PBTZ169, CFZ and the combination PBTZ169/CFZ, respectively [230]. In a similar experiment, the combination PBTZ169/BDQ/PZA was also shown to be more efficient in reducing lung and spleen burden of chronically *M. tuberculosis*-infected mice than the standard INH/RIF/PZA treatment [231].

Mode of action. PBTZ169 mode of action is similar to BTZ043. PBTZ169 is a prodrug, activated by DprE1 to yield a nitroso metabolite that covalently binds to the Cys387 residue of DprE1.

Clinical trial. A phase I study (NCT03423030) in healthy male subjects receiving increasing doses of PBTZ169 showed that PBTZ169 was well tolerated and safe. No result has been posted for a phase 2a EBA study (NCT03334734) where PBTZ169 was given as a single dose to TB patients.

4.2.4. OPC167832

Activity. OPC167832 was developed by Otsuka pharmaceuticals. OPC167832 is a lipophilic, bactericidal compound, active against drug-susceptible, MDR and XDR *M. tuberculosis* strains, with a MIC ranging from 0.5 to 5 nM. OPC167832 is active in macrophage and in mouse model of *M. tuberculosis* infection at a dose of 1.25 mg/kg. OPC167832 is not antagonistic with other TB drugs. The combination of OPC167832 with delamanid and other TB drugs showed better efficacy than standard TB treatment in a mouse model of chronic drug-susceptible and MDR *M. tuberculosis* infections [232].

Mode of action. OPC167832 is a DprE1 inhibitor. OPC167832 does not contain a nitro group, and so it inhibits DprE1 via other interactions than a covalent bond with DprE1 Cys387 (see below).

Resistance. Spontaneous OPC167832-resistant mutants isolated at 16 × MIC occurred at a frequency of 2.6×10^{-9} to 1.5×10^{-7} in *M. tuberculosis* H37Rv. Mutations in *dprE1* and *Rv0678* encoding a transcriptional regulator of the efflux pumps MmpS5 and Mmpl5 [233] drive resistance to OPC167832.

Clinical trial. A phase 1/2 clinical trial (NCT03678688) is recruiting to test the safety, tolerability and pharmacokinetics of OPC167832 given at increasing concentrations to patients infected with uncomplicated, smear-positive, drug-susceptible TB (phase I). A phase II study will compare delamanid and OPC167832 vs. delamanid only or INH/RIF/EMB/PZA to demonstrate that this new regimen, which uses only oral drugs, is safer and can shorten TB treatment [232].

4.2.5. TBA-7371

Discovery. TBA-7371 was developed in collaboration between Astra Zeneca and the TB Global Alliance. TBA-7371 is a derivative of 1,4-azaindoles developed through scaffold morphing of an imidazopyridine compound with good MIC but low minimum bactericidal activity (MBC) against *M. tuberculosis* [234]. SAR on the hit azaindole compound led to TBA-7371.

Activity. TBA-7371 is bactericidal against *M. tuberculosis*, with a MIC ranging from 0.78 to 3.12 μm against drug-susceptible and drug-resistant clinical isolates of *M. tuberculosis* [235]. Solubility of TBA-7371 is 170 μm. TBA-7371 does not exhibit toxicity in THP1 cell line up to concentration of 0.1 mM [235].

Mode of action. TBA-7371 inhibits DprE1 with an IC_{50} of 10–30 nm but, unlike BTZ043 and PBTZ169, TBA-7371 is a non-covalent inhibitor of DprE1 [234].

Resistance. Resistance to TBA-7371 was observed in strains overexpressing *M. tuberculosis dprE1* or carrying the mutation Y314H in *dprE1*, but not in strains with a mutated Cys387 in *dprE1* [234].

Clinical trial. TBA-7371 is in phase I clinical trial (NCT03199339) for safety, tolerability and pharmacokinetic studies in healthy adults. No results have been posted.

DprE1 is a successful drug target. Since the discovery of BTZ043 and its target DprE1 [225], numerous groups have uncovered novel molecules that target DprE1. DprE1 inhibitors can be divided into two groups depending on whether or not they bind covalently to DprE1. The presence of the nitro group is a requirement for the covalent binding to the Cys387 amino acid of DprE1. Among the covalent inhibitors are benzothiazinethione [236], dinitrobenzamides [237], nitroquinoxalines [238] and nitroimidazoles [239]. Several non-covalent DprE1 inhibitors have been identified with various chemical structures: azaindoles [234], benzothiazoles [240], pyrazolopyridones [241], aminoquinolone piperidine amides [242], carboxyquinoxalines [243], pyrrole-benzothiazinones [244], pyridobenzimidazole [245], sulfonylpiperazin-benzothiazinones [246] and piperidinopyrimidines [247]. These compounds inhibit DprE1 through electrostatic or hydrophobic interactions with different sites of DprE1. Most of these compounds have low MICs (in the nM range), are active against MDR and XDR TB as well as in a mouse model of *M. tuberculosis* infection. Considering the promiscuity of the DprE1 target, better DprE1 inhibitors might still be discovered to improve TB treatment.

4.3. Target: Peptidoglycan

4.3.1. β-Lactams and Clavulanic Acid

Discovery. The first class of synthetic drugs that successfully treated bacterial infections were the sulfa drugs in the early 1930s [163]; however, by the early 1940s, penicillin had quickly eclipsed the sulfa drugs. Penicillin, a β-lactam compound, was active against Gram-positive and Gram-negative, yet penicillin, like the sulfa drugs, had no activity against *M. tuberculosis*. For penicillin, the lack of activity was due to the presence in *M. tuberculosis* of a potent β-lactamase BlaC. The innate resistance of *M. tuberculosis* to β-lactams decreased when blaC was deleted from *M. tuberculosis* or when the β-lactamase irreversible inhibitor clavulanic acid [248] was used in combination with β-lactams [249]. Hugonnet and colleagues used this knowledge to validate the combination of meropenem, a carbapenem with poor affinity for BlaC, and clavulanic acid as an efficacious *M. tuberculosis* inhibitor [250].

Activity. The combination meropenem plus clavulanate is part of the group D3 (add-on agents) of second-line TB drugs. Meropenem is a potent inhibitor of drug-susceptible and XDR *M. tuberculosis* strains (MIC range from 0.6 μm to 3.3 μm) when combined with clavulanate [250]. Interestingly, this combination also inhibits non-replicating *M. tuberculosis* [250]. In a chronic model of *M. tuberculosis* infection in mice, meropenem had a modest activity in reducing lung burdens [251]. Surprisingly, the addition of clavulanate to the meropenem treatment did not increase meropenem activity in mice. In mice intravenously infected with *M. tuberculosis*, the lung burden had increased by almost a 1 $\log_{10}$ after 4 weeks of treatment with meropenem and clavulanate showing no inhibitory activity of this combination in vivo [252]. However, studies in mice are problematic since the mice dehydropeptidase (DHP) rapidly cleaves the β-lactam ring.

Meropenem is administered via intravenous injection and new regimens are steering away from injectables. Therefore, Dhar and colleagues tested faropenem, an *oral* β-lactam, and demonstrated activity against *M. tuberculosis* (MIC 4.6 μm) without clavulanate addition (meropenem MIC is 8-fold higher without clavulanate) [253]. Faropenem is also more soluble in water than meropenem (solubility 14.7 g/L vs. 8 g/L for meropenem). Using the dehydropeptidase inhibitor, probenecid, Dhar and colleagues demonstrated that *M. tuberculosis*-infected mice had a small but significant reduction in lung burden after 9 days of treatment with a combination faropenem/clavulanate/probenecid. In a study using DHP-I knockout mice [254], mice were infected intratracheally with *M. tuberculosis* H37Rv and treated for 8 consecutive days, 10 to 12 days post infection. Meropenem was given subcutaneously three times a day in combination with amoxicillin and clavulanate. Another group of mice received faropenem orally with amoxicillin and clavulanate. A control group was treated with moxifloxacin. Treatment also consisted of a dose of probenecid prior to drug administration. While the lung burden was 7.7 $\log_{10}$ CFU/mouse at the beginning of the treatment, the lung burdens at the end of the treatment were 7.8 and 7.5 $\log_{10}$ CFU/mouse for the meropenem and faropenem groups, respectively. Only the mice receiving moxifloxacin had a 2-$\log_{10}$ reduction in lung burden [254].

Experiments performed in other animal models (rabbits and monkeys) revealed that intravenous injections of meropenem with or without clavulanate and dehydropeptidase inhibitor led to serious adverse effects (diarrhea, weight loss and death) [251].

Mode of action. Meropenem binds to and inhibits the L,D-transpeptidase Ldt$_{Mt1}$ (Rv0116c), which catalyzes the 3→3 cross-linkage in the peptidoglycan [255,256]. Faropenem is 14-fold more potent in inactivating Ldt$_{Mt1}$ than meropenem, but this did not correlate with increasing killing against intracellular *M. tuberculosis* [253]. Dhar and colleagues demonstrated that the addition of faropenem to *M. tuberculosis* quickly arrested growth, which did not translate into an immediate lysis of the cells. This is opposite to the current understanding that β-lactams induce cell lysis by inhibiting peptidoglycan cross-linking while cells are still actively dividing. Dhar and colleagues concluded that the mechanism of *M. tuberculosis* killing by β-lactam should be revisited [253].

Clinical trial. A clinical study on MDR and XDR TB patients receiving meropenem/clavulanate (intravenous injections three times a day for an average of 85 days) in combination with an MDR/XDR drug regimen showed that the addition of meropenem/clavulanate did not improve: (1) sputum smear or culture conversion; (2) treatment outcome; or (3) treatment success [257]. A phase 2 clinical trial evaluated the early bactericidal activity, safety and tolerability of intravenously administered meropenem with amoxicillin and clavulanate versus orally-administered faropenem with amoxicillin and clavulanate (NCT02349841). Results have not been posted yet.

4.3.2. Sanfetrinem

Activity. Sanfetrinem is a tricyclic β-lactam developed by GlaxoSmithKline in the 1990s. Sanfetrinem has broad-spectrum activity against Gram-negative and Gram-positive bacteria. Sanfetrinem is also very stable to β-lactamases and to human renal dehydropeptidase (DHP) [258]. The MICs for sanfetrinem against *M. tuberculosis* H37Rv ranged from 1.25 to 5 μm with clavulanate (2.5 to 7.5 μm without clavulanate). In a checkerboard assay, the combination of sanfetrinem with RIF or amoxicillin was synergistic, but not with delamanid or ethambutol. Sanfetrinem cilexetil is an oral prodrug ester of sanfetrinem and has an MIC ranging from 5 to 20 μm with clavulanate (7.5 to 20 μm without clavulanate) against *M. tuberculosis*. The compounds performed better against intracellular *M. tuberculosis*. In THP1 monocytes, the MIC for sanfetrinem and sanfetrinem cilexetil ranged from 2.1 to 7.7 μm without clavulanate (6.1 to 8.8 μm with clavulanate) and 3.4 to 7.0 μm without clavulanate (4.6 to 5.3 μm with clavulanate), respectively. In a mouse model of acute *M. tuberculosis* infection, DHP-1 KO mice were infected intratracheally with *M. tuberculosis* H37Rv. Treatment started 9 days post infection, twice a day, for 5 days. Sanfetrinem was given subcutaneously, while sanfetrinem cilexetil and clavulanate were given orally. Growth arrest but not killing of *M. tuberculosis* was observed during this five-day treatment [259].

Clinical trial. A phase 2a clinical study is projected to start the first quarter of 2021 [260].

5. Conclusions

In 1882, when Robert Koch discovered the causative agent of TB, there was no valid treatment for the disease. In the past 70 years, with the introduction of multidrug chemotherapy, the cure rate for drug-susceptible *M. tuberculosis* infection has reached up to 95%. The treatment is effective but very long, with drugs that may not be well tolerated. New chemotherapy regimens are needed with drugs that are less toxic and more efficacious, so that a shorter treatment can be achieved. The pursuit of mycobacterial cell wall inhibitors has offered several hits with novel modes of action and remarkable potency that have reached clinical trials. Those pioneering compounds can lead the way to more valuable therapeutics. Numerous new compounds are being studied that either derive from the known hits or target other biosynthetic pathways of the cell wall [261,262]. In the past 20 years, *M. tuberculosis* has been unraveling many of its secrets: its niches in the host, its metabolism and its way of fighting drugs. New tools were developed to study *M. tuberculosis* and to search for new TB drugs more efficiently combining target and whole-cell screenings. We should be closer to a new TB regimen for a shorter treatment of drug-susceptible TB and a more successful treatment of MDR and XDR TB. The collaboration between academic and pharmaceutical laboratories and the involvement of funding foundations have opened the doors to the clinical testing of new drugs and new regimens [263].

In Selman Waksman's book "The Conquest of Tuberculosis" [264], a quote from Georges Canetti's speech at the 1961 Sixteenth International Conference on Tuberculosis, as reported by the chairman, stated: "We are not concerned with eradication in the absolutely literal sense of the word because this is something that many of us believe to be biologically impossible. What we are talking about, however, is tuberculosis 'eradication' in the sense of reducing the problem to the point where the disease is a scientific curiosity. This is a biologic possibility". With better diagnostic tools, a more efficient vaccine and sterilizing therapeutics, this might become a biologic reality.

A summary of the cell wall inhibitors discussed herein is presented in Table 1 with their chemical structures and targets.

Table 1. TB drugs targeting the mycobacterial cell wall.

Name	Structure	Cell Wall Component Inhibited	Prodrug/Activator	Target
Isoniazid		Mycolic acids	+ KatG	InhA
Ethionamide Prothionamide	ETH PTH	Mycolic acids	+ EthA, MymA, Rv0565c	InhA
Thiacetazone		Mycolic acids	+ EthA	HadAB
Pretonamid		Keto-mycolic acids	+ Ddn	?

Table 1. Cont.

Name	Structure	Cell Wall Component Inhibited	Prodrug/Activator	Target
Delamanid		Methoxy- and keto-mycolic acids	+ Ddn	?
SQ109		Mycolic acid transport	-	Mmpl3
CPZEN45		Arabinogalactan	-	WecA
Ethambutol		Arabinogalactan LAM	-	EmbCAB
BTZ043		Arabinogalactan LAM	+ DprE1	DprE1
PBTZ169		Arabinogalactan LAM	+ DprE1	DprE1
OPC167832		Arabinogalactan LAM	-	DprE1
TBA7371		Arabinogalactan LAM	-	DprE1
Cycloserine		Peptidoglycan	-	DdlA

Table 1. Cont.

Name	Structure	Cell Wall Component Inhibited	Prodrug/Activator	Target
Meropenem/ clavulanate		Peptidoglycan	-	LdtM1
sanfetrinem		Peptidoglycan	-	

Funding: CV acknowledges support through grant AI21670 from the US National Institutes of Health.

Acknowledgments: I thank Lawrence Leung, Claire Mulholland and Saranathan Rajagopalan for their critical reading, comments and feedback on this manuscript.

Conflicts of Interest: The author declares no conflict of interest.

References

1. Daniel, T.M. The history of tuberculosis. *Respir. Med.* **2006**, *100*, 1862–1870. [CrossRef] [PubMed]
2. Keers, R.Y. Two forgotten pioneers. James Carson and George Bodington. *Thorax* **1980**, *35*, 483–489. [CrossRef] [PubMed]
3. Bodington, G. Reviews of Books. An essay on the treatment and cure of pulmunory consumption, on principles natural, ratinal and succesful. *Lancet* **1840**, *34*, 575–577.
4. George Bodington, M.D. Obituary. *Lancet* **1882**, *119*, 477.
5. McCarthy, O.R. The key to the sanatoria. *J. R. Soc. Med.* **2001**, *94*, 413–417. [CrossRef] [PubMed]
6. Latour, A. Review of Books. On the preservative and curative treatment of pulmunory consumption. *Lancet* **1840**, *34*, 616–619.
7. Cox, G.L. Sanatorium treatment contrasted with home treatment. After-histories of 4067 cases. *Br. J. Tuberc.* **1923**, *17*, 27–30.
8. Schatz, A.; Bugie, E.; Waksman, S.A. Streptomycin, a substance exhibiting antibiotic activity against gram-positive and gram-negative bacteria. *Proc. Exp. Biol. Med.* **1944**, *55*, 66–69. [CrossRef]
9. Schatz, A.; Waksman, S.A. Effect of streptomycin upon *Mycobacterium tuberculosis* and related organisms. *Proc. Soc. Exp. Biol. Med.* **1944**, *57*, 244–248. [CrossRef]
10. Pfuetze, K.H.; Pyle, M.M.; Hinshaw, H.C.; Feldman, W.H. The first clinical trial of streptomycin in human tuberculosis. *Am. Rev. Tuberc.* **1955**, *71*, 752–754. [CrossRef]
11. Waksman, S.A. Tenth anniversary of the discovery of streptomycin, the first chemotherapeutic agent found to be effective against tuberculosis in humans. *Am. Rev. Tuberc.* **1954**, *70*, 1–8. [CrossRef]
12. Hinshaw, H.C.; Pyle, M.M.; Feldman, W.H. Streptomycin in tuberculosis. *Am. J. Med.* **1947**, *2*, 429–435. [CrossRef]
13. Crofton, J.; Mitchison, D.A. Streptomycin resistance in pulmonary tuberculosis. *Br. Med. J.* **1948**, *2*, 1009–1015. [CrossRef] [PubMed]
14. Lehmann, J. *Para*-Aminosalicylic acid in the treatment of tuberculosis. *Lancet* **1946**, *247*, 15. [CrossRef]
15. Medical Research Council Investigation. Treatment of pulmonary tuberculosis with streptomycin and *para*-amino-salicylic acid. *Br. Med. J.* **1950**, *2*, 4688.
16. Crofton, J. Chemotherapy of pulmonary tuberculosis. *Br. Med. J.* **1959**, *1*, 1610–1614. [CrossRef]
17. Koul, A.; Arnoult, E.; Lounis, N.; Guillemont, J.; Andries, K. The challenge of new drug discovery for tuberculosis. *Nature* **2011**, *469*, 483–490. [CrossRef]

18. Payne, D.J.; Gwynn, M.N.; Holmes, D.J.; Pompliano, D.L. Drugs for bad bugs: Confronting the challenges of antibacterial discovery. *Nat. Rev. Drug Discov.* **2007**, *6*, 29–40. [CrossRef]
19. World Health Organization. Global Tuberculosis Report 2019. Available online: https://apps.who.int/iris/bitstream/handle/10665/329368/9789241565714-eng.pdf?ua=1 (accessed on 15 February 2020).
20. Abrahams, K.A.; Besra, G.S. Mycobacterial cell wall biosynthesis: A multifaceted antibiotic target. *Parasitology* **2018**, *145*, 116–133. [CrossRef]
21. Konyarikova, Z.; Savkova, K.; Kozmon, S.; Mikusova, K. Biosynthesis of Galactan in *Mycobacterium tuberculosis* as a Viable TB Drug Target? *Antibiotics* **2020**, *9*, 20. [CrossRef]
22. Maitra, A.; Munshi, T.; Healy, J.; Martin, L.T.; Vollmer, W.; Keep, N.H.; Bhakta, S. Cell wall peptidoglycan in *Mycobacterium tuberculosis*: An Achilles' heel for the TB-causing pathogen. *FEMS Microbiol. Rev.* **2019**, *43*, 548–575. [CrossRef] [PubMed]
23. Lederer, E.; Adam, A.; Ciorbaru, R.; Petit, J.F.; Wietzerbin, J. Cell walls of Mycobacteria and related organisms; chemistry and immunostimulant properties. *Mol. Cell. Biochem.* **1975**, *7*, 87–104. [CrossRef] [PubMed]
24. Mahapatra, S.; Crick, D.C.; Brennan, P.J. Comparison of the UDP-*N*-acetylmuramate:L-alanine ligase enzymes from *Mycobacterium tuberculosis* and *Mycobacterium leprae*. *J. Bacteriol.* **2000**, *182*, 6827–6830. [CrossRef] [PubMed]
25. Mahapatra, S.; Scherman, H.; Brennan, P.J.; Crick, D.C. *N*-Glycolylation of the nucleotide precursors of peptidoglycan biosynthesis of *Mycobacterium* spp. is altered by drug treatment. *J. Bacteriol.* **2005**, *187*, 2341–2347. [CrossRef] [PubMed]
26. Raymond, J.B.; Mahapatra, S.; Crick, D.C.; Pavelka, M.S., Jr. Identification of the *namH* gene, encoding the hydroxylase responsible for the *N*-glycolylation of the mycobacterial peptidoglycan. *J. Biol. Chem.* **2005**, *280*, 326–333. [CrossRef]
27. McNeil, M.; Wallner, S.J.; Hunter, S.W.; Brennan, P.J. Demonstration that the galactosyl and arabinosyl residues in the cell-wall arabinogalactan of *Mycobacterium leprae* and *Mycobacterium tuberculosis* are furanoid. *Carbohydr. Res.* **1987**, *166*, 299–308. [CrossRef]
28. McNeil, M.; Daffe, M.; Brennan, P.J. Evidence for the nature of the link between the arabinogalactan and peptidoglycan of mycobacterial cell walls. *J. Biol. Chem.* **1990**, *265*, 18200–18206.
29. Marrakchi, H.; Laneelle, M.A.; Daffe, M. Mycolic acids: Structures, biosynthesis, and beyond. *Chem. Biol.* **2014**, *21*, 67–85. [CrossRef]
30. Brindley, D.N.; Matsumura, S.; Bloch, K. *Mycobacterium phlei* Fatty Acid Synthetase—A Bacterial Multienzyme Complex. *Nature* **1969**, *224*, 666–669. [CrossRef]
31. Cantaloube, S.; Veyron-Churlet, R.; Haddache, N.; Daffe, M.; Zerbib, D. The *Mycobacterium tuberculosis* FAS-II dehydratases and methyltransferases define the specificity of the mycolic acid elongation complexes. *PLoS ONE* **2011**, *6*, e29564. [CrossRef]
32. Fukuda, T.; Matsumura, T.; Ato, M.; Hamasaki, M.; Nishiuchi, Y.; Murakami, Y.; Maeda, Y.; Yoshimori, T.; Matsumoto, S.; Kobayashi, K.; et al. Critical roles for lipomannan and lipoarabinomannan in cell wall integrity of mycobacteria and pathogenesis of tuberculosis. *MBio* **2013**, *4*, e00472-12. [CrossRef] [PubMed]
33. Nigou, J.; Gilleron, M.; Rojas, M.; Garcia, L.F.; Thurnher, M.; Puzo, G. Mycobacterial lipoarabinomannans: Modulators of dendritic cell function and the apoptotic response. *Microbes Infect.* **2002**, *4*, 945–953. [CrossRef]
34. Patterson, J.H.; Waller, R.F.; Jeevarajah, D.; Billman-Jacobe, H.; McConville, M.J. Mannose metabolism is required for mycobacterial growth. *Biochem. J.* **2003**, *372*, 77–86. [CrossRef] [PubMed]
35. Stanley, S.A.; Cox, J.S. Host-pathogen interactions during *Mycobacterium tuberculosis* infections. *Curr. Top. Microbiol. Immunol.* **2013**, *374*, 211–241. [CrossRef] [PubMed]
36. McDermott, W. The story of INH. *J. Infect. Dis.* **1969**, *119*, 678–683. [CrossRef]
37. Hinshaw, H.C.; Mc, D.W. Thiosemicarbazone therapy of tuberculosis in humans. *Am. Rev. Tuberc.* **1950**, *61*, 145–157.
38. Donovick, R.; Pansy, F.; Stryker, G.; Bernstein, J. The chemotherapy of experimental tuberculosis. I. The in vitro activity of thiosemicarbazides, thiosemicarbazones, and related compounds. *J. Bacteriol.* **1950**, *59*, 667–674. [CrossRef]
39. Fox, H.H. The chemical approach to the control of tuberculosis. *Science* **1952**, *116*, 129–134. [CrossRef]
40. Hamre, D.; Bernstein, J.; Donovick, R. The chemotherapy of experimental tuberculosis. II. Thiosemicarbazones and analogues in experimental tuberculosis in the mouse. *J. Bacteriol.* **1950**, *59*, 675–680. [CrossRef]

41. Bernstein, J.W.; Lott, A.; Steinberg, B.A.; Yale, H.L. Chemotherapy of experimental tuberculosis. *Am. Rev. Tuberc.* **1952**, *65*, 357–374.
42. Domagk, G.; Offe, H.A.; Siefken, W. [Therapy of experimental tuberculosis with neoteben]. *Med. Colon.* **1952**, *20*, 517–528. [PubMed]
43. Jain, P.; Weinrick, B.C.; Kalivoda, E.J.; Yang, H.; Munsamy, V.; Vilcheze, C.; Weisbrod, T.R.; Larsen, M.H.; O'Donnell, M.R.; Pym, A.; et al. Dual-Reporter Mycobacteriophages (Phi2DRMs) Reveal Preexisting *Mycobacterium tuberculosis* Persistent Cells in Human Sputum. *MBio* **2016**, *7*. [CrossRef] [PubMed]
44. Karakousis, P.C.; Williams, E.P.; Bishai, W.R. Altered expression of isoniazid-regulated genes in drug-treated dormant *Mycobacterium tuberculosis*. *J. Antimicrob. Chemother.* **2008**, *61*, 323–331. [CrossRef] [PubMed]
45. Ohno, M.; Yamaguchi, I.; Yamamoto, I.; Fukuda, T.; Yokota, S.; Maekura, R.; Ito, M.; Yamamoto, Y.; Ogura, T.; Maeda, K.; et al. Slow N-acetyltransferase 2 genotype affects the incidence of isoniazid and rifampicin-induced hepatotoxicity. *Int. J. Tuberc. Lung Dis.* **2000**, *4*, 256–261. [PubMed]
46. Wang, P.; Pradhan, K.; Zhong, X.B.; Ma, X. Isoniazid metabolism and hepatotoxicity. *Acta Pharm. Sin. B* **2016**, *6*, 384–392. [CrossRef]
47. Mitchell, J.R.; Zimmerman, H.J.; Ishak, K.G.; Thorgeirsson, U.P.; Timbrell, J.A.; Snodgrass, W.R.; Nelson, S.D. Isoniazid liver injury: Clinical spectrum, pathology, and probable pathogenesis. *Ann. Intern. Med.* **1976**, *84*, 181–192. [CrossRef]
48. Nelson, S.D.; Mitchell, J.R.; Timbrell, J.A.; Snodgrass, W.R.; Corcoran, G.B., 3rd. Isoniazid and iproniazid: Activation of metabolites to toxic intermediates in man and rat. *Science* **1976**, *193*, 901–903. [CrossRef]
49. Lauterburg, B.H.; Smith, C.V.; Todd, E.L.; Mitchell, J.R. Pharmacokinetics of the toxic hydrazino metabolites formed from isoniazid in humans. *J. Pharmacol. Exp. Ther.* **1985**, *235*, 566–570.
50. Gajdacs, M. The Concept of an Ideal Antibiotic: Implications for Drug Design. *Molecules* **2019**, *24*, 892. [CrossRef]
51. Laborde, J.; Deraeve, C.; Bernardes-Genisson, V. Update of Antitubercular Prodrugs from a Molecular Perspective: Mechanisms of Action, Bioactivation Pathways, and Associated Resistance. *ChemMedChem* **2017**, *12*, 1657–1676. [CrossRef]
52. Zhang, Y.; Heym, B.; Allen, B.; Young, D.; Cole, S. The catalase-peroxidase gene and isoniazid resistance of *Mycobacterium tuberculosis*. *Nature* **1992**, *358*, 591–593. [CrossRef] [PubMed]
53. Johnsson, K.; Schultz, P.G. Mechanistic Studies of the Oxidation of Isoniazid by the Catalase Peroxidase from *Mycobacterium tuberculosis*. *J. Am. Chem. Soc.* **1994**, *116*, 7425–7426. [CrossRef]
54. Lei, B.; Wei, C.J.; Tu, S.C. Action mechanism of antitubercular isoniazid. Activation by *Mycobacterium tuberculosis* KatG, isolation, and characterization of *inhA* inhibitor. *J. Biol. Chem.* **2000**, *275*, 2520–2526. [CrossRef] [PubMed]
55. Rozwarski, D.A.; Grant, G.A.; Barton, D.H.; Jacobs, W.R., Jr.; Sacchettini, J.C. Modification of the NADH of the isoniazid target (InhA) from *Mycobacterium tuberculosis*. *Science* **1998**, *279*, 98–102. [CrossRef]
56. Wilming, M.; Johnsson, K. Spontaneous Formation of the Bioactive Form of the Tuberculosis Drug Isoniazid. *Angew. Chem. Int. Ed. Engl.* **1999**, *38*, 2588–2590. [CrossRef]
57. Banerjee, A.; Dubnau, E.; Quemard, A.; Balasubramanian, V.; Um, K.S.; Wilson, T.; Collins, D.; de Lisle, G.; Jacobs, W.R., Jr. *inhA*, a gene encoding a target for isoniazid and ethionamide in *Mycobacterium tuberculosis*. *Science* **1994**, *263*, 227–230. [CrossRef]
58. Nguyen, M.; Quemard, A.; Broussy, S.; Bernadou, J.; Meunier, B. Mn(III) pyrophosphate as an efficient tool for studying the mode of action of isoniazid on the InhA protein of *Mycobacterium tuberculosis*. *Antimicrob. Agents Chemother.* **2002**, *46*, 2137–2144. [CrossRef]
59. Rawat, R.; Whitty, A.; Tonge, P.J. The isoniazid-NAD adduct is a slow, tight-binding inhibitor of InhA, the *Mycobacterium tuberculosis* enoyl reductase: Adduct affinity and drug resistance. *Proc. Natl. Acad. Sci. USA* **2003**, *100*, 13881–13886. [CrossRef]
60. Vilcheze, C.; Wang, F.; Arai, M.; Hazbon, M.H.; Colangeli, R.; Kremer, L.; Weisbrod, T.R.; Alland, D.; Sacchettini, J.C.; Jacobs, W.R., Jr. Transfer of a point mutation in *Mycobacterium tuberculosis inhA* resolves the target of isoniazid. *Nat. Med.* **2006**, *12*, 1027–1029. [CrossRef]
61. Dessen, A.; Quemard, A.; Blanchard, J.S.; Jacobs, W.R., Jr.; Sacchettini, J.C. Crystal structure and function of the isoniazid target of *Mycobacterium tuberculosis*. *Science* **1995**, *267*, 1638–1641. [CrossRef]

62. Quemard, A.; Sacchettini, J.C.; Dessen, A.; Vilcheze, C.; Bittman, R.; Jacobs, W.R., Jr.; Blanchard, J.S. Enzymatic characterization of the target for isoniazid in *Mycobacterium tuberculosis*. *Biochemistry* **1995**, *34*, 8235–8241. [CrossRef] [PubMed]
63. Marrakchi, H.; Laneelle, G.; Quemard, A. InhA, a target of the antituberculous drug isoniazid, is involved in a mycobacterial fatty acid elongation system, FAS-II. *Microbiology* **2000**, *146 Pt 2*, 289–296. [CrossRef]
64. Takayama, K.; Wang, L.; David, H.L. Effect of isoniazid on the in vivo mycolic acid synthesis, cell growth, and viability of *Mycobacterium tuberculosis*. *Antimicrob. Agents Chemother.* **1972**, *2*, 29–35. [CrossRef]
65. Vilcheze, C.; Morbidoni, H.R.; Weisbrod, T.R.; Iwamoto, H.; Kuo, M.; Sacchettini, J.C.; Jacobs, W.R., Jr. Inactivation of the *inhA*-encoded fatty acid synthase II (FASII) enoyl-acyl carrier protein reductase induces accumulation of the FASI end products and cell lysis of *Mycobacterium smegmatis*. *J. Bacteriol.* **2000**, *182*, 4059–4067. [CrossRef]
66. Winder, F.G.; Collins, P.B. Inhibition by isoniazid of synthesis of mycolic acids in *Mycobacterium tuberculosis*. *J. Gen. Microbiol.* **1970**, *63*, 41–48. [CrossRef]
67. Vilcheze, C.; Jacobs, W.R., Jr. The mechanism of isoniazid killing: Clarity through the scope of genetics. *Annu. Rev. Microbiol.* **2007**, *61*, 35–50. [CrossRef] [PubMed]
68. Middlebrook, G.; Cohn, M.L. Some observations on the pathogenicity of isoniazid-resistant variants of tubercle bacilli. *Science* **1953**, *118*, 297–299. [CrossRef] [PubMed]
69. Middlebrook, G. Isoniazid resistance and catalase activity of tubercle bacilli. *Am. Rev. Tuberc.* **1954**, *69*, 471–472. [PubMed]
70. Zhang, Y.; Garbe, T.; Young, D. Transformation with *katG* restores isoniazid-sensitivity in *Mycobacterium tuberculosis* isolates resistant to a range of drug concentrations. *Mol. Microbiol.* **1993**, *8*, 521–524. [CrossRef]
71. Vilcheze, C.; Jacobs, W.R., Jr. Resistance to Isoniazid and Ethionamide in *Mycobacterium tuberculosis*: Genes, Mutations, and Causalities. *Microbiol. Spectr.* **2014**, *2*. [CrossRef]
72. Ando, H.; Kondo, Y.; Suetake, T.; Toyota, E.; Kato, S.; Mori, T.; Kirikae, T. Identification of *katG* mutations associated with high-level isoniazid resistance in *Mycobacterium tuberculosis*. *Antimicrob. Agents Chemother.* **2010**, *54*, 1793–1799. [CrossRef] [PubMed]
73. Sekiguchi, J.; Miyoshi-Akiyama, T.; Augustynowicz-Kopec, E.; Zwolska, Z.; Kirikae, F.; Toyota, E.; Kobayashi, I.; Morita, K.; Kudo, K.; Kato, S.; et al. Detection of multidrug resistance in *Mycobacterium tuberculosis*. *J. Clin. Microbiol.* **2007**, *45*, 179–192. [CrossRef] [PubMed]
74. Wei, C.J.; Lei, B.; Musser, J.M.; Tu, S.C. Isoniazid activation defects in recombinant *Mycobacterium tuberculosis* catalase-peroxidase (KatG) mutants evident in InhA inhibitor production. *Antimicrob. Agents Chemother.* **2003**, *47*, 670–675. [CrossRef]
75. Brossier, F.; Veziris, N.; Truffot-Pernot, C.; Jarlier, V.; Sougakoff, W. Performance of the genotype MTBDR line probe assay for detection of resistance to rifampin and isoniazid in strains of *Mycobacterium tuberculosis* with low- and high-level resistance. *J. Clin. Microbiol.* **2006**, *44*, 3659–3664. [CrossRef] [PubMed]
76. Ando, H.; Kitao, T.; Miyoshi-Akiyama, T.; Kato, S.; Mori, T.; Kirikae, T. Downregulation of *katG* expression is associated with isoniazid resistance in *Mycobacterium tuberculosis*. *Mol. Microbiol.* **2011**, *79*, 1615–1628. [CrossRef] [PubMed]
77. Muller, B.; Streicher, E.M.; Hoek, K.G.; Tait, M.; Trollip, A.; Bosman, M.E.; Coetzee, G.J.; Chabula-Nxiweni, E.M.; Hoosain, E.; Gey van Pittius, N.C.; et al. *inhA* promoter mutations: A gateway to extensively drug-resistant tuberculosis in South Africa? *Int. J. Tuberc. Lung Dis.* **2011**, *15*, 344–351. [PubMed]
78. Shaw, D.J.; Robb, K.; Vetter, B.V.; Tong, M.; Molle, V.; Hunt, N.T.; Hoskisson, P.A. Disruption of key NADH-binding pocket residues of the *Mycobacterium tuberculosis* InhA affects DD-CoA binding ability. *Sci. Rep.* **2017**, *7*, 4714. [CrossRef]
79. Machado, D.; Perdigao, J.; Ramos, J.; Couto, I.; Portugal, I.; Ritter, C.; Boettger, E.C.; Viveiros, M. High-level resistance to isoniazid and ethionamide in multidrug-resistant *Mycobacterium tuberculosis* of the Lisboa family is associated with *inhA* double mutations. *J Antimicrob. Chemother.* **2013**, *68*, 1728–1732. [CrossRef]
80. Morlock, G.P.; Metchock, B.; Sikes, D.; Crawford, J.T.; Cooksey, R.C. *ethA, inhA*, and *katG* loci of ethionamide-resistant clinical *Mycobacterium tuberculosis* isolates. *Antimicrob. Agents Chemother.* **2003**, *47*, 3799–3805. [CrossRef]
81. Nimmo, C.; Doyle, R.; Burgess, C.; Williams, R.; Gorton, R.; McHugh, T.D.; Brown, M.; Morris-Jones, S.; Booth, H.; Breuer, J. Rapid identification of a *Mycobacterium tuberculosis* full genetic drug resistance profile through whole genome sequencing directly from sputum. *Int. J. Infect. Dis.* **2017**, *62*, 44–46. [CrossRef]

82. Encinas, L.; O'Keefe, H.; Neu, M.; Remuinan, M.J.; Patel, A.M.; Guardia, A.; Davie, C.P.; Perez-Macias, N.; Yang, H.; Convery, M.A.; et al. Encoded library technology as a source of hits for the discovery and lead optimization of a potent and selective class of bactericidal direct inhibitors of *Mycobacterium tuberculosis* InhA. *J. Med. Chem.* **2014**, *57*, 1276–1288. [CrossRef] [PubMed]
83. Hartkoorn, R.C.; Sala, C.; Neres, J.; Pojer, F.; Magnet, S.; Mukherjee, R.; Uplekar, S.; Boy-Rottger, S.; Altmann, K.H.; Cole, S.T. Towards a new tuberculosis drug: Pyridomycin—Nature's isoniazid. *EMBO Mol. Med.* **2012**, *4*, 1032–1042. [CrossRef] [PubMed]
84. Lu, X.Y.; You, Q.D.; Chen, Y.D. Recent progress in the identification and development of InhA direct inhibitors of *Mycobacterium tuberculosis*. *Mini Rev. Med. Chem.* **2010**, *10*, 181–192. [CrossRef] [PubMed]
85. Manjunatha, U.H.; SP, S.R.; Kondreddi, R.R.; Noble, C.G.; Camacho, L.R.; Tan, B.H.; Ng, S.H.; Ng, P.S.; Ma, N.L.; Lakshminarayana, S.B.; et al. Direct inhibitors of InhA are active against *Mycobacterium tuberculosis*. *Sci. Transl. Med.* **2015**, *7*, 269ra263. [CrossRef] [PubMed]
86. Pan, P.; Tonge, P.J. Targeting InhA, the FASII enoyl-ACP reductase: SAR studies on novel inhibitor scaffolds. *Curr. Top. Med. Chem.* **2012**, *12*, 672–693. [CrossRef] [PubMed]
87. Shirude, P.S.; Madhavapeddi, P.; Naik, M.; Murugan, K.; Shinde, V.; Nandishaiah, R.; Bhat, J.; Kumar, A.; Hameed, S.; Holdgate, G.; et al. Methyl-thiazoles: A novel mode of inhibition with the potential to develop novel inhibitors targeting InhA in *Mycobacterium tuberculosis*. *J. Med. Chem.* **2013**, *56*, 8533–8542. [CrossRef]
88. Sink, R.; Sosic, I.; Zivec, M.; Fernandez-Menendez, R.; Turk, S.; Pajk, S.; Alvarez-Gomez, D.; Lopez-Roman, E.M.; Gonzales-Cortez, C.; Rullas-Triconado, J.; et al. Design, synthesis, and evaluation of new thiadiazole-based direct inhibitors of enoyl acyl carrier protein reductase (InhA) for the treatment of tuberculosis. *J. Med. Chem.* **2015**, *58*, 613–624. [CrossRef]
89. Vilcheze, C.; Baughn, A.D.; Tufariello, J.; Leung, L.W.; Kuo, M.; Basler, C.F.; Alland, D.; Sacchettini, J.C.; Freundlich, J.S.; Jacobs, W.R., Jr. Novel inhibitors of InhA efficiently kill *Mycobacterium tuberculosis* under aerobic and anaerobic conditions. *Antimicrob. Agents Chemother.* **2011**, *55*, 3889–3898. [CrossRef]
90. Martinez-Hoyos, M.; Perez-Herran, E.; Gulten, G.; Encinas, L.; Alvarez-Gomez, D.; Alvarez, E.; Ferrer-Bazaga, S.; Garcia-Perez, A.; Ortega, F.; Angulo-Barturen, I.; et al. Antitubercular drugs for an old target: GSK693 as a promising InhA direct inhibitor. *EBioMedicine* **2016**, *8*, 291–301. [CrossRef]
91. Robertson, G.T.; Ektnitphong, V.A.; Scherman, M.S.; McNeil, M.B.; Dennison, D.; Korkegian, A.; Smith, A.J.; Halladay, J.; Carter, D.S.; Xia, Y.; et al. Efficacy and Improved Resistance Potential of a Cofactor-Independent InhA Inhibitor of *Mycobacterium tuberculosis* in the C3HeB/FeJ Mouse Model. *Antimicrob. Agents Chemother.* **2019**, *63*. [CrossRef]
92. Xia, Y.; Zhou, Y.; Carter, D.S.; McNeil, M.B.; Choi, W.; Halladay, J.; Berry, P.W.; Mao, W.; Hernandez, V.; O'Malley, T.; et al. Discovery of a cofactor-independent inhibitor of *Mycobacterium tuberculosis* InhA. *Life Sci. Alliance* **2018**, *1*, e201800025. [CrossRef] [PubMed]
93. Shepherd, R.G.; Wilkinson, R.G. Antituberculous Agents. Ii. N,N'-Diisopropylethylenediamine and Analogs. *J. Med. Chem.* **1962**, *91*, 823–835. [CrossRef] [PubMed]
94. Thomas, J.P.; Baughn, C.O.; Wilkinson, R.G.; Shepherd, R.G. A new synthetic compound with antituberculous activity in mice: Ethambutol (dextro-2,2'-(ethylenediimino)-di-l-butanol). *Am. Rev. Respir. Dis.* **1961**, *83*, 891–893. [CrossRef] [PubMed]
95. Wilkinson, R.G.; Shepherd, R.G.; Thomas, J.P.; Baughn, C.O. Stereospecificity in a new type of synthetic antituberculous agent. *J. Am. Chem. Soc.* **1961**, *83*, 2212. [CrossRef]
96. Kuck, N.A.; Peets, E.A.; Forbes, M. Mode of action of ethambutol on *Mycobacterium tuberculosis*, strain H37R V. *Am. Rev. Respir. Dis.* **1963**, *87*, 905–906. [CrossRef]
97. Koul, P.A. Ocular toxicity with ethambutol therapy: Timely recaution. *Lung India* **2015**, *32*, 1–3. [CrossRef]
98. Forbes, M.; Kuck, N.A.; Peets, E.A. Effect of Ethambutol on Nucleic Acid Metabolism in *Mycobacterium Smegmatis* and Its Reversal by Polyamines and Divalent Cations. *J. Bacteriol.* **1965**, *89*, 1299–1305. [CrossRef]
99. Takayama, K.; Armstrong, E.L.; Kunugi, K.A.; Kilburn, J.O. Inhibition by ethambutol of mycolic acid transfer into the cell wall of *Mycobacterium smegmatis*. *Antimicrob. Agents Chemother.* **1979**, *16*, 240–242. [CrossRef]
100. Kilburn, J.O.; Takayama, K. Effects of ethambutol on accumulation and secretion of trehalose mycolates and free mycolic acid in *Mycobacterium smegmatis*. *Antimicrob. Agents Chemother.* **1981**, *20*, 401–404. [CrossRef]
101. Takayama, K.; Kilburn, J.O. Inhibition of synthesis of arabinogalactan by ethambutol in *Mycobacterium smegmatis*. *Antimicrob. Agents Chemother.* **1989**, *33*, 1493–1499. [CrossRef]

102. Wolucka, B.A.; McNeil, M.R.; de Hoffmann, E.; Chojnacki, T.; Brennan, P.J. Recognition of the lipid intermediate for arabinogalactan/arabinomannan biosynthesis and its relation to the mode of action of ethambutol on mycobacteria. *J. Biol. Chem.* **1994**, *269*, 23328–23335. [PubMed]
103. Mikusova, K.; Huang, H.; Yagi, T.; Holsters, M.; Vereecke, D.; D'Haeze, W.; Scherman, M.S.; Brennan, P.J.; McNeil, M.R.; Crick, D.C. Decaprenylphosphoryl arabinofuranose, the donor of the D-arabinofuranosyl residues of mycobacterial arabinan, is formed via a two-step epimerization of decaprenylphosphoryl ribose. *J. Bacteriol.* **2005**, *187*, 8020–8025. [CrossRef] [PubMed]
104. Deng, L.; Mikusova, K.; Robuck, K.G.; Scherman, M.; Brennan, P.J.; McNeil, M.R. Recognition of multiple effects of ethambutol on metabolism of mycobacterial cell envelope. *Antimicrob. Agents Chemother.* **1995**, *39*, 694–701. [CrossRef] [PubMed]
105. Mikusova, K.; Slayden, R.A.; Besra, G.S.; Brennan, P.J. Biogenesis of the mycobacterial cell wall and the site of action of ethambutol. *Antimicrob. Agents Chemother.* **1995**, *39*, 2484–2489. [CrossRef]
106. Pawar, A.; Jha, P.; Konwar, C.; Chaudhry, U.; Chopra, M.; Saluja, D. Ethambutol targets the glutamate racemase of *Mycobacterium tuberculosis*-an enzyme involved in peptidoglycan biosynthesis. *Appl. Microbiol. Biotechnol.* **2019**, *103*, 843–851. [CrossRef]
107. Sreevatsan, S.; Stockbauer, K.E.; Pan, X.; Kreiswirth, B.N.; Moghazeh, S.L.; Jacobs, W.R., Jr.; Telenti, A.; Musser, J.M. Ethambutol resistance in *Mycobacterium tuberculosis*: Critical role of *embB* mutations. *Antimicrob. Agents Chemother.* **1997**, *41*, 1677–1681. [CrossRef] [PubMed]
108. Telenti, A.; Philipp, W.J.; Sreevatsan, S.; Bernasconi, C.; Stockbauer, K.E.; Wieles, B.; Musser, J.M.; Jacobs, W.R., Jr. The *emb* operon, a gene cluster of *Mycobacterium tuberculosis* involved in resistance to ethambutol. *Nat. Med.* **1997**, *3*, 567–570. [CrossRef]
109. Belanger, A.E.; Besra, G.S.; Ford, M.E.; Mikusova, K.; Belisle, J.T.; Brennan, P.J.; Inamine, J.M. The *embAB* genes of *Mycobacterium avium* encode an arabinosyl transferase involved in cell wall arabinan biosynthesis that is the target for the antimycobacterial drug ethambutol. *Proc. Natl. Acad. Sci. USA* **1996**, *93*, 11919–11924. [CrossRef]
110. Safi, H.; Sayers, B.; Hazbon, M.H.; Alland, D. Transfer of *embB* codon 306 mutations into clinical *Mycobacterium tuberculosis* strains alters susceptibility to ethambutol, isoniazid, and rifampin. *Antimicrob. Agents Chemother.* **2008**, *52*, 2027–2034. [CrossRef]
111. Starks, A.M.; Gumusboga, A.; Plikaytis, B.B.; Shinnick, T.M.; Posey, J.E. Mutations at *embB* codon 306 are an important molecular indicator of ethambutol resistance in *Mycobacterium tuberculosis*. *Antimicrob. Agents Chemother.* **2009**, *53*, 1061–1066. [CrossRef]
112. Hazbon, M.H.; Bobadilla del Valle, M.; Guerrero, M.I.; Varma-Basil, M.; Filliol, I.; Cavatore, M.; Colangeli, R.; Safi, H.; Billman-Jacobe, H.; Lavender, C.; et al. Role of *embB* codon 306 mutations in *Mycobacterium tuberculosis* revisited: A novel association with broad drug resistance and IS6110 clustering rather than ethambutol resistance. *Antimicrob. Agents Chemother.* **2005**, *49*, 3794–3802. [CrossRef] [PubMed]
113. Lee, A.S.; Othman, S.N.; Ho, Y.M.; Wong, S.Y. Novel mutations within the *embB* gene in ethambutol-susceptible clinical isolates of *Mycobacterium tuberculosis*. *Antimicrob. Agents Chemother.* **2004**, *48*, 4447–4449. [CrossRef] [PubMed]
114. Mokrousov, I.; Otten, T.; Vyshnevskiy, B.; Narvskaya, O. Detection of *embB306* mutations in ethambutol-susceptible clinical isolates of *Mycobacterium tuberculosis* from Northwestern Russia: Implications for genotypic resistance testing. *J. Clin. Microbiol.* **2002**, *40*, 3810–3813. [CrossRef] [PubMed]
115. Shen, X.; Shen, G.M.; Wu, J.; Gui, X.H.; Li, X.; Mei, J.; DeRiemer, K.; Gao, Q. Association between *embB* codon 306 mutations and drug resistance in *Mycobacterium tuberculosis*. *Antimicrob. Agents Chemother.* **2007**, *51*, 2618–2620. [CrossRef]
116. Safi, H.; Lingaraju, S.; Amin, A.; Kim, S.; Jones, M.; Holmes, M.; McNeil, M.; Peterson, S.N.; Chatterjee, D.; Fleischmann, R.; et al. Evolution of high-level ethambutol-resistant tuberculosis through interacting mutations in decaprenylphosphoryl-beta-D-arabinose biosynthetic and utilization pathway genes. *Nat. Genet.* **2013**, *45*, 1190–1197. [CrossRef]
117. Tulyaprawat, O.; Chaiprasert, A.; Chongtrakool, P.; Suwannakarn, K.; Ngamskulrungroj, P. Association of *ubiA* mutations and high-level of ethambutol resistance among *Mycobacterium tuberculosis* Thai clinical isolates. *Tuberculosis* **2019**, *114*, 42–46. [CrossRef]
118. Gardner, T.S.; Wenis, E.; Lee, J.H. The synthesis of compounds for the chemotherapy of tuberculosis. IV. The amide function. *J. Org. Chem.* **1954**, *19*, 753–757. [CrossRef]

119. Grumbach, F.; Rist, N.; Libermann, D.; Moyeux, M.; Cals, S.; Clavel, S. Experimental antituberculous activity of certain isonicotinic thioamides substituted on the nucleus. *C. R. Hebd. Seances Acad. Sci.* **1956**, *242*, 2187–2189.
120. Scardigli, A.; Caminero, J.A.; Sotgiu, G.; Centis, R.; D'Ambrosio, L.; Migliori, G.B. Efficacy and tolerability of ethionamide versus prothionamide: A systematic review. *Eur. Respir. J.* **2016**, *48*, 946–952. [CrossRef]
121. Winder, F.G.; Collins, P.B.; Whelan, D. Effects of ethionamide and isoxyl on mycolic acid synthesis in *Mycobacterium tuberculosis* BCG. *J. Gen. Microbiol.* **1971**, *66*, 379–380. [CrossRef]
122. Hok, T.T. A comparative study of the susceptibility to ethionamide, thiosemicarbazone, and isoniazid of tubercle bacilli from patients never treated with ethionamide or thiosemicarbazone. *Am. Rev. Respir. Dis.* **1964**, *90*, 468–469. [PubMed]
123. Lefford, M.J. The ethionamide sensitivity of British pre-treatment strains of *Mycobacterium tuberculosis*. *Tubercle* **1966**, *47*, 198–206. [CrossRef]
124. Stewart, S.M.; Hall, E.; Riddell, R.W.; Somner, A.R. Bacteriological aspects of the use of ethionamide, pyrazinamide and cycloserine in the treatment of chronic pulmonary tuberculosis. *Tubercle* **1962**, *43*, 417–431. [CrossRef]
125. DeBarber, A.E.; Mdluli, K.; Bosman, M.; Bekker, L.G.; Barry, C.E., 3rd. Ethionamide activation and sensitivity in multidrug-resistant *Mycobacterium tuberculosis*. *Proc. Natl. Acad. Sci. USA* **2000**, *97*, 9677–9682. [CrossRef]
126. Vannelli, T.A.; Dykman, A.; Ortiz de Montellano, P.R. The antituberculosis drug ethionamide is activated by a flavoprotein monooxygenase. *J. Biol. Chem.* **2002**, *277*, 12824–12829. [CrossRef]
127. Wang, F.; Langley, R.; Gulten, G.; Dover, L.G.; Besra, G.S.; Jacobs, W.R., Jr.; Sacchettini, J.C. Mechanism of thioamide drug action against tuberculosis and leprosy. *J. Exp. Med.* **2007**, *204*, 73–78. [CrossRef]
128. Ang, M.L.T.; Zainul Rahim, S.Z.; de Sessions, P.F.; Lin, W.; Koh, V.; Pethe, K.; Hibberd, M.L.; Alonso, S. EthA/R-Independent Killing of *Mycobacterium tuberculosis* by Ethionamide. *Front. Microbiol.* **2017**, *8*, 710. [CrossRef]
129. Grant, S.S.; Wellington, S.; Kawate, T.; Desjardins, C.A.; Silvis, M.R.; Wivagg, C.; Thompson, M.; Gordon, K.; Kazyanskaya, E.; Nietupski, R.; et al. Baeyer-Villiger Monooxygenases EthA and MymA Are Required for Activation of Replicating and Non-replicating *Mycobacterium tuberculosis* Inhibitors. *Cell Chem. Biol.* **2016**, *23*, 666–677. [CrossRef]
130. Hicks, N.D.; Carey, A.F.; Yang, J.; Zhao, Y.; Fortune, S.M. Bacterial Genome-Wide Association Identifies Novel Factors That Contribute to Ethionamide and Prothionamide Susceptibility in *Mycobacterium tuberculosis*. *MBio* **2019**, *10*. [CrossRef]
131. Brossier, F.; Veziris, N.; Truffot-Pernot, C.; Jarlier, V.; Sougakoff, W. Molecular investigation of resistance to the antituberculous drug ethionamide in multidrug-resistant clinical isolates of *Mycobacterium tuberculosis*. *Antimicrob. Agents Chemother.* **2011**, *55*, 355–360. [CrossRef]
132. Zhang, H.N.; Xu, Z.W.; Jiang, H.W.; Wu, F.L.; He, X.; Liu, Y.; Guo, S.J.; Li, Y.; Bi, L.J.; Deng, J.Y.; et al. Cyclic di-GMP regulates *Mycobacterium tuberculosis* resistance to ethionamide. *Sci. Rep.* **2017**, *7*, 5860. [CrossRef] [PubMed]
133. Vilcheze, C.; Av-Gay, Y.; Attarian, R.; Liu, Z.; Hazbon, M.H.; Colangeli, R.; Chen, B.; Liu, W.; Alland, D.; Sacchettini, J.C.; et al. Mycothiol biosynthesis is essential for ethionamide susceptibility in *Mycobacterium tuberculosis*. *Mol. Microbiol.* **2008**, *69*, 1316–1329. [CrossRef] [PubMed]
134. Vilcheze, C.; Weisbrod, T.R.; Chen, B.; Kremer, L.; Hazbon, M.H.; Wang, F.; Alland, D.; Sacchettini, J.C.; Jacobs, W.R., Jr. Altered NADH/NAD$^+$ ratio mediates coresistance to isoniazid and ethionamide in mycobacteria. *Antimicrob. Agents Chemother.* **2005**, *49*, 708–720. [CrossRef] [PubMed]
135. Baulard, A.R.; Betts, J.C.; Engohang-Ndong, J.; Quan, S.; McAdam, R.A.; Brennan, P.J.; Locht, C.; Besra, G.S. Activation of the pro-drug ethionamide is regulated in mycobacteria. *J. Biol. Chem.* **2000**, *275*, 28326–28331. [CrossRef] [PubMed]
136. Willand, N.; Dirie, B.; Carette, X.; Bifani, P.; Singhal, A.; Desroses, M.; Leroux, F.; Willery, E.; Mathys, V.; Deprez-Poulain, R.; et al. Synthetic EthR inhibitors boost antituberculous activity of ethionamide. *Nat. Med.* **2009**, *15*, 537–544. [CrossRef]
137. Flipo, M.; Desroses, M.; Lecat-Guillet, N.; Villemagne, B.; Blondiaux, N.; Leroux, F.; Piveteau, C.; Mathys, V.; Flament, M.P.; Siepmann, J.; et al. Ethionamide boosters. 2. Combining bioisosteric replacement and structure-based drug design to solve pharmacokinetic issues in a series of potent 1,2,4-oxadiazole EthR inhibitors. *J. Med. Chem.* **2012**, *55*, 68–83. [CrossRef]

138. Blondiaux, N.; Moune, M.; Desroses, M.; Frita, R.; Flipo, M.; Mathys, V.; Soetaert, K.; Kiass, M.; Delorme, V.; Djaout, K.; et al. Reversion of antibiotic resistance in *Mycobacterium tuberculosis* by spiroisoxazoline SMARt-420. *Science* **2017**, *355*, 1206–1211. [CrossRef]
139. Harris, D.A.; Ruger, M.; Reagan, M.A.; Wolf, F.J.; Peck, R.L.; Wallick, H.; Woodruff, H.B. Discovery, development, and antimicrobial properties of D-4-amino-3-isoxazolidone (oxamycin), a new antibiotic produced by *Streptomyces garyphalus* n. sp. *Antibiot. Chemother.* **1955**, *5*, 183–190.
140. Shull, G.M.; Sardinas, J.L. PA-94, an antibiotic identical with D-4-amino-3-isoxazolidinone (cycloserine, oxamycin). *Antibiot. Chemother.* **1955**, *5*, 398–399.
141. Kurosawa, H. The isolation of an antibiotic produced by a strain of streptomyces K-300. *Yokohama Med. Bull.* **1952**, *3*, 386–399.
142. Harned, R.L.; Hidy, P.H.; La Baw, E.K. Cycloserine. I. A preliminary report. *Antibiot. Chemother.* **1955**, *5*, 204–205.
143. Steenken, W., Jr.; Wolinsky, E. Cycloserine: Antituberculous activity in vitro and in the experimental animal. *Am. Rev. Tuberc.* **1956**, *73*, 539–546. [CrossRef] [PubMed]
144. Cuckler, A.C.; Frost, B.M.; Mc, C.L.; Solotorovsky, M. The antimicrobial evaluation of oxamycin (D-4-amino-3-isoxazolidone), a new broad-spectrum antibiotic. *Antibiot. Chemother.* **1955**, *5*, 191–197.
145. Patnode, R.A.; Hudgins, P.C.; Cummings, M.M. Effect of cycloserine on experimental tuberculosis in guinea pigs. *Am. Rev. Tuberc.* **1955**, *72*, 117–118. [CrossRef]
146. Conzelman, G.M., Jr.; Jones, R.K. On the physiologic disposition of cycloserine in experimental animals. *Am. Rev. Tuberc.* **1956**, *74*, 802–806. [CrossRef]
147. Epstein, I.G.; Nair, K.G.; Boyd, L.J. Cycloserine, a new antibiotic, in the treatment of human pulmonary tuberculosis: A preliminary report. *Antibiot. Med. Clin. Ther.* **1955**, *1*, 80–93.
148. Lester, W., Jr.; Salomon, A.; Reimann, A.F.; Shulruff, E.; Berg, G.S. Cycloserine therapy in tuberculosis in humans. *Am. Rev. Tuberc.* **1956**, *74*, 121–127. [CrossRef]
149. Hoeprich, P.D. Alanine: Cycloserine Antagonism. Vi. Demonstration of D-Alanine in the Serum of Guinea Pigs and Mice. *J. Biol. Chem.* **1965**, *240*, 1654–1660.
150. Li, Y.; Wang, F.; Wu, L.; Zhu, M.; He, G.; Chen, X.; Sun, F.; Liu, Q.; Wang, X.; Zhang, W. Cycloserine for treatment of multidrug-resistant tuberculosis: A retrospective cohort study in China. *Infect. Drug Resist.* **2019**, *12*, 721–731. [CrossRef]
151. Walker, W.C.; Murdoch, J.M. Cycloserine in the treatment of pulmonary tuberculosis; a report on toxicity. *Tubercle* **1957**, *38*, 297–302. [CrossRef]
152. Bondi, A.; Kornblum, J.; Forte, C. Inhibition of antibacterial activity of cycloserine by alpha-alanine. *Proc. Soc. Exp. Biol. Med.* **1957**, *96*, 270–272. [CrossRef] [PubMed]
153. Strominger, J.L.; Ito, E.; Threnn, R.H. Competitive inhibition of enzymatic reactions by oxamycin. *J. Am. Chem. Soc.* **1960**, *82*, 998–999. [CrossRef]
154. Halouska, S.; Fenton, R.J.; Zinniel, D.K.; Marshall, D.D.; Barletta, R.G.; Powers, R. Metabolomics analysis identifies d-Alanine-d-Alanine ligase as the primary lethal target of d-Cycloserine in mycobacteria. *J. Proteome Res.* **2014**, *13*, 1065–1076. [CrossRef] [PubMed]
155. Prosser, G.A.; de Carvalho, L.P. Reinterpreting the mechanism of inhibition of *Mycobacterium tuberculosis* D-alanine:D-alanine ligase by D-cycloserine. *Biochemistry* **2013**, *52*, 7145–7149. [CrossRef]
156. Prosser, G.A.; de Carvalho, L.P. Kinetic mechanism and inhibition of *Mycobacterium tuberculosis* D-alanine:D-alanine ligase by the antibiotic D-cycloserine. *FEBS J.* **2013**, *280*, 1150–1166. [CrossRef]
157. Viallier, J.; Cayre, R.M.; Biot, N. Sensitivity and resistance of *Mycobacterium tuberculosis* to cycloserine; study of 115 strains isolated from human pathological materials. *Ann. Inst. Pasteur (Paris)* **1957**, *93*, 127–131.
158. Chen, J.M.; Uplekar, S.; Gordon, S.V.; Cole, S.T. A point mutation in *cycA* partially contributes to the D-cycloserine resistance trait of *Mycobacterium bovis* BCG vaccine strains. *PLoS ONE* **2012**, *7*, e43467. [CrossRef]
159. Desjardins, C.A.; Cohen, K.A.; Munsamy, V.; Abeel, T.; Maharaj, K.; Walker, B.J.; Shea, T.P.; Almeida, D.V.; Manson, A.L.; Salazar, A.; et al. Genomic and functional analyses of *Mycobacterium tuberculosis* strains implicate *ald* in D-cycloserine resistance. *Nat. Genet.* **2016**, *48*, 544–551. [CrossRef]
160. Nakatani, Y.; Opel-Reading, H.K.; Merker, M.; Machado, D.; Andres, S.; Kumar, S.S.; Moradigaravand, D.; Coll, F.; Perdigao, J.; Portugal, I.; et al. Role of Alanine Racemase Mutations in *Mycobacterium tuberculosis* d-Cycloserine Resistance. *Antimicrob. Agents Chemother.* **2017**, *61*. [CrossRef]

161. Chen, J.; Zhang, S.; Cui, P.; Shi, W.; Zhang, W.; Zhang, Y. Identification of novel mutations associated with cycloserine resistance in *Mycobacterium tuberculosis*. *J. Antimicrob. Chemother.* **2017**, *72*, 3272–3276. [CrossRef]
162. Evangelopoulos, D.; Prosser, G.A.; Rodgers, A.; Dagg, B.M.; Khatri, B.; Ho, M.M.; Gutierrez, M.G.; Cortes, T.; de Carvalho, L.P.S. Comparative fitness analysis of D-cycloserine resistant mutants reveals both fitness-neutral and high-fitness cost genotypes. *Nat. Commun.* **2019**, *10*, 4177. [CrossRef] [PubMed]
163. Hager, T. *The Demon under the Microscope. From Battlefield Hospitals to Nazi Labs, One Doctor's Heroic Search for the World's First Miracle Drug*, 1st ed.; Three River Press: New York, NY, USA, 2006; p. 340.
164. Domagk, G. Investigations on the antituberculous activity of the thiosemicarbazones in vitro and in vivo. *Am. Rev. Tuberc.* **1950**, *61*, 8–19. [PubMed]
165. Mertens, A.; Bunge, R. The present status of the chemotherapy of tuberculosis with conteben a substance of the thiosemicarbazone series; a review. *Am. Rev. Tuberc.* **1950**, *61*, 20–38. [CrossRef] [PubMed]
166. Cavanagh, P.; McPherson, K. The thiacetazone sensitivity of *Mycobacterium tuberculosis*. *J. Med. Microbiol.* **1969**, *2*, 237–242. [CrossRef]
167. Rieder, H.L.; Enarson, D.A. Rebuttal: Time to call a halt to emotions in the assessment of thioacetazone. *Tuber. Lung Dis.* **1996**, *77*, 109–111. [CrossRef]
168. Falzon, D.; Hill, G.; Pal, S.N.; Suwankesawong, W.; Jaramillo, E. Pharmacovigilance and tuberculosis: Applying the lessons of thioacetazone. *Bull. World Health Organ.* **2014**, *92*, 918–919. [CrossRef]
169. Dover, L.G.; Alahari, A.; Gratraud, P.; Gomes, J.M.; Bhowruth, V.; Reynolds, R.C.; Besra, G.S.; Kremer, L. EthA, a common activator of thiocarbamide-containing drugs acting on different mycobacterial targets. *Antimicrob. Agents Chemother.* **2007**, *51*, 1055–1063. [CrossRef]
170. Belardinelli, J.M.; Morbidoni, H.R. Mutations in the essential FAS II beta-hydroxyacyl ACP dehydratase complex confer resistance to thiacetazone in *Mycobacterium tuberculosis* and *Mycobacterium kansasii*. *Mol. Microbiol.* **2012**, *86*, 568–579. [CrossRef]
171. Grzegorzewicz, A.E.; Kordulakova, J.; Jones, V.; Born, S.E.; Belardinelli, J.M.; Vaquie, A.; Gundi, V.A.; Madacki, J.; Slama, N.; Laval, F.; et al. A common mechanism of inhibition of the *Mycobacterium tuberculosis* mycolic acid biosynthetic pathway by isoxyl and thiacetazone. *J. Biol. Chem.* **2012**, *287*, 38434–38441. [CrossRef]
172. Dong, Y.; Qiu, X.; Shaw, N.; Xu, Y.; Sun, Y.; Li, X.; Li, J.; Rao, Z. Molecular basis for the inhibition of beta-hydroxyacyl-ACP dehydratase HadAB complex from *Mycobacterium tuberculosis* by flavonoid inhibitors. *Protein Cell* **2015**, *6*, 504–517. [CrossRef]
173. Grzegorzewicz, A.E.; Eynard, N.; Quemard, A.; North, E.J.; Margolis, A.; Lindenberger, J.J.; Jones, V.; Kordulakova, J.; Brennan, P.J.; Lee, R.E.; et al. Covalent modification of the *Mycobacterium tuberculosis* FAS-II dehydratase by Isoxyl and Thiacetazone. *ACS Infect. Dis.* **2015**, *1*, 91–97. [CrossRef] [PubMed]
174. Coxon, G.D.; Craig, D.; Corrales, R.M.; Vialla, E.; Gannoun-Zaki, L.; Kremer, L. Synthesis, antitubercular activity and mechanism of resistance of highly effective thiacetazone analogues. *PLoS ONE* **2013**, *8*, e53162. [CrossRef] [PubMed]
175. Grzegorzewicz, A.E.; Gee, C.; Das, S.; Liu, J.; Belardinelli, J.M.; Jones, V.; McNeil, M.R.; Lee, R.E.; Jackson, M. Mechanisms of Resistance Associated with the Inhibition of the Dehydration Step of Type II Fatty Acid Synthase in *Mycobacterium tuberculosis*. *ACS Infect. Dis.* **2019**. [CrossRef]
176. Alahari, A.; Alibaud, L.; Trivelli, X.; Gupta, R.; Lamichhane, G.; Reynolds, R.C.; Bishai, W.R.; Guerardel, Y.; Kremer, L. Mycolic acid methyltransferase, MmaA4, is necessary for thiacetazone susceptibility in *Mycobacterium tuberculosis*. *Mol. Microbiol.* **2009**, *71*, 1263–1277. [CrossRef] [PubMed]
177. Sacksteder, K.A.; Protopopova, M.; Barry, C.E., 3rd; Andries, K.; Nacy, C.A. Discovery and development of SQ109: A new antitubercular drug with a novel mechanism of action. *Future Microbiol.* **2012**, *7*, 823–837. [CrossRef] [PubMed]
178. Alland, D.; Steyn, A.J.; Weisbrod, T.; Aldrich, K.; Jacobs, W.R., Jr. Characterization of the *Mycobacterium tuberculosis iniBAC* promoter, a promoter that responds to cell wall biosynthesis inhibition. *J. Bacteriol.* **2000**, *182*, 1802–1811. [CrossRef] [PubMed]
179. Protopopova, M.; Hanrahan, C.; Nikonenko, B.; Samala, R.; Chen, P.; Gearhart, J.; Einck, L.; Nacy, C.A. Identification of a new antitubercular drug candidate, SQ109, from a combinatorial library of 1,2-ethylenediamines. *J. Antimicrob. Chemother.* **2005**, *56*, 968–974. [CrossRef]
180. Chen, P.; Gearhart, J.; Protopopova, M.; Einck, L.; Nacy, C.A. Synergistic interactions of SQ109, a new ethylene diamine, with front-line antitubercular drugs in vitro. *J. Antimicrob. Chemother.* **2006**, *58*, 332–337. [CrossRef]

181. Jia, L.; Tomaszewski, J.E.; Hanrahan, C.; Coward, L.; Noker, P.; Gorman, G.; Nikonenko, B.; Protopopova, M. Pharmacodynamics and pharmacokinetics of SQ109, a new diamine-based antitubercular drug. *Br. J. Pharmacol.* **2005**, *144*, 80–87. [CrossRef]
182. Boshoff, H.I.; Myers, T.G.; Copp, B.R.; McNeil, M.R.; Wilson, M.A.; Barry, C.E., 3rd. The transcriptional responses of *Mycobacterium tuberculosis* to inhibitors of metabolism: Novel insights into drug mechanisms of action. *J. Biol. Chem.* **2004**, *279*, 40174–40184. [CrossRef]
183. Tahlan, K.; Wilson, R.; Kastrinsky, D.B.; Arora, K.; Nair, V.; Fischer, E.; Barnes, S.W.; Walker, J.R.; Alland, D.; Barry, C.E., 3rd; et al. SQ109 targets MmpL3, a membrane transporter of trehalose monomycolate involved in mycolic acid donation to the cell wall core of *Mycobacterium tuberculosis*. *Antimicrob. Agents Chemother.* **2012**, *56*, 1797–1809. [CrossRef] [PubMed]
184. Grzegorzewicz, A.E.; Pham, H.; Gundi, V.A.; Scherman, M.S.; North, E.J.; Hess, T.; Jones, V.; Gruppo, V.; Born, S.E.; Kordulakova, J.; et al. Inhibition of mycolic acid transport across the *Mycobacterium tuberculosis* plasma membrane. *Nat. Chem. Biol.* **2012**, *8*, 334–341. [CrossRef] [PubMed]
185. Zhang, B.; Li, J.; Yang, X.; Wu, L.; Zhang, J.; Yang, Y.; Zhao, Y.; Zhang, L.; Cheng, X.; Liu, Z.; et al. Crystal Structures of Membrane Transporter MmpL3, an Anti-TB Drug Target. *Cell* **2019**, *176*, 636–648.e13. [CrossRef] [PubMed]
186. Heinrich, N.; Dawson, R.; du Bois, J.; Narunsky, K.; Horwith, G.; Phipps, A.J.; Nacy, C.A.; Aarnoutse, R.E.; Boeree, M.J.; Gillespie, S.H.; et al. Early phase evaluation of SQ109 alone and in combination with rifampicin in pulmonary TB patients. *J. Antimicrob. Chemother.* **2015**, *70*, 1558–1566. [CrossRef]
187. Niemi, M.; Backman, J.T.; Fromm, M.F.; Neuvonen, P.J.; Kivisto, K.T. Pharmacokinetic interactions with rifampicin: Clinical relevance. *Clin. Pharmacokinet.* **2003**, *42*, 819–850. [CrossRef]
188. Boeree, M.J.; Heinrich, N.; Aarnoutse, R.; Diacon, A.H.; Dawson, R.; Rehal, S.; Kibiki, G.S.; Churchyard, G.; Sanne, I.; Ntinginya, N.E.; et al. High-dose rifampicin, moxifloxacin, and SQ109 for treating tuberculosis: A multi-arm, multi-stage randomised controlled trial. *Lancet Infect. Dis.* **2017**, *17*, 39–49. [CrossRef]
189. Tiberi, S.; du Plessis, N.; Walzl, G.; Vjecha, M.J.; Rao, M.; Ntoumi, F.; Mfinanga, S.; Kapata, N.; Mwaba, P.; McHugh, T.D.; et al. Tuberculosis: Progress and advances in development of new drugs, treatment regimens, and host-directed therapies. *Lancet Infect. Dis.* **2018**, *18*, e183–e198. [CrossRef]
190. Poce, G.; Consalvi, S.; Biava, M. MmpL3 Inhibitors: Diverse Chemical Scaffolds Inhibit the Same Target. *Mini Rev. Med. Chem.* **2016**, *16*, 1274–1283. [CrossRef]
191. Williams, J.T.; Haiderer, E.R.; Coulson, G.B.; Conner, K.N.; Ellsworth, E.; Chen, C.; Alvarez-Cabrera, N.; Li, W.; Jackson, M.; Dick, T.; et al. Identification of New MmpL3 Inhibitors by Untargeted and Targeted Mutant Screens Defines MmpL3 Domains with Differential Resistance. *Antimicrob. Agents Chemother.* **2019**, *63*. [CrossRef]
192. Samuelson, J. Why metronidazole is active against both bacteria and parasites. *Antimicrob. Agents Chemother.* **1999**, *43*, 1533–1541. [CrossRef]
193. Brooks, J.V.; Furney, S.K.; Orme, I.M. Metronidazole therapy in mice infected with tuberculosis. *Antimicrob. Agents Chemother.* **1999**, *43*, 1285–1288. [CrossRef] [PubMed]
194. Ashtekar, D.R.; Costa-Perira, R.; Nagrajan, K.; Vishvanathan, N.; Bhatt, A.D.; Rittel, W. In vitro and in vivo activities of the nitroimidazole CGI 17341 against *Mycobacterium tuberculosis*. *Antimicrob. Agents Chemother.* **1993**, *37*, 183–186. [CrossRef] [PubMed]
195. Stover, C.K.; Warrener, P.; VanDevanter, D.R.; Sherman, D.R.; Arain, T.M.; Langhorne, M.H.; Anderson, S.W.; Towell, J.A.; Yuan, Y.; McMurray, D.N.; et al. A small-molecule nitroimidazopyran drug candidate for the treatment of tuberculosis. *Nature* **2000**, *405*, 962–966. [CrossRef] [PubMed]
196. Li, S.Y.; Tasneen, R.; Tyagi, S.; Soni, H.; Converse, P.J.; Mdluli, K.; Nuermberger, E.L. Bactericidal and Sterilizing Activity of a Novel Regimen with Bedaquiline, Pretomanid, Moxifloxacin, and Pyrazinamide in a Murine Model of Tuberculosis. *Antimicrob. Agents Chemother.* **2017**, *61*. [CrossRef] [PubMed]
197. Xu, J.; Li, S.Y.; Almeida, D.V.; Tasneen, R.; Barnes-Boyle, K.; Converse, P.J.; Upton, A.M.; Mdluli, K.; Fotouhi, N.; Nuermberger, E.L. Contribution of Pretomanid to Novel Regimens Containing Bedaquiline with either Linezolid or Moxifloxacin and Pyrazinamide in Murine Models of Tuberculosis. *Antimicrob. Agents Chemother.* **2019**, *63*. [CrossRef] [PubMed]
198. Singh, R.; Manjunatha, U.; Boshoff, H.I.; Ha, Y.H.; Niyomrattanakit, P.; Ledwidge, R.; Dowd, C.S.; Lee, I.Y.; Kim, P.; Zhang, L.; et al. PA-824 kills nonreplicating *Mycobacterium tuberculosis* by intracellular NO release. *Science* **2008**, *322*, 1392–1395. [CrossRef]

199. Manjunatha, U.; Boshoff, H.I.; Barry, C.E. The mechanism of action of PA-824: Novel insights from transcriptional profiling. *Commun. Integr. Biol.* **2009**, *2*, 215–218. [CrossRef]
200. Haver, H.L.; Chua, A.; Ghode, P.; Lakshminarayana, S.B.; Singhal, A.; Mathema, B.; Wintjens, R.; Bifani, P. Mutations in genes for the F420 biosynthetic pathway and a nitroreductase enzyme are the primary resistance determinants in spontaneous in vitro-selected PA-824-resistant mutants of *Mycobacterium tuberculosis*. *Antimicrob. Agents Chemother.* **2015**, *59*, 5316–5323. [CrossRef]
201. Diacon, A.H.; Dawson, R.; von Groote-Bidlingmaier, F.; Symons, G.; Venter, A.; Donald, P.R.; van Niekerk, C.; Everitt, D.; Winter, H.; Becker, P.; et al. 14-day bactericidal activity of PA-824, bedaquiline, pyrazinamide, and moxifloxacin combinations: A randomised trial. *Lancet* **2012**, *380*, 986–993. [CrossRef]
202. Tweed, C.D.; Dawson, R.; Burger, D.A.; Conradie, A.; Crook, A.M.; Mendel, C.M.; Conradie, F.; Diacon, A.H.; Ntinginya, N.E.; Everitt, D.E.; et al. Bedaquiline, moxifloxacin, pretomanid, and pyrazinamide during the first 8 weeks of treatment of patients with drug-susceptible or drug-resistant pulmonary tuberculosis: A multicentre, open-label, partially randomised, phase 2b trial. *Lancet Respir. Med.* **2019**, *7*, 1048–1058. [CrossRef]
203. Conradie, F.; Diacon, A.; Everitt, D.; Mendel, C.M.; Crook, A.M.; Howell, P.; Comins, K.; Spigelman, M. Sustained High Rate of Successful Treatment Outcomes: Interim Results of 75 Patients in the Nix-TB Clinical Study of Pretomanid, Bedaquiline and Linezolid. Available online: https://www.dropbox.com/s/gu8l27grq38psul/Nix%20TB%20interim%20results%20-%2010-25-18.pdf?dl=0 (accessed on 30 January 2020).
204. Sasaki, H.; Haraguchi, Y.; Itotani, M.; Kuroda, T.; Hashizume, H.; Tomishige, T.; Kawasaki, M.; Matsumoto, M.; Komatsu, M.; Tsubouchi, H. Synthesis and antituberculosis activity of a novel series of optically active 6-nitro-2,3-dihydroimidazo[2,1-b]oxazoles. *J. Med. Chem.* **2006**, *49*, 7854–7860. [CrossRef] [PubMed]
205. Liu, Y.; Matsumoto, M.; Ishida, H.; Ohguro, K.; Yoshitake, M.; Gupta, R.; Geiter, L.; Hafkin, J. Delamanid: From discovery to its use for pulmonary multidrug-resistant tuberculosis (MDR-TB). *Tuberculosis* **2018**, *111*, 20–30. [CrossRef]
206. Matsumoto, M.; Hashizume, H.; Tomishige, T.; Kawasaki, M.; Tsubouchi, H.; Sasaki, H.; Shimokawa, Y.; Komatsu, M. OPC-67683, a nitro-dihydro-imidazooxazole derivative with promising action against tuberculosis in vitro and in mice. *PLoS Med.* **2006**, *3*, e466. [CrossRef] [PubMed]
207. Upton, A.M.; Cho, S.; Yang, T.J.; Kim, Y.; Wang, Y.; Lu, Y.; Wang, B.; Xu, J.; Mdluli, K.; Ma, Z.; et al. In vitro and in vivo activities of the nitroimidazole TBA-354 against *Mycobacterium tuberculosis*. *Antimicrob. Agents Chemother.* **2015**, *59*, 136–144. [CrossRef] [PubMed]
208. Fujiwara, M.; Kawasaki, M.; Hariguchi, N.; Liu, Y.; Matsumoto, M. Mechanisms of resistance to delamanid, a drug for *Mycobacterium tuberculosis*. *Tuberculosis* **2018**, *108*, 186–194. [CrossRef] [PubMed]
209. Gler, M.T.; Skripconoka, V.; Sanchez-Garavito, E.; Xiao, H.; Cabrera-Rivero, J.L.; Vargas-Vasquez, D.E.; Gao, M.; Awad, M.; Park, S.K.; Shim, T.S.; et al. Delamanid for multidrug-resistant pulmonary tuberculosis. *N. Engl. J. Med.* **2012**, *366*, 2151–2160. [CrossRef]
210. Gupta, R.; Geiter, L.J.; Wells, C.D.; Gao, M.; Cirule, A.; Xiao, H. Delamanid for Extensively Drug-Resistant Tuberculosis. *N. Engl. J. Med.* **2015**, *373*, 291–292. [CrossRef]
211. Wells, C.D.; Gupta, R.; Hittel, N.; Geiter, L.J. Long-term mortality assessment of multidrug-resistant tuberculosis patients treated with delamanid. *Eur. Respir. J.* **2015**, *45*, 1498–1501. [CrossRef]
212. Von Groote-Bidlingmaier, F.; Patientia, R.; Sanchez, E.; Balanag, V., Jr.; Ticona, E.; Segura, P.; Cadena, E.; Yu, C.; Cirule, A.; Lizarbe, V.; et al. Efficacy and safety of delamanid in combination with an optimised background regimen for treatment of multidrug-resistant tuberculosis: A multicentre, randomised, double-blind, placebo-controlled, parallel group phase 3 trial. *Lancet Respir. Med.* **2019**, *7*, 249–259. [CrossRef]
213. Kumar, P.; Capodagli, G.C.; Awasthi, D.; Shrestha, R.; Maharaja, K.; Sukheja, P.; Li, S.G.; Inoyama, D.; Zimmerman, M.; Ho Liang, H.P.; et al. Synergistic Lethality of a Binary Inhibitor of *Mycobacterium tuberculosis* KasA. *MBio* **2018**, *9*. [CrossRef]
214. Aggarwal, A.; Parai, M.K.; Shetty, N.; Wallis, D.; Woolhiser, L.; Hastings, C.; Dutta, N.K.; Galaviz, S.; Dhakal, R.C.; Shrestha, R.; et al. Development of a Novel Lead that Targets *M. tuberculosis* Polyketide Synthase 13. *Cell* **2017**, *170*, 249–259.e25. [CrossRef] [PubMed]
215. Dal Molin, M.; Selchow, P.; Schafle, D.; Tschumi, A.; Ryckmans, T.; Laage-Witt, S.; Sander, P. Identification of novel scaffolds targeting *Mycobacterium tuberculosis*. *J. Mol. Med.* **2019**, *97*, 1601–1613. [CrossRef]

216. Wilson, R.; Kumar, P.; Parashar, V.; Vilcheze, C.; Veyron-Churlet, R.; Freundlich, J.S.; Barnes, S.W.; Walker, J.R.; Szymonifka, M.J.; Marchiano, E.; et al. Antituberculosis thiophenes define a requirement for Pks13 in mycolic acid biosynthesis. *Nat. Chem. Biol.* **2013**, *9*, 499–506. [CrossRef] [PubMed]
217. Zhang, W.; Lun, S.; Wang, S.H.; Jiang, X.W.; Yang, F.; Tang, J.; Manson, A.L.; Earl, A.M.; Gunosewoyo, H.; Bishai, W.R.; et al. Identification of Novel Coumestan Derivatives as Polyketide Synthase 13 Inhibitors against *Mycobacterium tuberculosis*. *J. Med. Chem.* **2018**, *61*, 791–803. [CrossRef] [PubMed]
218. Li, W.; Sanchez-Hidalgo, A.; Jones, V.; de Moura, V.C.; North, E.J.; Jackson, M. Synergistic Interactions of MmpL3 Inhibitors with Antitubercular Compounds In Vitro. *Antimicrob. Agents Chemother.* **2017**, *61*. [CrossRef] [PubMed]
219. Poce, G.; Bates, R.H.; Alfonso, S.; Cocozza, M.; Porretta, G.C.; Ballell, L.; Rullas, J.; Ortega, F.; De Logu, A.; Agus, E.; et al. Improved BM212 MmpL3 inhibitor analogue shows efficacy in acute murine model of tuberculosis infection. *PLoS ONE* **2013**, *8*, e56980. [CrossRef] [PubMed]
220. Igarashi, M.; Nakagawa, N.; Doi, N.; Hattori, S.; Naganawa, H.; Hamada, M. Caprazamycin B, a novel anti-tuberculosis antibiotic, from *Streptomyces* sp. *J. Antibiot.* **2003**, *56*, 580–583. [CrossRef]
221. Takahashi, Y.; Igarashi, M.; Miyake, T.; Soutome, H.; Ishikawa, K.; Komatsuki, Y.; Koyama, Y.; Nakagawa, N.; Hattori, S.; Inoue, K.; et al. Novel semisynthetic antibiotics from caprazamycins A-G: Caprazene derivatives and their antibacterial activity. *J. Antibiot.* **2013**, *66*, 171–178. [CrossRef]
222. Ishizaki, Y.; Hayashi, C.; Inoue, K.; Igarashi, M.; Takahashi, Y.; Pujari, V.; Crick, D.C.; Brennan, P.J.; Nomoto, A. Inhibition of the first step in synthesis of the mycobacterial cell wall core, catalyzed by the GlcNAc-1-phosphate transferase WecA, by the novel caprazamycin derivative CPZEN-45. *J. Biol. Chem.* **2013**, *288*, 30309–30319. [CrossRef]
223. Takahashi, Y.; Igarashi, M.; Okada, M. Anti-XDR-TB, Anti-MDR-TB Drug, and Combination Anti-Tuberculoses Drug. U.S. Patent 9040502 B2, 26 May 2015.
224. Huszar, S.; Singh, V.; Polcicova, A.; Barath, P.; Barrio, M.B.; Lagrange, S.; Leblanc, V.; Nacy, C.A.; Mizrahi, V.; Mikusova, K. N-Acetylglucosamine-1-Phosphate Transferase, WecA, as a Validated Drug Target in *Mycobacterium tuberculosis*. *Antimicrob. Agents Chemother.* **2017**, *61*. [CrossRef]
225. Makarov, V.; Manina, G.; Mikusova, K.; Mollmann, U.; Ryabova, O.; Saint-Joanis, B.; Dhar, N.; Pasca, M.R.; Buroni, S.; Lucarelli, A.P.; et al. Benzothiazinones kill *Mycobacterium tuberculosis* by blocking arabinan synthesis. *Science* **2009**, *324*, 801–804. [CrossRef]
226. Lechartier, B.; Hartkoorn, R.C.; Cole, S.T. In vitro combination studies of benzothiazinone lead compound BTZ043 against *Mycobacterium tuberculosis*. *Antimicrob. Agents Chemother.* **2012**, *56*, 5790–5793. [CrossRef]
227. New TB Drugs. BTZ-043. Available online: https://www.newtbdrugs.org/pipeline/compound/btz-043 (accessed on 15 February 2020).
228. Trefzer, C.; Skovierova, H.; Buroni, S.; Bobovska, A.; Nenci, S.; Molteni, E.; Pojer, F.; Pasca, M.R.; Makarov, V.; Cole, S.T.; et al. Benzothiazinones are suicide inhibitors of mycobacterial decaprenylphosphoryl-beta-D-ribofuranose 2′-oxidase DprE1. *J. Am. Chem. Soc.* **2012**, *134*, 912–915. [CrossRef] [PubMed]
229. Lupien, A.; Vocat, A.; Foo, C.S.; Blattes, E.; Gillon, J.Y.; Makarov, V.; Cole, S.T. Optimized Background Regimen for Treatment of Active Tuberculosis with the Next-Generation Benzothiazinone Macozinone (PBTZ169). *Antimicrob. Agents Chemother.* **2018**, *62*. [CrossRef] [PubMed]
230. Lechartier, B.; Cole, S.T. Mode of Action of Clofazimine and Combination Therapy with Benzothiazinones against *Mycobacterium tuberculosis*. *Antimicrob. Agents Chemother.* **2015**, *59*, 4457–4463. [CrossRef] [PubMed]
231. Makarov, V.; Lechartier, B.; Zhang, M.; Neres, J.; van der Sar, A.M.; Raadsen, S.A.; Hartkoorn, R.C.; Ryabova, O.B.; Vocat, A.; Decosterd, L.A.; et al. Towards a new combination therapy for tuberculosis with next generation benzothiazinones. *EMBO Mol. Med.* **2014**, *6*, 372–383. [CrossRef] [PubMed]
232. Workshop "Critical Path to TB Drug Regimens". Available online: http://www.cptrinitiative.org/wp-content/uploads/2017/05/Jeffrey_Hafkin_CPTR2017_JH.pdf (accessed on 26 January 2020).
233. Hartkoorn, R.C.; Uplekar, S.; Cole, S.T. Cross-Resistance between Clofazimine and Bedaquiline through Upregulation of MmpL5 in *Mycobacterium tuberculosis*. *Antimicrob. Agents Chemother.* **2014**, *58*, 2979–2981. [CrossRef]
234. Shirude, P.S.; Shandil, R.; Sadler, C.; Naik, M.; Hosagrahara, V.; Hameed, S.; Shinde, V.; Bathula, C.; Humnabadkar, V.; Kumar, N.; et al. Azaindoles: Noncovalent DprE1 inhibitors from scaffold morphing efforts, kill *Mycobacterium tuberculosis* and are efficacious in vivo. *J. Med. Chem.* **2013**, *56*, 9701–9708. [CrossRef]

235. Chatterji, M.; Shandil, R.; Manjunatha, M.R.; Solapure, S.; Ramachandran, V.; Kumar, N.; Saralaya, R.; Panduga, V.; Reddy, J.; Prabhakar, K.R.; et al. 1,4-azaindole, a potential drug candidate for treatment of tuberculosis. *Antimicrob. Agents Chemother.* **2014**, *58*, 5325–5331. [CrossRef] [PubMed]
236. Gao, C.; Peng, C.; Shi, Y.; You, X.; Ran, K.; Xiong, L.; Ye, T.H.; Zhang, L.; Wang, N.; Zhu, Y.; et al. Benzothiazinethione is a potent preclinical candidate for the treatment of drug-resistant tuberculosis. *Sci. Rep.* **2016**, *6*, 29717. [CrossRef] [PubMed]
237. Christophe, T.; Jackson, M.; Jeon, H.K.; Fenistein, D.; Contreras-Dominguez, M.; Kim, J.; Genovesio, A.; Carralot, J.P.; Ewann, F.; Kim, E.H.; et al. High content screening identifies decaprenyl-phosphoribose 2′ epimerase as a target for intracellular antimycobacterial inhibitors. *PLoS Pathog.* **2009**, *5*, e1000645. [CrossRef] [PubMed]
238. Magnet, S.; Hartkoorn, R.C.; Szekely, R.; Pato, J.; Triccas, J.A.; Schneider, P.; Szantai-Kis, C.; Orfi, L.; Chambon, M.; Banfi, D.; et al. Leads for antitubercular compounds from kinase inhibitor library screens. *Tuberculosis* **2010**, *90*, 354–360. [CrossRef] [PubMed]
239. Stanley, S.A.; Grant, S.S.; Kawate, T.; Iwase, N.; Shimizu, M.; Wivagg, C.; Silvis, M.; Kazyanskaya, E.; Aquadro, J.; Golas, A.; et al. Identification of novel inhibitors of *M. tuberculosis* growth using whole cell based high-throughput screening. *ACS Chem. Biol.* **2012**, *7*, 1377–1384. [CrossRef] [PubMed]
240. Wang, F.; Sambandan, D.; Halder, R.; Wang, J.; Batt, S.M.; Weinrick, B.; Ahmad, I.; Yang, P.; Zhang, Y.; Kim, J.; et al. Identification of a small molecule with activity against drug-resistant and persistent tuberculosis. *Proc. Natl. Acad. Sci. USA* **2013**, *110*, E2510–E2517. [CrossRef] [PubMed]
241. Panda, M.; Ramachandran, S.; Ramachandran, V.; Shirude, P.S.; Humnabadkar, V.; Nagalapur, K.; Sharma, S.; Kaur, P.; Guptha, S.; Narayan, A.; et al. Discovery of pyrazolopyridones as a novel class of noncovalent DprE1 inhibitor with potent anti-mycobacterial activity. *J. Med. Chem.* **2014**, *57*, 4761–4771. [CrossRef] [PubMed]
242. Naik, M.; Humnabadkar, V.; Tantry, S.J.; Panda, M.; Narayan, A.; Guptha, S.; Panduga, V.; Manjrekar, P.; Jena, L.K.; Koushik, K.; et al. 4-aminoquinolone piperidine amides: Noncovalent inhibitors of DprE1 with long residence time and potent antimycobacterial activity. *J. Med. Chem.* **2014**, *57*, 5419–5434. [CrossRef]
243. Neres, J.; Hartkoorn, R.C.; Chiarelli, L.R.; Gadupudi, R.; Pasca, M.R.; Mori, G.; Venturelli, A.; Savina, S.; Makarov, V.; Kolly, G.S.; et al. 2-Carboxyquinoxalines kill *Mycobacterium tuberculosis* through noncovalent inhibition of DprE1. *ACS Chem. Biol.* **2015**, *10*, 705–714. [CrossRef]
244. Makarov, V.; Neres, J.; Hartkoorn, R.C.; Ryabova, O.B.; Kazakova, E.; Sarkan, M.; Huszar, S.; Piton, J.; Kolly, G.S.; Vocat, A.; et al. The 8-Pyrrole-Benzothiazinones Are Noncovalent Inhibitors of DprE1 from *Mycobacterium tuberculosis*. *Antimicrob. Agents Chemother.* **2015**, *59*, 4446–4452. [CrossRef]
245. Warrier, T.; Kapilashrami, K.; Argyrou, A.; Ioerger, T.R.; Little, D.; Murphy, K.C.; Nandakumar, M.; Park, S.; Gold, B.; Mi, J.; et al. N-methylation of a bactericidal compound as a resistance mechanism in *Mycobacterium tuberculosis*. *Proc. Natl. Acad. Sci. USA* **2016**, *113*, E4523–E4530. [CrossRef]
246. Piton, J.; Vocat, A.; Lupien, A.; Foo, C.S.; Riabova, O.; Makarov, V.; Cole, S.T. Structure-Based Drug Design and Characterization of Sulfonyl-Piperazine Benzothiazinone Inhibitors of DprE1 from *Mycobacterium tuberculosis*. *Antimicrob. Agents Chemother.* **2018**, *62*. [CrossRef]
247. Borthwick, J.A.; Alemparte, C.; Wall, I.; Whitehurst, B.C.; Argyrou, A.; Burley, G.; de Dios-Anton, P.; Guijarro, L.; Monteiro, M.C.; Ortega, F.; et al. *Mycobacterium tuberculosis* Decaprenylphosphoryl-beta-d-ribose Oxidase Inhibitors: Expeditious Reconstruction of Suboptimal Hits into a Series with Potent in Vivo Activity. *J. Med. Chem.* **2020**. [CrossRef] [PubMed]
248. Hugonnet, J.E.; Blanchard, J.S. Irreversible inhibition of the *Mycobacterium tuberculosis* beta-lactamase by clavulanate. *Biochemistry* **2007**, *46*, 11998–12004. [CrossRef] [PubMed]
249. Flores, A.R.; Parsons, L.M.; Pavelka, M.S. Genetic analysis of the beta-lactamases of *Mycobacterium tuberculosis* and *Mycobacterium smegmatis* and susceptibility to beta-lactam antibiotics. *Microbiology* **2005**, *151*, 521–532. [CrossRef] [PubMed]
250. Hugonnet, J.E.; Tremblay, L.W.; Boshoff, H.I.; Barry, C.E., 3rd; Blanchard, J.S. Meropenem-clavulanate is effective against extensively drug-resistant *Mycobacterium tuberculosis*. *Science* **2009**, *323*, 1215–1218. [CrossRef]
251. England, K.; Boshoff, H.I.; Arora, K.; Weiner, D.; Dayao, E.; Schimel, D.; Via, L.E.; Barry, C.E., 3rd. Meropenem-clavulanic acid shows activity against *Mycobacterium tuberculosis* in vivo. *Antimicrob. Agents Chemother.* **2012**, *56*, 3384–3387. [CrossRef]

252. Veziris, N.; Truffot, C.; Mainardi, J.L.; Jarlier, V. Activity of carbapenems combined with clavulanate against murine tuberculosis. *Antimicrob. Agents Chemother.* **2011**, *55*, 2597–2600. [CrossRef]
253. Dhar, N.; Dubee, V.; Ballell, L.; Cuinet, G.; Hugonnet, J.E.; Signorino-Gelo, F.; Barros, D.; Arthur, M.; McKinney, J.D. Rapid cytolysis of *Mycobacterium tuberculosis* by faropenem, an orally bioavailable beta-lactam antibiotic. *Antimicrob. Agents Chemother.* **2014**, *59*, 1308–1319. [CrossRef]
254. Rullas, J.; Dhar, N.; McKinney, J.D.; Garcia-Perez, A.; Lelievre, J.; Diacon, A.H.; Hugonnet, J.E.; Arthur, M.; Angulo-Barturen, I.; Barros-Aguirre, D.; et al. Combinations of beta-Lactam Antibiotics Currently in Clinical Trials Are Efficacious in a DHP-I-Deficient Mouse Model of Tuberculosis Infection. *Antimicrob. Agents Chemother.* **2015**, *59*, 4997–4999. [CrossRef]
255. Dubee, V.; Triboulet, S.; Mainardi, J.L.; Etheve-Quelquejeu, M.; Gutmann, L.; Marie, A.; Dubost, L.; Hugonnet, J.E.; Arthur, M. Inactivation of *Mycobacterium tuberculosis* L,D-transpeptidase Ldt$_{Mt1}$ by carbapenems and cephalosporins. *Antimicrob. Agents Chemother.* **2012**, *56*, 4189–4195. [CrossRef]
256. Lavollay, M.; Arthur, M.; Fourgeaud, M.; Dubost, L.; Marie, A.; Veziris, N.; Blanot, D.; Gutmann, L.; Mainardi, J.L. The peptidoglycan of stationary-phase *Mycobacterium tuberculosis* predominantly contains cross-links generated by L,D-transpeptidation. *J. Bacteriol.* **2008**, *190*, 4360–4366. [CrossRef]
257. Tiberi, S.; Payen, M.C.; Sotgiu, G.; D'Ambrosio, L.; Alarcon Guizado, V.; Alffenaar, J.W.; Abdo Arbex, M.; Caminero, J.A.; Centis, R.; De Lorenzo, S.; et al. Effectiveness and safety of meropenem/clavulanate-containing regimens in the treatment of MDR- and XDR-TB. *Eur. Respir. J.* **2016**, *47*, 1235–1243. [CrossRef] [PubMed]
258. Di Modugno, E.; Erbetti, I.; Ferrari, L.; Galassi, G.; Hammond, S.M.; Xerri, L. In vitro activity of the tribactam GV104326 against gram-positive, gram-negative, and anaerobic bacteria. *Antimicrob. Agents Chemother.* **1994**, *38*, 2362–2368. [CrossRef] [PubMed]
259. Barros Aguirre, D.; Bates, R.H.; Gonzalez Del Rio, R.; Mendoza Losana, A.; Ramón García, S. Sanfetrinem or a Salt or Ester Thereof for Use in Treating Mycobacterial Infection. Patent No. WO/2018/206466, 15 November 2018.
260. New TB Drugs. Sanfetrinem. Available online: https://www.newtbdrugs.org/pipeline/compound/sanfetrinem (accessed on 31 January 2020).
261. Campanico, A.; Moreira, R.; Lopes, F. Drug discovery in tuberculosis. New drug targets and antimycobacterial agents. *Eur. J. Med. Chem.* **2018**, *150*, 525–545. [CrossRef] [PubMed]
262. Libardo, M.; Boshoff, H.I.; Barry, C.E., 3rd. The present state of the tuberculosis drug development pipeline. *Curr. Opin. Pharmacol.* **2018**, *42*, 81–94. [CrossRef] [PubMed]
263. Tiberi, S.; Munoz-Torrico, M.; Duarte, R.; Dalcolmo, M.; D'Ambrosio, L.; Migliori, G.B. New drugs and perspectives for new anti-tuberculosis regimens. *Pulmonology* **2018**, *24*, 86–98. [CrossRef]
264. Waksman, S.A. *The Conquest of Tuberculosis*; University of California Press: Berkeley, CA, USA, 1964; p. 211.

© 2020 by the author. Licensee MDPI, Basel, Switzerland. This article is an open access article distributed under the terms and conditions of the Creative Commons Attribution (CC BY) license (http://creativecommons.org/licenses/by/4.0/).

Review

Promiscuous Targets for Antitubercular Drug Discovery: The Paradigm of DprE1 and MmpL3

Giulia Degiacomi [1,†], Juan Manuel Belardinelli [2,†], Maria Rosalia Pasca [1], Edda De Rossi [1], Giovanna Riccardi [1,*] and Laurent Roberto Chiarelli [1,*]

1. Department of Biology and Biotechnology "Lazzaro Spallanzani", University of Pavia, 27100 Pavia, Italy; giulia.degiacomi@unipv.it (G.D.); mariarosalia.pasca@unipv.it (M.R.P.); edda.derossi@unipv.it (E.D.R.)
2. Mycobacteria Research Laboratories, Department of Microbiology, Immunology and Pathology, Colorado State University, Fort Collins, CO 80523-1682, USA; juan.belardinelli@colostate.edu
* Correspondence: giovanna.riccardi@unipv.it (G.R.); laurent.chiarelli@unipv.it (L.R.C.); Tel.: +39-382-985574 (G.R.); +39-382-987241 (L.R.C.)
† These authors contributed equally to this work.

Received: 18 December 2019; Accepted: 11 January 2020; Published: 15 January 2020

Abstract: The development and spread of *Mycobacterium tuberculosis* multi-drug resistant strains still represent a great global health threat, leading to an urgent need for novel anti-tuberculosis drugs. Indeed, in the last years, several efforts have been made in this direction, through a number of high-throughput screenings campaigns, which allowed for the identification of numerous hit compounds and novel targets. Interestingly, several independent screening assays identified the same proteins as the target of different compounds, and for this reason, they were named "promiscuous" targets. These proteins include DprE1, MmpL3, QcrB and Psk13, and are involved in the key pathway for *M. tuberculosis* survival, thus they should represent an Achilles' heel which could be exploited for the development of novel effective drugs. Indeed, among the last molecules which entered clinical trials, four inhibit a promiscuous target. Within this review, the two most promising promiscuous targets, the oxidoreductase DprE1 involved in arabinogalactan synthesis and the mycolic acid transporter MmpL3 are discussed, along with the latest advancements in the development of novel inhibitors with anti-tubercular activity.

Keywords: mycobacteria; tuberculosis; multi-drug resistance; drug discovery; promiscuous targets

---

1. The Added Value of Promiscuous Targets for Antitubercular Drug Discovery

One of the biggest problems in fighting *Mycobacterium tuberculosis* (Mtb), the etiologic agent of tuberculosis (TB), is the development and spread of multi-drug resistant strains [1]. Therefore, there is an urgent need to identify novel druggable targets for the development of more efficient anti-TB agents [2,3]. In this context, the medicinal chemistry efforts made in the last years led to the discovery of new antimycobacterial compounds, and the identification of novel targets [3–5].

High-throughput screenings (HTS), based on Mtb whole cells, were developed to identify hit compounds with potent inhibitory activity, and consequently, new targets emerged. The HTS had the merit of fueling the scarce TB drug development pipeline, since many molecules under preclinical and clinical development came out from this approach. Indeed, the first new TB drug, Bedaquiline, approved by the Food and Drug Administration (FDA) since 1971, was discovered in a whole-cell HTS campaign [6], and the same origin had both Delamanid [7] and Pretomanid [8], other two drugs recently approved for multidrug resistant tuberculosis (MDR-TB) and extensively drug-resistant tuberculosis (XDR-TB) treatment [9,10].

Curiously, an unforeseen outcome of many phenotypic screens against Mtb is the finding that the same targets have been frequently found in many different screening assays, despite the use of different compound libraries [11]. These targets were named "promiscuous" targets, for their nonspecific susceptibility to being inhibited by different scaffolds [2,12]. Although this term was initially given with a negative connotation, the value of these potential pharmacological targets has been reevaluated. Indeed, the high vulnerability of these essential targets, reflecting the biggest vulnerabilities of Mtb, can provide new opportunities to be explored for the development of TB drugs [12]. Among them, there are DprE1, MmpL3, QcrB and Psk13 [3,13].

DprE1 is an enzyme that works in concert with DprE2 to synthesize the unique arabinose precursor for lipoarabinomannan and arabinogalactan, essential building blocks of the mycobacterial cell-wall [14]. To date, more than 15 pharmacophores were found to inhibit DprE1 activity.

MmpL3 is the only Mtb transporter of trehalose monomycolate, required for the formation of the mycolic acid layer of the cell wall [15], and has been found to be affected by several molecules. Recently, direct inhibition of MmpL3 by BM212, the first compound found to hit MmpL3, was shown using spheroplasts [15], while the dissipation of proton motive force is the proposed mechanism for the other molecules [16].

QcrB encodes the *b* subunit of ubiquinol cytochrome C reductase, involved in the electron transport. Among the other compounds, Q203 is a promising clinical candidate, which blocks Mtb cell growth by targeting the respiratory cytochrome bc1 complex [17,18].

Finally, polyketide synthase Psk13, essential for the formation of the mycolic acid precursors, has been targeted by thiophenes and benzofuran containing molecules [19,20].

It is noteworthy that several research groups independently identified these promiscuous targets. These proteins are embedded in pathways that have key roles for Mtb growth and survival under the screening conditions, but even more important during infection. Their essentiality represents an Mtb Achilles' heel that is important to exploit for the development of new effective drug regimens able to shorten the TB treatment. Interestingly, all these promiscuous targets are localized within the cell envelope, emphasizing that the cell wall still represents a fruitful source of drug targets.

In recent years, several innovative strategies have been implemented to avoid the rediscovery of promiscuous targets [17,21–23]. Nevertheless, it is noteworthy that promiscuous targets proved to be successful in medicinal chemistry: four inhibitors Q203 (QcrB), SQ-109 (MmpL3), Macozinone (MCZ) and TBA-7371 (DprE1) entered clinical trials [24–29]. Thus, in parallel, a rational exploitation of already available promiscuous targets could lead to optimized scaffolds that retain the ability to attack the most vulnerable weakness of tubercle bacilli.

2. The First Discovery of DprE1 and MmpL3 as Drug Target

Dr. Vadim Makarov and collaborators, including our Laboratory, published in 2009 the first report in which DprE1 was described as the target of 1,3-benzothiazin-4-ones (BTZ), a new class of agents endowed with antimycobacterial activity [30]. To find the BTZ target, two genetic approaches were performed in parallel using both a *Mycobacterium smegmatis* cosmid library for resistance to BTZ, and the selection and characterization of *M. smegmatis*, *Mycobacterium bovis* BCG and *M. tuberculosis* spontaneous resistant mutants. Both methods revealed that *dprE1* gene was responsible for BTZ resistance. It is noteworthy that all the mutants carried mutations in the same codon of *dprE1*, leading to the replacement of Cys387 with a Ser or Gly residue [30]. DprE1 was known to be involved in the arabinogalactan biosynthesis, a key precursor required for the synthesis of the cell-wall [14,31], and it was demonstrated that its inhibition abolishes the formation of decaprenylphosphoryl arabinose (DPA), thus provoking cell lysis and bacterial death [30]. Later, new antimycobacterial compounds targeting DprE1 were identified by independent whole cell-based screens, like dinitrobenzamides (DNB) and bromoquinoxaline (VI-9376) [32,33], and afterward, numerous DprE1 inhibitors have come out since that groundbreaking report in 2009.

In this context, it has been demonstrated that DprE1 is essential for mycobacterial cell growth and survival [34,35]. Moreover, using *M. tuberculosis* conditional knockdown mutants, it has been shown that depletion of the DprE1 mRNA and protein level leads to a rapid cell death in vitro and ex vivo, suggesting that some enzymatic reactions are more sensitive to inhibition [35], providing the proof-of-principle of why it is so successful hitting DprE1 enzyme.

MmpL3 was firstly identified in our laboratory as the cellular target of BM212, the hit compound of a pyrrole-derivative class of antimycobacterial agents by screening the *M. smegmatis* cosmid library for BM212 resistance, as well as the selection and characterization of spontaneous resistant mutants [36]. MmpL3 is a transporter belonging to the mycobacterial membrane protein large (MmpL) family, specific for Actinobacteria. Since MmpL3 protein belongs to the resistance-nodulation-cell division (RND) permease superfamily, involved mainly in drug efflux, there was concern that the BM212 resistance could have been due to an enhanced efflux of the drug. Using efflux pump inhibitors, the authors provided evidence that mutations in *mmpL3* gene were instead responsible for resistance to BM212 per se and did not mediate resistance to BM212 via drug efflux [36]. Moreover, *M. smegmatis* spheroplasts assay showed that BM212 inhibits the MmpL3 transport activity, corroborating the evidence that MmpL3 was the BM212 cellular target [15].

Immediately after this first publication, MmpL3 was demonstrated to be the target of SQ109, an antitubercular in clinical trials, for which the mechanism of action was still unclear, and adamantyl urea AU1235, a compound displaying potent bactericidal activity against Mtb [37,38]. Genetic and biochemical studies indicated a clear effect of inhibiting MmpL3 on the translocation of trehalose monomycolate (TMM) in Mtb, thus abolishing the trehalose dimycolate (TDM) formation and mycolic acid transfer onto arabinogalactan [37,38]. Subsequently, several papers have reported the discovery of new MmpL3 inhibitors.

MmpL3 is essential for both mycobacterial growth and intracellular survival [16,39]. Strikingly, its broad promiscuity can be explained not only for its essentiality for tubercle bacilli, but also because the proton motive force needed to TMM transport across the membrane is dissipated by different MmpL3 inhibitors [16], a possible indirect mechanism.

The importance of these promiscuous targets can be exemplified by the numerous papers that report novel compound(s) able to exert antimycobacterial activity targeting these proteins, but the presence of drug candidates in the antitubercular pipeline better highlight the relevance of promiscuous targets. Indeed, among the new drug candidates, Macozinone (MCZ), BTZ-043 and TBA-7371 (or AZ 7371), present in Phase I or II, inhibit DprE1, while SQ109, hitting MmpL3, has completed Phase IIb study. Moreover, the Phase IIa of telacebec (Q203) has given positive results: telacebec has a high specificity for the cytochrome bc1 complex of Mtb, particularly for QcrB, another promiscuous target.

3. DprE1: The Hot TB Target of the Moment

3.1. Why DprE1 Is a Promiscuous Target

DprE1 (Decaprenylphosphoryl-β-D-ribofuranose 2-oxidoreductase, EC 1.1.98.3), is involved along with DprE2 (Decaprenylphosphoryl-2-keto-β-D-erythro-pentose reductase, EC 1.1.1.333), in the two-step epimerization of decaprenylphosphoribose (DPR) to decaprenylphosphoarabinose (DPA), a key arabinosyl donor essential for the biosynthesis of cell-wall arabinan polymers [40].

As a flavoprotein, after the first half-reaction in which DPR is oxidized into decaprenylphosphorylerythropentose (DPX), DprE1 needs to reoxidize the reduced flavin to begin a new catalytic cycle, thus, to complete this half-reaction the enzyme must use an electron acceptor. Initially, DprE1 was considered an oxidase, which uses molecular oxygen as an electron acceptor [41]. However, the enzyme was found to utilize more efficiently organic compounds, including bacterial membrane-embedded quinones, such as menaquinone, which was postulated as a physiological electron acceptor [41]. For this reason, the enzyme should be more precisely defined as an oxidoreductase.

In this context, DprE1 was found to have a very broad specificity for the electron acceptor, being able to use several compounds belonging to very different chemical classes. Indeed, the enzyme can reduce not only quinones, but also indophenols such as 2,6-dichlorophenolindophenol, or the phenoxazine moiety of resazurin, both compounds used in DprE1 activity assays [41,42], as well as different nitroaromatic moieties that characterize several covalent inhibitors of this enzyme [30,32,33,43,44].

To date, more than 15 chemical classes of compounds inhibiting DprE1 have been reported. These classes encompass both covalent and noncovalent inhibitors, and although the covalent ones are the most successful, several noncovalent inhibitors with significant antitubercular activity have actually been described [45,46].

The success of DprE1 covalent inhibitors in antitubercular drug discovery partly resides in this peculiar reactivity of the enzyme with the nitroaromatic moiety. Indeed, the most potent class of these inhibitors, the benzothiazinones, has been demonstrated to be suicide prodrugs that need activation to covalently bind DprE1 [47]. This activation consists of the reduction of the nitro group to nitroso, which rapidly forms a semimercaptal covalent adduct with a cysteine residue, located in a pocket just in front of the flavin cofactor. In this way, since the electron acceptor site is occupied, the enzyme is unable to reoxidize the flavin cofactor, resulting in being irreversibly inhibited [41,48] (Figure 1).

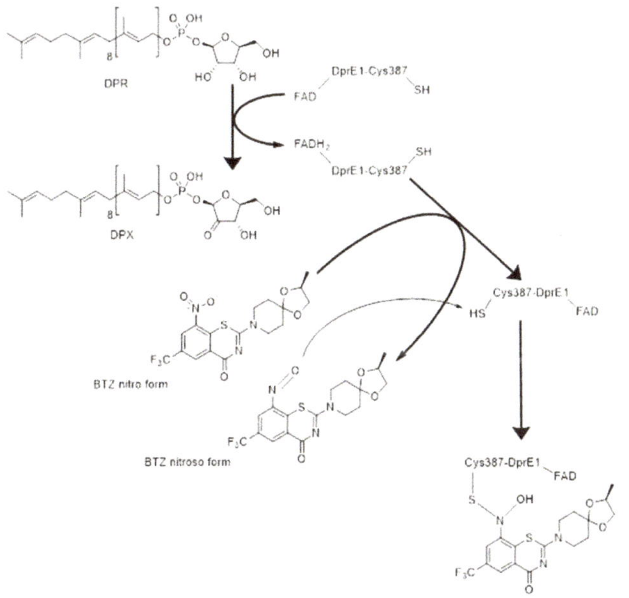

Figure 1. Mechanism of activation of the BTZ043 suicide inhibitor of DprE1.

This mechanism of action has been demonstrated for benzothiazinones and benzothiazinone analogs [41,43,49,50]. However, other different classes of nitro-compounds have been identified as potent DprE1 inhibitors, including dinitrobenzamides [32], nitroquinoxalines [33] and nitroimidazoles [51], that conceivably share the same mechanism of action of benzothiazinones [52].

Currently, 33 DprE1 structures from *M. tuberculosis* or *M. smegmatis* have been deposited in the Protein Data Bank, as apoenzyme [41,53] or in complex with both covalent [41,43,44,49,54] and noncovalent inhibitors [43,50,55–58]. The enzyme is characterized by the two-domain topology of the vanillyl-alcohol oxidase family of oxidoreductases, which includes a flavin adenine dinucleotide (FAD)-binding domain and the substrate-binding domain [41]. The isoalloxazine ring of FAD is located at the interface of the substrate-binding domain and the cofactor-binding domain, where are located in two disordered loops, which have been supposed to be involved in interactions with the DPR substrate, or with other proteins involved in the DPA biosynthesis [42].

In all structures of DprE1 in complex with inhibitors, all molecules showed binding sites that are significantly overlapped (Figure 2). This binding pocket, which conceivably constitutes the binding site of the physiological electron acceptor of DprE1, shows a very broad specificity, thus explaining the promiscuity of DprE1 for inhibitors.

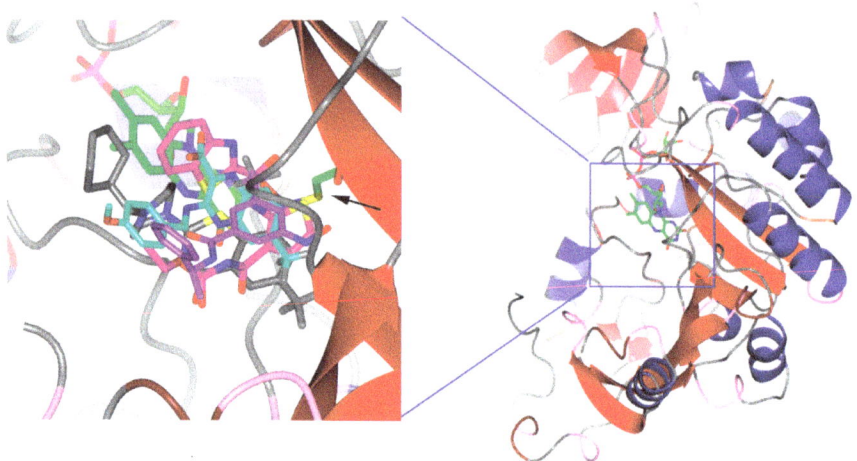

Figure 2. Crystal structure of *Mycobacterium tuberculosis* (Mtb) DprE1 and detail of the inhibitor binding sites. The superimposition of the structures of DprE1, cocrystallized with both covalent and noncovalent inhibitors (PBTZ-169, PDB: 4NCR [49] with carbon atoms in grey; Ty38c, PDB: 4P8K [58] with carbon atoms in cyan; TCA1, PDB: 4KWS [55] with carbon atoms in magenta; CT325, PDB: 4FF6 [43] with carbon atoms in violet), shows how the compounds have significantly overlapping binding sites, located in front of the isoalloxazine ring of the FAD cofactor (with carbon atoms in green). The black arrow points to the Cys387 side chain.

A broad specificity for inhibitors is a necessary but not sufficient condition for a target, to be considered promiscuous. Indeed, as promiscuous targets usually emerge during phenotypic screening, this implies that the inhibitors identified have antitubercular activity, so they can reach the cellular compartment in which the target resides. It is noteworthy that among the obstacles for the antitubercular drug discovery process, there is the difficulty of the compounds to cross the cell wall and the cellular membranes. Moreover, the presence of efflux pumps that can extrude the compounds outside the cells, and potential cytosolic inactivation processes of drugs lead to a lack of activity. In this context, it is noteworthy that one of the most successful pathways for antitubercular discovery is the cell wall compartment [59].

Recently, it was demonstrated that DPA biosynthesis partly occurs also outside the cytosolic membrane, and in particular, DprE1 was localized into the periplasm [60]. How DprE1 can reach the periplasmic compartment is still not clear, despite this protein, as well as its partner DprE2, do not possess any export signal for the translocation across the cytosolic membrane [60]. Anyhow, this localization allows drugs against DprE1 not only to easily reach the target, but also to escape several resistance mechanisms such as efflux pump or potential cytoplasmic inactivating enzymes.

3.2. DprE1 Inhibitors, Which Are the Current Status and Future Perspectives

Currently, at least 11 different scaffolds have been reported as effective DprE1 inhibitors, either covalent and noncovalent, showing different efficacy in vitro, ex vivo and in vivo [46]. It is noteworthy that the majority of these compounds have been identified through phenotypic screening, then subjected to optimization processes. The most representative compounds are reported in Table 1.

Table 1. Main covalent and not covalent DprE1 inhibitors showing antitubercular activity.

Chemical Class	Compound	Structure	*M. tuberculosis* MIC	Ref.
Covalent inhibitors				
Benzothiazinones	BTZ043		0.0023 µM	[30]
	PBTZ169		0.001 µM	[49]
Dinitrobenzamides	DNB1		0.69 µM	[32]
Nitro-quinoxalines	VI-9376		3.1 µM	[33]
Triazoles	377790		0.48 µM	[51]
Noncovalent inhibitors				
Benzothiazoles	TCA1		0.51 µM	[55]
	7a		0.08 µM	[61]
Thiadiazoles	GSK-710		4 µM	[42]
Carboxy-quinoxalines	Ty38c		3.1 µM	[58]
Aminoquinolones	3		0.8 µM	[62]

Table 1. Cont.

Chemical Class	Compound	Structure	M. tuberculosis MIC	Ref.
Pyrazolopyridones	19		0.1 μM	[63]
Azaindoles	TBA-7371		0.01 μM	[64]

The first DprE1 inhibitors identified is the BTZ043, belonging to the class of the benzothiazinones [30]. Moreover, nearly in the same period other classes of compounds, such as dinitrobenzamides, nitroquinoxaline and triazoles, have been described as covalent inhibitors [32,33,51].

As above reported, all these compounds possess a nitro group, that is converted to nitroso through the oxidation of the $FADH_2$ cofactor of DprE1, allowing the formation of a covalent bond with a cysteine residue. However, an alternative and similarly probable mechanism of formation of the nitroso group has been proposed. In this model, the thiol group of the cysteine is able to start the redox reaction which converts the nitro group into nitroso similarly to the von Richter reaction [65].

Despite BTZ043 was extremely potent against *M. tuberculosis* with an MIC of 1 ng/mL (2.3 nM) [30], its great efficacy was not translated into a mouse model, probably due to its relatively high lipophilicity, and low pharmacokinetic properties. Several attempts were then done to improve this class of compounds, and in particular, a structure activity relationship (SAR) study of piperazine benzothiazinone derivatives afforded the PBTZ-169 [49]. This compound, bearing a cyclohexylmethyl substituted piperazine showed improved in vitro and in vivo potency, and resulted in being less susceptible to the attack of nitroreductases [66]. PBTZ-169 is currently in a clinical trial under the name macozinone (MCZ) [29].

Although the nitro group is essential for the activity of the benzothiazinones [30,66], although potentially mutagenic, several attempts to substitute it, for instance by a pyrrole ring, or by an azide group were performed [67,68]. However, all these attempts led to noncovalent inhibitors, with much-reduced efficacy.

Also, noncovalent inhibitors, although less potent in vitro, can show in-vivo efficacy. For instance, the first discovered noncovalent inhibitor was the benzothiazole TCA1. This compound showed a relatively high MIC against *M. tuberculosis* of 0.5 μM, nevertheless, it proved to be effective in mice, both in acute and chronic infection models [55]. A series of pyrimidine conjugate of the benzothiazole give a further improvement of this compound, with the best derivatives showing MIC of 0.08–0.09 μM [61]. More recently a new class of inhibiting DprE1 benzothiazoles, benzothiazole-N-oxides, have been identified through a whole cells screening [44], but despite their promising bactericidal activity they showed toxicity issues.

The high vulnerability of DprE1 as an antitubercular drug target was confirmed by the number of screening campaigns that identified effective noncovalent inhibitors, including thiadiazoles [42], carboxy-quinoxalines [58], aminoquinolones [62], pyrazolopyridones [63], and azaindoles [69,70].

The latter are the most promising drug candidates among the noncovalent DprE1 inhibitors, the best inhibitor showing potent antitubercular activity in vitro and in in vivo chronic infection models, low toxicity and no antagonistic activity with other TB drugs [64]. Currently, the azaindole TBA-7371 has started phase I clinical trials [27], further confirming the great potential of DprE1 inhibitors in anti-tubercular drug discovery.

3.3. DprE2: A Promising Target for Future Drug Discovery

As above reported, the conversion of DPR to DPA is a two steps reaction, in which the product of DprE1 decaprenyl-phospho-2′-keto-D-arabinose must be reduced to finalize the epimerization of the substrate. This second step is catalyzed by the enzyme DprE2 (decaprenylphosphoryl-2-keto-β-D-erythro-pentose reductase) [14]. As DprE1, also DprE2 is essential for mycobacterial growth [35]. Moreover, the two enzymes have been supposed to strongly interact, and sometimes they were considered two subunits of a single enzyme named decaprenylphosphoryl-β-D-ribofuranose 2-epimerase. Indeed, DprE1 and DprE2 interaction in *Corynebacterium glutamicum*, strongly related to mycobacteria, have been demonstrated by a bacterial two-hybrid system [71]. An in silico study using homology modeling, protein threading and molecular dynamics proposed a model in which DprE1 and DprE2 form a complex [72], but currently, no structural and functional data on DprE2 have been reported yet. Anyhow, to fulfill their reactions it is conceivable that the two enzymes must work in concert, thus, reasonably a strong interaction occurs. In this context, the elucidation of the structural and functional behavior of the full decaprenylphosphoryl-β-D-ribofuranose 2-epimerase complex should be a useful tool for the development of further antitubercular inhibitors [73].

4. Mmpl3 Transporter: The Other Hot TB Target of the Moment

4.1. MmpL3: A Mycolic Acid Transporter

MmpL3 is a membrane protein, member of the resistance-nodulation-cell division (RND) superfamily of transporters. In mycobacteria, MmpL transporters have specialized in the export of several lipids and glycolipids across the plasma membrane to the cell surface, namely, trehalose monomycolates (TMM), di- and poly-acyltrehaloses, sulfolipids, phthiocerol dimycocerosates, monomycolyldiacylglycerol, glycopeptidolipids and mycobactins [38,74–78]. Genetic studies with transposon mutant libraries and the inability to knock-out the gene using different strategies suggested that MmpL3 was essential for the survival of *M. tuberculosis* [38,79]. An alternative strategy, utilizing knock-down strains both in vitro and in vivo had confirmed that MmpL3 is indeed essential for survival. Silencing of MmpL3 in mice, both during the acute or persistence phase of infection, led to a complete clearance of bacteria from lungs and spleens. These studies not only reinforce the idea of MmpL3 as an attractive drug target but also the potential of MmpL3 inhibitors to shorten TB treatment [80].

MmpL3 contains 12 transmembrane segments with two periplasmic loops and one C-terminal cytoplasmic domain. The periplasmic loops are essential for the MmpL3 function and are involved in substrate binding. On the other hand, the cytoplasmic loop, which is only present in MmpL3 and MmpL11 but not in other members of the MmpL family, is not essential for MmpL3 transport activity [81]. The exact role of the C-terminal domain is still not clear but it has been proposed to act as a signal for polar localization of the protein, since its removal caused MmpL3 to be more diffuse across the cells in contrast to the typical subpolar localization of the full protein [82]. In addition to that, this domain was shown to be involved in protein–protein interactions with itself as well as with other proteins, pointing to a role in oligomerization and/or coordination of synthesis of other cell wall components [83]. MmpL3 transports TMM across the cell membrane to the periplasmic space where it serves as a mycolic acid donor to another TMM molecule to form TDM, or to arabinogalactan as part of the mAGP core. Both of these reactions are catalyzed by the Ag85 family of enzymes. Mycolic acids are essential components of the mycobacterial cell wall and its biosynthesis has been shown as an effective drug target in mycobacteria, with several antibiotics targeting this pathway. The best example of this is isoniazid, which inhibits the FASII enoyl-ACP reductase InhA, and is one of the pillars in tuberculosis treatment [84].

MmpL3 relies on the proton motive force as an energy source to drive TMM transport. Thus, the flux of protons across the transmembrane region of MmpL3 would drive the conformational changes necessary for the translocation of TMM. This is evidenced by mutagenesis studies showing that charged residues lying on the transmembrane region are essential for MmpL3 function and are most likely involved in this proton relay pathway. The mechanism and extent of TMM transport mediated by MmpL3 is still not fully understood. Spheroplasts assays have suggested that MmpL3 acts as a flippase [15]. This conclusion was derived from the accumulation of TMM in the inner leaflet of the plasma membrane after MmpL3 inhibition. Although this would point to MmpL3 acting as a flippase, there is the possibility that MmpL3 inhibition would affect also other proteins acting upstream in the transport pathway and/or TMM biosynthetic enzymes. This holds true for other MmpL transporters where disruption of the transport leads to the accumulation of biosynthetic precursors and, in some cases, the complete abolition of the synthesis of the substrate [74,78,85,86].

The recently solved crystal structure of MmpL3 has shed some light as to the possible mechanism of TMM translocation [25,87]. Crystals of MmpL3 from *M. smegmatis* were diffracted to a 2.59 Å resolution. Surprisingly, the 3D structure of this transporter is different from other RND transporters like AcrB, MexB, CusA, MtrD, CmeB and HpnN. The structure shows that the N-terminal and C-terminal halves are assembled in a twofold pseudo symmetrical fashion and the two periplasmic domains (PD1 and PD2) interact with each other and are also flexible. MmpL3 possesses a cavity that extends from the outer leaflet of the inner membrane up to the periplasmic domain. This cavity is large enough to fit a TMM molecule. Fortuitously, MmpL3 co-purified with n-dodecyl-β-D-maltoside (the detergent used for solubilization) and phosphatidylethanolamine (PE) bound to it. The former was found in the hydrophobic pocket created by TMs 7–10 while the latter was bound to the periplasmic domain within the large space between PD1 and PD2. These results reinforce the idea of MmpL3 channeling its substrate from the outer leaflet of the inner membrane up to the periplasmic space. The authors further demonstrated that TMM binds to purified MmpL3 with a K_d of 3.7 ± 1.3 µM while there was no binding when TDM was assayed. In addition to PE and TMM, both phosphatidylglycerol and cardiolipin interacted with MmpL3. Whether these phospholipids act as substrates for MmpL3 in whole cells remains to be determined.

Native mass spectra, Small-angle X-ray scattering (SAXS) data and Blue-Native PAGE of the purified MmpL3 transporter indicate that it is a monomer in detergent solution [25,87]. It has to be noted that in both cases MmpL3 was purified devoid of the C-terminal domain due to its instability in solution and this domain was shown to be important for MmpL3 oligomerization [83]. This is in contrast to the quaternary structure of CmpL1, a *Corynebacterium glutamicum* ortholog, which has been shown to possess a trimeric quaternary structure [81]. There is a great variety of quaternary structures among the RND superfamily of transporters, while the efflux subfamily of RND transporters, namely, AcrB [88], MexB [89] and MtrD [90] exist as trimers, the hopanoid biosynthesis-associated transporter (HpnN) from *Burkholderia multivorans* is a dimer [91] and human NPC1 is a monomer [92]. Due to this high variability, we cannot rule out that MmpL3 has a higher oligomerization state in native conditions found in the mycobacterial cell membrane.

4.2. MmpL3: The Achilles Heel of Mycobacterium Tuberculosis

In recent years, a diversity of scaffolds have been reported to inhibit MmpL3 (Table 2): SQ109 (diamine), DA5 (SQ109 related compound), BM212 (diarylpyrrole), AU1235 (adamantyl urea), C215 (benzimidazole), NITD-349 (indolecarboxamides), THP P (tetrahydropyrazolo pyrimidine), Spiro (N-benzyl-6′,7′-dihydrospiro[piperidine-4,4′-thieno[3,2-c]pyrans]), PIPD1 (piperidinol), E11 (acetamide) and HC2091 (carboxamide) [36–38,51,93–97].

Table 2. Main MmpL3 inhibitors showing antitubercular activity.

Chemical Class	Compound	Structure	M. tuberculosis MIC	Ref.
Diarylpyrroles	BM212		3.6 μM	[36]
1,2-diamine derivatives	DA5		9.97 μM	[37]
	SQ109		0.78 μM	[37]
Adamantyl urea	AU1235		0.3 μM	[38]
Benzimidazoles	C215		16.0 μM	[51]
Indolcarboxamides	NIDT349		0.023 μM	[93]
Tetrahydropyrazolo pyrimidine	THP P		0.3 μM	[94]
Dihydrospiro[piperidine-4,4′-thieno[3,2-c]pyrans]	Spiro		0.3 μM	[94]
Piperidinols	PIPD1		1.28 μM	[95]
Acetamides	E11		8.0 μM	[96]
Carboxamides	HC2091		19.3 μM	[97]

Among these compounds, SQ109 is the most advanced in the TB drug pipeline and has completed Phase I clinical trials in the US and Phase II clinical trials in Africa. SQ109 was identified originally from a combinatorial library screening based on the 1,2-ethylenediamine structure of ethambutol. However, other structural dissimilarities between SQ109 and ethambutol, the significant activity of SQ109 against drug-resistant strains of *M. tuberculosis*, including those that were ethambutol-resistant, suggested that SQ109 had a different target than its parent molecule [98]. SQ109 showed bactericidal activity against *M. tuberculosis* in vitro, including MDR and XDR-TB. Furthermore, it had activity against intracellular bacilli as well as an extremely low spontaneous resistance rate [37]. In addition to that, it showed synergistic activity with isoniazid and rifampicin and additive effects with streptomycin in vitro [99].

SQ109 also increases the activity of the new TB drug bedaquiline in vitro [100]. Replacement of ethambutol by SQ109 in the standard regimen in mice led to a 1.5 n log decrease in colony-forming units (CFUs) [101]. Furthermore, SQ109 was safe and well-tolerated in Phase I and Phase IIa clinical trials [102].

MmpL3 seems to be a promiscuous target since a plethora of compounds with dissimilar scaffolds have been shown to affect its activity and every new library screening seems to lead to novel scaffolds targeting the same enzyme. However, the reason for this extreme susceptibility is not clear. One possibility is that the compounds, due to their high hydrophobicity, get vastly enriched in the membrane where they can bind readily to MmpL3. This, in addition to the extreme vulnerability of MmpL3, would explain the variety of compounds targeting MmpL3.

There has been some controversy as to the direct or indirect inhibition of these compounds against MmpL3, it has been suggested that some compounds may act indirectly on MmpL3 by disrupting the pH gradient and/or membrane potential which will disrupt the proton relay needed by MmpL3 to pump out TMM. Indeed, while SQ109 and BM212 display a broad spectrum of activity and are active against non-replicating Mtb bacilli, some others specifically target mycobacteria and do not kill nonreplicating bacilli [16,103,104]. In order to solve this controversy and to shed more light on the mechanism of action of these inhibitors, Li et al. have recently developed assays both in vitro and in whole cells to identify direct inhibitors of MmpL3. Their use of fluorescent competition assays and surface plasmon resonance experiments with MmpL3 purified protein has shown that SQ109, BM212, Au1235, NITD304, NITD349 and THPP1 all bind to MmpL3. The authors also showed that only SQ109 and BM212 dissipated both ΔpH and $\Delta\Psi$ (although the latter only at high concentrations). These assays could be a great tool for assessing the specificity of new and previously identified drugs that are believed to act against MmpL3 [105].

Recently, the crystal structure of *M. smegmatis* MmpL3 alone and in complex with four TB drug candidates has been solved [25]. Analysis of the transmembrane region has shown the existence of two pairs of hydrophilic residues (D256-Y646 and Y257-D645) on TM4 and TM 10 which links these two helices by forming hydrogen bonds. These residues are similar to the Asp-Asp-Lys triad found in AcrB and the Asp-Asp-Thr triad in SecDF, which are known to be involved in the proton relay pathway [106,107]. Consistently with this, three out of four of these residues have been shown to be essential for MmpL3 activity [81]. Furthermore, this catalytic tetrad is conserved in most MmpL transporters, with the exception of MmpL7, so it is likely that all MmpL transporters use the same mechanism to drive its substrate translocation [81].

The crystal structure of MmpL3 bound to SQ109, Au1235 and ICA38 has shown that these MmpL3 inhibitors bind to the same pocket in the center of the transmembrane region of the protein (Figure 3A), inducing conformational changes which in turn disrupt the interaction of the two Asp-Tyr pairs involved in proton translocation (Figure 3B,C). Furthermore, molecular docking assays with six other compounds (BM212, NITD-349, GSK2200150A, C215, PIPD1 and HC2091) showed that they can fit the same binding pocket within MmpL3. Thus, MmpL3 inhibitors seem to have an identical mechanism of action, binding to the same pocket and blocking the proton translocation relay pathway that is essential for substrate transport. This is also supported by the fact that the majority of resistance mutations found in MmpL3 lie in the TM region, close to the binding pocket. Zhang et al. purified 12 different mutant versions of MmpL3 and showed that most of the mutations affected the binding of the inhibitors to the pocket by disrupting hydrophobic/electrostatic interactions, inducing conformational changes or causing steric hindrance which leads to a decrease or abolition of the binding. On the other hand, there is still no clear evidence as to how mutations located further for the binding pocket induce resistance. Although MmpL3 is the only essential MmpL protein in *M. tuberculosis*, several other MmpL proteins have been shown to be required for virulence and/or drug efflux [108,109]. The fact that all inhibitors act by disrupting the proton relay pathway, a conserved feature of all MmpL transporters, opens the avenue for the development of molecules that target not only MmpL3 but all the MmpL family of transporters and thus affect virulence and resistance to other drugs.

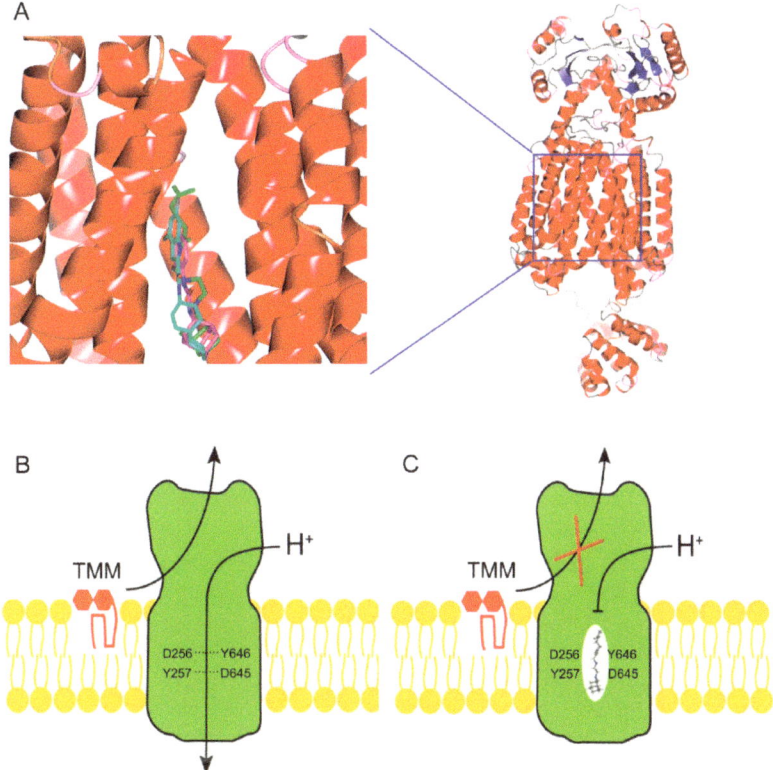

Figure 3. (**A**) Crystal structure of Mtb MmpL3 and detail of the inhibitors binding site (SQ109, PDB: 6AJG with carbon atoms in green; AU1235, PDB: 6AJH with carbon atoms in magenta; ICA38 PDB: 6AJJ with carbon atoms in cyan [25]). (**B**) Mechanism of transport of MmpL3: The flow of protons through charged residues in the transmembrane region of MmpL3 drives conformational changes that lead to the transport of TMM from the membrane to the periplasmic space. (**C**) MmpL3 inhibitors bind to a pocket in the transmembrane region of MmpL3, disrupting the electrostatic interactions between D256 and Y257 with Y646 and D645, respectively. This blocks the proton relay pathway and ultimately TMM translocation.

In summary, these findings will lead to a new era in MmpL3 inhibitor drug discovery since structure-guided molecules can now be designed with better anti-tubercular efficacy and pharmacokinetic/pharmacodynamic (PK/PD) properties.

5. Future Perspectives for DprE1 and MmpL3 Inhibitors in Clinical Therapy

Promiscuous targets could be an obstacle in the discovery of novel antitubercular drug targets, being recurrently found in HTS [110], nevertheless, they have been proven really useful for the development of novel drug candidates. This is well demonstrated by the fact that five compounds currently in clinical trials inhibit these targets, with DprE1 inhibitors TBA-7379, BTZ043 and Macozinone in Phase I, while the QcrB inhibitor Telacebec, the MmpL3 inhibitor SQ109 and again the Macozinone are in Phase II [26–29].

In particular, Macozinone has completed a dose-escalation phase I study in healthy male volunteers and a multiple ascending dose study in Russia, which demonstrated the good safety profile of the compound. Moreover, in 2018 it completed a Phase IIa Early Bactericidal Activity (EBA) study in

Russia and Belarus in drug sensitive-TB (DS-TB) patients, confirming the safety and efficacy of the drug. In parallel, a Phase I clinical study has been initiated in Switzerland, by the nonprofit Innovative Medicines for Tuberculosis (iM4TB) foundation [24].

The MmpL3 inhibitor SQ109 has completed a Phase 2b study in Russia which demonstrated its efficacy in MDR-TB patients, with a good tolerability profile [111]. Furthermore, SQ109 was demonstrated to enhance the activity of isoniazid, rifampicin and bedaquiline and shortened clearance of TB in a mice model. The drug now has also completed three Phase I studies in the USA and two Phase II studies in Africa in DS-TB patients.

These studies thus demonstrate the great potential of these targets for the development of a new drug to reach the objective to eradicate TB.

Author Contributions: Conceptualization, G.R. and L.R.C.; writing, G.D., J.M.B. and L.R.C., review and editing, M.R.P., E.D.R., G.R. and L.R.C.; visualization, J.M.B. and L.R.C.; project administration, G.R. All authors have read and agreed to the published version of the manuscript.

Funding: This research was funded by the Italian Ministry of Education, University and Research (MIUR): Dipartimenti di Eccellenza Program (2018–2022)—Dept. of Biology and Biotechnology "L. Spallanzani", University of Pavia (G.D., M.R.P., E.D.R., G.R. and L.R.C.).

Conflicts of Interest: The authors declare no conflict of interest.

References

1. World Health Organization. Global Tuberculosis Report 2019. Available online: https://www.who.int/tb/publications/global_report/en/ (accessed on 7 January 2020).
2. Lechartier, B.; Rybniker, J.; Zumla, A.; Cole, S.T. Tuberculosis drug discovery in the post-post-genomic era. *EMBO Mol. Med.* **2014**, *6*, 158–168. [CrossRef]
3. Chiarelli, L.R.; Mori, G.; Esposito, M.; Orena, B.S.; Pasca, M.R. New and old hot drug targets in tuberculosis. *Curr. Med. Chem.* **2016**, *23*, 3813–3846. [CrossRef]
4. Meneghetti, F.; Villa, S.; Gelain, A.; Barlocco, D.; Chiarelli, L.R.; Pasca, M.R.; Costantino, L. Iron acquisition pathways as targets for antitubercular drugs. *Curr. Med. Chem.* **2016**, *23*, 4009–4026. [CrossRef]
5. Mori, M.; Sammartino, J.C.; Costantino, L.; Gelain, A.; Meneghetti, F.; Villa, S.; Chiarelli, L.R. An Overview on the Potential Antimycobacterial Agents Targeting Serine/Threonine Protein Kinases from *Mycobacterium tuberculosis*. *Curr. Top. Med. Chem.* **2019**, *19*, 646–661. [CrossRef] [PubMed]
6. Andries, K.; Verhasselt, P.; Guillemont, J.; Göhlmann, H.W.; Neefs, J.M.; Winkler, H.; Van Gestel, J.; Timmerman, P.; Zhu, M.; Lee, E.; et al. A diarylquinoline drug active on the ATP synthase of *Mycobacterium tuberculosis*. *Science* **2005**, *307*, 223–227. [CrossRef] [PubMed]
7. Matsumoto, M.; Hashizume, H.; Tomishige, T.; Kawasaki, M.; Tsubouchi, H.; Sasaki, H.; Shimokawa, Y.; Komatsu, M. OPC-67683, a nitro-dihydro-imidazooxazole derivative with promising action against tuberculosis in vitro and in mice. *PLoS Med.* **2006**, *3*, e466. [CrossRef] [PubMed]
8. Stover, C.K.; Warrener, P.; VanDevanter, D.R.; Sherman, D.R.; Arain, T.M.; Langhorne, M.H.; Anderson, S.W.; Towell, J.A.; Yuan, Y.; McMurray, D.N.; et al. A small-molecule nitroimidazopyran drug candidate for the treatment of tuberculosis. *Nature* **2000**, *405*, 962–966. [CrossRef] [PubMed]
9. Ryan, N.J.; Lo, J.H. Delamanid: First global approval. *Drugs* **2014**, *74*, 1041–1045. [CrossRef] [PubMed]
10. Keam, S.J. Pretomanid: First Approval. *Drugs* **2019**, *79*, 1797–1803. [CrossRef]
11. Cole, S.T. Inhibiting Mycobacterium tuberculosis within and without. *Philos. Trans. R. Soc. Lond. B Biol. Sci.* **2016**, *371*, 20150506. [CrossRef]
12. Lee, B.S.; Pethe, K. Therapeutic potential of promiscuous targets in *Mycobacterium tuberculosis*. *Curr. Opin. Pharmacol.* **2018**, *42*, 22–26. [CrossRef] [PubMed]
13. Campaniço, A.; Moreira, R.; Lopes, F. Drug discovery in tuberculosis. New drug targets and antimycobacterial agents. *Eur. J. Med. Chem.* **2018**, *150*, 525–545. [CrossRef] [PubMed]
14. Mikušová, K.; Huang, H.; Yagi, T.; Holsters, M.; Vereecke, D.; D'Haeze, W.; Scherman, M.S.; Brennan, P.J.; McNeil, M.R.; Crick, D.C. Decaprenylphosphoryl arabinofuranose, the donor of the D-arabinofuranosyl residues of mycobacterial arabinan, is formed via a two-step epimerization of decaprenylphosphoryl ribose. *J. Bacteriol.* **2005**, *187*, 8020–8025. [CrossRef] [PubMed]

15. Xu, Z.; Meshcheryakov, V.A.; Poce, G.; Chng, S.S. MmpL3 is the flippase for mycolic acids in mycobacteria. *Proc. Natl. Acad. Sci. USA* **2017**, *114*, 7993–7998. [CrossRef] [PubMed]
16. Li, W.; Upadhyay, A.; Fontes, F.L.; North, E.J.; Wang, Y.; Crans, D.C.; Grzegorzewicz, A.E.; Jones, V.; Franzblau, S.G.; Lee, R.E.; et al. Novel insights into the mechanism of inhibition of MmpL3, a target of multiple pharmacophores in *Mycobacterium tuberculosis*. *Antimicrob. Agents Chemother.* **2014**, *58*, 6413–6423. [CrossRef] [PubMed]
17. Abrahams, K.A.; Cox, J.A.; Spivey, V.L.; Loman, N.J.; Pallen, M.J.; Constantinidou, C.; Fernández, R.; Alemparte, C.; Remuiñán, M.J.; Barros, D.; et al. Identification of novel imidazo[1,2-a]pyridine inhibitors targeting *M. tuberculosis* QcrB. *PLoS ONE* **2012**, *7*, e52951. [CrossRef]
18. Pethe, K.; Bifani, P.; Jang, J.; Kang, S.; Park, S.; Ahn, S.; Jiricek, J.; Jung, J.; Jeon, H.K.; Cechetto, J.; et al. Discovery of Q203, a potent clinical candidate for the treatment of tuberculosis. *Nat. Med.* **2013**, *19*, 1157–1160. [CrossRef]
19. Wilson, R.; Kumar, P.; Parashar, V.; Vilchèze, C.; Veyron-Churlet, R.; Freundlich, J.S.; Barnes, S.W.; Walker, J.R.; Szymonifka, M.J.; Marchiano, E.; et al. Antituberculosis thiophenes define a requirement for Pks13 in mycolic acid biosynthesis. *Nat. Chem. Biol.* **2013**, *9*, 499–506. [CrossRef]
20. Aggarwal, A.; Parai, M.K.; Shetty, N.; Wallis, D.; Woolhiser, L.; Hastings, C.; Dutta, N.K.; Galaviz, S.; Dhakal, R.C.; Shrestha, R.; et al. Development of a Novel Lead that Targets *M. tuberculosis* Polyketide Synthase 13. *Cell* **2017**, *170*, 249–259. [CrossRef]
21. Grant, S.S.; Kawate, T.; Nag, P.P.; Silvis, M.R.; Gordon, K.; Stanley, S.A.; Kazyanskaya, E.; Nietupski, R.; Golas, A.; Fitzgerald, M.; et al. Identification of novel inhibitors of nonreplicating *Mycobacterium tuberculosis* using a carbon starvation model. *ACS Chem. Biol.* **2013**, *8*, 2224–2234. [CrossRef]
22. Bonnett, S.A.; Dennison, D.; Files, M.; Bajpai, A.; Parish, T. A class of hydrazones are active against non-replicating *Mycobacterium tuberculosis*. *PLoS ONE* **2018**, *13*, e0198059. [CrossRef] [PubMed]
23. Boldrin, F.; Degiacomi, G.; Serafini, A.; Kolly, G.S.; Ventura, M.; Sala, C.; Provvedi, R.; Palù, G.; Cole, S.T.; Manganelli, R. Promoter mutagenesis for fine-tuning expression of essential genes in *Mycobacterium tuberculosis*. *Microb. Biotechnol.* **2018**, *11*, 238–247. [CrossRef] [PubMed]
24. Lupien, A.; Vocat, A.; Foo, C.S.; Blattes, E.; Gillon, J.Y.; Makarov, V.; Cole, S.T. Optimized background regimen for treatment of active tuberculosis with the next-generation benzothiazinone Macozinone (PBTZ169). *Antimicrob. Agents Chemother.* **2018**, *62*, e00840-18. [CrossRef] [PubMed]
25. Zhang, B.; Li, J.; Yang, X.; Wu, L.; Zhang, J.; Yang, Y.; Zhao, Y.; Zhang, L.; Cheng, X.; Liu, Z.; et al. Crystal structures of membrane transporter MmpL3, an anti-TB drug target. *Cell* **2019**, *176*, 636–648. [CrossRef] [PubMed]
26. Available online: https://www.newtbdrugs.org/pipeline/compound/sq109 (accessed on 4 December 2019).
27. Available online: https://www.newtbdrugs.org/pipeline/compound/tba-7371 (accessed on 4 December 2019).
28. Available online: https://www.newtbdrugs.org/pipeline/trials/phase-2-telacebec-q203-eba (accessed on 4 December 2019).
29. Available online: https://www.newtbdrugs.org/pipeline/compound/macozinone-mcz-pbtz-169 (accessed on 4 December 2019).
30. Makarov, V.; Manina, G.; Mikusova, K.; Möllmann, U.; Ryabova, O.; Saint-Joanis, B.; Dhar, N.; Pasca, M.R.; Buroni, S.; Lucarelli, A.P.; et al. Benzothiazinones kill *Mycobacterium tuberculosis* by blocking arabinan synthesis. *Science* **2009**, *324*, 801–804. [CrossRef] [PubMed]
31. Wolucka, B.A. Biosynthesis of D-arabinose in mycobacteria—A novel bacterial pathway with implications for antimycobacterial therapy. *FEBS J.* **2008**, *275*, 2691–2711. [CrossRef]
32. Christophe, T.; Jackson, M.; Jeon, H.K.; Fenistein, D.; Contreras-Dominguez, M.; Kim, J.; Genovesio, A.; Carralot, J.P.; Ewann, F.; Kim, E.H.; et al. High content screening identifies decaprenyl-phosphoribose 2′ epimerase as a target for intracellular antimycobacterial inhibitors. *PLoS Pathog.* **2009**, *5*, e1000645. [CrossRef]
33. Magnet, S.; Hartkoorn, R.C.; Székely, R.; Pató, J.; Triccas, J.A.; Schneider, P.; Szántai-Kis, C.; Orfi, L.; Chambon, M.; Banfi, D.; et al. Leads for antitubercular compounds from kinase inhibitor library screens. *Tuberculosis* **2010**, *90*, 354–360. [CrossRef]
34. Matsoso, L.G.; Kana, B.D.; Crellin, P.K.; Lea-Smith, D.J.; Pelosi, A.; Powell, D.; Dawes, S.S.; Rubin, H.; Coppel, R.L.; Mizrahi, V. Function of the cytochrome bc1-aa3 branch of the respiratory network in mycobacteria and network adaptation occurring in response to its disruption. *J. Bacteriol.* **2005**, *187*, 6300–6308. [CrossRef]

35. Kolly, G.S.; Boldrin, F.; Sala, C.; Dhar, N.; Hartkoorn, R.C.; Ventura, M.; Serafini, A.; McKinney, J.D.; Manganelli, R.; Cole, S.T. Assessing the essentiality of the decaprenyl-phospho-d-arabinofuranose pathway in *Mycobacterium tuberculosis* using conditional mutants. *Mol. Microbiol.* **2014**, *92*, 194–211. [CrossRef]
36. La Rosa, V.; Poce, G.; Canseco, J.O.; Buroni, S.; Pasca, M.R.; Biava, M.; Raju, R.M.; Porretta, G.C.; Alfonso, S.; Battilocchio, C.; et al. MmpL3 is the cellular target of the antitubercular pyrrole derivative BM212. *Antimicrob. Agents Chemother.* **2012**, *56*, 324–331. [CrossRef] [PubMed]
37. Tahlan, K.; Wilson, R.; Kastrinsky, D.B.; Arora, K.; Nair, V.; Fischer, E.; Barnes, S.W.; Walker, J.R.; Alland, D.; Barry, C.E.; et al. SQ109 targets MmpL3, a membrane transporter of trehalose monomycolate involved in mycolic acid donation to the cell wall core of *Mycobacterium tuberculosis*. *Antimicrob. Agents Chemother.* **2012**, *56*, 1797–1809. [CrossRef] [PubMed]
38. Grzegorzewicz, A.E.; Pham, H.; Gundi, V.A.; Scherman, M.S.; North, E.J.; Hess, T.; Jones, V.; Gruppo, V.; Born, S.E.; Korduláková, J.; et al. Inhibition of mycolic acid transport across the Mycobacterium tuberculosis plasma membrane. *Nat. Chem. Biol.* **2012**, *8*, 334–341. [CrossRef]
39. Degiacomi, G.; Benjak, A.; Madacki, J.; Boldrin, F.; Provvedi, R.; Palù, G.; Kordulakova, J.; Cole, S.T.; Manganelli, R. Essentiality of mmpL3 and impact of its silencing on *Mycobacterium tuberculosis* gene expression. *Sci. Rep.* **2017**, *7*, 43495. [CrossRef] [PubMed]
40. Riccardi, G.; Pasca, M.R.; Chiarelli, L.R.; Manina, G.; Mattevi, A.; Binda, C. The DprE1 enzyme, one of the most vulnerable targets of *Mycobacterium tuberculosis*. *Appl. Microbiol. Biotechnol.* **2013**, *97*, 8841–8848. [CrossRef]
41. Neres, J.; Pojer, F.; Molteni, E.; Chiarelli, L.R.; Dhar, N.; Boy-Rottger, S.; Buroni, S.; Fullam, E.; Degiacomi, G.; Lucarelli, A.P.; et al. Structural basis for benzothiazinone-mediated killing of *Mycobacterium tuberculosis*. *Sci. Trans. Med.* **2012**, *4*, 150ra121. [CrossRef]
42. Batt, S.M.; Cacho Izquierdo, M.; Castro Pichel, J.; Stubbs, C.J.; Del Peral, L.V.-G.; Pérez-Herrán, E.; Dhar, N.; Mouzon, B.; Rees, M.; Hutchinson, J.P.; et al. Whole cell target engagement identifies novel inhibitors of *Mycobacterium tuberculosis* decaprenylphosphoryl-β-D-ribose oxidase. *ACS Infect. Dis* **2015**, *1*, 615–626. [CrossRef]
43. Batt, S.M.; Jabeen, T.; Bhowruth, V.; Quill, L.; Lund, P.A.; Eggeling, L.; Alderwick, L.J.; Fütterer, K.; Besra, G.S. Structural basis of inhibition of *Mycobacterium tuberculosis* DprE1 by benzothiazinone inhibitors. *Proc. Natl. Acad. Sci. USA* **2012**, *109*, 11354–11359. [CrossRef]
44. Landge, S.; Mullick, A.B.; Nagalapur, K.; Neres, J.; Subbulakshmi, V.; Murugan, K.; Ghosh, A.; Sadler, C.; Fellows, M.D.; Humnabadkar, V.; et al. Discovery of benzothiazoles as antimycobacterial agents: Synthesis, structure-activity relationships and binding studies with *Mycobacterium tuberculosis* decaprenylphosphoryl-β-D-ribose 2′-oxidase. *Bioorg. Med. Chem.* **2015**, *23*, 7694–7710. [CrossRef]
45. Piton, J.; Foo, C.S.; Cole, S.T. Structural studies of *Mycobacterium tuberculosis* DprE1 interacting with its inhibitors. *Drug Discov. Today* **2017**, *22*, 526–533. [CrossRef]
46. Chikhale, R.V.; Barmade, M.A.; Murumkar, P.R.; Yadav, M.R. Overview of the development of dpre1 inhibitors for combating the menace of tuberculosis. *J. Med. Chem.* **2018**, *61*, 8563–8593. [CrossRef] [PubMed]
47. Mori, G.; Chiarelli, L.R.; Riccardi, G.; Pasca, M.R. New prodrugs against tuberculosis. *Drug Discov. Today* **2017**, *22*, 519–525. [CrossRef] [PubMed]
48. Trefzer, C.; Škovierová, H.; Buroni, S.; Bobovská, A.; Nenci, S.; Molteni, E.; Pojer, F.; Pasca, M.R.; Makarov, V.; Cole, S.T.; et al. Benzothiazinones are suicide inhibitors of mycobacterial decaprenylphosphoryl-β-D-ribofuranose 2′-oxidase DprE1. *J. Am. Chem. Soc.* **2012**, *134*, 912–915. [CrossRef] [PubMed]
49. Makarov, V.; Lechartier, B.; Zhang, M.; Neres, J.; van der Sar, A.M.; Raadsen, S.A.; Hartkoorn, R.C.; Ryabova, O.B.; Vocat, A.; Decosterd, L.A.; et al. Towards a new combination therapy for tuberculosis with next generation benzothiazinones. *EMBO Mol. Med.* **2014**, *6*, 372–383. [CrossRef]
50. Richter, A.; Rudolph, I.; Möllmann, U.; Voigt, K.; Chung, C.W.; Singh, O.M.P.; Rees, M.; Mendoza-Losana, A.; Bates, R.; Ballell, L.; et al. Novel insight into the reaction of nitro, nitroso and hydroxylamino benzothiazinones and of benzoxacinones with *Mycobacterium tuberculosis* DprE1. *Sci. Rep.* **2018**, *8*, 13473. [CrossRef]
51. Stanley, S.A.; Grant, S.S.; Kawate, T.; Iwase, N.; Shimizu, M.; Wivagg, C.; Silvis, M.; Kazyanskaya, E.; Aquadro, J.; Golas, A.; et al. Identification of novel inhibitors of *M. tuberculosis* growth using whole cell based high-throughput screening. *ACS Chem. Biol.* **2012**, *7*, 1377–1384. [CrossRef]

52. De Jesus Lopes Ribeiro, A.L.; Degiacomi, G.; Ewann, F.; Buroni, S.; Incandela, M.L.; Chiarelli, L.R.; Mori, G.; Kim, J.; Contreras-Dominguez, M.; Park, Y.-S.; et al. Analogous mechanisms of resistance to benzothiazinones and dinitrobenzamides in *Mycobacterium smegmatis*. *PLoS ONE* **2011**, *6*, e26675. [CrossRef]
53. Li, H.; Jogl, G. Crystal structure of decaprenylphosphoryl-β-D-ribose 2′-epimerase from *Mycobacterium smegmatis*. *Proteins* **2013**, *81*, 538–543. [CrossRef]
54. Landge, S.; Ramachandran, V.; Kumar, A.; Neres, J.; Murugan, K.; Sadler, C.; Fellows, M.D.; Humnabadkar, V.; Vachaspati, P.; Raichurkar, A.; et al. Nitroarenes as antitubercular agents: Stereoelectronic modulation to mitigate mutagenicity. *ChemMedChem* **2016**, *11*, 331–339. [CrossRef]
55. Wang, F.; Sambandan, D.; Halder, R.; Wang, J.; Batt, S.M.; Weinrick, B.; Ahmad, I.; Yang, P.; Zhang, Y.; Kim, J.; et al. Identification of a small molecule with activity against drug-resistant and persistent tuberculosis. *Proc. Natl. Acad. Sci. USA* **2013**, *110*, E2510–E2517. [CrossRef]
56. Liu, R.; Lyu, X.; Batt, S.M.; Hsu, M.H.; Harbut, M.B.; Vilchèze, C.; Cheng, B.; Ajayi, K.; Yang, B.; Yang, Y.; et al. Determinants of the inhibition of DprE1 and CYP2C9 by antitubercular thiophenes. *Angew. Chem. Int. Ed. Engl.* **2017**, *56*, 13011–13015. [CrossRef] [PubMed]
57. Piton, J.; Vocat, A.; Lupien, A.; Foo, C.S.; Riabova, O.; Makarov, V.; Cole, S.T. Structure-based drug design and characterization of sulfonyl-piperazine benzothiazinone inhibitors of DprE1 from *Mycobacterium tuberculosis*. *Antimicrob. Agents Chemother.* **2018**, *62*, e00681-18. [CrossRef] [PubMed]
58. Neres, J.; Hartkoorn, R.C.; Chiarelli, L.R.; Gadupudi, R.; Pasca, M.R.; Mori, G.; Venturelli, A.; Savina, S.; Makarov, V.; Kolly, G.S.; et al. 2-carboxyquinoxalines kill *Mycobacterium tuberculosis* through noncovalent inhibition of DprE1. *ACS Chem. Biol.* **2015**, *10*, 705–714. [CrossRef] [PubMed]
59. Abrahams, K.A.; Besra, G.S. Mycobacterial cell wall biosynthesis: A multifaceted antibiotic target. *Parasitology* **2018**, *145*, 116–133. [CrossRef] [PubMed]
60. Brecik, M.; Centárová, I.; Mukherjee, R.; Kolly, G.S.; Huszár, S.; Bobovská, A.; Kilacsková, E.; Mokošová, V.; Svetlíková, Z.; Šarkan, M.; et al. DprE1 is a vulnerable tuberculosis drug target due to its cell wall localization. *ACS Chem. Biol.* **2015**, *10*, 1631–1636. [CrossRef] [PubMed]
61. Chikhale, R.; Menghani, S.; Babu, R.; Bansode, R.; Bhargavi, G.; Karodia, N.; Rajasekharan, M.V.; Paradkar, A.; Khedekar, P. Development of selective DprE1 inhibitors: Design, synthesis, crystal structure and antitubercular activity of benzothiazolylpyrimidine-5-carboxamides. *Eur. J. Med. Chem.* **2015**, *96*, 30–46. [CrossRef]
62. Naik, M.; Humnabadkar, V.; Tantry, S.J.; Panda, M.; Narayan, A.; Guptha, S.; Panduga, V.; Manjrekar, P.; Jena, L.K.; Koushik, K.; et al. 4-Aminoquinolone piperidine amides: Noncovalent inhibitors of DprE1 with long residence time and potent antimycobacterial activity. *J. Med. Chem.* **2014**, *57*, 5419–5434. [CrossRef]
63. Panda, M.; Ramachandran, S.; Ramachandran, V.; Shirude, P.S.; Humnabadkar, V.; Nagalapur, K.; Sharma, S.; Kaur, P.; Guptha, S.; Narayan, A.; et al. Discovery of pyrazolopyridones as a novel class of noncovalent DprE1 inhibitor with potent anti-mycobacterial activity. *J. Med. Chem.* **2014**, *57*, 4761–4771. [CrossRef]
64. Chatterji, M.; Shandil, R.; Manjunatha, M.R.; Solapure, S.; Ramachandran, V.; Kumar, N.; Saralaya, R.; Panduga, V.; Reddy, J.; Prabhakar, K.R.; et al. 1,4-azaindole, a potential drug candidate for treatment of tuberculosis. *Antimicrob. Agents Chemother.* **2014**, *58*, 5325–5331. [CrossRef]
65. Tiwari, R.; Moraski, G.C.; Krchňák, V.; Miller, P.A.; Colon-Martinez, M.; Herrero, E.; Oliver, A.G.; Miller, M.J. Thiolates chemically induce redox activation of BTZ043 and related potent nitroaromatic anti-tuberculosis agents. *J. Am. Chem. Soc.* **2013**, *135*, 3539–3549. [CrossRef]
66. Manina, G.; Bellinzoni, M.; Pasca, M.R.; Neres, J.; Milano, A.; De Jesus Lopes Ribeiro, A.L.; Buroni, S.; Skovierová, H.; Dianišková, P.; Mikušová, K.; et al. Biological and structural characterization of the *Mycobacterium smegmatis* nitroreductase NfnB, and its role in benzothiazinone resistance. *Mol. Microbiol.* **2010**, *77*, 1172–1185. [CrossRef] [PubMed]
67. Makarov, V.; Neres, J.; Hartkoorn, R.C.; Ryabova, O.B.; Kazakova, E.; Šarkan, M.; Huszár, S.; Piton, J.; Kolly, G.S.; Vocat, A.; et al. The 8-Pyrrole-Benzothiazinones Are Noncovalent Inhibitors of DprE1 from Mycobacterium tuberculosis. *Antimicrob. Agents Chemother.* **2015**, *59*, 4446–4452. [CrossRef] [PubMed]
68. Tiwari, R.; Miller, P.A.; Chiarelli, L.R.; Mori, G.; Šarkan, M.; Centárová, I.; Cho, S.; Mikusová, K.; Franzblau, S.G.; Oliver, A.G.; et al. Design, syntheses, and anti-tb activity of 1,3-benzothiazinone azide and click chemistry products inspired by BTZ043. *ACS Med. Chem. Lett.* **2016**, *7*, 266–270. [CrossRef] [PubMed]

69. Shirude, P.S.; Shandil, R.; Sadler, C.; Naik, M.; Hosagrahara, V.; Hameed, S.; Shinde, V.; Bathula, C.; Humnabadkar, V.; Kumar, N.; et al. Azaindoles: Noncovalent DprE1 inhibitors from scaffold morphing efforts, kill *Mycobacterium tuberculosis* and are efficacious in vivo. *J. Med. Chem.* **2013**, *56*, 9701–9708. [CrossRef]
70. Shirude, P.S.; Shandil, R.K.; Manjunatha, M.R.; Sadler, C.; Panda, M.; Panduga, V.; Reddy, J.; Saralaya, R.; Nanduri, R.; Ambady, A.; et al. Lead optimization of 1,4-azaindoles as antimycobacterial agents. *J. Med. Chem.* **2014**, *57*, 5728–5737. [CrossRef]
71. Jankute, M.; Byng, C.V.; Alderwick, L.J.; Besra, G.S. Elucidation of a protein-protein interaction network involved in *Corynebacterium glutamicum* cell wall biosynthesis as determined by bacterial two-hybrid analysis. *Glycoconj. J.* **2014**, *31*, 475–483. [CrossRef]
72. Bhutani, I.; Loharch, S.; Gupta, P.; Madathil, R.; Parkesh, R. Structure, dynamics, and interaction of *Mycobacterium tuberculosis* (Mtb) DprE1 and DprE2 examined by molecular modeling, simulation, and electrostatic studies. *PLoS ONE* **2015**, *10*, e0119771. [CrossRef]
73. Gawad, J.; Bonde, C. Decaprenyl-phosphoryl-ribose 2'-epimerase (DprE1): Challenging target for antitubercular drug discovery. *Chem. Cent. J.* **2018**, *12*, 72. [CrossRef]
74. Camacho, L.R.; Constant, P.; Raynaud, C.; Laneelle, M.A.; Triccas, J.A.; Gicquel, B.; Daffe, M.; Guilhot, C. Analysis of the phthiocerol dimycocerosate locus of *Mycobacterium tuberculosis*. Evidence that this lipid is involved in the cell wall permeability barrier. *J. Biol. Chem.* **2001**, *276*, 19845–19854. [CrossRef]
75. Converse, S.E.; Mougous, J.D.; Leavell, M.D.; Leary, J.A.; Bertozzi, C.R.; Cox, J.S. MmpL8 is required for sulfolipid-1 biosynthesis and *Mycobacterium tuberculosis* virulence. *Proc. Natl. Acad. Sci. USA* **2003**, *100*, 6121–6126. [CrossRef]
76. Pacheco, S.A.; Hsu, F.F.; Powers, K.M.; Purdy, G.E. MmpL11 protein transports mycolic acid-containing lipids to the mycobacterial cell wall and contributes to biofilm formation in *Mycobacterium smegmatis*. *J. Biol. Chem.* **2013**, *288*, 24213–24222. [CrossRef] [PubMed]
77. Wells, R.M.; Jones, C.M.; Xi, Z.; Speer, A.; Danilchanka, O.; Doornbos, K.S.; Sun, P.; Wu, F.; Tian, C.; Niederweis, M. Discovery of a siderophore export system essential for virulence of *Mycobacterium tuberculosis*. *PLoS Pathog.* **2013**, *9*, e1003120. [CrossRef] [PubMed]
78. Belardinelli, J.M.; Larrouy-Maumus, G.; Jones, V.; de Carvalho, L.P.S.; McNeil, M.R.; Jackson, M. Biosynthesis and translocation of unsulfated acyltrehaloses in *Mycobacterium tuberculosis*. *J. Biol. Chem.* **2014**, *289*, 27952–27965. [CrossRef] [PubMed]
79. Domenech, P.; Reed, M.B.; Barry, C.E. Contribution of the *Mycobacterium tuberculosis* MmpL protein family to virulence and drug resistance. *Infect. Immun.* **2005**, *73*, 3492–3501. [CrossRef]
80. Li, W.; Obregón-Henao, A.; Wallach, J.B.; North, E.J.; Lee, R.E.; Gonzalez-Juarrero, M.; Schnappinger, D.; Jackson, M. Therapeutic potential of the *Mycobacterium tuberculosis* mycolic acid transporter, MmpL3. *Antimicrob. Agents Chemother.* **2016**, *60*, 5198–5207. [CrossRef]
81. Belardinelli, J.M.; Yazidi, A.; Yang, L.; Fabre, L.; Li, W.; Jacques, B.; Angala, S.K.; Rouiller, I.; Zgurskaya, H.I.; Sygusch, J.; et al. Structure-function profile of MmpL3, the essential mycolic acid transporter from *Mycobacterium tuberculosis*. *ACS Infect. Dis.* **2016**, *2*, 702–713. [CrossRef]
82. Carel, C.; Nukdee, K.; Cantaloube, S.; Bonne, M.; Diagne, C.T.; Laval, F.; Daffé, M.; Zerbib, D. Mycobacterium tuberculosis proteins involved in mycolic acid synthesis and transport localize dynamically to the old growing pole and septum. *PLoS ONE* **2014**, *9*, e97148. [CrossRef]
83. Belardinelli, J.M.; Stevens, C.M.; Li, W.; Tan, Y.Z.; Jones, V.; Mancia, F.; Zgurskaya, H.I.; Jackson, M. The MmpL3 interactome reveals a complex crosstalk between cell envelope biosynthesis and cell elongation and division in mycobacteria. *Sci. Rep.* **2019**, *9*, 10728. [CrossRef]
84. Vilchèze, C.; Jacobs, W.R. The mechanism of isoniazid killing: Clarity through the scope of genetics. *Annu. Rev. Microbiol.* **2007**, *61*, 35–50. [CrossRef]
85. Medjahed, H.; Reyrat, J.M. Construction of *Mycobacterium abscessus* defined glycopeptidolipid mutants: Comparison of genetic tools. *Appl. Environ. Microbiol.* **2009**, *75*, 1331–1338. [CrossRef]
86. Seeliger, J.C.; Holsclaw, C.M.; Schelle, M.W.; Botyanszki, Z.; Gilmore, S.A.; Tully, S.E.; Niederweis, M.; Cravatt, B.F.; Leary, J.A.; Bertozzi, C.R. Elucidation and chemical modulation of sulfolipid-1 biosynthesis in *Mycobacterium tuberculosis*. *J. Biol. Chem.* **2012**, *287*, 7990–8000. [CrossRef] [PubMed]

87. Su, C.C.; Klenotic, P.A.; Bolla, J.R.; Purdy, G.E.; Robinson, C.V.; Yu, E.W. MmpL3 is a lipid transporter that binds trehalose monomycolate and phosphatidylethanolamine. *Proc. Natl. Acad. Sci. USA* **2019**, *116*, 11241–11246. [CrossRef] [PubMed]
88. Murakami, S.; Nakashima, R.; Yamashita, E.; Yamaguchi, A. Crystal structure of bacterial multidrug efflux transporter AcrB. *Nature* **2002**, *419*, 587–593. [CrossRef]
89. Sennhauser, G.; Bukowska, M.A.; Briand, C.; Grütter, M.G. Crystal structure of the multidrug exporter MexB from *Pseudomonas aeruginosa*. *J. Mol. Biol.* **2009**, *389*, 134–145. [CrossRef] [PubMed]
90. Bolla, J.R.; Su, C.C.; Do, S.V.; Radhakrishnan, A.; Kumar, N.; Long, F.; Chou, T.H.; Delmar, J.A.; Lei, H.T.; Rajashankar, K.R.; et al. Crystal structure of the *Neisseria gonorrhoeae* MtrD inner membrane multidrug efflux pump. *PLoS ONE* **2014**, *9*, e97903. [CrossRef] [PubMed]
91. Kumar, N.; Su, C.C.; Chou, T.H.; Radhakrishnan, A.; Delmar, J.A.; Rajashankar, K.R.; Yu, E.W. Crystal structures of the *Burkholderia multivorans* hopanoid transporter HpnN. *Proc. Natl. Acad. Sci. USA* **2017**, *114*, 6557–6562. [CrossRef]
92. Gong, X.; Qian, H.; Zhou, X.; Wu, J.; Wan, T.; Cao, P.; Huang, W.; Zhao, X.; Wang, X.; Wang, P.; et al. Structural insights into the Niemann-Pick C1 (NPC1)-mediated cholesterol transfer and ebola infection. *Cell* **2016**, *165*, 1467–1478. [CrossRef]
93. Rao, S.P.; Lakshminarayana, S.B.; Kondreddi, R.R.; Herve, M.; Camacho, L.R.; Bifani, P.; Kalapala, S.K.; Jiricek, J.; Ma, N.L.; Tan, B.H.; et al. Indolcarboxamide is a preclinical candidate for treating multidrug-resistant tuberculosis. *Sci. Transl. Med.* **2013**, *5*, 214ra168. [CrossRef]
94. Remuiñán, M.J.; Pérez-Herrán, E.; Rullás, J.; Alemparte, C.; Martínez-Hoyos, M.; Dow, D.J.; Afari, J.; Mehta, N.; Esquivias, J.; Jiménez, E.; et al. Tetrahydropyrazolo[1,5-a]pyrimidine-3-carboxamide and N-benzyl-6′,7′-dihydrospiro[piperidine-4,4′-thieno[3,2-c]pyran] analogues with bactericidal efficacy against *Mycobacterium tuberculosis* targeting MmpL3. *PLoS ONE* **2013**, *8*, e60933. [CrossRef]
95. Dupont, C.; Viljoen, A.; Dubar, F.; Blaise, M.; Bernut, A.; Pawlik, A.; Bouchier, C.; Brosch, R.; Guérardel, Y.; Lelièvre, J.; et al. A new piperidinol derivative targeting mycolic acid transport in Mycobacterium abscessus. *Mol. Microbiol.* **2016**, *101*, 515–529. [CrossRef]
96. Shetty, A.; Xu, Z.; Lakshmanan, U.; Hill, J.; Choong, M.L.; Chng, S.S.; Yamada, Y.; Poulsen, A.; Dick, T.; Gengenbacher, M. Novel acetamide indirectly targets mycobacterial transporter MmpL3 by proton motive force disruption. *Front. Microbiol.* **2018**, *9*, 2960. [CrossRef] [PubMed]
97. Zheng, H.; Williams, J.T.; Coulson, G.B.; Haiderer, E.R.; Abramovitch, R.B. HC2091 Kills *Mycobacterium tuberculosis* by targeting the MmpL3 mycolic acid transporter. *Antimicrob. Agents Chemother.* **2018**, *62*, e02459-17. [CrossRef] [PubMed]
98. Protopopova, M.; Hanrahan, C.; Nikonenko, B.; Samala, R.; Chen, P.; Gearhart, J.; Einck, L.; Nacy, C.A. Identification of a new antitubercular drug candidate, SQ109, from a combinatorial library of 1,2-ethylenediamines. *J. Antimicrob. Chemother.* **2005**, *56*, 968–974. [CrossRef] [PubMed]
99. Chen, P.; Gearhart, J.; Protopopova, M.; Einck, L.; Nacy, C.A. Synergistic interactions of SQ109, a new ethylene diamine, with front-line antitubercular drugs in vitro. *J. Antimicrob. Chemother.* **2006**, *58*, 332–337. [CrossRef] [PubMed]
100. Reddy, V.M.; Einck, L.; Andries, K.; Nacy, C.A. In vitro interactions between new antitubercular drug candidates SQ109 and TMC207. *Antimicrob. Agents Chemother.* **2010**, *54*, 2840–2846. [CrossRef] [PubMed]
101. Nikonenko, B.V.; Protopopova, M.; Samala, R.; Einck, L.; Nacy, C.A. Drug therapy of experimental tuberculosis (TB): Improved outcome by combining SQ109, a new diamine antibiotic, with existing TB drugs. *Antimicrob. Agents Chemother.* **2007**, *51*, 1563–1565. [CrossRef]
102. Heinrich, N.; Dawson, R.; du Bois, J.; Narunsky, K.; Horwith, G.; Phipps, A.J.; Nacy, C.A.; Aarnoutse, R.E.; Boeree, M.J.; Gillespie, S.H.; et al. Early phase evaluation of SQ109 alone and in combination with rifampicin in pulmonary TB patients. *J. Antimicrob. Chemother.* **2015**, *70*, 1558–1566. [CrossRef]
103. Veiga-Santos, P.; Li, K.; Lameira, L.; de Carvalho, T.M.; Huang, G.; Galizzi, M.; Shang, N.; Li, Q.; Gonzalez-Pacanowska, D.; Hernandez-Rodriguez, V.; et al. SQ109, a new drug lead for Chagas disease. *Antimicrob. Agents Chemother.* **2015**, *59*, 1950–1961. [CrossRef]
104. García-García, V.; Oldfield, E.; Benaim, G. Inhibition of *Leishmania mexicana* growth by the tuberculosis drug SQ109. *Antimicrob. Agents Chemother.* **2016**, *60*, 6386–6389. [CrossRef]

105. Li, W.; Stevens, C.M.; Pandya, A.N.; Darzynkiewicz, Z.; Bhattarai, P.; Tong, W.; Gonzalez-Juarrero, M.; North, E.J.; Zgurskaya, H.I.; Jackson, M. Direct inhibition of MmpL3 by novel antitubercular compounds. *ACS Infect. Dis.* **2019**, *5*, 1001–1012. [CrossRef]
106. Seeger, M.A.; Diederichs, K.; Eicher, T.; Brandstätter, L.; Schiefner, A.; Verrey, F.; Pos, K.M. The AcrB efflux pump: Conformational cycling and peristalsis lead to multidrug resistance. *Curr. Drug Targets* **2008**, *9*, 729–749. [CrossRef]
107. Tsukazaki, T.; Mori, H.; Echizen, Y.; Ishitani, R.; Fukai, S.; Tanaka, T.; Perederina, A.; Vassylyev, D.G.; Kohno, T.; Maturana, A.D.; et al. Structure and function of a membrane component SecDF that enhances protein export. *Nature* **2011**, *474*, 235–238. [CrossRef] [PubMed]
108. Lamichhane, G.; Tyagi, S.; Bishai, W.R. Designer arrays for defined mutant analysis to detect genes essential for survival of *Mycobacterium tuberculosis* in mouse lungs. *Infect. Immun.* **2005**, *73*, 2533–2540. [CrossRef] [PubMed]
109. Hartkoorn, R.C.; Uplekar, S.; Cole, S.T. Cross-resistance between clofazimine and bedaquiline through upregulation of MmpL5 in *Mycobacterium tuberculosis*. *Antimicrob. Agents Chemother.* **2014**, *58*, 2979–2981. [CrossRef]
110. Oh, S.; Park, Y.; Engelhart, C.A.; Wallach, J.B.; Schnappinger, D.; Arora, K.; Manikkam, M.; Gac, B.; Wang, H.; Murgolo, N.; et al. Discovery and structure-activity-relationship study of N-Alkyl-5-hydroxypyrimidinone carboxamides as novel antitubercular agents targeting decaprenylphosphoryl-β-D-ribose 2′-oxidase. *J. Med. Chem.* **2018**, *61*, 9952–9965. [CrossRef] [PubMed]
111. Borisov, S.E.; Bogorodskaya, E.M.; Volchenkov, G.V.; Kulchavenya, E.V.; Maryandyshev, A.O.; Skornyakov, S.N.; Talibov, O.B.; Tikhonov, A.M.; Vasilyeva, I.A. Efficiency and safety of chemotherapy regimen with SQ109 in those suffering from multiple drug resistant tuberculosis. *Tuberc. Lung Dis.* **2018**, *96*, 6–18. [CrossRef]

© 2020 by the authors. Licensee MDPI, Basel, Switzerland. This article is an open access article distributed under the terms and conditions of the Creative Commons Attribution (CC BY) license (http://creativecommons.org/licenses/by/4.0/).

Review

Development of Macozinone for TB treatment: An Update

Vadim Makarov [1],* and Katarína Mikušová [2]

1. Federal Research Center "Fundamentals of Biotechnology RAS", Leninsky prospect 33, bld. 2, 119071 Moscow, Russia
2. Department of Biochemistry, Faculty of Natural Sciences, Comenius University in Bratislava, Mlynská dolina, Ilkovičova 6, 84215 Bratislava, Slovakia; katarina.mikusova@uniba.sk
* Correspondence: makarov@inbi.ras.ru

Received: 5 March 2020; Accepted: 25 March 2020; Published: 26 March 2020

Abstract: Macozinone, a piperazine-benzothiazinone PBTZ169, is currently undergoing Phase 1/2 clinical studies for the treatment of tuberculosis (TB). In this review we summarize the key findings that led to the development of this compound and to identification of its target, decaprenylphospohoryl ribose oxidase DprE1, which is involved in the synthesis of the essential arabinan polymers of the cell wall in a TB pathogen, *Mycobacterium tuberculosis*. We present the results of the pilot clinical studies, which raise optimism regarding its further development towards more efficient TB drug regimens.

Keywords: macozinone; DprE1 inhibitor; clinical studies

1. Introduction

Tuberculosis (TB) as both an acute and chronic infectious disease is still unbeaten and represents an important social, medical, and biological challenge for the healthcare worldwide [1]. The infection caused by *Mycobacterium tuberculosis* strains is most often located in the respiratory tract from which it is transmitted in the population by aerosols. At the same time, mycobacteria can be disseminated also to other organs. However, it is now well acknowledged that even in its primary location, the TB pathogen is present in different microenvironments and thus in various metabolic states, which substantially complicates the treatment [2]. According to the last Global Tuberculosis Report issued by the World Health Organization (WHO) [3], in 2018 the death rate was estimated to be about 1.45 million people. Among the estimated 10 million new cases of TB, 3.4% had multidrug-resistant TB or rifampicin-resistant TB (MDR/RR-TB). This number increased to 18% in previously treated cases, with the highest proportions (>50% in previously treated cases) in countries of the former Soviet Union. Although efficient TB treatment is available, only 7 million new cases were treated for the disease in 2018, while an estimated 3 million patients did not have access to quality care or were not reported, and only one in three people with drug-resistant TB accessed care [3].

In addition to interventions in the social area and point-of-care management of TB, one of the key challenges in making the fight against TB more effective is to improve treatment options. The current chemotherapy for the drug-sensitive TB takes about six months and requires administration of four different medicines (isoniazid, rifampicin, ethambutol, pyrazinamide) to avoid development of drug resistance. The length of the treatment and its adverse effects result in bad compliance and thus it is imperative to find novel drugs and regimens that would be less demanding on patients [4].

During recent years, three novel drugs against TB were approved. These are bedaquiline [5], delamanid [6] and pretomanid [7], which are currently undergoing several clinical trials aimed at treatment of the drug-resistant TB, as well as at shortening standard TB treatment [8]. The efforts and hope for better TB drugs are exemplified also by the pipeline of several new molecules, that are currently at different stages of clinical development. Among them there are inhibitors of protein synthesis from the oxazolidinone family (delpazolid, sutezolid, contezolid, TBI-223), inhibitor of leucyl-tRNA synthetase (GSK3036656), inhibitor of DNA gyrase B (SPR720), compounds interfering with the respiratory chain of a TB pathogen (telacebec, TBI-166), a cell wall inhibitor targeting MmpL3 transporter (SQ109) and four compounds targeting another cell wall target, DprE1. These are BTZ043, TBA-7371, OPC-167832, and macozinone (Figure 1) [4,8]. In this review we will present the key findings towards the discovery and development of macozinone, an optimized benzothiazinone, which currently appears to be the most advanced DprE1 inhibitor.

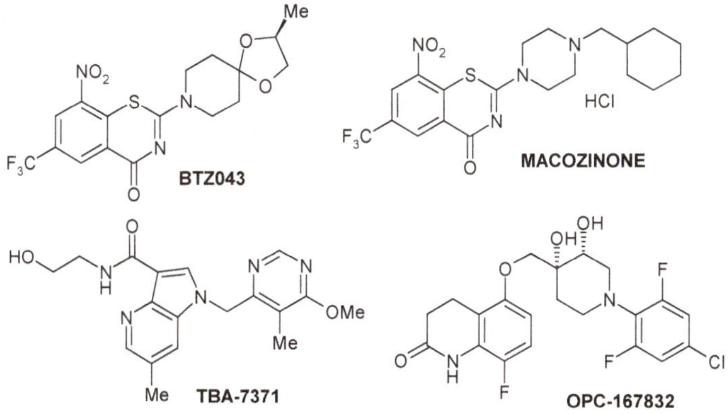

Figure 1. DprE1 inhibitors in clinical development.

2. Discovery of Benzothiazinones as Highly Potent Killers of *Mycobacterium Tuberculosis*

The very first scientific report on benzothiazinones (BTZs) with extremely high antimycobacterial activity was published in 2009 [9]. These compounds emerged as a result of several years of efforts focused on investigation of a family of ditiocarbamate (DTC) derivatives with antimicrobial properties, which was carried out at the Research Center for Antibiotics in Moscow and the Hans Knöll Institute in Jena [10]. DTCs gained their fame primarily as additives in industry and pesticides, which have been in use since 1930. However, some of them were tested as pharmacological agents for treatment of alcoholism, human immunodeficiency virus (HIV) or cancer [11]. It occurred that selected DTC derivatives had minimal inhibitory concentrations (MICs) between 3–30 µg/ml for *M. tuberculosis*, but were active also on several species of fast-growing mycobacteria and multiresistant staphylococci [10]. Thorough structure-activity studies focused on improvement of the activity and selectivity of these compounds for mycobacteria was performed. Selected compounds derived from the basic active ortho nitro-dialkyldithiocarbamate structure showed enhanced activity towards fast growing mycobacteria, with *M. vaccae* being the most sensitive reaching MICs 0.4–1.56 µg/ml [12]. In addition, 3 out of 7 tested molecules were efficient and well tolerated also in the mouse foot pad model of leprosy [12]. Nevertheless, it was clear that reactivity of DTCs poses a significant problem in further development of these compounds despite their activity and rather unexpected low level of toxicity as exemplified by studies in mice [12]. Consequently, a thorough investigation of the metabolic conversion of the most active DTCs containing spiroamine moiety, was undertaken. The drug was incubated with the non-pathogenic strain *M. smegmatis* and all isolated metabolites were tested for antimycobacterial activity. One of these metabolites was particularly active and its structure established by nuclear magnetic resonance (NMR) revealed the principal BTZ scaffold. A set of novel BTZ derivatives was synthesized and tested against both fast-growing and pathogenic mycobacteria in collaboration with Prof. Stewart Cole at Pasteur Institute in Paris [13]. Surprisingly, these compounds were highly active and selective against mycobacteria in vitro, with MIC values 0.195–1.56 µg/ml for *M. tuberculosis*, including the resistant strains. The lowest MIC value, 0.023 pg/ml, was obtained for *M. fortuitum* and 2,2-[2-methyl-1,4-dioxa-8-azaspiro[4.5]dec-8-yl]-8-nitro- 6-(trifluoromethyl)-4H-1,3-benzothiazin-4-one, later named BTZ038 [9]. The pilot in vivo studies with this compound revealed its potency in the murine model of TB, which was comparable to isoniazid used in the same dose of 25 mg/kg [13]. These encouraging results warranted further exploration of BTZs as perspective compounds for development of a new TB drug.

3. Mechanism of Action of Benzothiazinones

An investigation of mechanism of action of BTZ was carried out within the European Union (EU)-funded consortium New Medicines for Tuberculosis (NM4TB), which was formed in 2006 under the leadership of Prof. Stewart Cole. The compound BTZ038 was initially chosen as a lead molecule. Since it contains a chiral center, pure enantiomers BTZ043 (S) and BTZ044 (R) were prepared and subjected to MIC determination. Both forms showed comparable activities in vitro in *M. tuberculosis*. Accordingly, only one of them, BTZ043, was subjected to mechanistic studies [9].

Genetic experiments aimed at identification of the BTZ target were carried out by Prof. Giovanna Riccardi and her team at University of Pavia. The two approaches which were chosen for this purpose pointed out to the role of the gene *rv3790* in conferring resistance against the drug. In the first approach, *M. smegmatis* was transformed with a cosmid library containing DNA fragments from the same organism. An ortholog of the *rv3790* gene from *M. smegmatis* mc^{2}155, *MSMEG_6382*, was identified on a cosmid, which resulted in an increase in MICs towards BTZ043. In the second approach, BTZ-resistant strains of *M. smegmatis*, *M. bovis* BCG and *M. tuberculosis* were isolated and their resistance was traced to mutations in the same codon in *rv3790* and its orthologs. In all mutant strains codons corresponding to Cys387 in *rv3790* gene product were replaced by Ser or Gly, which increased the MICs of the resistant strains up to 10,000 times. Accordingly, the mycobacterial species *M. avium* and *M. aurum*, in which this position was occupied by Ala or Ser, respectively, were naturally resistant against BTZ, which supported identification of *rv3790* as a target [9]. Coincidentally, the function of the protein encoded by this gene was described only a short while before the data on the BTZ target became available. In the course of studies focused on the metabolism of arabinose in mycobacteria carried out by Prof. Michael McNeil with collaborators at Colorado State University in Fort Collins, the *rv3790* and *rv3791* gene products were proposed to be involved in a unique epimerisation of decaprenylphosphoryl ribose (DPR) to decaprenylphosphoryl arabinose (DPA) [14]. The latter molecule serves as a sole donor of arabinose residues for synthesis of the essential arabinan polymers of the mycobacterial cell wall [15]. It was shown that the conversion takes place via oxidation of the DPR substrate, followed by reduction of the resulting keto-intermediate and thus the two oxidoreductases, *rv3790* and *rv3791*, from the so-called arabinogalactan biosynthesis cluster [16] were ideal candidates to catalyse these reactions. However, at the time it was not possible to conclude which of the two is responsible for oxidation and which for the reduction step. Examination of the effects of BTZ043 on enzymatic conversion of DPR to DPA proved that the compound inhibited the oxidation step. Given the confirmed roles of Rv3790 and Rv3791 in epimerization of DPR, we named them DprE1 and DprE2, respectively (Figure 2) [9].

An important step towards understanding the mechanism of action of BTZs was isolation of a covalent adduct of BTZ043 and DprE1 by Trefzer et al. in 2010 [17]. The authors concluded that reduction of the nitro-group to nitroso derivative is critical for formation of the semimercaptal adduct of BTZ043 and DprE1 through the Cys387, which was verified by mass spectrometry. The following study proved that DprE1 itself is responsible for the reductive activation of DprE1. The electrons needed for the reaction are provided by the enzyme's prosthetic group, FAD, which gets reduced during the catalysis of DPR oxidation (Figure 2) [18]. Accordingly, BTZ represents a typical example of a suicide inhibitor. The proposed mechanism of action was confirmed by crystallographic studies in which structures of DprE1 from *M. smegmatis* were obtained in native and BTZ043-bound forms. As expected, the semimercaptal adduct was formed only in the presence of the substrate analog, farnesylphosphoryl ribose, which provided electrons for reduction of the FAD cofactor necessary for BTZ activation [19].

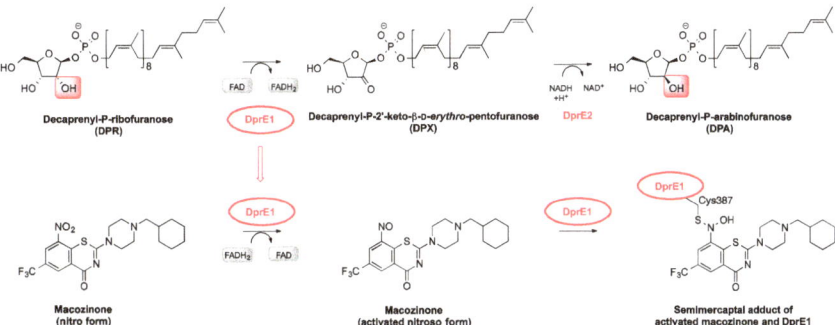

Figure 2. Mechanism of action of macozinone. *Upper line:* Reactions catalyzed by DprE1 and DprE2. *Bottom line:* Activation of macozinone and its covalent attachment to DprE1 from *M. tuberculosis* through Cys387 of the enzyme's active site. The scheme is based on mechanistic studies performed initially with benzothiazinone BTZ043 [17–19] and subsequently confirmed for piperazine-benzothiazinone PBTZ169 (macozinone) [20].

4. From BTZ043 to PBTZ169

At the time of its discovery, BTZ043 was the most efficient inhibitor of in vitro mycobacterial growth ever described. While MICs for first line drugs isoniazid or ethambutol for *M. tuberculosis* were 0.02–0.2 µg/ml or 1–5 µg/ml, respectively, the value for BTZ043 was just 1 ng/ml. The compound proved to be active also in intracellular *M. tuberculosis* infection in Raw 264.7 macrophages and in a mouse model of chronic tuberculosis and it did not show any signs of toxicity or mutagenicity in concentrations well above the amounts used for the experiments. Nevertheless, the extremely low MIC did not translate to comparably high efficiencies in animal studies [9].

In the effort to improve pharmacological properties of benzothiazinone series, structure-activity relationship (SAR) analysis was performed, with main focus on position 2 of BTZ scaffold [20]. It was previously shown that the rest of the positions harbor substituents critical for the antimycobacterial activity, e.g., replacement of 8-nitro-group with any other group, including hydroxylamino-, amino- and nitryl-, led to at least a 500-fold increase in the MIC against *M. tuberculosis* [9]. The new series contained a piperazine at the position 2 of the BTZ structure, which allowed extensive modifications of the N-4 position of this substituent. Evaluation of the 60 prepared piperazine-benzothiazinones (PBTZ) for in vitro activity on *M. tuberculosis* pointed to a strong correlation between the MICs and lipophilicity. The alkyl-PBTZ derivatives were most active and some of them reached the MICs between 0.19–0.75 ng/ml for *M. tuberculosis*. Five of these derivatives were tested in the chronic TB mouse model, which allowed for selection of PBTZ169 as the most promising candidate [20].

Comparison of BTZ043 with PBTZ169 confirmed superior properties of the latter in all tested parameters. It was three to seven times more active in vitro against *M. tuberculosis*, *M. bovis* BCG, *M. marinum*, *M. smegmatis*, *Corynebacterium diphtheriae* and *C. glutamicum* as established by resazurin microtiter assay (REMA). The molecule was also highly active against a panel of 9 MDR- and XDR-clinical isolates of *M. tuberculosis* [20]. Cross-resistance between BTZ043 and PBTZ169 was confirmed for BTZ-resistant strains of *M. tuberculosis*, *M. bovis* BCG and *M. smegmatis*, indicating the common mechanism of action. Interestingly, PBTZ169 was less susceptible to inactivation by nitro-reduction by the nitroreductase NfnB from *M. smegmatis* compared to BTZ043, as confirmed by the liquid chromatography-mass spectrometry analyses. This property could be critical for in vivo activity, due to inevitable actions of host nitroreductases or microbial nitroreductases in the digestive tract of the patients, which would potentially reduce the amount of the available active drug. Enzymologic studies confirmed that PBTZ169 is more efficient inhibitor of its DprE1 target than BTZ043; the enzyme was completely inactivated after 5 min of incubation with 5 µM PBTZ169, while four times higher concentration was needed to achieve the same result with BTZ043. The crystal structure of

M. tuberculosis DprE1 with covalently bound PBTZ169 showed similarities with the BTZ043-bound structure of the *M. smegmatis* enzyme. However, higher flexibility of the methyl-cyclohexyl group on piperazine, could account for better affinity of this drug for its target and explain its higher capacity to inactivate the enzyme [20].

Efficacies of PBTZ169 and BTZ043 were also evaluated against *M. marinum* in a zebrafish embryo model. Although after 5 days of treatment both drugs efficiently decreased the bacterial burden at concentrations of 25 nM and 50 nM, examination of BTZ043-treated embryos pointed to developmental defects. By contrast, application of PBTZ169 up to 10 μM did not cause any pathological changes. In vivo studies in the mouse chronic model of TB proved the superiority of PBTZ169 over BTZ043 in lowering colony forming unit counts in the spleen and lungs, which, however, could not be attributed to better pharmacokinetics, since the two compounds had similar properties except for faster uptake of PBTZ169 [20]. Aiming at the use of the novel PBTZ derivative in designing a new regimen for TB, combination studies with selection of approved and experimental drugs were performed. The combination of PBTZ169 and bedaquiline, which proved to be synergistic in the in vitro REMA assay, was tested in the mouse chronic model of TB. Simultaneous administration of PBTZ, bedaquiline and pyrazinamide was more efficient in reducing the mycobacterial counts both in the lungs and spleen compared to the standard mixture of isoniazid, rifampicin and pyrazinamide [20].

Among favorable properties of PBTZ169 over BTZ043 is the lack of the chiral center, which simplified the production of the drug and significantly decreased its price. This is critical for developing treatment for a disease greatly affecting low-income and middle-income countries in particular. Taken together, PBTZ169 was identified as a promising candidate for development as a novel TB medicine [20].

5. From PBTZ169 to Macozinone: First Results of the Clinical Studies

Establishment of a non-profit Innovative Medicines for Tuberculosis Foundation (iM4TB Foundation) at Ecole Polytechnique Fédérale de Lausanne (EPFL) in March 2014 was a principal milestone for the further development of PBTZ169 and its transformation from a valuable TB drug candidate to the compound reaching promising results in the pilot clinical studies (Figure 3) [21]. This approach was necessary to overcome the lack of interest from pharmaceutical companies to address the disease primarily affecting low-income countries. The mission of the foundation is "to develop better and faster-acting medicines to fight tuberculosis and bridge the gap between scientific discovery and the market in order to provide affordable TB treatment to anyone in the world". In July 2014 the foundation joined in this effort with the Russian pharmaceutical company Nearmedic Plus LLC, which bought a license covering the use of PBTZ169 in most countries of the former Soviet Union, while iM4TB retained the rights for the rest of the world. Investments of Nearmedic, along with funds raised by iM4TB, including the generous support of the Bill and Melinda Gates foundation, enabled performing all necessary pre-clinical trials with PBTZ169, such as ADME (absorption, distribution, metabolism, excretion) profiling, toxicology studies, as well as production of clinical supplies of the drug, and preparation of regulatory documents necessary for entering the first clinical trials [21,22].

Clinical studies were initiated by both iM4TB and Nearmedic. In the randomized, double-blind, placebo-controlled Phase 1a study performed in Switzerland (NCT03423030) [23] the aim was to assess safety, tolerability and pharmacokinetic profile of PBTZ169 formulated as a spray-dried dispersion versus a native crystal powder. In addition, antitubercular activity of single escalating doses of PBTZ169 ex vivo was evaluated. The study included 32 healthy male volunteers, grouped to 4 panels of 8 subjects, each undergoing 2 investigation periods, during which they received either single doses of PBTZ169 at increasing dose levels up to 320 mg or a matching placebo. Promising results of this study encouraged the iM4TB team to design of the Phase 1b trial (NCT03776500) [24] aimed at evaluation of the safety, tolerability and pharmacokinetics of PBTZ169 in multiple dosing up to 1200 mg/day in healthy volunteers (32 subjects grouped in 4 consecutive panels) receiving PBTZ169 for 14 consecutive days. An additional goal is to evaluate interactions of PBTZ16 with the human cytochrome P-450 enzyme family [24].

Figure 3. Pre-clinical and clinical development of macozinone.

The first Phase 1 clinical trial in Russia (Nearmedic Plus LLC) (NCT03036163, PBTZ169-Z00-C01-1) [25] with an official title "Open-label Prospective Non-comparative Study of Safety, Tolerability and Pharmacokinetics of PBTZ169 after Single and Multiple Fasting Oral Administration in Increasing Doses in Healthy Volunteers" was carried out between January and November 2016. The drug was administered in the form of capsules containing 40 mg of the drug. The study included single escalating doses (40, 80, 160, 320, and 640 mg) in fasting conditions, as well as multiple fasting doses (320 and 640 mg) for 14 days. Overall the study included 40 participants [22].

A completed open-label prospective Phase 1b study initiated by Nearmedic Plus LLC (NCT04150224, PBTZ169-Z00-C01-3) [26] was designed to assess the safety, tolerability, pharmacokinetics, and food effects on single, double, and multiple escalating doses in healthy volunteers. During the first part of the study, safety, tolerability, and pharmacokinetics of PBTZ169 in 80 mg capsules were studied in healthy volunteers who received single or double (twice daily) fasting doses, which increased sequentially (640, 960, and 1280 mg once a day and 640 mg twice a day). The food effects were studied for a single 640 mg dose. The second part of the study involved the safety, tolerability, and pharmacokinetics of PBTZ169 in 80 mg capsules in healthy volunteers who received 1280 mg daily in fed conditions, for 14 days. In total, 60 healthy volunteers (10 in each cohort) received the investigational drug and completed the study according to the protocol [22].

During these clinical trials, safety and tolerability of PBTZ169 in capsules were studied, including assessment of adverse effects, vital signs (blood pressure, heart rate, body temperature, respiratory rate), laboratory and instrumental parameters (hematology, blood chemistry, coagulation parameters, urinalysis, electrocardiography) and physical examination. In these studies (NCT03036163 and NCT04150224), good tolerability and favorable safety profile were demonstrated for PBTZ169 in the studied dose range. One event of dose-limiting toxicity (increase in the glucose level 2 hours after the drug administration) was documented after a single 80 mg dose of PBTZ169 in 40 mg capsules (NCT03036163). There were no other cases of dose-limiting toxicity in healthy volunteers in these studies. There were no serious adverse effects. There was no increase in the frequency of adverse effects associated with increased dose or with the administration regimen (fasting or in fed conditions), for single or multiple doses. Changes in some mean vital signs, laboratory, instrumental parameters, or physical examination data were not associated with a trend to increase with a growing dose [22].

Efficacy of PBTZ169 based on the early bactericidal activity (EBA), was studied in a Phase 2a clinical trial (PBTZ169-A15-C2A-1, NCT03334734) [27] by Nearmedic Plus LLC in patients with newly diagnosed smear-positive tuberculosis of the respiratory tract, with preserved sensitivity to isoniazid and rifampicin. Although the trial was terminated early because of slow enrollment, relevant information was obtained from 16 patients who participated in the study. The drug was administered in 80 mg capsules, orally as a monotherapy for 14 days, as follows: 160 mg/day dose—4 patients; 320 mg/day dose—4 patients; 640 mg/day dose—7 patients. Isoniazid tablets were used as a control treatment at 600 mg/day dose in one patient.

The primary efficacy of the treatment was evaluated by two methods:

(i) According to quantification of colony forming units in sputum by its inoculation on agar plates, in the group of patients receiving PBTZ169 in the 640 mg dose, EBA 0–14 (mean of the two measurements) and EBA 0–14 (the higher of the two measurements) were 0.071 log10 cells/ml/day [95% confidence interval (CI) 0; 0.143] and 0.080 log10 cells/ml/day [95% CI 0.002; 0.158], respectively.

(ii) According to quantitative polymerase chain reaction, in the same group of patients, EBA 0–14 (mean of the two measurements) and EBA 0–14 (the higher of the two measurements) were 0.098 log10 cells/ml/day [95% CI 0.021; 0.175] and 0.100 log10 cells/ml/day [95% CI 0.021; 0.180], respectively.

Thus, the analysis of efficacy based on the data obtained in a pilot Phase 2a clinical study revealed statistically significant early bactericidal activity of PBTZ169 14 days after the start of monotherapy (EBA0-14) in the group of PBTZ169 with the 640 mg dose. During this clinical study, there were no cases of death or severe adverse effects related to the PBTZ169. No adverse effects were considered as definitely related to the drug and, at the study completion, all adverse effects were resolved. There was no increase in adverse effects frequency with growing dose [22].

During the clinical studies it was confirmed that PBTZ169 after single, double, or multiple doses was detectable in the blood plasma of all volunteers and patients who had received the drug. However, it should be noted that pharmacokinetic studies in animals and in healthy volunteers (in Phase 1 clinical studies) demonstrated that PBTZ169, being more soluble in an acidic environment, is absorbed primarily in the stomach. These data agree with the results of in vitro studies, which were conducted by Nearmedic Plus LLC, assessing cellular permeability and solubility in biorelevant media, which mimicked native gastric and intestinal juices. They confirmed that the highest solubility of PBTZ169 is observed at the pH range of 1 to 2, which coincides with the pH in the lower part of the stomach, and its solubility gradually decreases with the growing pH. In the media with pH higher than 5, PBTZ169 was virtually insoluble. According to the test of solubility in biorelevant media, the drug level in the gastric juice was much higher than in the intestinal juice. So, one of the possible methods to increase bioavailability and exposure of PBTZ169 is to prolong its presence in the stomach.

After a single fasting dose, PBTZ169 absorbed rapidly, and the speed of absorption was independent of the administered dose (median T_{max} was 1.5–2.5 hours for the dose range 40–640 mg (NCT03036163), and 1.5–1.75 hours for the dose range 640–1280 mg (NCT04150224). Meantime of retention of the drug in the body was 13.02–19.30 hours and was independent of the administered dose (NCT03036163) [22].

According to the obtained pharmacokinetic data, dose proportionality of C_{max} and AUC_{0-t} was demonstrated for single, as well as for multiple administrations of the investigational drug in the dose range 40–640 mg (NCT03036163, NCT03334734). After single fasting escalating doses (640, 960, and 1280 mg), C_{max}, AUC_{0-24}, $AUC_{0-\infty}$ were less than dose proportional. Thus, it can be concluded that there is a trend for linear pharmacokinetics for single and multiple doses up to 640 mg.

When pharmacokinetics of different administration regimens for 1280 mg daily dose was studied (640 mg twice daily or 1280 mg once in fasting conditions), the pharmacokinetic parameters $AUC_{0-\infty}$ and AUC_{0-t} increased 1.5–1.6-fold (NCT04150224). The relation of mean $AUC_{0-\infty}$ and AUC_{0-t} in cohorts receiving the investigational drug as 640 mg twice a day and 1280 mg once a day, was 153.68% and 160.32%. For C_{max}, a slight decrease (1.1-fold) was observed for the 640 mg twice a day (the relation of means was 90.84%). All observed differences were statistically significant.

According to the results of analysis of relative bioavailability and relative degree of absorption, the main pharmacokinetic parameters of PBTZ169 substantially and statistically significantly increased when 640 mg dose was administered in fed conditions: $AUC_{0-\infty}$ increased 3.45-fold, AUC_{0-t} (f') increased 3.50-fold, C_{max} increased 2.29-fold. The drug absorbed and cleared from the body significantly more slowly, when it was administered in fed conditions compared to the fasting conditions (median T_{max}

increased 2-fold, mean $T_{1/2}$ increased 1.44-fold). These data allow to conclude that it is preferable to administer PBTZ169 in fed conditions [22].

In October 2018, the international non-proprietary name "macozinone" was given to PBTZ169 by WHO [28]. In addition to the clinical studies, macozinone is undergoing further in vitro and in vivo testing focused on improvement of its pharmacological properties by structure-based design [29], or on the interactions with anti-TB drugs [30]. Importantly, an efficient method for monitoring metabolism of macozinone in human plasma was recently developed. It enables simultaneous measurement of concentrations of macozinone and its five active metabolites in the clinical samples, which is important for comprehensive pharmacokinetic/pharmacodynamic analyses [31].

6. Other DprE1 Inhibitors under Active Clinical Development

Following the discovery of DprE1 as a target of BTZ043, numerous molecules acting on this enzyme emerged, particularly from the whole-cell based phenotypic screening campaigns (for a recent review see [32]). As a result, DprE1 was given an unflattering attribute—a promiscuous target [33]. Nevertheless, DprE1 remains to be one of the best understood and the most vulnerable novel targets. One of the reasons for its high sensitivity against a number of different pharmacophores could be its periplasmic localization, which makes it easily accessible for the drugs [34]. Consequently, in addition to macozinone, there are currently three other DprE1 inhibitors progressing down the clinical studies pipeline. BTZ043, sponsored by the University of Munich; the Hans Knöll Institute, Jena; and the German Center for Infection Research, successfully completed a single dose escalation Phase 1 study (NCT03590600) [35] and is currently recruiting for a combined Phase 1 and 2 study, which will evaluate safety, tolerability, pharmacokinetics and early bactericidal activity of the multiple ascending doses (NCT04044001) [36]. A non-covalent DprE1 inhibitor, TBA-7371 [37] developed by TB Alliance, Bill and Melinda Gates Medical Research Institute and Foundation for Neglected Disease Research, completed a Phase 1 study (NCT03199339) [38]—a partially blind, placebo-controlled study of (i) a combined single ascending dose with a food effect cohort, (ii) multiple ascending dose group, and (iii) a cohort to investigate interactions between TBA-7371 with midazolam and bupropion. Currently, recruitment of adult patients with rifampicin-sensitive pulmonary tuberculosis is taking place for the participation in a Phase 2 study (NCT04176250) [39] to assess safety, early bactericidal activity, and pharmacokinetics of escalating doses of TBA-7371. A combined Phase 1 and 2 study of OPC-167832 by Otsuka is in progress (NCT03678688) [40], aimed at evaluation of the safety, tolerability, pharmacokinetics, and efficacy of multiple oral doses of the drug in patients with uncomplicated drug-sensitive pulmonary tuberculosis with a positive smear. In a parallel group, low dose or high dose OPC-167832 will be combined with delamanid and compared to delamanid only, or standard treatment with isoniazid, rifampin, pyrazinamide, and ethambutol.

7. Conclusions

The ultimate goal of the TB drug-development efforts is a design of new, more effective regimen, which could replace the current combination therapy. One of the most promising DprE1 inhibitors is macozinone. Its safety and tolerability profiles, pharmacokinetics, and efficacy in terms of early bactericidal activity were evaluated in the course of three clinical trials conducted by Nearmedic Plus LLC in the Russian Federation, two clinical trials by iM4TB foundation are in progress. High tolerability and a favorable safety profile of the drug in the studied dose range were demonstrated both in healthy volunteers and in patients with newly diagnosed pulmonary tuberculosis with bacterial excretion and preserved sensitivity to isoniazid and rifampicin. The main pharmacokinetics parameters of macozinone after single and multiple administration in the dosage range up to 1280 mg were studied, and a statistically significant efficacy of the drug after monotherapy at a dose of 640 mg a day was established, which allowed the preferred regimen of its intake to be determined.

A thorough study aimed at interactions of macozinone with a spectrum of clinically used and experimental TB drugs was recently performed [30]. It revealed that macozinone does not have synergistic or antagonistic interactions with the tested first-line (rifampin, isoniazid, ethambutol) or second-line (amikacin, levofloxacin, moxifloxacin, D-cycloserin, ethionamide, *para*-aminosalicyc acid) drugs. Among the tested re-purposed or new drugs, clarithromycin, delamanid, lansoprazole sulfide, linezolid, meropenem or sutezolid did not show synergistic effects, when tested individually with macozinone by the checkerboard assay. Synergism was observed for macozinone and clofazimine or bedaquiline, confirming previous findings [20,41]. At the same time, enumeration of colony forming units after drug exposure revealed the next two potentially synergistic candidates—delamanid and sutezolid. In the mouse model of TB, the combination of macozinone-delamanid-sutezolid was more active in the lungs of *M. tuberculosis*-infected animals compared to the regimen of rifampin-isoniazid-pyrazinamide, while the bacterial burden in spleen remained comparable. However, increased toxicity of the combination for HepG2 cells was observed, so further studies are needed to evaluate the potential of this drug combination [30]. Nevertheless, neither antagonism nor increased toxicity was found for most combinations, which paves the way for using macozinone in the development of more efficient TB regimens [30].

Author Contributions: Both authors contributed to writing of the paper. All authors have read and agreed to the published version of the manuscript.

Funding: K.M. acknowledges current support by Slovak Research and Development Agency (APVV-15-0515).

Acknowledgments: Authors would like to thank Nearmedic Plus LLC, and personally to R. Bolgarin for information concerning clinical trials and preclinical studies including unpublished data.

Conflicts of Interest: V.M. is a named inventor on patents pertaining to work on benzothiazinones mentioned herein. K.M. declares no conflict of interest.

References

1. Bloom, B.R.; Atun, R.; Cohen, T.; Dye, C.; Fraser, H.; Gomez, G.B.; Knight, G.; Murray, M.; Nardell, E.; Rubin, E.; et al. Chapter 11. Tuberculosis. In *Major Infectious Diseases. Disease Control Priorities*, 3rd ed.; Holmes, K.K., Bertozzi, S., Bloom, B., Jha, P., Eds.; World Bank: Washington, DC, USA, 2017; Volume 6, pp. 233–313.
2. Baer, C.E.; Rubin, E.J.; Sassetti, C.M. New insights into TB physiology suggest untapped therapeutic opportunities. *Immunol. Rev.* **2015**, *264*, 327–343. [CrossRef]
3. *Global TB Report 2019*; WHO: Geneva, Switzerland, 2019; Available online: https://apps.who.int/iris/bitstream/handle/10665/329368/9789241565714-eng.pdf?ua=1 (accessed on 20 February 2020).
4. Tiberi, S.; du Plessis, N.; Walzl, G.; Vjecha, M.J.; Rao, M.; Ntoumi, F.; Mfinanga, S.; Kapata, N.; Mwaba, P.; McHugh, T.D.; et al. Tuberculosis: Progress and advances in development of new drugs, treatment regimens, and host-directed therapies. *Lancet Infect. Dis.* **2018**, *18*, e183–e198. [CrossRef]
5. Andries, K.; Verhasselt, P.; Guillemont, J.; Gohlmann, H.W.; Neefs, J.M.; Winkler, H.; Van Gestel, J.; Timmerman, P.; Zhu, M.; Lee, E.; et al. A diarylquinoline drug active on the ATP synthase of *Mycobacterium tuberculosis*. *Science* **2005**, *307*, 223–227. [CrossRef] [PubMed]
6. Matsumoto, M.; Hashizume, H.; Tomishige, T.; Kawasaki, M.; Tsubouchi, H.; Sasaki, H.; Shimokawa, Y.; Komatsu, M. OPC-67683, a nitro-dihydro-imidazooxazole derivative with promising action against tuberculosis in vitro and in mice. *PLoS Med.* **2006**, *3*, e466. [CrossRef] [PubMed]
7. Stover, C.K.; Warrener, P.; VanDevanter, D.R.; Sherman, D.R.; Arain, T.M.; Langhorne, M.H.; Anderson, S.W.; Towell, J.A.; Yuan, Y.; McMurray, D.N.; et al. A small-molecule nitroimidazopyran drug candidate for the treatment of tuberculosis. *Nature* **2000**, *405*, 962–966. [CrossRef] [PubMed]
8. WGND Global TB Drug Pipeline. Available online: https://www.newtbdrugs.org/pipeline/clinical (accessed on 20 February 2020).
9. Makarov, V.; Manina, G.; Mikusova, K.; Mollmann, U.; Ryabova, O.; Saint-Joanis, B.; Dhar, N.; Pasca, M.R.; Buroni, S.; Lucarelli, A.P.; et al. Benzothiazinones kill *Mycobacterium tuberculosis* by blocking arabinan synthesis. *Science* **2009**, *324*, 801–804. [CrossRef] [PubMed]
10. Makarov, V.; Moellmann, U. New Dithiocarbamate Derivatives, Useful as Antibacterial Agents, for Treating e.g., Tuberculosis and Infections Caused by Mycobacteria and Staphylococci. U.S. Patent WO2003042186-A, 22 May 2003.

11. Szolar, O.H. Environmental and pharmaceutical analysis of dithiocarbamates. *Anal. Chim. Acta* **2007**, *582*, 191–200. [CrossRef] [PubMed]
12. Makarov, V.; Riabova, O.B.; Yuschenko, A.; Urlyapova, N.; Daudova, A.; Zipfel, P.F.; Mollmann, U. Synthesis and antileprosy activity of some dialkyldithiocarbamates. *J.Antimicrob. Chemother.* **2006**, *57*, 1134–1138. [CrossRef]
13. Makarov, V.A.; Cole, S.T.; Moellmann, U. New Benzothiazinone Derivatives Useful for the Treatment of Tuberculosis Infection or Leprosy Infection in Mammals. U.S. Patent WO2007134625-A1, 29 November 2007.
14. Mikusova, K.; Huang, H.; Yagi, T.; Holsters, M.; Vereecke, D.; D'Haeze, W.; Scherman, M.S.; Brennan, P.J.; McNeil, M.R.; Crick, D.C. Decaprenylphosphoryl arabinofuranose, the donor of the D-arabinofuranosyl residues of mycobacterial arabinan, is formed via a two-step epimerization of decaprenylphosphoryl ribose. *J. Bacteriol.* **2005**, *187*, 8020–8025. [CrossRef]
15. Wolucka, B.A.; McNeil, M.R.; de Hoffmann, E.; Chojnacki, T.; Brennan, P.J. Recognition of the lipid intermediate for arabinogalactan/arabinomannan biosynthesis and its relation to the mode of action of ethambutol on mycobacteria. *J. Biol. Chem.* **1994**, *269*, 23328–23335.
16. Belanger, A.E.; Inamine, J.M. Genetics of Cell Wall Biosynthesis. In *Molecular Genetics of Mycobacteria*; Hatfull, G.F., Jacobs, W.R., Jr., Eds.; ASM Press: Washington, DC, USA, 2000; pp. 191–202.
17. Trefzer, C.; Rengifo-Gonzalez, M.; Hinner, M.J.; Schneider, P.; Makarov, V.; Cole, S.T.; Johnsson, K. Benzothiazinones: Prodrugs that covalently modify the decaprenylphosphoryl-beta-D-ribose 2′-epimerase DprE1 of *Mycobacterium tuberculosis*. *J. Am. Chem. Soc.* **2010**, *132*, 13663–13665. [CrossRef] [PubMed]
18. Trefzer, C.; Skovierova, H.; Buroni, S.; Bobovska, A.; Nenci, S.; Molteni, E.; Pojer, F.; Pasca, M.R.; Makarov, V.; Cole, S.T.; et al. Benzothiazinones are suicide inhibitors of mycobacterial decaprenylphosphoryl-beta-D-ribofuranose 2′-oxidase DprE1. *J. Am. Chem. Soc.* **2012**, *134*, 912–915. [CrossRef] [PubMed]
19. Neres, J.; Pojer, F.; Molteni, E.; Chiarelli, L.R.; Dhar, N.; Boy-Rottger, S.; Buroni, S.; Fullam, E.; Degiacomi, G.; Lucarelli, A.P.; et al. Structural basis for benzothiazinone-mediated killing of *Mycobacterium tuberculosis*. *Sci. Transl. Med.* **2012**, *4*, 150ra121. [CrossRef]
20. Makarov, V.; Lechartier, B.; Zhang, M.; Neres, J.; van der Sar, A.M.; Raadsen, S.A.; Hartkoorn, R.C.; Ryabova, O.B.; Vocat, A.; Decosterd, L.A.; et al. Towards a new combination therapy for tuberculosis with next generation benzothiazinones. *EMBO Mol. Med.* **2014**, *6*, 372–383. [CrossRef] [PubMed]
21. Innovative Medicines for Tuberculosis. Available online: http://im4tb.org (accessed on 24 February 2020).
22. Mariandyshev, A.O.; Khokhlov, A.L.; Smerdin, S.V.; Shcherbakova, V.S.; Igumnova, O.V.; Ozerova, I.V.; Bolgarina, A.A.; Nikitina, N.A. The main results of clinical trials of the efficacy, safety and pharmacokinetics of the perspective anti-tuberculosis drug Macozinone (PBTZ169). *Terapevt. Arkh.* **2020**, *92*, 61–72.
23. Study to Evaluate the Safety, Tolerability, Pharmacokinetics and Ex-Vivo Antitubercular Activity of PBTZ169 Formulation. Available online: https://clinicaltrials.gov/ct2/show/NCT03423030 (accessed on 25 February 2020).
24. Study to Evaluate the Safety, Tolerability and Pharmacokinetics of PBTZ169 in Multiple Dosing. Available online: https://clinicaltrials.gov/ct2/show/NCT03776500 (accessed on 25 February 2020).
25. Phase 1 Study of PBTZ169. Available online: https://clinicaltrials.gov/ct2/show/NCT03036163 (accessed on 25 February 2020).
26. Safety, Tolerability, Pharmacokinetics and Food Effects Study of PBTZ169. Available online: https://clinicaltrials.gov/ct2/show/study/NCT04150224 (accessed on 25 February 2020).
27. Phase 2a Study of PBTZ169. Available online: https://clinicaltrials.gov/ct2/show/NCT03334734 (accessed on 25 February 2020).
28. WHO Drug Information Volume 32, N° 3. 2018. Available online: https://www.who.int/medicines/publications/druginformation/issues/DrugInformation2018_Vol32-3/en/ (accessed on 25 February 2020).
29. Piton, J.; Vocat, A.; Lupien, A.; Foo, C.S.; Riabova, O.; Makarov, V.; Cole, S.T. Structure-Based drug design and characterization of sulfonyl-piperazine benzothiazinone inhibitors of DprE1 from *Mycobacterium tuberculosis*. *Antimicrob. Agents Chemot.* **2018**, *62*. [CrossRef] [PubMed]
30. Lupien, A.; Vocat, A.; Foo, C.S.; Blattes, E.; Gillon, J.Y.; Makarov, V.; Cole, S.T. Optimized background regimen for treatment of active tuberculosis with the next-generation benzothiazinone macozinone (PBTZ169). *Antimicrob. Agents Chemot.* **2018**, *62*, e00840-18. [CrossRef]

31. Spaggiari, D.; Desfontaine, V.; Cruchon, S.; Guinchard, S.; Vocat, A.; Blattes, E.; Pitteloud, J.; Ciullini, L.; Bardinet, C.; Ivanyuk, A.; et al. Development and validation of a multiplex UHPLC-MS/MS method for the determination of the investigational antibiotic against multi-resistant tuberculosis macozinone (PBTZ169) and five active metabolites in human plasma. *PLoS ONE* **2019**, *14*, e0217139. [CrossRef]
32. Degiacomi, G.; Belardinelli, J.M.; Pasca, M.R.; De Rossi, E.; Riccardi, G.; Chiarelli, L.R. Promiscuous targets for antitubercular drug discovery: The paradigm of DprE1 and MmpL3. *Appl. Sci.* **2020**, *10*, 623. [CrossRef]
33. Lechartier, B.; Rybniker, J.; Zumla, A.; Cole, S.T. Tuberculosis drug discovery in the post-post-genomic era. *EMBO Mol. Med.* **2014**, *6*, 158–168. [CrossRef]
34. Brecik, M.; Centarova, I.; Mukherjee, R.; Kolly, G.S.; Huszar, S.; Bobovska, A.; Kilacskova, E.; Mokosova, V.; Svetlikova, Z.; Sarkan, M.; et al. DprE1 is a vulnerable tuberculosis drug target due to its cell wall localization. *ACS Chem. Biol.* **2015**, *10*, 1631–1636. [CrossRef] [PubMed]
35. A Single Ascending Dose Study of BTZ043. Available online: https://clinicaltrials.gov/ct2/show/NCT03590600 (accessed on 25 February 2020).
36. BTZ-043—Multiple Ascending Dose (MAD) to Evaluate Safety, Tolerability and Early Bactericidal Activity (EBA). Available online: https://clinicaltrials.gov/ct2/show/NCT04044001 (accessed on 25 February 2020).
37. Chatterji, M.; Shandil, R.; Manjunatha, M.R.; Solapure, S.; Ramachandran, V.; Kumar, N.; Saralaya, R.; Panduga, V.; Reddy, J.; Prabhakar, K.R.; et al. 1,4-azaindole, a potential drug candidate for treatment of tuberculosis. *Antimicrob. Agents Chemother.* **2014**, *58*, 5325–5331. [CrossRef] [PubMed]
38. A Phase 1 Study to Evaluate Safety, Tolerability, PK, and PK Interactions of TBA-7371. Available online: https://clinicaltrials.gov/ct2/show/NCT03199339 (accessed on 25 February 2020).
39. Early Bactericidal Activity of TBA-7371 in Pulmonary Tuberculosis. Available online: https://clinicaltrials.gov/ct2/show/NCT04176250 (accessed on 25 February 2020).
40. A Phase 1/2 Trial of Multiple oral Doses of OPC-167832 for Uncomplicated Pulmonary Tuberculosis. Available online: https://clinicaltrials.gov/ct2/show/NCT03678688 (accessed on 25 February 2020).
41. Lechartier, B.; Cole, S.T. Mode of action of clofazimine and combination therapy with benzothiazinones against *Mycobacterium tuberculosis*. *Antimicrob. Agents Chemother.* **2015**, *59*, 4457–4463. [CrossRef] [PubMed]

© 2020 by the authors. Licensee MDPI, Basel, Switzerland. This article is an open access article distributed under the terms and conditions of the Creative Commons Attribution (CC BY) license (http://creativecommons.org/licenses/by/4.0/).

Review

Oxidative Phosphorylation—an Update on a New, Essential Target Space for Drug Discovery in *Mycobacterium tuberculosis*

Caroline Shi-Yan Foo [1], Kevin Pethe [2,*] and Andréanne Lupien [3,4,*]

1. Laboratory of Virology and Chemotherapy, Rega Institute, Leuven 3000, Belgium; caroline.foo@kuleuven.be
2. Lee Kong Chian School of Medicine and School of Biological Sciences, Nanyang Technological University, Singapore 636921, Singapore
3. Infectious Diseases and Immunity in Global Health Program, Research Institute of the McGill University Health Centre, Montreal, QC H4A 3J1, Canada
4. McGill International TB Centre, Montreal, QC H4A 3J1, Canada
* Correspondence: kevin.pethe@ntu.edu.sg (K.P.); andreanne.lupien@mail.mcgill.ca (A.L.)

Received: 29 February 2020; Accepted: 24 March 2020; Published: 29 March 2020

Abstract: New drugs with new mechanisms of action are urgently required to tackle the global tuberculosis epidemic. Following the FDA-approval of the ATP synthase inhibitor bedaquiline (Sirturo®), energy metabolism has become the subject of intense focus as a novel pathway to exploit for tuberculosis drug development. This enthusiasm stems from the fact that oxidative phosphorylation (OxPhos) and the maintenance of the transmembrane electrochemical gradient are essential for the viability of replicating and non-replicating *Mycobacterium tuberculosis* (*M. tb*), the etiological agent of human tuberculosis (TB). Therefore, new drugs targeting this pathway have the potential to shorten TB treatment, which is one of the major goals of TB drug discovery. This review summarises the latest and key findings regarding the OxPhos pathway in *M. tb* and provides an overview of the inhibitors targeting various components. We also discuss the potential of new regimens containing these inhibitors, the flexibility of this pathway and, consequently, the complexity in targeting it. Lastly, we discuss opportunities and future directions of this drug target space.

Keywords: *Mycobacterium tuberculosis*; energy metabolism; electron transport chain; oxidative phosphorylation; tuberculosis; drug discovery; bedaquiline; Q203

1. Introduction

Tuberculosis (TB) continues to have a significant global burden by remaining one of the top ten killers worldwide and the leading cause of death due to a single infectious agent [1]. The causative agent, *Mycobacterium tuberculosis* (*M. tb*), has evolved to adapt to a range of environmental stresses encountered during infection, giving rise to sub-populations of bacteria in heterogenous metabolic states, from actively replicating, to slow-growing, and dormant, non-replicating bacteria [2]. Conventional anti-TB drugs primarily target processes required for cell growth and replication, and are less efficacious against non-replicating bacteria, resulting in lengthy treatments in both drug-susceptible (DS-TB) and drug-resistant TB (DR-TB) cases [1,3]. In addition, current treatments for DR-TB are associated with low cure rates and toxicity [4–6]. Therefore, the successful elimination of the disease requires the development of new anti-TB drugs ideally active against non-replicating bacterial subpopulations and drug-resistant strains.

A new target space that has garnered increasing interest is energy metabolism, and in particular, the oxidative phosphorylation (OxPhos) pathway. The first clinical validation of this pathway as a mycobacterial drug target occurred in 2012, with the FDA approval of bedaquiline (BDQ) for treating

DR-TB by targeting the mycobacterial F_1F_0 ATP synthase [7]. This has been followed by the regulatory approvals of nitroimidazoles delamanid (Deltyba; DEL) and pretomanid (PA-824) [8,9], with the demonstration of PA-824 killing non-replicating *M. tb* by inhibition of the respiratory cytochromes through the release of nitric oxide (NO) [10]. Additionally, Telacebec (Q203), which is currently in Phase 2 clinical trials, targets the primary terminal oxidase cytochrome bcc-aa3 complex (Cyt-bcc-aa3) of *M. tb* [11,12]. Altogether, this highlights the vulnerability of energy metabolism in this pathogen and the sensitivity of components of the OxPhos pathway to specific chemical inhibition. Despite the high degree of conservation of this pathway between prokaryotes and eukaryotes, the ability to specifically target mycobacterial components without affecting the human mitochondrial counterparts has alleviated safety concerns, opening up a new and attractive target space for drug discovery.

From a biological viewpoint, this pathway is promising as a drug target space for several reasons. Whereas substrate-level phosphorylation can provide a sufficient amount of energy for replication in many bacteria, *M. tb* is dependent on the more energetically-efficient OxPhos to sustain growth [13–15]. This is probably linked to the absence of a NADH-dependent lactate dehydrogenase in mycobacteria, which would limit efficient fermentation [16,17]. During the OxPhos process, electrons are transferred from electron donors of central metabolism to oxygen through the electron transport chain (ETC). The energy released in this transfer is conserved by the proton-pumping components of the ETC, generating an electrochemical gradient in the form of a proton motive force (PMF). This stored energy is then mined through the ATP synthase as protons flow back down the gradient to yield ATP. Since a sustained PMF and ATP replenishment are essential even for the viability of non-replicating *M. tb* [18–21], it is believed that inhibition of the OxPhos pathway is an effective strategy to eradicate non-replicating subpopulations, thereby shortening future treatment durations. Moreover, since drug efflux pumps are energy-dependent in actively transporting drugs out of the bacterial cell, perturbation of the PMF and ATP levels may interfere with their function as well [13]. It has been demonstrated that efflux pump activities play a significant role in drug susceptibility of *M. tb* [22–27] and, therefore, perturbation of the OxPhos pathway may indirectly aid in overcoming issues of efflux-pump mediated drug resistance.

In light of these significant interests in targeting energy metabolism in mycobacteria, this review aims to (1) summarise key findings of the main components comprising the OxPhos pathway of *M. tb*, (2) provide an overview of validated inhibitors of this pathway which are either in the discovery stage, in clinical development, or have been approved, (3) discuss the potential of new regimens containing these inhibitors, (4) highlight possible concerns in targeting this pathway, and (5) discuss future opportunities and directions for drug development in this space.

2. Main Components of the OxPhos Pathway in *M. tb*

The ETC of *M. tb* comprises nine primary dehydrogenases, which fuel the respiratory chain as electron donors, and four-terminal oxidoreductases, which catalyse the transfer of electrons to terminal electron acceptors such as oxygen [14]. The lipoquinone, menaquinone (MK), serves as the intermediary electron carrier between the primary dehydrogenases and terminal oxidoreductases. In the following section, we summarise the main components of the pathway against which inhibitors have been discovered and those which are potentially suitable drug targets. In particular, this includes the NADH dehydrogenases, succinate dehydrogenase, the MK biosynthesis pathway, the two terminal oxidases, ATP synthase and the PMF (Figure 1).

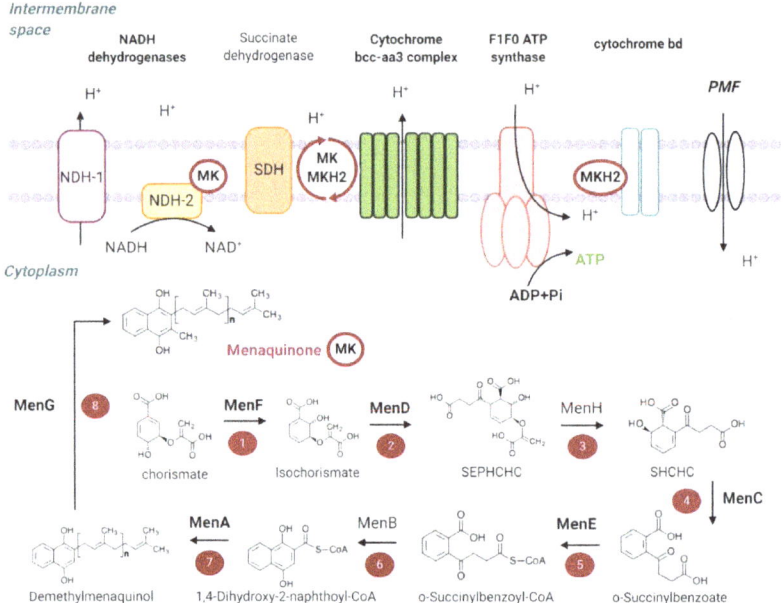

Figure 1. Schematic representation of the *Mycobacterium tuberculosis* (*M. tb*) electron transport chain (ETC) and menaquinone (MK) biosynthesis pathway. Electrons provided by NADH are fuelled to the ETC by the NADH dehydrogenases (mainly NDH-2 in *M. tb*), leading to the reduction of menaquinone (MKH2). MK is the sole carrier of electrons across the ETC and eight main enzymes are involved in its biosynthesis. The succinate dehydrogenases Sdh1 and Sdh2 also contribute to the reduction of the menaquinone pool. Electrons from the menaquinone pool are then transferred to one of the terminal oxidases, the proton-pumping Cyt-bcc-aa3 or the Cyt-bd oxidase, both of which catalyse the reduction of oxygen to water. Along the respiratory chain, protons are pumped across the membrane, which generate the proton motive force (PMF) that is then used by the F_1F_0 ATP synthase to produce ATP.

2.1. NADH Dehydrogenases

M. tb possesses three membrane-bound NADH dehydrogenases that are capable of transferring electrons to MK with the oxidation of NADH to NAD+. There is a proton-translocating type I NADH dehydrogenase (NDH-1) and two non-proton-translocating type II NADH dehydrogenases (NDH-2) [28]. NDH-2 has been chemically validated as a drug target in *M. tb* [28–30]. Unlike NDH-1, it is absent in the mammalian genome, which is an advantage to develop specific inhibitors. NDH-1 is encoded by the nuo operon, NuoA-NuoN, (rv3145-rv3158) and the two NDH-2 are identified as Ndh (encoded by ndh = rv1854c) and NdhA (encoded by ndhA = rv0392c) [17,28]. High-throughput transposon mutagenesis studies suggested that NDH-1 and NdhA are dispensable for growth, whereas Ndh is essential in vitro and in vivo [31–33]. However, recent studies have clarified the redundancy and essentiality of the NADH dehydrogenases. Detailed gene-deletion and gene-silencing studies of nuo, ndh, and ndhA performed by Vilcheze et al., established that individually, none of the NADH dehydrogenases are essential in vitro or in vivo [34]. Beites et al. further demonstrated that NDH-2 enzymes are jointly essential for growth in the presence of long-chain fatty acids, but are dispensable for growth on some carbon sources such as glycerol [35]. Moreover, a strain deficient for NDH-2 activity is only mildly attenuated in mice [35]. Both studies agree that while *M. tb* requires at least one non-proton-pumping NDH-2 enzyme for growth, chemically targeting NDH-2 alone will likely be inefficacious, but is conceivable as part of a rational drug combination.

2.2. Succinate Dehydrogenase

Succinate dehydrogenase (SDH), also known as complex II of the ETC, couples the oxidation of succinate to fumarate with the reduction of MK to menaquinol (MKH2) in *M. tb* [36]. This complex directly connects the tricarboxylic acid cycle and the OxPhos pathway, thereby playing a critical role in bacterial carbon metabolism and respiration [37]. Despite its significance, there has been a general reluctance to develop inhibitors against this complex due to the presence of a mammalian counterpart and the presence of two isotypes [38]. *M. tb* possesses two canonical SDH enzymes, Sdh1 and Sdh2, and an additional reversible fumarate reductase that can catalyse the reaction in both directions [39]. Independent transposon-site hybridization screens in *M. tb* suggest that Sdh1 is essential for optimal growth under aerobic conditions [32], whereas Sdh2 seems to play a more significant role in the growth arrest and dormancy of *M. tb* during its transition from aerobic to hypoxic conditions [40]. SDH enzymes also appear to function as an important regulator of respiration in *M. tb*. The deletion of sdh1 results in perturbation of respiration, leading to an increased rate of oxygen consumption which prevented recovery of *M. tb* from the stationary phase [41]. While the functional redundancy of these enzymes remains to be fully clarified, it is established that the SDH enzymes in *M. tb* are phylogenetically and biochemically distinct from each other, as well as from the mammalian complex [14,42,43]. On this basis, it would be theoretically possible to selectively inhibit the mycobacterial SDHs without off-target effects, although functional redundancy and structural differences between the mycobacterial SDHs would pose a challenge.

2.3. Menaquinone (MK) Biosynthesis

As the sole lipoquinone in mycobacteria, MK has a central and essential role in OxPhos [44,45]. Its biosynthetic pathway is a highly attractive drug target space as it is absent in humans, and has been demonstrated to be critical for maintaining mycobacterial survival under aerobic and anaerobic conditions [46]. In *M. tb*, the biosynthesis of MK mirrors that of *Escherichia coli*, in which the pathway has been better characterised [45]. Eight enzymes (MenF, MenD, MenH, MenC, MenE, MenB, MenA and MenG) are involved in the pathway, with MenF catalysing the first committed step of the conversion of chorismate [14]. Not all of the genes encoding for the MK synthesis pathway seem to be essential. Through initial transposon studies and confirmation by high-resolution phenotypic profiling experiments, most MK biosynthesis enzymes have been reported to be essential via transposon mutagenesis in H37Rv for *M. tb* growth, except for MenF, whose deletion resulted in a growth defect in *M. tb* [31–33].

Small molecules inhibiting MenD, MenE, MenB, MenA and MenG, and have been reported [47–51]. MenD is a thiamine diphosphate-dependent enzyme which catalyses the second step of MK biosynthesis. It uses 2-oxoglutarate and isochorismate for the synthesis of 2-Succinyl-5-enolpyruvyl-6-hydroxy-3-cyclohexene-1-carboxylate (SEPHCHC). MenE encodes for an o-succinylbenzoate-CoA ligase that converts o-succinylbenzoate (OSB) to OSB-CoA, and is the fifth step of MK biosynthesis in *M. tb*. It is responsible for the addition of a CoA group via an OSB-AMP intermediate [50]. MenB encodes for a 1,4-dihydroxy-2-naphthoyl-CoA (DHNA-CoA) synthase that catalyses the formation of DHNA-CoA from o-succinylbenzoate-CoA and is the sixth step of the MK biosynthesis [52]. MenA encodes for a DHNA-octaprenyltransferase, which is involved in the seventh step of the MK biosynthesis. It catalyses the conversion of 1,4-dihydroxy-2-naphthoate (DHNA) to demethylmenaquinone (DMK), and uses a variety of allylic isoprenyl diphosphates as substrates, requiring at least three isoprene units [46,49]. Bacterial MenA was reported to be homologous to the eukaryotic UbiA [53]. Additionally known as UbiE, MenG (rv0558) catalyses the methylation of DMK-9 using S-adenosylmethionine to form MK [54]. This reaction is the last step of MK biosynthesis in *M. tb*. That these enzymes of the MK biosynthesis pathway can be pharmacologically inhibited demonstrate their essentiality and highlight their potential as drug targets.

2.4. Terminal Oxidases

MKH2 is reoxidised with the transfer of electrons to branched routes of terminal respiratory oxidases or reductases. In *M. tb*, there are two terminal oxidases present which catalyse the four-electron reduction of oxygen to water, namely the proton-pumping Cyt-bcc-aa3 supercomplex and an alternative cytochrome bd oxidase (Cyt-bd) [39].

Cytochromes bcc and aa3 (also known as complexes III and IV) are encoded by qcrCAB and ctaB, ctaC, ctaD, ctaE, respectively, and form a supercomplex of Cyt-bcc-aa3 in mycobacteria [55–57]. As this complex is proton-pumping, it is energetically more favourable for optimum growth and seems to act as the primary respiratory route in *M. tb* under aerobic conditions [15]. Through inducible repression and genetic deletion, elegant studies revealed that, while the Cyt-bcc-aa3 complex is required for optimal *M. tb* growth rates in vitro and in mice, it is not strictly essential for growth nor persistence as long as the alternate Cyt-bd oxidase is expressed [35].

Following the first purification and initial characterisation of a hybrid respiratory complex consisting of *M. tb* cytochrome bcc and *Mycobacterium smegmatis* (*M. smeg*) cytochrome aa3 [56], the crystal structure of the Cyt-bcc-aa3 supercomplex of *M. smeg* has been recently solved by cryo-EM by two independent groups to a resolution of 3.3–3.5 Å [57,58]. This has enabled the full visualization of the supercomplex with its associated subunits, including cytochrome bcc subunits QcrC, QcrA (the Rieske iron-sulphur protein) and QcrB in a dimeric form with their respective prosthetic groups [57,58]. These studies shed light on the direct internal transfer of electrons from MKH2 to oxygen without the need for a separate cytochrome c electron shuttle as in other respiratory systems, and identifies novel subunits that contribute to the stability of the supercomplex, including an enzymatically active superoxide dismutase (SOD1). The presence of SOD1 in the bcc-aa3 supercomplex suggests the need for efficient ROS detoxification at a site dealing with a high amount of oxygen. Altogether, the structural information obtained from these studies is possibly relevant for further rationale-based TB drug discovery studies for inhibitors of this complex due to the high degree of sequence similarity of the quinol oxidation (Qp) site between *M. tb* and *M. smeg*, the site of inhibition of the TB drug candidate Q203 and other known QcrB inhibitors discovered thus far.

The bacterial-specific Cyt-bd is less characterised in *M. tb* [59]. The pathogen harbours genes encoding the two main subunits, CydA and CydB, and a putative ABC transporter CydDC, which has been proposed to be important for the assembly of the cytochrome in *E. coli* [60–63]. Genetic inactivation of the Cyt-bd-encoding genes in *M. tb* presented no loss of bacterial fitness and did not significantly impact ATP homeostasis under standard aerobic growth conditions, nor did it affect growth and persistence of *M. tb* in mice [35,64,65]. To date, two structures of the Cyt-bd have been reported in *Geobacillus thermodenitrificans* and *E. coli* [66,67], and these studies have highlighted the structural diversity within this family of enzymes. Given the low sequence homology of mycobacterial Cyt-bd to that of *G. thermodenitrificans* and *E. coli* [68], further structural studies of these enzymes in mycobacteria would be necessary to aid in drug development against the *M. tb* Cyt-bd oxidase. While structural and biochemical information are currently lacking for the *M. tb* Cyt-bd, similarities can be drawn with the *E. coli* homologue, which exhibits a very high affinity to oxygen and is non-proton-pumping, making it less efficient energetically compared to proton-pumping oxidases [69–71]. These characteristics would indicate a role of this terminal oxidase in bacterial survival in environments of low oxygen tension and protection against oxidative stress, even though it appears to function efficiently under normoxic conditions as well [65]. The upregulation of the cydAB operon has been reported for *M. tb* under hypoxia and in the presence of NO, as well as during the chronic phase of infection in mice [72–74], and when the function of the Cyt-bcc-aa3 is compromised [32,59–61,75–77].

2.5. ATP Synthase

M. tb possesses a F_1F_0 ATP synthase consisting of two functional domains, a membrane-embedded F0 unit and an external F1 domain, which is linked by central and periphery stalks on which F0 rotates [78]. This rotation is powered by a flow of protons down the electrochemical gradient of the PMF, and drives a cycle of conformational changes in the catalytic F1 domain, resulting in successive ADP binding and entrapment, ADP phosphorylation to form ATP and ATP release [79]. Conventionally, the ATP synthase is also capable of ATPase activity when intracellular ATP levels are high and the PMF is low, hydrolysing ATP while pumping protons across the cytoplasm to re-energise the PMF [80]. In mycobacteria however, the ATP synthase has been characteristically observed to have suppressed ATP hydrolysis activity, which has been postulated to be an adaptation to conserving ATP under low oxygen tensions [39,78,81,82].

In *M. tb*, the F_1F_0 ATP synthase is encoded by the atpBEFHAGDC operon and is essential for the viability of replicating and non-growing *M. tb* [31–33,83], highlighting its critical role in ATP production and in maintaining a respiratory electron flow in all metabolic states [84]. Even though intracellular ATP levels in non-replicating *M. tb* are significantly reduced compared to replicating bacilli, basal levels of ATP are still maintained in non-replicating states [19,21]. The essentiality of the ATP synthase in these conditions is underlined by the fact that its pharmacological inhibition by BDQ, a specific ATP synthase inhibitor, kills hypoxic, non-replicating *M. tb* [83].

Interestingly, the mycobacterial ATP synthase is deprived of efficient ATP hydrolysis activity [82]. As a step towards uncovering the molecular basis of the extreme latency of ATP hydrolysis, a crystal structure of the *M. smeg* catalytic F1 domain revealed similarities with the *Caldalkalibacillus thermarum* F1-ATPase, which also hydrolyses ATP poorly [85]. This is likely due to an arrest in the catalytic rotary cycle of the F1 component, resulting in the inability to release products of ATP hydrolysis. The ε-subunit of the F1 module of *M. tb*, whose structure has been recently solved by nuclear magnetic resonance (NMR), has been implicated in the regulation of ATP hydrolysis. The removal of its C-terminal resulted in an increased ATP hydrolysis rate and decreased ATP synthesis [86]. Furthermore, Saw et al., demonstrated that this site is amenable to chemical inhibition in *M. smeg* by epigallocatechin gallate, the most abundant catechin in green tea [87]. Other subunits involved in the regulation of ATP hydrolysis include the γ- and α- subunits of the F1 module [88,89]. Altogether, these advances in the understanding of the F_1F_0 ATP synthase aid in identifying new, specific inhibitors that either block *de novo* ATP synthesis and/or activate ATP hydrolysis, with the aim of depleting the residual pool of ATP in *M. tb*.

2.6. Proton Motive Force (PMF)

As in all other bacteria, an energised cell membrane is essential for the viability and growth of *M. tb* in all metabolic states [19]. The generation and maintenance of a PMF, consisting of an electrical potential due to charge separation across the membrane and a chemical potential of protons, is, therefore, vital and occurs mainly through the proton-pumping components of the ETC [14]. A PMF of about - 110 mV has been measured both in aerobic, replicating and in hypoxic, non-growing *M. tb*, indicating that even in a non-growing state, the bacteria maintain a similarly energized membrane [19]. This PMF is less than the typical PMF range of −180 to −200 mV observed in other bacteria, a difference which has been postulated to be an adaptation to host physiological environments of low nutrient levels and/or terminal electron acceptors [39,78,90].

3. Overview of the Inhibitors Targeting the OxPhos Pathway

3.1. NADH Dehydrogenases

NADH dehydrogenases play a major role in maintaining mycobacterial respiratory chain energization by using MK as an electron carrier. NDH-1 is non-essential for mycobacterial growth and for persistence [19,34]. Therefore, no drug development is currently in progress for inhibitors of mycobacterial NDH-1 [39]. NDH-2 is widespread in bacteria and the mitochondria of fungi, plants and some protists. In some organisms, more than one copy is present [91]. Notably, NDH-2 (Ndh and NdhA) are absent in mammalian genomes, suggesting a potential leading target for anti-mycobacterial drug development.

Several inhibitors have been found to target NDH-2 in mycobacteria, including the phenothiazines, quinolinyl pyrimidines and 2-mercapto-quinazolinones (Table 1). It is also thought that NDH-2 plays a role in the reduction of the anti-leprosy drug clofazimine (CFZ), leading to the production of reactive oxygen species (ROS) and, consequently, its bactericidal activity against *M. tb* [92]. However, CFZ is still potent against a *M. tb* Δndh-2 mutant strain, indicating that CFZ does not require activation by NDH-2 to exert its anti-mycobacterial potency [35]. Further studies will be needed to understand the mechanism of action of CFZ against mycobacteria. The CFZ analogue TBI-166 is more potent than CFZ against *M. tb in vitro*, and it is at least as potent as CFZ in acute and chronic murine models of TB while being potentially associated with less skin discoloration [93]. A study by Beites et al. demonstrated that the inactivation of both NDH-2 (Δndh-2) can be achieved in the absence of fatty acids in the growth media. Interestingly, rotenone, an inhibitor of NDH-1, has bactericidal activity against Δndh-2, which suggests that NDH-1 and NDH-2 have a redundant role in *M. tb* [35].

The phenothiazines, a class of antipsychotic drugs, including thioridazine, chlorpromazine and trifluoperazine, have been widely studied against *M. tb*. They have activity in vitro against *M. tb* with minimum inhibitory concentration (MIC) of thioridazine, chlorpromazine and trifluoperazine ranging between 5 to 32 µg/mL [94–98]. However, their limited potency coupled with an unfavourable toxicological profile preclude clinical use for TB treatment [99]. However, phenothiazines accumulate in macrophages and are particularly potent against intracellular *M. tb* [100,101]. Current efforts aim at developing new phenothiazine derivatives with improved activity against *M. tb* and reduced toxicity [102–105].

Another class of NDH-2 inhibitors is the quinolinyl pyrimidines [106]. This class of compounds was identified during an NDH-2 target-based high-throughput screening of more than 100 000 compounds. The series showed a good inhibitory range against *M. tb* in vitro, with MIC_{50} below 1 µg/mL [106]. In contrast to the phenothiazines, the quinolinyl pyrimidines series is not hemolytic.

CBR-1825 and CBR-4032 are two representatives compounds of the thioquinazoline (TQZ) and tetrahydroindazole scaffolds, respectively, that were identified in a high-throughput screen of over 800 000 compounds in mycobacterial inverted membrane vesicles (IMVs) [29]. The assay was designed to identify small-molecules interfering with ATP production. CBR-1825 and CBR-4032 had an MIC_{50} of 0.43 µM and 6.6 µM, respectively, when tested, and no apparent cytotoxicity [29]. Both compounds are bactericidal against *M. tb* in vitro and seem to perturb NADH turnover. Sequencing of three escape mutants resistant to TQZ revealed a non-synonymous mutation in the promoter region of ndhA (rv0392c), suggesting that these compounds target NDH-2.

A 2-mercapto-quinazolinones cluster of hits (1, 2, and 3) was identified in a screening of a commercial diversity library [30]. Compound 1 potently inhibiting *M. tb* growth in vitro in the low micromolar range without cytotoxicity against HepG2 cells [30]. Bioenergetic analyses conducted in cells in which the membrane potential was uncoupled from ATP production revealed a decrease in oxygen consumption rates (OCR) in response to the inhibitor, while IMVs showed mercapto-quinazolinone-dependent inhibition of ATP production when NADH was the primary electron donor of the respiratory chain. Enzyme kinetic studies further demonstrated non-competitive inhibition of recombinant *M. tb* Ndh protein [30]. Resistance to the compounds in *M. tb* was conferred by promoter mutations in ndhA. Interestingly, hypersusceptibility to this class of compounds was observed when ndhA was deleted, suggesting that NDH-1 and other electron donors cannot compensate for the inhibition of NDH-2 by 2-mercapto-quinazolinones [30]. Chemical optimisation is required to improve the pharmacokinetics (PK) properties of this interesting series [30].

Quinolinequinones (QQ) exert mycobactericidal activity through NDH-2 inhibition. Interestingly, QQ prevent the emergence of persistent cells in a time- and dose-dependant manner in *Mycobacterium bovis* BCG in vitro [107]. This scaffold is known for its anti-cancer and anti-inflammatory properties, and can be modified to give rise to derivatives with anti-mycobacterial activity [108–110]. Mulchin BJ et al. synthesised a range of 6,7-substituted-5,8-quinolinequinones, and their anti-mycobacterial activities were assessed against *M. bovis* BCG, in addition to their anti-tumour and anti-inflammatory properties [111]. Several compounds containing amine and halogen functionality exhibited tuberculostatic activity in the range 3–12.5 µg/mL [111]. The derivative QQ8c stimulated NADH oxidation in *M. tb* and *M. smeg* IMVs. Ndh overexpression enhanced the stimulation of NADH oxidation [107]. This increase in NDH-2 catalytic activity has been associated with the production of lethal concentrations of ROS. Recently, Santoso et al. synthesised a library of 32 new QQ derivatives, with one derivative (16b) showing enhanced in vitro activity against *M. tb* [112]. Activation of NADH-dependent oxygen consumption in *M. smeg* IMVs in the presence of 16b suggests that this new inhibitor has the same anti-mycobacterial mechanism of action than QQ8c [112].

Table 1. Structures of the NADH dehydrogenase inhibitors discussed in this review.

Chemical Class	Represented by	Structure	References
Riminophenazines	Clofazimine		[113]
	TBI-166		[114]
Phenothiazines	Chlorpromazine		[99]
	Thioridazine		
	Trifluoperazine		

Table 1. *Cont.*

Chemical Class	Represented by	Structure	References
Quinolinyl pyrimidines	13a		[106]
Thioquinazoline	CBR-1825		[29]
Tetrahydroindazole	CBR-4032		[29]
2-mercapto-quinazolinone	1		[30]
Quinolinequinones	QQ8c		[107,111]
	16b		[112]

3.2. Menaquinone (MK) Biosynthesis

The essentiality of this pathway makes it an attractive target for anti-TB drug development. To date, chemical inhibitors of MenA, MenB, MenG, MenD and MenE have proven efficacious in inhibiting *M. tb* growth. Inhibitors of the MK biosynthesis discussed in this review are listed in Table 2.

Table 2. Menaquinone (MK) biosynthesis inhibitors discussed in this review.

Chemical Class	Represented by	Structure	Ref
MenD inhibitors			
methyl succinylphosphonate	2		[115]
succinylphosphonate	1		[115]
MenE inhibitors			
vinyl sulphonamide MeOSB-AVSN	3		[50]
MenA inhibitors			
benzophenone O-methyl oxime derivatives	(R)-13		[49]
4-bromophenyl)[2-fluoro-4-[[6-(methyl-2-propenylamino)hexyl]oxy]phenyl]-methanone	Ro-48-8071		[46]
aminoalkoxydiphenylmethane	CSU-20		[46,116]
bicyclic inhibitors	NM-4		[117]
MenG inhibitors			
diphenylborinic acid quinoline esters	4b		[118]
biphenyl amide	DG70		[51]

3.2.1. MenD

Studies by Fang et al. identified two compounds, methyl succinylphosphonate (2) and succinylphosphonate (1), that inhibit *M. tb* MenD at a Ki of 0.7 and 16 μM, respectively [115]. These inhibitors are structural analogues of the substrate 2-oxoglutarate and bind covalently to MenD. The resolution of the three-dimensional structure of *M. tb* MenD by NMR will support target-based drug design [119].

3.2.2. MenE

Due to the instability of OSB, efforts to create MenE inhibitors were mainly focused on the inhibition of the assembly of the OSB-AMP intermediate. Lu et al. identified three potential MenE inhibitors using the OSB as a scaffold [50,120]. In these inhibitors, the AMP moiety was replaced with a bioisosteres sulphamate, a sulphamide, and a vinyl sulphonamide group [120]. The vinyl sulphonamide compound was the most potent with IC_{50} of 5.7 µM against the purified *M. tb* MenE enzyme [50].

3.2.3. MenA

Several inhibitors have been identified to target MenA. Aurachin RE is a natural product from *Rhodococcus erythropolis* JCM6824, and harbours potent activity against a wide range of Gram-positive and Gram-negative bacteria, including *Corynebacterium glutamicum* [121]. Using the aurachin RE scaffold, Debnath et al. synthesised a series of more than 400 derivatives that were assessed for their ability to specifically inhibit the *M. tb* MenA enzyme [49]. Compounds with IC_{50} lower than 20 µM against *M. tb* and *M. smeg*, and without activity against *Staphylococcus aureus* MenA enzyme, were identified [49]. The lead compound of this series, (R)-13, had a MIC of 2.3 µg/mL against *M. tb*. The compound is also bactericidal against hypoxic, non-replicating mycobacteria, suggesting a high MK turnover rate during persistence.

Ro 48-8071, a (4-bromophenyl)[2-fluoro-4-[[6-(methyl-2-propenylamino)hexyl]oxy] phenyl]-methanone developed by Hoffman La Roche Inc is a known inhibitor of oxidosqualene cyclase, which is involved in cholesterol biosynthesis in mammalians. It is effective against *M. tb*, *M. bovis* BCG and *M. smeg* [46,53]. The target of Ro 48-8071 was inferred from metabolic labelling experiment in *M. bovis* BCG [46]. The synthesis of a neutral, apolar liquid, which was biochemically identified as MK through the incorporation of radiolabelled isoprenoid precursors and methionine, was inhibited in the presence of Ro 48-8071 [46]. Moreover, when *M. smeg* was treated with Ro 48-8071, a reduction of 2.5–3.3-fold in the concentrations of DMK-9 and MK-9 was observed, suggesting that the compound inhibits the late steps of MK biosynthesis in mycobacteria. Ro 48-8071 decreased the oxygen consumption in *M. smeg* and *M. tb*, a phenotype that was rescued by MK supplementation. To validate MenA as the direct target of Ro 48-8071, MenA activity was assessed in membranes prepared from a recombinant *E.coli* strain expressing the *M. tb* MenA [46]. An eight-fold decrease in MenA activity was observed upon treatment with Ro 48-8071, indicating that MenA is the target of this inhibitor. Structure-activity relationship (SAR) studies led to the derivative CSU-20 which has improved MIC against *M. tb* and was bactericidal against hypoxic, non-replicating mycobacteria [46]. Dhiman et al. showed that Ro 48-8071 has dual activity against MenA by behaving as a non-competitive inhibitor of the substrate DHNA, but also as a competitive inhibitor of isoprenyl diphosphate [122].

The bicyclic long-chain fatty acids 7-methoxy-2-naphthol-based inhibitors also target MenA [117,123,124]. Berube et al. tested four compounds of this series (NM-1 to NM-4) for their ability to inhibit MenA and *M. tb* growth. The most active compound, NM-4, had an MIC_{90} of 4.5 µM and was bactericidal against replicating and non-replicating *M. tb* [117]. NM-4 was synergistic with BDQ, CFZ and imidazopyridine ND-10885 in vitro against *M. tb*, with culture sterilisation observed after 21 days of incubation. This new lead is non-cytotoxic, specific for Gram-positive bacteria, and amenable to chemical optimisation, properties that could be exploited to optimise the chemical series [124]. Further characterisation work on the PK/PD properties and in vivo activity remains to be undertaken.

3.2.4. MenG

Benkovic et al. identified inhibitors of CcrM, an essential DNA methyltransferase in the α-proteobacteria *Caulobacter crescentus*, which were later demonstrated to inhibit MenH in *B. subtilis*-an orthologue of *M. tb* MenG [118]. The derivative compound 12a, a diphenylborinic acid quinoline ester, was the most active MenH inhibitor in *Bacillus subtillis*. In *M. tb*, the MICs of the derivatives range from 0.31 to 0.64 µg/mL, and to our knowledge, no further SAR and target validation studies have been performed.

Another chemical class of MenG inhibitors, the biphenyl amides, were also identified in a whole-cell screen targeting the mycobacterial respiratory pathway. 168 small molecules with anti-TB activity from GlaxoSmithKline were screened against a *M. bovis* BCG strain containing a mWasabi reporter fused to the putative promoter of cydAB, a reporter system used to identify drugs targeting respiration [51,125]. The compound GSK1733953A, also known as DG70, induced the expression of the reporter strain compared to untreated control. This bactericidal compound showed specific activity against drug-susceptible and drug-resistant *M. tb*. DG70 was bactericidal against nutrient-starved *M. tb*, which mimics a sub-population that is particularly difficult to kill [51]. MK-4 supplementation rescued the bactericidal potency of DG70, implying that the MK biosynthesis pathway is involved in the mechanism of action of this compound. Resistant mutants selected in *M. tb* H37Rv with DG70 harboured mutations V20A, F118L, W75A, and S188A in MenG, leading to high-level resistance to DG70. DG70 was also shown to block *de novo* biosynthesis of MK, highlighting its role as an inhibitor of MenG. Interestingly, this compound showed synergism with BDQ, isoniazid (INH) and rifampicin (RIF) in time-kill assays, suggesting that MenG inhibitors could be part of a potent drug combination [51].

3.3. Inhibitors of the Cyt-bcc-aa3 Complex

A number of chemically-diverse and distinct scaffolds target the *M. tb* cytochrome bcc oxidase by binding at the Qp site of the QcrB subunit, otherwise known as the stigmatellin pocket. The Qp site is one of two quinol catalytic sites on QcrB. The promiscuous nature of this target has been attributed to its localization in the bacterial membrane, as with other membrane-associated targets [54,68]. In this section, we provide an overview of chemical classes reported to inhibit the Cyt-bcc-aa3 terminal oxidase (Table 3).

Table 3. Inhibitors of the Cyt-bcc-aa3 complex and the bd oxidase discussed in this review.

Chemical Class	Represented by	Structure	Ref
Cytochrome bcc inhibitors			
Imidazopyridines	Q203 (Telacebec)		[12]
Imidazo[2,1-b]thiazole-5-carboxamide	ND-11543		[126]
Imidazopyridine ethers	19e		[127]
Pyrazolo[1,5-a]pyridine-3-carboxamide	TB47		[128]

Table 3. *Cont.*

Chemical Class	Represented by	Structure	Ref
Lansoprazole sulphide (LPZS)			[129]
Phenoxyalkylbenzimidazole	54		[130]
Pyrrolo[3,4-c]pyridine(2H)-dione	5h		[131]
2-(quinolin-4-yloxy)acetamides	5s		[132]
	12 n		[133]
Arylvinylpiperazine amides	AX-35		[76]
Morpholino thiophenes	37		[134]
4-Amino-thieno[2,3-d]pyrimidines	CWHM-1023		[135]

Table 3. Cont.

Chemical Class	Represented by	Structure	Ref
2-Ethylthio-4-methylaminoquinazolines	11726148		[77]
Cyt-bd inhibitor			
Aurachin D			[136]

Several promising hits belonging to the chemical class of imidazopyridines (IPs) have been identified in independent screening programs [12,75,137–139]. Extensive efforts have been undertaken to optimise and to explore the SAR of this scaffold [126,140–143]. To date, the most advanced QcrB inhibitor is Q203, an IP derivative in Phase 2 clinical trials. The drug candidate was initially discovered in a phenotypic high-content screen in *M. tb*-infected macrophages and further optimised to achieve an MIC_{50} of 2.7 nM against *M. tb* H37Rv in vitro. Its activity is comparable to that of BDQ and INH in a mouse model of chronic TB infection [12]. In addition to its potency, the positive results from the Phase 1 and Phase 2a early bactericidal activity clinical trial indicate its safety and tolerability in TB patients, emphasizing the specificity of bacterial inhibition without affecting the human counterpart [144]. Resistant mutants generated to Q203 revealed amino acid mutations T313A or T313I in QcrB [12]. This interaction between Q203 and QcrB has recently been biochemically demonstrated by in-cell NMR using whole *M. smeg* cells expressing the *M. tb* QcrCAB complex and a Q203 derivative, indicating target engagement at the Qp site of QcrB [145]. While bacteriostatic on its own in *M. tb*, Q203 becomes highly bactericidal when the alternate Cyt-bd oxidase is absent, as demonstrated in *M. tb*-infected macrophages and in a mouse model of acute TB infection [65,146]. Q203 is also highly bactericidal and potent against *Mycobacterium ulcerans*, reducing bacterial load in a mouse footpad infection model of Buruli ulcer by 99.99% at a dose of 0.5 mg/kg administered three times per week for four weeks [147]. This highlights additional opportunities for repurposing QcrB inhibitors against other organisms that do not harbour functional alternative terminal oxidases, such as *M. ulcerans* and *Mycobacterium leprae*.

Extended SAR studies of IPs imidazopyridine carboxamides led to the novel scaffold imidazothiazole carboxamides (ITA), of which analogues have nanomolar potency (MIC < 10 nM) against replicating and drug-resistant *M. tb* [143]. Derivatives of this class also demonstrate intracellular activity in macrophages and in vivo [143]. Additionally, cross-resistance studies with strains harbouring QcrB mutations and dose-response studies in a *M. tb* strain without Cyt-bd oxidase indicate that these compounds target QcrB [143]. This series is promising for its tolerability and good oral bioavailability in mice [126].

Imidazopyridine ethers (IPEs) were identified from a biochemical screen of AstraZeneca's corporate compound collection aimed at identifying inhibitors of ATP homeostasis [127]. The extent of ATP synthesis inhibition, as measured in IMV of *M. smeg*, correlated well with anti-tubercular potency in *M. tb*, with a MIC of 0.03 µM for some optimised derivatives [127]. While generating resistant mutants to the series was unsuccessful, target site deconvolution through a series of biochemical tests and cell-based assays indicated that these chemical entities target the Cyt-bcc-aa3 complex. These series have been deprioritised by the group due to their poor solubility and poor PK profile [127].

Structurally similar to the IPs, the pyrazolopyridine carboxamide series were designed as novel anti-tubercular agents through a scaffold hopping strategy [148,149]. The lead compound, TB-47, is currently in early-stage development. It has activity against replicating *M. tb* in vitro with MICs of 0.016 to 0.5 µg/mL when tested against a panel of *M. tb* drug-susceptible and drug-resistant clinical isolates, and a high-selective index. Furthermore, TB-47 is well-tolerated and has good oral bioavailability in rats. While it did not display bactericidal activity alone in mouse models of acute and chronic TB infection, TB-47 potentiated the activity of pyrazinamide (PZA) and RIF during an acute TB infection in mice [128]. The authors speculated that alteration of the NADH/NAD$^+$ ratio triggered by QcrB inhibition sensitized the bacteria to the front-line drugs. While this hypothesis is interesting, it remains to be further explored since no positive interactions between RIF and the QcrB inhibitor Q203 in vitro have been observed [150]. Resistant mutants to TB-47 were generated in a *M. smeg* strain with a deletion of Cyt-bd oxidase, and whole-genome sequencing of the isolates revealed mutation H190Y in QcrB of *M. smeg* [128]. This residue, equivalent to H195 in *M. tb*, is located in the cd2 loop of QcrB, which is thought to interact with the Qp site [128].

A host cell-based screen of FDA-approved drugs of the Prestwick chemical library for compounds that protect lung fibroblasts from *M. tb*-induced cytotoxicity identified the gastric proton-pump inhibitor, lansoprazole (LPZ, Prevacid), as a potent hit compound [129]. LPZ is a prodrug that requires the host cell environment for conversion into lansoprazole sulphide (LPZS). Oral administration of LPZS significantly reduced bacterial burden during an acute TB infection in mice, and the compound lacks cytotoxicity, being well-tolerated in mice at a dose of 300 mg/kg. Whole-genome sequencing of resistant mutants generated to LPZS revealed mutation L176P in QcrB, and cross-resistance studies with imidazopyridine amide compounds and the T313A mutant strain indicated a distinct binding mode [129]. A follow-up study demonstrated that when LPZ is administered orally or intraperitoneally (i.p.) to rats, conversion to the active metabolite LPZS is undetectable in plasma and lung tissue, whereas LPZS was highly stable after i.p. administration [151]. In spite of this, LPZ intake has been found to significantly protect against TB incidence in individuals based on a cohort study conducted using the United Kingdom Clinical Practice Research Datalink [152].

The phenoxyalkylbenzimidazole (PAB) series, with an excellent growth-inhibitory potency and low cytotoxicity, is another promising QcrB inhibitor [130,153]. The lead compound, 54, demonstrated good efficacy against intracellular *M. tb*. Resistant mutants to PAB isolated in *M. tb* revealed mutations in qcrB and rv1339, a gene of unknown function. Through additional cross-resistance studies, QcrB was implicated as the main target [130]. The mutations identified in QcrB were A179P, M342T, W312C, and W312G, residues of the Qp binding site. Interestingly, while PAB compounds are bacteriostatic against actively growing *M. tb*, as with other QcrB inhibitors such as Q203 and LPZS, they are bactericidal against *M. tb* grown under conditions of nutrient starvation, in contrast to the lack of activity of other QcrB inhibitors in non-replicating models [129]. This discrepancy may be due to a potential involvement of a secondary target of PAB, for instance, rv1339, the second gene in which mutations were identified from resistant mutants.

The screening of a library of small polar molecules in a distinctive chemical space conducted at the Novartis Institute for Tropical Diseases led to the identification of hit compound pyrrolo[3,4-c]pyridine-1,3(2H)-dione, which was further optimised to increase its stability by replacing the ester moiety with a methyl oxadiazole bioisostere [131]. Lead compound 5h has an MIC$_{90}$ below 0.156 µM in replicating *M. tb*, and is non-cytotoxic. The high clearance and poor plasma exposure of 5h in mouse PK studies could be improved by co-dosing with a pan-CYP inhibitor, such that a compound concentration above that of the MIC could be achieved in the blood for 6 h at a dose of 20 mg/kg. While attempts at isolating resistant mutants to this novel chemical class were unsuccessful, *M. tb* with a Cyt-bd oxidase deletion was hypersusceptible to the analogues. Moreover, the presence of mutation A317T in QcrB in this strain rendered *M. tb* resistant to the compounds, thus implicating QcrB in the mechanism of action of this class. Further studies on the pyrrolo[3,4-c]pyridine-1,3(2H)-diones will be aimed at improving their plasma concentration in vivo.

In 2013, GlaxoSmithKline published the results of a large phenotypic screening campaign against *M. bovis* BCG and *M. tb*, in which 177 hits belonging to several structurally distinct groups were identified [125]. To date, these chemical starting points have led to the discovery of two new classes of QcrB inhibitors, the 2-quinolin-4-yloxyacetamides (QOA) and the arylvinylpiperazine amides. Several

SAR studies of the QOAs have been reported and are in agreement regarding the potency of this chemical entity against *M. tb*, the lack of toxicity in a zebrafish model and the narrow-spectrum activity specific for *M. tb* [132,154,155]. Independent work through cross-resistance studies and isolation of resistance mutants harbouring the mutation T313A in QcrB confirmed cytochrome bcc as the target of QOAs [155,156]. The solubility and stability of this series, as well as their in vivo potency, remain to be addressed [133,154]. The arylvinylpiperazine amides were generated through lead optimisation and SAR studies conducted with GW861072X from the GSK screen, which was an attractive starting point for its structural simplicity and potent activity against *M. bovis* and *M. tb* [76]. The most potent analogues have improved MICs compared to the parent scaffold, displaying low cytotoxicity and activity against intracellular *M. tb*. The lead compound, AX-35 and two other analogues, AX-37 and AX-39, were also active in an acute TB infection model in mice. While isolation of *M. tb*-resistant mutants on agar proved unsuccessful, continual and increasing exposure of *M. tb* to AX-35 over several passages in liquid culture resulted in the selection of resistant clones harbouring the mutations S182P, M342V and M342I in QcrB. Cross-resistance studies provided insight on residues which were important for the binding of the compound to QcrB, indicating a slightly different interaction with the quinol binding pocket than Q203 and LPZS. Improving the metabolic stability of this series would likely improve the in vivo activity in mice, although these compounds appear to be more stable in human microsomes.

The morpholino-thiophenes were identified from the Lilly corporate library that was screened in an aerobic whole-cell phenotypic setting against *M. tb*. The series was extensively profiled and optimised, with the lead of this series having an MIC_{90} of 0.24 µM against *M. tb*, a high-selective index and improved microsomal stability [134]. In an acute model of TB infection, the lead compound reduced the bacterial burden by 0.8 log CFUs in mouse lungs. Cross-resistance studies of the compounds to *M. tb* strains harbouring QcrB T313I and M342T mutations indicated that the compounds interact with the quinol binding site of QcrB.

4-amino-thieno[2,3-d] pyrimidines were found to inhibit the growth of *M. smeg* in a screen of small-molecule nucleotide mimetics from ChemBridge Corporation. The most potent compound of the series, CWHM-1023, had an IC_{50} of 0.083 µM against *M. tb* [135]. Resistant mutants raised to CWHM-1023 in *M. tb* had mutations A178T, A178V, V338G, G175S or G315S in QcrB, and this target was further confirmed by the increased susceptibility of a *M. tb* strain lacking Cyt-bd oxidase to the compounds.

The most recent class of Cyt-bcc-aa3 inhibitors identified are the quinazoline derivatives, 2-ethylthio-4-methylaminoquinazolines [77]. The most potent derivatives have activity against in vitro and intracellular *M. tb* in the micromolar range, and have low cytotoxicity in human hepatocytes. The lead compound 11726148 has a low clearance in human microsomes and was active in an acute TB infection model. Whole-genome sequencing of *M. tb* resistant mutants to the quinazoline derivatives revealed mutations Trp312Gly and Gly175Ser in QcrB. Interestingly, one escape mutant has a SNP in qcrA leading to the substitution of Leu356Val in QcrA, the Rieske iron-sulphur protein of the Cyt-bcc-aa3. Mapping of the mutated residues on a model of the *M. smeg* Cyt-bcc-aa3 complex [58] indicated that all of the mutated residues, including Leu356Val in QcrA, map to the Qp site of QcrB. This is in line with the Q-cycle model, in which quinol oxidation occurs at the interface of cytochrome b and the 2Fe-2S cluster domain of the Rieske protein, which make up the catalytic Qp site [57,157]. Cross-resistance studies confirmed the target of QcrB, and also reveal the role of Leu356Val in QcrA in its interaction with other QcrB inhibitors, namely Q203, AX-35, and LPZS. With the implication of QcrA in the pharmacological inhibition of Cyt-bcc-aa3, novel inhibitors can be generated which target both QcrB and QcrA subunits using structure-assisted drug-design to decrease the likelihood of resistance to QcrB inhibitors. Within this novel class, lead compound 11726148 appears the most amenable for further optimisation, as its phenyl moiety can be changed on different heterocycles to improve pharmacological properties.

The Qp site of the Cyt-bcc-aa3 complex is particularly susceptible to chemical inhibition, as evident from the multitude of structurally diverse and distinct compounds detailed above. It has been consistently observed across several of the studies that characteristic consequences of cytochrome bcc inhibition include a general depletion of intracellular bacterial ATP levels, an upregulation of Cyt-bd oxidase, and an increase in OCR and bacterial respiration due to Cyt-bd [12,76,77,128–130,134,135,155,158]. This compensation by the alternate terminal oxidase leads to an incomplete respiratory shutdown in *M. tb*, resulting in the bacteriostatic nature of cytochrome bcc inhibitors alone. However, with the deletion of Cyt-bd, mycobacterial respiration is effectively blocked, and cytochrome bcc inhibitors become bactericidal [65,76,77]. Taken altogether, these concerted efforts have generated a potential pool of backup cytochrome bcc inhibitors as Q203 continues in clinical trials. Additionally, these findings have highlighted the extreme vulnerability of *M. tb* without both of its terminal oxidases, and the immense potential of both respiratory branches as drug targets.

3.4. Inhibitors of Cyt-bd Oxidase

Aurachin D is a quinone analogue of the aurachin class that inhibits Cyt-bd oxidase of *E. coli* [136] (Table 3). When tested against membrane vesicles of *M. smeg*, aurachin D demonstrated dose-dependent inhibition of oxygen consumption of up to 50% [159]. Oxygen consumption was further inhibited by 90% in membrane vesicles without QcrCAB, indicating the likelihood of aurachin D targeting Cyt-bd in mycobacteria as well. On its own, aurachin D did not display activity against replicating *M. smeg* nor *M. tb* [159,160]. In spite of the lack of observable activity alone, aurachin D in combination with Q203 resulted in a ~10-fold decrease of the MIC of Q203 against *M. tb*, as well as enhanced bactericidal killing of *M. tb* by >2 log10 CFUs, demonstrating that aurachin D does indeed potentiate the activity of Q203 [160]. However, it remains to be demonstrated that aurachin D is a specific Cyt-bd inhibitor in mycobacteria as given its structural relation with MK, it may interfere with other respiratory complexes. This is particularly important since it was recently demonstrated that small-molecules are able to enhance the bactericidal potency of QcrB inhibitors without targeting the Cyt-bd [161].

It was shown that genetically inactivating both terminal oxidases completely abolished the in vitro growth of *M. tb* [35]. This also dramatically impacted bacterial fitness in vivo in mice, with a strain having a knockdown of Cyt-bcc-aa3 and a knockout of Cyt-bd being unable to establish an initial infection [35]. Even if an initial infection could be first achieved by regulating the expression of one of the terminal oxidases, the absence of both terminal oxidases led to severe persistence defects, with a decrease of five orders of magnitude in bacterial load in 35 days [35]. These findings are consistent with previous reports, whereby the joint inactivation, either pharmacologically or genetically, of both cytochrome bcc and bd oxidases resulted in bactericidal effects on *M. tb* in vitro and in vivo [65,76,77,160]. Exploiting the synthetic lethal interaction between these two terminal oxidases is an extremely attractive approach to eradicate *M. tb*. Inhibitors of this alternate terminal oxidase may have been missed in previous screens, which may be due to the fact that many screens have been performed under conditions in which the Cyt-bd is non-essential. In light of this, screening under stressed conditions may lead to the identification of Cyt-bd inhibitors amenable to chemical optimisation.

3.5. Inhibitors of the F_1F_0 ATP Synthase

The F_1F_0 ATP synthase is a clinically-validated drug target in *M. tb*, since the approval of BDQ by the US FDA for the treatment of multi-drug resistant TB. The mechanism of action of BDQ has been extensively studied in mycobacteria [20,83,162–164]. Briefly, BDQ binds to the c-subunit and the ε-subunit of the ATP synthase by mimicking key residues in the proton transfer chain and blocking the rotary movement of the c-subunit during the catalysis of ATP [165,166], disrupting a fundamental process for bacterial survival in both actively growing and non-growing states [20,162]. The inhibition of the ATP synthase leads to ATP depletion and also dissipation of the membrane potential, as BDQ was shown to have uncoupling properties mediated by the H+/K+ antiporter [167]. Unfortunately, drug resistance to BDQ has already emerged in MDR-TB patients, with several mechanisms of resistance having been reported [27,168–170]. Predominantly, mutations in atpE, the transcriptional repressor of the MmpL5-MmpS5, and the coding region of this efflux pump have been identified in humans [170–173]. In light of the emergence of resistance and the cardiotoxicity (QT prolongation) associated with BDQ treatment [174], efforts are currently ongoing to develop optimised diarylquinoline analogues (Table 4). The development of BDQ analogues 3,5-dialkoxypyridines is particularly interesting [175–180]. The preclinical candidate of this series, TBAJ-876, is less cardiotoxic than BDQ and 10 times more potent [180]. Through the isolation of resistant mutants and NMR studies, TBAJ-876 was shown to target the c- and ε-subunits of the ATP synthase at the same binding site as BDQ, indicating that this new analogue retains the mechanism of action of BDQ [180]. Kumar et al. screened a set of 700 compounds from the CSIR-IIIM repository for inhibition of mycobacterial ATP synthase activity using IMVs of *M. smeg*, leading to the identification of two compounds, the thiazolidine 5228485 and the cyclohexanediones 5220632. Both compounds inhibited ATP synthesis in IMVs and had MICs in the low micromolar range [181]. These bactericidal compounds were active against drug-resistant *M. tb* strains and non-replicating *M. tb* without apparent cytotoxicity [181]. Mutants selected against each compound were cross-resistant to BDQ, although resistance studies additionally suggest that these compounds may inhibit a secondary target besides the ATP synthase [181]. Further SAR and mechanistic studies are needed to fully understand the mechanism of action and for the development of these two new classes of ATP synthase inhibitors.

Table 4. Inhibitors of the ATP synthase discussed in this review.

Chemical Class	Represented by	Structure	Ref
Diarylquinoline	BDQ		[83]
	TBAJ-587		[175–177]
	TBAJ-876		[175–177]
Thiazolidines	5228485		[181]

Table 4. *Cont.*

Chemical Class	Represented by	Structure	Ref
Cyclohexanediones	5220632		[181]
Squaramides	31f		[127]

Squaramides were identified in the same screen as imidazo[1,2-a]pyridine ethers (IPE) (see section inhibitors of the Cyt-bcc-aa3) [127], with both compounds identified as inhibitors of ATP synthesis. Isolation of mutants resistant to squaramides revealed the presence of mutations in the α- and c- subunits of the ATP synthase. Interestingly, these mutants showed no cross-resistance to BDQ, suggesting a different mode of interaction with the ATP synthase than BDQ. Docking analysis of the most potent squaramide derivative, 31f (MIC *M. tb* 0.8 μM), showed that squaramides bind at the interface of the α and c -subunits of the ATP synthase. Compound 31f shows good accumulation in the serum in mice at a concentration above the MIC for over 15 h when administered with 100 mg/kg ABT, and was bacteriostatic in an acute model of TB. Further optimisation of the PK properties and in vivo safety is needed to advance this compound as a potential pre-clinical candidate for TB.

3.6. PMF

A wide range of compounds currently exists that target the PMF in bacteria, including rotenone which inhibits major proton pumps and protonophores (e.g., carbonyl cyanide m-chlorophenyl hydrazone, also known as CCCP), which translocate protons across the cell membranes [14] (Table 5). More specifically to mycobacteria, pyrazinoic acid, the active form of the first-line TB drug PZA, was demonstrated to decrease PMF and ATP levels in *M. bovis* BCG [182]. Several other compounds active against *M. tb* including SQ109, BDQ and CFZ were also found to be multi-targeting by behaving as uncouplers in addition to targeting enzymes [183]. While such an intrinsic and critical characteristic of the ETC is highly attractive therapeutically, such an approach would necessitate identifying compounds that are specific to the perturbation of mycobacterial PMF.

Recently, 2-aminoimidazoles (2-AI), a class of molecules with anti-biofilm activity, was shown to revert drug tolerance in an in vitro *M. tb* biofilm model [184]. This class of compounds potentiate the activity of β-lactams by altering protein secretion and lipid export, suggesting that 2-AI may perturb membrane energization [185]. Derivative 2B8 rapidly depolarized the membrane potential of live *M. smeg*, and collapsed the ΔpH generated by *M. smeg* IMVs energized with NADH, similar to CCCP and other mycobacterial uncouplers. In addition, a decrease in the OCR and intracellular ATP levels were observed in *M. tb* upon exposure to 2B8. Taken together, the perturbations of 2B8 on the PMF, OCR and ATP synthesis validates its uncoupling activity in mycobacteria [184].

Table 5. Inhibitors of the mycobacterial proton motive force (PMF) discuss in this review.

Chemical Class	Represented by	Structure	Ref
Ethylenediamine	SQ109		[186,187]
Pyrazinamide			[188]
2-Aminoimidazoles	2B8		[184]

3.7. Respiratory Poisoning

NO is a key component of the innate immune response against intracellular pathogens like *M. tb* [189]. Two drugs, PA-824 and DEL (OPC-67683), were shown to release NO when activated by the deazaflavin-dependant nitroreductase Ddn (rv3547) of *M. tb* [10,190]. PA-824 and DEL are bicyclic nitroimidazoles that are approved for the treatment of DR-TB as part of a drug combination. Both drugs kill replicating and non-replicating *M. tb*. Under aerobic conditions, PA-824 and DEL inhibit mycolic acid synthesis [191,192], whereas, under anaerobic conditions, Singh et al. identified that the release of NO correlated with the formation of des-nitroimidazole metabolites, leading to the antimicrobial activity of PA-824 [10]. Transcriptomic analysis of *M. tb* treated with DEL and PA-824 revealed that respiratory poisoning by NO is fundamental for the activity of the drug in mycobacteria [193]. The transcriptomic profile of bacteria exposed to DEL and PA-824 is similar to potassium cyanide, a cytochrome c oxidase-specific inhibitor, which suggests that NO poisoning in *M. tb* may lead to the inhibition of the terminal oxidases [190].

4. Combinations Including ETC Inhibitors

TB treatment relies on the combination of several antibacterial agents. New regimens for TB need to be (1) effective against DS- and DR-TB, (2) contain drugs with new mechanisms of action, (3) are suitable for oral administration, and (4) do not interfere with drugs used to treat chronic conditions or chronic infections [194]. An effective regimen should combine drugs that preserve or even potentiate their activity (additivity or synergism) when given as a regimen. Due to their ability to perturb the energy metabolism of replicating and non-replicating *M. tb*, including drugs that target the ETC may shorten treatments against DS- and DR-TB. The recent approval of the BPaL regimen (BDQ-PA824-Linezolid; Nix-TB trial) for DR-TB highlights that inhibiting components of the ETC is key in developing new regimens against *M. tb*, even though the relative contribution of each drug to the sterilizing potency of the BPaL regimen remains to be further investigated in humans.

Several inhibitors in lead optimisation were tested in combination with other anti-TB drugs. PAB, a Cyt-bcc oxidase inhibitor, resulted in the synergistic killing of *M. tb* under both replicating and non-replicating conditions when combined with CFZ [195]. PABs in combination with BDQ demonstrated antagonism at early time points, particularly under non-replicating conditions. However, this antagonistic effect disappeared within three weeks, with PAB-BDQ combinations becoming highly bactericidal [195]. The specificity of the PAB series needs to be further studied since it kills nutrient-starved *M. tb* while remaining bacteriostatic against replicating mycobacteria, a property not shared with other specific Cyt-bcc inhibitors [130,153]. A MenA inhibitor, NM-4, was synergistic even at low doses together with sub-bactericidal concentrations of BDQ, CFZ, and the QcrB inhibitor ND-10885 [196], causing enhanced and efficient killing of *M. tb* in a time-kill curve assay [117]. These preliminary results underline the need to assess the efficacy of candidates in combination for TB therapy in an early stage of development.

Several ETC inhibitors are currently in ongoing pre-clinical and clinical trials to assess their efficacy against TB as part of new regimens [11]. These include BDQ, TBAJ-587 and TBAJ-876 (ATP synthase inhibitors), the first-line anti-TB drug PZA and SQ109 (inhibitors of the PMF), CFZ and its analogue TBI-166, and Q203 (inhibitor of Cyt-bcc-aa3 oxidase) [11]. SQ109 was reported to enhance the activity of anti-tuberculosis drugs INH, RIF and BDQ, and to shorten the time required to cure *M. tb*-infected mice [197]. Additionally, several inhibitors of the ETC compounds have a multi-target activity against the bacilli. The uncoupling effects of SQ109, PZA, BDQ, and CFZ suggest that these inhibitors may have a critical role due to their multi-targeting activity and should be taken into consideration for the further development of regimens against *M. tb*, particularly against DR-TB [183]. Assessing the in vitro or in vivo combinatory effects with repurposed or approved drugs for the treatment of TB has been another fruitful avenue to find new potential regimens comprising drug candidates that target the ETC. An example is TBI-166, an analogue of clofazimine with excellent potency alone or in combination with BDQ, PZA and linezolid in vitro and in vivo [198].

Q203 and BDQ were the first-in-class, orally-available representatives of the Cyt-bcc-aa3 and ATP synthase inhibitors, respectively. Q203 is currently in Phase 2b clinical trials and demonstrated potency in a 14-day, proof-of-concept design study of early bactericidal activity. It was safe and well-tolerated throughout the different dose strengths (100, 200 and 300 mg). Preliminary studies in vivo using a mouse footpad infection model of Buruli ulcer (*M. ulcerans*) showed that the addition of Q203 to the two-drug regimen of RIF and CFZ, or the three-drug regimen rifamycin, CFZ, and BDQ can decrease the treatment duration from four to two weeks, without any relapse after 12 weeks from the completion of treatment [199]. A similar approach could be developed to find an effective Q203-based regimen against *M. tb* with the addition of a chemical inhibitor of the Cyt-bd. Nonetheless, the plasticity and the possible re-routing of the mycobacterial ETC through chemical inhibition can be used against mycobacteria to conceive an effective regimen. Bioenergetics and *ex vivo* efficacy studies revealed that a combination of BDQ, CFZ and Q203 killed *M. tb* synergistically, with BDQ and Q203 potentiating CFZ's ROS production [158]. These results suggest that the potentiation of Q203 can be achieved without a Cyt-bd oxidase inhibitor as well.

With the number of BDQ-containing regimens undergoing clinical trials, BDQ seems a drug of choice for further development of TB therapy. Several Phase 1 and 2 clinical trials which include BDQ are currently in progress [194]. However, the emergence of resistance and its high cardiotoxicity may jeopardize the potency of this antitubercular drug. The development of new ATP synthase inhibitors that harbour less cardiotoxicity and are potentially less prone to the development of resistance, such as TBAJ-876 and TBAJ-587, will most likely lead to the development of new regimens including these second generation of diaryquinolines [179,180].

5. Conclusions and Perspectives

This review has highlighted the current efforts made to find new inhibitors against components of the ETC. To date, tackling the mycobacterial OxPhos pathway has been a prolific avenue in finding new inhibitors against *M. tb*, as well as other mycobacteria such as *M. ulcerans*. Deciphering the role of each new scaffold targeting ETC components will enable a further understanding of this pivotal pathway in *M. tb* survival and metabolism. Several components are still poorly exploited as potential targets for chemical inhibition, either due to their homology to eukaryotic components or their regulation as a specific bacterial response to the environment/stress (e.g., Cyt-bd). Redundancy of several components of the ETC further complicates the establishment of a background regimen comprising of ETC inhibitors. A deeper understanding of the intricately-linked energy metabolism processes would be required to target the various metabolic pathways which *M. tb* can reroute to, such that a successful regimen can be developed. Most screens to identify inhibitors of the OxPhos have been conducted using whole-cell assays, target-based assays or phenotypic screening using ATP as a readout. However, these screening methods may not be suitable for the identification of inhibitors of some components such as the Cyt-bd, which are conditionally essential under specific conditions. Therefore, a better understanding of the modulation of the ETC under the host physiological conditions encountered by *M. tb* would be of great interest for further development of energy metabolism inhibitors. Lastly, ETC inhibitors have the potential to revolutionise future TB treatments by contributing to efficacious regimens which are simpler and shorter, as evident from BDQ's role in the novel, three-drug, all-oral BPaL regimen.

Author Contributions: Conceptualization, C.S.-Y.F., K.P. and A.L.; writing—original draft preparation, C.S.-Y.F., K.P. and A.L.; writing—review and editing, C.S.-Y.F., K.P. and A.L.; visualization, C.S.-Y.F. and A.L.; project administration, K.P. and A.L.; funding acquisition, K.P. All authors have read and agreed to the published version of the manuscript.

Funding: This work was supported in part by the National Research Foundation (NRF) Singapore, NRF Competitive Research Programme (CRP), Grant Award Number NRF–CRP18–2017–01 (K.P.)

Acknowledgments: We would like to thank Marcel Behr and Jean-Yves Dubé for their contributions to the revision of the manuscript.

Conflicts of Interest: The authors declare no conflict of interest. The funders had no role in the design of the study; in the collection, analyses, or interpretation of data; in the writing of the manuscript, or in the decision to publish the results

References

1. WHO | Global Tuberculosis Report 2019. Available online: http://www.who.int/tb/publications/global_report/en/ (accessed on 17 February 2020).
2. McKinney, J.D. In vivo veritas: The search for TB drug targets goes live. *Nat. Med.* **2000**, *6*, 1330–1333. [CrossRef] [PubMed]
3. WHO | WHO Consolidated Guidelines on Drug-Resistant Tuberculosis Treatment. Available online: http://www.who.int/tb/publications/2019/consolidated-guidelines-drug-resistant-TB-treatment/en/ (accessed on 17 February 2020).
4. Sotgiu, G.; Centis, R.; D'ambrosio, L.; Migliori, G.B. Tuberculosis treatment and drug regimens. *Cold Spring Harb. Perspect. Med.* **2015**, *5*, a017822. [CrossRef] [PubMed]
5. Borisov, S.; Danila, E.; Maryandyshev, A.; Dalcolmo, M.; Miliauskas, S.; Kuksa, L.; Manga, S.; Skrahina, A.; Diktanas, S.; Codecasa, L.R.; et al. Surveillance of adverse events in the treatment of drug-resistant tuberculosis: First global report. *Eur. Respir. J.* **2019**, *54*, 1901522. [CrossRef] [PubMed]
6. Akkerman, O.; Aleksa, A.; Alffenaar, J.-W.; Al-Marzouqi, N.H.; Arias-Guillén, M.; Belilovski, E.; Bernal, E.; Boeree, M.J.; Borisov, S.E.; Bruchfeld, J.; et al. Surveillance of adverse events in the treatment of drug-resistant tuberculosis: A global feasibility study. *Int. J. Infect. Dis.* **2019**, *83*, 72–76. [CrossRef]
7. Janssen Therapeutics, Division of Janssen Products, LP. (2019) Sirturo ®: HIGHLIGHTS OF PRESCRIBING INFORMATION. Available online: https://www.accessdata.fda.gov/drugsatfda_docs/label/2019/204384s010lbl.pdf (accessed on 26 March 2020).
8. European Medicines Agency-Find Medicine-Deltyba. Available online: http://www.ema.europa.eu/ema/index.jsp?curl=pages/medicines/human/medicines/002552/human_med_001699.jsp&mid=WC0b01ac058001d124 (accessed on 31 May 2018).
9. U.S. Food and Drug Administration FDA Approves New Drug for Treatment-Resistant Forms of Tuberculosis that Affects the Lungs. Available online: http://www.fda.gov/news-events/press-announcements/fda-approves-new-drug-treatment-resistant-forms-tuberculosis-affects-lungs (accessed on 31 January 2020).
10. Singh, R.; Manjunatha, U.; Boshoff, H.I.M.; Ha, Y.H.; Niyomrattanakit, P.; Ledwidge, R.; Dowd, C.S.; Lee, I.Y.; Kim, P.; Zhang, L.; et al. PA-824 kills nonreplicating Mycobacterium tuberculosis by intracellular NO release. *Science* **2008**, *322*, 1392–1395. [CrossRef]
11. Pipeline | Working Group for New TB Drugs. Available online: https://www.newtbdrugs.org/pipeline/clinical (accessed on 17 February 2020).

12. Pethe, K.; Bifani, P.; Jang, J.; Kang, S.; Park, S.; Ahn, S.; Jiricek, J.; Jung, J.; Jeon, H.K.; Cechetto, J.; et al. Discovery of Q203, a potent clinical candidate for the treatment of tuberculosis. *Nat. Med.* **2013**, *19*, 1157–1160. [CrossRef]
13. Bald, D.; Villellas, C.; Lu, P.; Koul, A. Targeting energy metabolism in mycobacterium tuberculosis, a new paradigm in antimycobacterial drug discovery. *mBio* **2017**, *8*, e00272-17. [CrossRef]
14. Cook, G.M.; Hards, K.; Dunn, E.; Heikal, A.; Nakatani, Y.; Greening, C.; Crick, D.C.; Fontes, F.L.; Pethe, K.; Hasenoehrl, E.; et al. Oxidative phosphorylation as a target space for tuberculosis: Success, caution, and future directions. *Microbiol. Spectr.* **2017**, *5*. [CrossRef]
15. Matsoso, L.G.; Kana, B.D.; Crellin, P.K.; Lea-Smith, D.J.; Pelosi, A.; Powell, D.; Dawes, S.S.; Rubin, H.; Coppel, R.L.; Mizrahi, V. Function of the cytochrome bc1-aa3 branch of the respiratory network in mycobacteria and network adaptation occurring in response to its disruption. *J. Bacteriol.* **2005**, *187*, 6300–6308. [CrossRef]
16. Billig, S.; Schneefeld, M.; Huber, C.; Grassl, G.A.; Eisenreich, W.; Bange, F.-C. Lactate oxidation facilitates growth of Mycobacterium tuberculosis in human macrophages. *Sci. Rep.* **2017**, *7*, 6484. [CrossRef] [PubMed]
17. Cole, S.T.; Brosch, R. Deciphering the biology of Mycobacterium tuberculosis from the complete genome sequence. *Trends Biochem. Sci.* **1997**, *22*, 28–31.
18. Boshoff, H.I.M.; Barry 3rd, C.E. Tuberculosis—Metabolism and respiration in the absence of growth. *Nat. Rev. Microbiol.* **2005**, *3*, 70–80. [CrossRef] [PubMed]
19. Rao, S.P.S.; Alonso, S.; Rand, L.; Dick, T.; Pethe, K. The protonmotive force is required for maintaining ATP homeostasis and viability of hypoxic, nonreplicating mycobacterium tuberculosis. *Proc. Natl. Acad. Sci. USA* **2008**, *105*, 11945–11950. [CrossRef] [PubMed]
20. Koul, A.; Vranckx, L.; Dendouga, N.; Balemans, W.; den Wyngaert, I.V.; Vergauwen, K.; Göhlmann, H.W.H.; Willebrords, R.; Poncelet, A.; Guillemont, J.; et al. Diarylquinolines are bactericidal for dormant mycobacteria as a result of disturbed ATP homeostasis. *J. Biol. Chem.* **2008**, *283*, 25273–25280. [CrossRef] [PubMed]
21. Gengenbacher, M.; Rao, S.P.S.; Pethe, K.; Dick, T. Nutrient-starved, non-replicating Mycobacterium tuberculosis requires respiration, ATP synthase and isocitrate lyase for maintenance of ATP homeostasis and viability. *Microbiol. Read. Engl.* **2010**, *156*, 81–87. [CrossRef] [PubMed]
22. Adams, K.N.; Takaki, K.; Connolly, L.E.; Wiedenhoft, H.; Winglee, K.; Humbert, O.; Edelstein, P.H.; Cosma, C.L.; Ramakrishnan, L. Drug tolerance in replicating mycobacteria mediated by a macrophage-induced efflux mechanism. *Cell* **2011**, *145*, 39–53. [CrossRef]
23. Machado, D.; Couto, I.; Perdigão, J.; Rodrigues, L.; Portugal, I.; Baptista, P.; Veigas, B.; Amaral, L.; Viveiros, M. Contribution of efflux to the emergence of isoniazid and multidrug resistance in Mycobacterium tuberculosis. *PLoS ONE* **2012**, *7*, e34538. [CrossRef]
24. Adams, K.N.; Szumowski, J.D.; Ramakrishnan, L. Verapamil, and its metabolite norverapamil, inhibit macrophage-induced, bacterial efflux pump-mediated tolerance to multiple anti-tubercular drugs. *J. Infect. Dis.* **2014**, *210*, 456–466. [CrossRef]
25. Coelho, T.; Machado, D.; Couto, I.; Maschmann, R.; Ramos, D.; von Groll, A.; Rossetti, M.L.; Silva, P.A.; Viveiros, M. Enhancement of antibiotic activity by efflux inhibitors against multidrug resistant Mycobacterium tuberculosis clinical isolates from Brazil. *Front. Microbiol.* **2015**, *6*, 330. [CrossRef]
26. Li, G.; Zhang, J.; Guo, Q.; Jiang, Y.; Wei, J.; Zhao, L.; Zhao, X.; Lu, J.; Wan, K. Efflux pump gene expression in multidrug-resistant Mycobacterium tuberculosis clinical isolates. *PLoS ONE* **2015**, *10*, e0119013. [CrossRef] [PubMed]
27. Andries, K.; Villellas, C.; Coeck, N.; Thys, K.; Gevers, T.; Vranckx, L.; Lounis, N.; de Jong, B.C.; Koul, A. Acquired resistance of Mycobacterium tuberculosis to bedaquiline. *PLoS ONE* **2014**, *9*, e102135. [CrossRef] [PubMed]
28. Weinstein, E.A.; Yano, T.; Li, L.-S.; Avarbock, D.; Avarbock, A.; Helm, D.; McColm, A.A.; Duncan, K.; Lonsdale, J.T.; Rubin, H. Inhibitors of type II NADH:menaquinone oxidoreductase represent a class of antitubercular drugs. *Proc. Natl. Acad. Sci. USA* **2005**, *102*, 4548–4553. [CrossRef] [PubMed]
29. Harbut, M.B.; Yang, B.; Liu, R.; Yano, T.; Vilchèze, C.; Cheng, B.; Lockner, J.; Guo, H.; Yu, C.; Franzblau, S.G.; et al. Small molecules targeting mycobacterium tuberculosis Type II NADH dehydrogenase exhibit antimycobacterial activity. *Angew. Chem. Int. Ed. Engl.* **2018**, *57*, 3478–3482. [CrossRef] [PubMed]

30. Murugesan, D.; Ray, P.C.; Bayliss, T.; Prosser, G.A.; Harrison, J.R.; Green, K.; Soares de Melo, C.; Feng, T.-S.; Street, L.J.; Chibale, K.; et al. 2-Mercapto-Quinazolinones as inhibitors of Type II NADH Dehydrogenase and Mycobacterium tuberculosis: Structure-activity relationships, mechanism of action and absorption, distribution, metabolism, and excretion characterization. *ACS Infect. Dis.* **2018**, *4*, 954–969. [CrossRef] [PubMed]
31. Sassetti, C.M.; Boyd, D.H.; Rubin, E.J. Genes required for mycobacterial growth defined by high density mutagenesis. *Mol. Microbiol.* **2003**, *48*, 77–84. [CrossRef] [PubMed]
32. Griffin, J.E.; Gawronski, J.D.; Dejesus, M.A.; Ioerger, T.R.; Akerley, B.J.; Sassetti, C.M. High-resolution phenotypic profiling defines genes essential for mycobacterial growth and cholesterol catabolism. *PLoS Pathog.* **2011**, *7*, e1002251. [CrossRef]
33. DeJesus, M.A.; Gerrick, E.R.; Xu, W.; Park, S.W.; Long, J.E.; Boutte, C.C.; Rubin, E.J.; Schnappinger, D.; Ehrt, S.; Fortune, S.M.; et al. Comprehensive essentiality analysis of the mycobacterium tuberculosis genome via saturating transposon mutagenesis. *mBio* **2017**, *8*, e02133-16. [CrossRef]
34. Vilchèze, C.; Weinrick, B.; Leung, L.W.; Jacobs, W.R. Plasticity of Mycobacterium tuberculosis NADH dehydrogenases and their role in virulence. *Proc. Natl. Acad. Sci. USA* **2018**, *115*, 1599–1604. [CrossRef]
35. Beites, T.; O'Brien, K.; Tiwari, D.; Engelhart, C.A.; Walters, S.; Andrews, J.; Yang, H.-J.; Sutphen, M.L.; Weiner, D.M.; Dayao, E.K.; et al. Plasticity of the mycobacterium tuberculosis respiratory chain and its impact on tuberculosis drug development. *Nat. Commun.* **2019**, *10*, 4970. [CrossRef]
36. Maklashina, E.; Cecchini, G.; Dikanov, S.A. Defining a direction: Electron transfer and catalysis in Escherichia coli complex II enzymes. *Biochim. Biophys. Acta* **2013**, *1827*, 668–678. [CrossRef] [PubMed]
37. Hards, K.; Adolph, C.; Harold, L.K.; McNeil, M.B.; Cheung, C.-Y.; Jinich, A.; Rhee, K.Y.; Cook, G.M. Two for the price of one: Attacking the energetic-metabolic hub of mycobacteria to produce new chemotherapeutic agents. *Prog. Biophys. Mol. Biol.* **2019**, S0079610719302111. [CrossRef] [PubMed]
38. Rutter, J.; Winge, D.R.; Schiffman, J.D. Succinate dehydrogenase—Assembly, regulation and role in human disease. *Mitochondrion* **2010**, *10*, 393–401. [CrossRef] [PubMed]
39. Cook, G.M.; Hards, K.; Vilchèze, C.; Hartman, T.; Berney, M. Energetics of respiration and oxidative phosphorylation in mycobacteria. *Microbiol. Spectr.* **2014**, *2*, 389–409. [CrossRef] [PubMed]
40. Baek, S.-H.; Li, A.H.; Sassetti, C.M. Metabolic regulation of mycobacterial growth and antibiotic sensitivity. *PLoS Biol.* **2011**, *9*, e1001065. [CrossRef] [PubMed]
41. Hartman, T.; Weinrick, B.; Vilchèze, C.; Berney, M.; Tufariello, J.; Cook, G.M.; Jacobs, W.R.J., Jr. Succinate Dehydrogenase is the Regulator of Respiration in Mycobacterium tuberculosis. *PLoS Pathog.* **2014**, *10*, e1004510. [CrossRef]
42. Lemos, R.S.; Fernandes, A.S.; Pereira, M.M.; Gomes, C.M.; Teixeira, M. Quinol:fumarate oxidoreductases and succinate:quinone oxidoreductases: Phylogenetic relationships, metal centres and membrane attachment. *Biochim. Biophys. Acta* **2002**, *1553*, 158–170. [CrossRef]
43. Pecsi, I.; Hards, K.; Ekanayaka, N.; Berney, M.; Hartman, T.; Jacobs, W.R.; Cook, G.M. Essentiality of succinate dehydrogenase in mycobacterium smegmatis and its role in the generation of the membrane potential under hypoxia. *mBio* **2014**, *5*, e01093-14. [CrossRef]
44. Collins, M.D.; Jones, D. Distribution of isoprenoid quinone structural types in bacteria and their taxonomic implication. *Microbiol. Rev.* **1981**, *45*, 316–354. [CrossRef]
45. Meganathan, R. Biosynthesis of menaquinone (vitamin K2) and ubiquinone (coenzyme Q): A perspective on enzymatic mechanisms. In *Vitamins & Hormones*; Cofactor Biosynthesis; Academic Press: Cambridge, MA, USA, 2001; Volume 61, pp. 173–218.
46. Dhiman, R.K.; Mahapatra, S.; Slayden, R.A.; Boyne, M.E.; Lenaerts, A.; Hinshaw, J.C.; Angala, S.K.; Chatterjee, D.; Biswas, K.; Narayanasamy, P.; et al. Menaquinone synthesis is critical for maintaining mycobacterial viability during exponential growth and recovery from non-replicating persistence. *Mol. Microbiol.* **2009**, *72*, 85–97. [CrossRef]
47. Kurosu, M.; Crick, D. MenA is a promising drug target for developing novel lead molecules to combat mycobacterium tuberculosis. *Med. Chem.* **2009**, *5*, 197–207. [CrossRef] [PubMed]
48. Li, X.; Liu, N.; Zhang, H.; Knudson, S.E.; Li, H.-J.; Lai, C.-T.; Simmerling, C.; Slayden, R.A.; Tonge, P.J. CoA Adducts of 4-Oxo-4-phenylbut-2-enoates: Inhibitors of MenB from the M. tuberculosis Menaquinone Biosynthesis Pathway. *ACS Med. Chem. Lett.* **2011**, *2*, 818–823. [CrossRef] [PubMed]

49. Debnath, J.; Siricilla, S.; Wan, B.; Crick, D.C.; Lenaerts, A.J.; Franzblau, S.G.; Kurosu, M. Discovery of selective menaquinone biosynthesis inhibitors against Mycobacterium tuberculosis. *J. Med. Chem.* **2012**, *55*, 3739–3755. [CrossRef] [PubMed]
50. Lu, X.; Zhang, H.; Tonge, P.J.; Tan, D.S. Mechanism-based inhibitors of MenE, an acyl-CoA synthetase involved in bacterial menaquinone biosynthesis. *Bioorg. Med. Chem. Lett.* **2008**, *18*, 5963–5966. [CrossRef]
51. Sukheja, P.; Kumar, P.; Mittal, N.; Li, S.-G.; Singleton, E.; Russo, R.; Perryman, A.L.; Shrestha, R.; Awasthi, D.; Husain, S.; et al. A novel small-molecule inhibitor of the mycobacterium tuberculosis demethylmenaquinone methyltransferase meng is bactericidal to both growing and nutritionally deprived persister cells. *mBio* **2017**, *8*, e02022-16. [CrossRef]
52. Truglio, J.J.; Theis, K.; Feng, Y.; Gajda, R.; Machutta, C.; Tonge, P.J.; Kisker, C. Crystal structure of Mycobacterium tuberculosis MenB, a key enzyme in vitamin K2 biosynthesis. *J. Biol. Chem.* **2003**, *278*, 42352–42360. [CrossRef]
53. Morand, O.H.; Aebi, J.D.; Dehmlow, H.; Ji, Y.H.; Gains, N.; Lengsfeld, H.; Himber, J. Ro 48-8.071, a new 2,3-oxidosqualene:lanosterol cyclase inhibitor lowering plasma cholesterol in hamsters, squirrel monkeys, and minipigs: Comparison to simvastatin. *J. Lipid Res.* **1997**, *38*, 373–390.
54. Goldman, R.C. Why are membrane targets discovered by phenotypic screens and genome sequencing in Mycobacterium tuberculosis? *Tuberc. Edinb. Scotl.* **2013**, *93*, 569–588. [CrossRef]
55. Megehee, J.A.; Hosler, J.P.; Lundrigan, M.D. Evidence for a cytochrome bcc-aa3 interaction in the respiratory chain of Mycobacterium smegmatis. *Microbiol. Read. Engl.* **2006**, *152*, 823–829. [CrossRef]
56. Kim, M.-S.; Jang, J.; Ab Rahman, N.B.; Pethe, K.; Berry, E.A.; Huang, L.-S. Isolation and characterization of a hybrid respiratory supercomplex consisting of mycobacterium tuberculosis cytochrome bcc and mycobacterium smegmatis cytochrome aa3. *J. Biol. Chem.* **2015**, *290*, 14350–14360. [CrossRef]
57. Gong, H.; Li, J.; Xu, A.; Tang, Y.; Ji, W.; Gao, R.; Wang, S.; Yu, L.; Tian, C.; Li, J.; et al. An electron transfer path connects subunits of a mycobacterial respiratory supercomplex. *Science* **2018**, *362*, eaat8923. [CrossRef] [PubMed]
58. Wiseman, B.; Nitharwal, R.G.; Fedotovskaya, O.; Schäfer, J.; Guo, H.; Kuang, Q.; Benlekbir, S.; Sjöstrand, D.; Ädelroth, P.; Rubinstein, J.L.; et al. Structure of a functional obligate complex III 2 IV 2 respiratory supercomplex from Mycobacterium smegmatis. *Nat. Struct. Mol. Biol.* **2018**, *25*, 1128–1136. [CrossRef] [PubMed]
59. Borisov, V.B.; Gennis, R.B.; Hemp, J.; Verkhovsky, M.I. The cytochrome bd respiratory oxygen reductases. *Biochim. Biophys. Acta* **2011**, *1807*, 1398–1413. [CrossRef] [PubMed]
60. Allen, R.J.; Brenner, E.P.; VanOrsdel, C.E.; Hobson, J.J.; Hearn, D.J.; Hemm, M.R. Conservation analysis of the CydX protein yields insights into small protein identification and evolution. *BMC Genom.* **2014**, *15*, 946. [CrossRef] [PubMed]
61. Poole, R.K.; Hatch, L.; Cleeter, M.W.J.; Gibson, F.; Cox, G.B.; Wu, G. Cytochrome bd biosynthesis in Escherichia coli: The sequences of the cydC and cydD genes suggest that they encode the components of an ABC membrane transporter. *Mol. Microbiol.* **1993**, *10*, 421–430. [CrossRef]
62. Shepherd, M. The CydDC ABC transporter of Escherichia coli: New roles for a reductant efflux pump. *Biochem. Soc. Trans.* **2015**, *43*, 908–912. [CrossRef]
63. Mascolo, L.; Bald, D. Cytochrome bd in Mycobacterium tuberculosis: A respiratory chain protein involved in the defense against antibacterials. *Prog. Biophys. Mol. Biol.* **2019**. [CrossRef]
64. Berney, M.; Hartman, T.E.; Jacobs, W.R. A Mycobacterium tuberculosis Cytochrome bd Oxidase Mutant Is Hypersensitive to Bedaquiline. *mBio* **2014**, *5*, e01275-14. [CrossRef]
65. Kalia, N.P.; Hasenoehrl, E.J.; Rahman, N.B.A.; Koh, V.H.; Ang, M.L.T.; Sajorda, D.R.; Hards, K.; Grüber, G.; Alonso, S.; Cook, G.M.; et al. Exploiting the synthetic lethality between terminal respiratory oxidases to kill Mycobacterium tuberculosis and clear host infection. *Proc. Natl. Acad. Sci. USA* **2017**, *114*, 7426–7431. [CrossRef]
66. Safarian, S.; Rajendran, C.; Müller, H.; Preu, J.; Langer, J.D.; Ovchinnikov, S.; Hirose, T.; Kusumoto, T.; Sakamoto, J.; Michel, H. Structure of a bd oxidase indicates similar mechanisms for membrane-integrated oxygen reductases. *Science* **2016**, *352*, 583–586. [CrossRef]
67. Safarian, S.; Hahn, A.; Mills, D.J.; Radloff, M.; Eisinger, M.L.; Nikolaev, A.; Meier-Credo, J.; Melin, F.; Miyoshi, H.; Gennis, R.B.; et al. Active site rearrangement and structural divergence in prokaryotic respiratory oxidases. *Science* **2019**, *366*, 100–104. [CrossRef] [PubMed]

68. Lee, B.S.; Sviriaeva, E.; Pethe, K. Targeting the cytochrome oxidases for drug development in mycobacteria. *Prog. Biophys. Mol. Biol.* **2020**. [CrossRef] [PubMed]
69. D'mello, R.; Hill, S.; Poole, R.K. The cytochrome bd quinol oxidase in Escherichia coli has an extremely high oxygen affinity and two oxygen-binding haems: Implications for regulation of activity in vivo by oxygen inhibition. *Microbiology* **1996**, *142*, 755–763. [CrossRef] [PubMed]
70. Miller, M.J.; Gennis, R.B. The cytochrome d complex is a coupling site in the aerobic respiratory chain of Escherichia coli. *J. Biol. Chem.* **1985**, *260*, 14003–14008. [PubMed]
71. Jasaitis, A.; Borisov, V.B.; Belevich, N.P.; Morgan, J.E.; Konstantinov, A.A.; Verkhovsky, M.I. Electrogenic reactions of cytochrome bd. *Biochemistry* **2000**, *39*, 13800–13809. [CrossRef]
72. Gopinath, V.; Raghunandanan, S.; Gomez, R.L.; Jose, L.; Surendran, A.; Ramachandran, R.; Pushparajan, A.R.; Mundayoor, S.; Jaleel, A.; Kumar, R.A. Profiling the proteome of mycobacterium tuberculosis during dormancy and reactivation. *Mol. Cell. Proteom.* **2015**, *14*, 2160–2176. [CrossRef]
73. Cortes, T.; Schubert, O.T.; Banaei-Esfahani, A.; Collins, B.C.; Aebersold, R.; Young, D.B. Delayed effects of transcriptional responses in Mycobacterium tuberculosis exposed to nitric oxide suggest other mechanisms involved in survival. *Sci. Rep.* **2017**, *7*, 1–9. [CrossRef]
74. Shi, L.; Sohaskey, C.D.; Kana, B.D.; Dawes, S.; North, R.J.; Mizrahi, V.; Gennaro, M.L. Changes in energy metabolism of Mycobacterium tuberculosis in mouse lung and under in vitro conditions affecting aerobic respiration. *Proc. Natl. Acad. Sci. USA* **2005**, *102*, 15629–15634. [CrossRef]
75. Arora, K.; Ochoa-Montano, B.; Tsang, P.S.; Blundell, T.L.; Dawes, S.S.; Mizrahi, V.; Bayliss, T.; Mackenzie, C.J.; Cleghorn, L.A.T.; Ray, P.C.; et al. Respiratory flexibility in response to inhibition of cytochrome c oxidase in mycobacterium tuberculosis. *Antimicrob. Agents Chemother.* **2014**, *58*, 6962–6965. [CrossRef]
76. Foo, C.S.; Lupien, A.; Kienle, M.; Vocat, A.; Benjak, A.; Sommer, R.; Lamprecht, D.A.; Steyn, A.J.C.; Pethe, K.; Piton, J.; et al. Arylvinylpiperazine Amides, a New Class of Potent Inhibitors Targeting QcrB of Mycobacterium tuberculosis. *mBio* **2018**, *9*. [CrossRef]
77. Lupien, A.; Foo, C.S.-Y.; Savina, S.; Vocat, A.; Piton, J.; Monakhova, N.; Benjak, A.; Lamprecht, D.A.; Steyn, A.J.C.; Pethe, K.; et al. New 2-Ethylthio-4-methylaminoquinazoline derivatives inhibiting two subunits of cytochrome bc1 in Mycobacterium tuberculosis. *PLoS Pathog.* **2020**, *16*, e1008270. [CrossRef] [PubMed]
78. Lu, P.; Lill, H.; Bald, D. ATP synthase in mycobacteria: Special features and implications for a function as drug target. *Biochim. Biophys. Acta BBA-Bioenerg.* **2014**, *1837*, 1208–1218. [CrossRef] [PubMed]
79. Walker, J.E. The ATP synthase: The understood, the uncertain and the unknown. *Biochem. Soc. Trans.* **2013**, *41*, 1–16. [CrossRef] [PubMed]
80. von Ballmoos, C.; Cook, G.M.; Dimroth, P. Unique rotary atp synthase and its biological diversity. *Annu. Rev. Biophys.* **2008**, *37*, 43–64. [CrossRef]
81. Higashi, T.; Kalra, V.K.; Lee, S.H.; Bogin, E.; Brodie, A.F. Energy-transducing membrane-bound coupling factor-ATPase from Mycobacterium phlei. I. Purification, homogeneity, and properties. *J. Biol. Chem.* **1975**, *250*, 6541–6548.
82. Haagsma, A.C.; Driessen, N.N.; Hahn, M.-M.; Lill, H.; Bald, D. ATP synthase in slow- and fast-growing mycobacteria is active in ATP synthesis and blocked in ATP hydrolysis direction. *FEMS Microbiol. Lett.* **2010**, *313*, 68–74. [CrossRef]
83. Andries, K.; Verhasselt, P.; Guillemont, J.; Göhlmann, H.W.H.; Neefs, J.-M.; Winkler, H.; Gestel, J.V.; Timmerman, P.; Zhu, M.; Lee, E.; et al. A Diarylquinoline drug active on the ATP synthase of mycobacterium tuberculosis. *Science* **2005**, *307*, 223–227. [CrossRef]
84. Bald, D.; Koul, A. Respiratory ATP synthesis: The new generation of mycobacterial drug targets? *FEMS Microbiol. Lett.* **2010**, *308*, 1–7. [CrossRef]
85. Zhang, A.T.; Montgomery, M.G.; Leslie, A.G.W.; Cook, G.M.; Walker, J.E. The structure of the catalytic domain of the ATP synthase from Mycobacterium smegmatis is a target for developing antitubercular drugs. *Proc. Natl. Acad. Sci. USA* **2019**, *116*, 4206–4211. [CrossRef]
86. Joon, S.; Ragunathan, P.; Sundararaman, L.; Nartey, W.; Kundu, S.; Manimekalai, M.S.S.; Bogdanović, N.; Dick, T.; Grüber, G. The NMR solution structure of *Mycobacterium tuberculosis* F- ATP synthase subunit ε provides new insight into energy coupling inside the rotary engine. *FEBS J.* **2018**, *285*, 1111–1128. [CrossRef]

87. Saw, W.-G.; Wu, M.-L.; Ragunathan, P.; Biuković, G.; Lau, A.-M.; Shin, J.; Harikishore, A.; Cheung, C.-Y.; Hards, K.; Sarathy, J.P.; et al. Disrupting coupling within mycobacterial F-ATP synthases subunit ε causes dysregulated energy production and cell wall biosynthesis. *Sci. Rep.* **2019**, *9*, 1–15. [CrossRef] [PubMed]
88. Hotra, A.; Suter, M.; Biuković, G.; Ragunathan, P.; Kundu, S.; Dick, T.; Grüber, G. Deletion of a unique loop in the mycobacterial F-ATP synthase γ subunit sheds light on its inhibitory role in ATP hydrolysis-driven H+ pumping. *FEBS J.* **2016**, *283*, 1947–1961. [CrossRef] [PubMed]
89. Ragunathan, P.; Sielaff, H.; Sundararaman, L.; Biuković, G.; Manimekalai, M.S.S.; Singh, D.; Kundu, S.; Wohland, T.; Frasch, W.; Dick, T.; et al. The uniqueness of subunit α of mycobacterial F-ATP synthases: An evolutionary variant for niche adaptation. *J. Biol. Chem.* **2017**, *292*, 11262–11279. [CrossRef] [PubMed]
90. Kinoshita, N.; Unemoto, T.; Kobayashi, H. Proton motive force is not obligatory for growth of Escherichia coli. *J. Bacteriol.* **1984**, *160*, 1074–1077. [CrossRef] [PubMed]
91. Melo, A.M.P.; Bandeiras, T.M.; Teixeira, M. New insights into type II NAD (P) H:quinone oxidoreductases. *Microbiol. Mol. Biol. Rev.* **2004**, *68*, 603–616. [CrossRef]
92. Yano, T.; Kassovska-Bratinova, S.; Teh, J.S.; Winkler, J.; Sullivan, K.; Isaacs, A.; Schechter, N.M.; Rubin, H. Reduction of clofazimine by mycobacterial type 2 NADH: Quinone oxidoreductase: A pathway for the generation of bactericidal levels of reactive oxygen species. *J. Biol. Chem.* **2011**, *286*, 10276–10287. [CrossRef]
93. Xu, J.; Wang, B.; Fu, L.; Zhu, H.; Guo, S.; Huang, H.; Yin, D.; Zhang, Y.; Lu, Y. In vitro and in vivo activities of the Riminophenazine TBI-166 against Mycobacterium tuberculosis. *Antimicrob. Agents Chemother.* **2019**, *63*, e02155-18. [CrossRef]
94. Bourdon, J.L. Contribution to the study of the antibiotic properties of chlorpromazine or 4560 RP. *Ann. Inst. Pasteur* **1961**, *101*, 876–886.
95. Amaral, L.; Kristiansen, J.E.; Abebe, L.S.; Millett, W. Inhibition of the respiration of multi-drug resistant clinical isolates of Mycobacterium tuberculosis by thioridazine: Potential use for initial therapy of freshly diagnosed tuberculosis. *J. Antimicrob. Chemother.* **1996**, *38*, 1049–1053. [CrossRef]
96. Bettencourt, M.V.; Bosne-David, S.; Amaral, L. Comparative in vitro activity of phenothiazines against multidrug-resistant Mycobacterium tuberculosis. *Int. J. Antimicrob. Agents* **2000**, *16*, 69–71. [CrossRef]
97. Ratnakar, P.; Murthy, P.S. Antitubercular activity of trifluoperazine, a calmodulin antagonist. *FEMS Microbiol. Lett.* **1992**, *76*, 73–76. [CrossRef] [PubMed]
98. Gadre, D.V.; Talwar, V. In vitro susceptibility testing of Mycobacterium tuberculosis strains to trifluoperazine. *J. Chemother. Florence Italy* **1999**, *11*, 203–206. [CrossRef] [PubMed]
99. Amaral, L.; Kristiansen, J.E.; Viveiros, M.; Atouguia, J. Activity of phenothiazines against antibiotic-resistant Mycobacterium tuberculosis: A review supporting further studies that may elucidate the potential use of thioridazine as anti-tuberculosis therapy. *J. Antimicrob. Chemother.* **2001**, *47*, 505–511. [CrossRef] [PubMed]
100. Amaral, L.; Viveiros, M. Thioridazine: A non-antibiotic drug highly effective, in combination with first line anti-tuberculosis drugs, against any form of antibiotic resistance of mycobacterium tuberculosis due to its multi-mechanisms of action. *Antibiotics* **2017**, *6*, 3. [CrossRef]
101. Crowle, A.J.; Douvas, G.S.; May, M.H. Chlorpromazine: A drug potentially useful for treating mycobacterial infections. *Chemotherapy* **1992**, *38*, 410–419. [CrossRef]
102. Salie, S.; Hsu, N.-J.; Semenya, D.; Jardine, A.; Jacobs, M. Novel non-neuroleptic phenothiazines inhibit Mycobacterium tuberculosis replication. *J. Antimicrob. Chemother.* **2014**, *69*, 1551–1558. [CrossRef]
103. He, C.-X.; Meng, H.; Zhang, X.; Cui, H.-Q.; Yin, D.-L. Synthesis and bio-evaluation of phenothiazine derivatives as new anti-tuberculosis agents. *Chin. Chem. Lett.* **2015**, *26*, 951–954. [CrossRef]
104. Jardine, M.A.; Jacobs, M. Phenothiazine Derivatives and Their Use against Tuberculosis. WO2014080378A1, 30 May 2014.
105. Trivedi, A.R.; Siddiqui, A.B.; Shah, V.H. Design, synthesis, characterization and antitubercular activity of some 2-heterocycle-substituted phenothiazines. *Arkivoc* **2008**, *2008*, 210–217.
106. Shirude, P.S.; Paul, B.; Roy Choudhury, N.; Kedari, C.; Bandodkar, B.; Ugarkar, B.G. Quinolinyl Pyrimidines: Potent Inhibitors of NDH-2 as a Novel Class of Anti-TB Agents. *ACS Med. Chem. Lett.* **2012**, *3*, 736–740. [CrossRef]
107. Heikal, A.; Hards, K.; Cheung, C.-Y.; Menorca, A.; Timmer, M.S.M.; Stocker, B.L.; Cook, G.M. Activation of type II NADH dehydrogenase by quinolinequinones mediates antitubercular cell death. *J. Antimicrob. Chemother.* **2016**, *71*, 2840–2847. [CrossRef]

108. Bringmann, G.; Reichert, Y.; Kane, V.V. The total synthesis of streptonigrin and related antitumor antibiotic natural products. *Tetrahedron* **2004**, *60*, 3539–3574. [CrossRef]
109. Colucci, M.A.; Moody, C.J.; Couch, G.D. Natural and synthetic quinones and their reduction by the quinone reductase enzyme NQO1: From synthetic organic chemistry to compounds with anticancer potential. *Org. Biomol. Chem.* **2008**, *6*, 637–656. [CrossRef] [PubMed]
110. Pearce, A.N.; Chia, E.W.; Berridge, M.V.; Clark, G.R.; Harper, J.L.; Larsen, L.; Maas, E.W.; Page, M.J.; Perry, N.B.; Webb, V.L.; et al. Anti-inflammatory thiazine alkaloids isolated from the New Zealand ascidian Aplidium sp.: Inhibitors of the neutrophil respiratory burst in a model of gouty arthritis. *J. Nat. Prod.* **2007**, *70*, 936–940. [CrossRef] [PubMed]
111. Mulchin, B.J.; Newton, C.G.; Baty, J.W.; Grasso, C.H.; Martin, W.J.; Walton, M.C.; Dangerfield, E.M.; Plunkett, C.H.; Berridge, M.V.; Harper, J.L.; et al. The anti-cancer, anti-inflammatory and tuberculostatic activities of a series of 6,7-substituted-5,8-quinolinequinones. *Bioorg. Med. Chem.* **2010**, *18*, 3238–3251. [CrossRef]
112. Santoso, K.T.; Menorca, A.; Cheung, C.-Y.; Cook, G.M.; Stocker, B.L.; Timmer, M.S.M. The synthesis and evaluation of quinolinequinones as anti-mycobacterial agents. *Bioorg. Med. Chem.* **2019**, *27*, 3532–3545. [CrossRef]
113. Barry, V.C.; Belton, J.G.; Conalty, M.L.; Denneny, J.M.; Edward, D.W.; O'sullivan, J.F.; Twomey, D.; Winder, F. A new series of phenazines (rimino-compounds) with high antituberculosis activity. *Nature* **1957**, *179*, 1013–1015. [CrossRef]
114. Zhang, D.; Liu, Y.; Zhang, C.; Zhang, H.; Wang, B.; Xu, J.; Fu, L.; Yin, D.; Cooper, C.B.; Ma, Z.; et al. Synthesis and biological evaluation of novel 2-methoxypyridylamino-substituted riminophenazine derivatives as antituberculosis agents. *Molecules* **2014**, *19*, 4380–4394. [CrossRef]
115. Fang, M.; Toogood, R.D.; Macova, A.; Ho, K.; Franzblau, S.G.; McNeil, M.R.; Sanders, D.A.R.; Palmer, D.R.J. Succinylphosphonate esters are competitive inhibitors of mend that show active-site discrimination between homologous α-ketoglutarate-decarboxylating enzymes. *Biochemistry* **2010**, *49*, 2672–2679. [CrossRef]
116. Kurosu, M.; Narayanasamy, P.; Biswas, K.; Dhiman, R.; Crick, D.C. Discovery of 1,4-dihydroxy-2-naphthoate [corrected] prenyltransferase inhibitors: New drug leads for multidrug-resistant gram-positive pathogens. *J. Med. Chem.* **2007**, *50*, 3973–3975. [CrossRef]
117. Berube, B.J.; Russell, D.; Castro, L.; Choi, S.; Narayanasamy, P.; Parish, T. Novel mena inhibitors are bactericidal against mycobacterium tuberculosis and synergize with electron transport chain inhibitors. *Antimicrob. Agents Chemother.* **2019**, *63*, e02661-18. [CrossRef]
118. Benkovic, S.J.; Baker, S.J.; Alley, M.R.K.; Woo, Y.-H.; Zhang, Y.-K.; Akama, T.; Mao, W.; Baboval, J.; Rajagopalan, P.T.R.; Wall, M.; et al. Identification of borinic esters as inhibitors of bacterial cell growth and bacterial methyltransferases, CcrM and MenH. *J. Med. Chem.* **2005**, *48*, 7468–7476. [CrossRef] [PubMed]
119. Jirgis, E.N.M.; Bashiri, G.; Bulloch, E.M.M.; Johnston, J.M.; Baker, E.N. Structural views along the mycobacterium tuberculosis MenD reaction pathway illuminate key aspects of thiamin diphosphate-dependent enzyme mechanisms. *Structure* **2016**, *24*, 1167–1177. [CrossRef] [PubMed]
120. Lu, X.; Zhou, R.; Sharma, I.; Li, X.; Kumar, G.; Swaminathan, S.; Tonge, P.J.; Tan, D.S. Stable analogues of OSB-AMP: Potent inhibitors of MenE, the o-succinylbenzoate-CoA synthetase from bacterial menaquinone biosynthesis. *Chembiochem Eur. J. Chem. Biol.* **2012**, *13*, 129–136. [CrossRef] [PubMed]
121. Kitagawa, W.; Tamura, T. A Quinoline Antibiotic from Rhodococcus erythropolis JCM 6824. *J. Antibiot. (Tokyo)* **2008**, *61*, 680–682. [CrossRef]
122. Dhiman, R.K.; Pujari, V.; Kincaid, J.M.; Ikeh, M.A.; Parish, T.; Crick, D.C. Characterization of MenA (isoprenyl diphosphate:1,4-dihydroxy-2-naphthoate isoprenyltransferase) from Mycobacterium tuberculosis. *PLoS ONE* **2019**, *14*, e0214958. [CrossRef]
123. Choi, S.; Frandsen, J.; Narayanasamy, P. Novel long-chain compounds with both immunomodulatory and MenA inhibitory activities against Staphylococcus aureus and its biofilm. *Sci. Rep.* **2017**, *7*, 40077. [CrossRef]
124. Choi, S.; Larson, M.A.; Hinrichs, S.H.; Bartling, A.M.; Frandsen, J.; Narayanasamy, P. Discovery of bicyclic inhibitors against menaquinone biosynthesis. *Future Med. Chem.* **2016**, *8*, 11–16. [CrossRef]
125. Ballell, L.; Bates, R.H.; Young, R.J.; Alvarez-Gomez, D.; Alvarez-Ruiz, E.; Barroso, V.; Blanco, D.; Crespo, B.; Escribano, J.; González, R.; et al. Fueling open-source drug discovery: 177 small-molecule leads against tuberculosis. *ChemMedChem* **2013**, *8*, 313–321. [CrossRef]

126. Moraski, G.C.; Deboosère, N.; Marshall, K.L.; Weaver, H.A.; Vandeputte, A.; Hastings, C.; Woolhiser, L.; Lenaerts, A.J.; Brodin, P.; Miller, M.J. Intracellular and in vivo evaluation of imidazo[2,1-b]thiazole-5-carboxamide anti-tuberculosis compounds. *PLoS ONE* **2020**, *15*, e0227224. [CrossRef]
127. Tantry, S.J.; Markad, S.D.; Shinde, V.; Bhat, J.; Balakrishnan, G.; Gupta, A.K.; Ambady, A.; Raichurkar, A.; Kedari, C.; Sharma, S.; et al. Discovery of Imidazo[1,2-a]pyridine ethers and squaramides as selective and potent inhibitors of mycobacterial adenosine triphosphate (ATP) synthesis. *J. Med. Chem.* **2017**, *60*, 1379–1399. [CrossRef]
128. Lu, X.; Williams, Z.; Hards, K.; Tang, J.; Cheung, C.-Y.; Aung, H.L.; Wang, B.; Liu, Z.; Hu, X.; Lenaerts, A.; et al. Pyrazolo[1,5- a]pyridine inhibitor of the respiratory cytochrome bcc complex for the treatment of drug-resistant tuberculosis. *ACS Infect. Dis.* **2019**, *5*, 239–249. [CrossRef] [PubMed]
129. Rybniker, J.; Vocat, A.; Sala, C.; Busso, P.; Pojer, F.; Benjak, A.; Cole, S.T. Lansoprazole is an antituberculous prodrug targeting cytochrome bc1. *Nat. Commun.* **2015**, *6*, 7659. [CrossRef] [PubMed]
130. Chandrasekera, N.S.; Berube, B.J.; Shetye, G.; Chettiar, S.; O'Malley, T.; Manning, A.; Flint, L.; Awasthi, D.; Ioerger, T.R.; Sacchettini, J.; et al. Improved phenoxyalkylbenzimidazoles with activity against Mycobacterium tuberculosis appear to target QcrB. *ACS Infect. Dis.* **2017**, *3*, 898–916. [CrossRef] [PubMed]
131. van der Westhuyzen, R.; Winks, S.; Wilson, C.R.; Boyle, G.A.; Gessner, R.K.; Soares de Melo, C.; Taylor, D.; de Kock, C.; Njoroge, M.; Brunschwig, C.; et al. Pyrrolo[3,4-c]pyridine-1,3(2H)-diones: A novel antimycobacterial class targeting mycobacterial respiration. *J. Med. Chem.* **2015**, *58*, 9371–9381. [CrossRef] [PubMed]
132. Pissinate, K.; Villela, A.D.; Rodrigues-Junior, V.; Giacobbo, B.C.; Grams, E.S.; Abbadi, B.L.; Trindade, R.V.; Roesler Nery, L.; Bonan, C.D.; Back, D.F.; et al. 2-(Quinolin-4-yloxy)acetamides are active against drug-susceptible and drug-resistant mycobacterium tuberculosis strains. *ACS Med. Chem. Lett.* **2016**, *7*, 235–239. [CrossRef] [PubMed]
133. Giacobbo, B.C.; Pissinate, K.; Rodrigues-Junior, V.; Villela, A.D.; Grams, E.S.; Abbadi, B.L.; Subtil, F.T.; Sperotto, N.; Trindade, R.V.; Back, D.F.; et al. New insights into the SAR and drug combination synergy of 2-(quinolin-4-yloxy)acetamides against Mycobacterium tuberculosis. *Eur. J. Med. Chem.* **2017**, *126*, 491–501. [CrossRef]
134. Cleghorn, L.A.T.; Ray, P.C.; Odingo, J.; Kumar, A.; Wescott, H.; Korkegian, A.; Masquelin, T.; Lopez Moure, A.; Wilson, C.; Davis, S.; et al. Identification of morpholino thiophenes as novel mycobacterium tuberculosis inhibitors, targeting QcrB. *J. Med. Chem.* **2018**, *61*, 6592–6608. [CrossRef]
135. Harrison, G.A.; Bridwell, A.E.M.; Singh, M.; Jayaraman, K.; Weiss, L.A.; Kinsella, R.L.; Aneke, J.S.; Flentie, K.; Schene, M.E.; Gaggioli, M.; et al. Identification of 4-Amino-Thieno[2,3-d]Pyrimidines as QcrB Inhibitors in Mycobacterium tuberculosis. *mSphere* **2019**, *4*, e00606-19. [CrossRef]
136. Meunier, B.; Madgwick, S.A.; Reil, E.; Oettmeier, W.; Rich, P.R. New inhibitors of the quinol oxidation sites of bacterial cytochromes bo and bd. *Biochemistry* **1995**, *34*, 1076–1083. [CrossRef]
137. Moraski, G.C.; Markley, L.D.; Hipskind, P.A.; Boshoff, H.; Cho, S.; Franzblau, S.G.; Miller, M.J. Advent of Imidazo[1,2-a]pyridine-3-carboxamides with potent multi- and extended drug resistant antituberculosis activity. *ACS Med. Chem. Lett.* **2011**, *2*, 466–470. [CrossRef]
138. Abrahams, K.A.; Cox, J.A.G.; Spivey, V.L.; Loman, N.J.; Pallen, M.J.; Constantinidou, C.; Fernández, R.; Alemparte, C.; Remuiñán, M.J.; Barros, D.; et al. Identification of novel imidazo[1,2-a]pyridine inhibitors targeting M. tuberculosis QcrB. *PLoS ONE* **2012**, *7*, e52951. [CrossRef] [PubMed]
139. Mak, P.A.; Rao, S.P.S.; Ping Tan, M.; Lin, X.; Chyba, J.; Tay, J.; Ng, S.H.; Tan, B.H.; Cherian, J.; Duraiswamy, J.; et al. A high-throughput screen to identify inhibitors of atp homeostasis in non-replicating mycobacterium tuberculosis. *ACS Chem. Biol.* **2012**, *7*, 1190–1197. [CrossRef]
140. Moraski, G.C.; Markley, L.D.; Cramer, J.; Hipskind, P.A.; Boshoff, H.; Bailey, M.; Alling, T.; Ollinger, J.; Parish, T.; Miller, M.J. Advancement of Imidazo[1,2-a]pyridines with improved pharmacokinetics and nanomolar activity against mycobacterium tuberculosis. *ACS Med. Chem. Lett.* **2013**, *4*, 675–679. [CrossRef] [PubMed]
141. Moraski, G.C.; Oliver, A.G.; Markley, L.D.; Cho, S.; Franzblau, S.G.; Miller, M.J. Scaffold-switching: An exploration of 5,6-fused bicyclic heteroaromatics systems to afford antituberculosis activity akin to the imidazo[1,2-a]pyridine-3-carboxylates. *Bioorg. Med. Chem. Lett.* **2014**, *24*, 3493–3498. [CrossRef] [PubMed]

142. Kang, S.; Kim, R.Y.; Seo, M.J.; Lee, S.; Kim, Y.M.; Seo, M.; Seo, J.J.; Ko, Y.; Choi, I.; Jang, J.; et al. Lead optimization of a novel series of imidazo[1,2-a]pyridine amides leading to a clinical candidate (Q203) as a multi- and extensively-drug-resistant anti-tuberculosis agent. *J. Med. Chem.* **2014**, *57*, 5293–5305. [CrossRef] [PubMed]
143. Moraski, G.C.; Seeger, N.; Miller, P.A.; Oliver, A.G.; Boshoff, H.I.; Cho, S.; Mulugeta, S.; Anderson, J.R.; Franzblau, S.G.; Miller, M.J. Arrival of Imidazo[2,1-b]thiazole-5-carboxamides: Potent anti-tuberculosis agents that target QcrB. *ACS Infect. Dis.* **2016**, *2*, 393–398. [CrossRef] [PubMed]
144. Telacebec (Q203) | Working Group for New TB Drugs. Available online: https://www.newtbdrugs.org/pipeline/compound/telacebec-q203 (accessed on 3 February 2020).
145. Bouvier, G.; Simenel, C.; Jang, J.; Kalia, N.P.; Choi, I.; Nilges, M.; Pethe, K.; Izadi-Pruneyre, N. Target engagement and binding mode of an antituberculosis drug to its bacterial target deciphered in whole living cells by NMR. *Biochemistry* **2019**, *58*, 526–533. [CrossRef]
146. Moosa, A.; Lamprecht, D.A.; Arora, K.; Barry, C.E.; Boshoff, H.I.M.; Ioerger, T.R.; Steyn, A.J.C.; Mizrahi, V.; Warner, D.F. Susceptibility of mycobacterium tuberculosis cytochrome bd oxidase mutants to compounds targeting the terminal respiratory oxidase, cytochrome c. *Antimicrob. Agents Chemother.* **2017**, *61*, e01338-17. [CrossRef]
147. Scherr, N.; Bieri, R.; Thomas, S.S.; Chauffour, A.; Kalia, N.P.; Schneide, P.; Ruf, M.-T.; Lamelas, A.; Manimekalai, M.S.S.; Grüber, G.; et al. Targeting the Mycobacterium ulcerans cytochrome bc1:aa3 for the treatment of Buruli ulcer. *Nat. Commun.* **2018**, *9*, 5370. [CrossRef]
148. Tang, J.; Wang, B.; Wu, T.; Wan, J.; Tu, Z.; Njire, M.; Wan, B.; Franzblauc, S.G.; Zhang, T.; Lu, X.; et al. Design, synthesis, and biological evaluation of Pyrazolo[1,5-a]pyridine-3-carboxamides as novel antitubercular agents. *ACS Med. Chem. Lett.* **2015**, *6*, 814–818. [CrossRef]
149. Lu, X.; Tang, J.; Cui, S.; Wan, B.; Franzblauc, S.G.; Zhang, T.; Zhang, X.; Ding, K. Pyrazolo[1,5-a]pyridine-3-carboxamide hybrids: Design, synthesis and evaluation of anti-tubercular activity. *Eur. J. Med. Chem.* **2017**, *125*, 41–48. [CrossRef] [PubMed]
150. Lee, B.S.; Kalia, N.P.; Jin, X.E.F.; Hasenoehrl, E.J.; Berney, M.; Pethe, K. Inhibitors of energy metabolism interfere with antibiotic-induced death in mycobacteria. *J. Biol. Chem.* **2019**, *294*, 1936–1943. [CrossRef] [PubMed]
151. Mdanda, S.; Baijnath, S.; Shobo, A.; Singh, S.D.; Maguire, G.E.M.; Kruger, H.G.; Arvidsson, P.I.; Naicker, T.; Govender, T. Lansoprazole-sulfide, pharmacokinetics of this promising anti-tuberculous agent. *Biomed. Chromatogr.* **2017**, *31*, e4035. [CrossRef] [PubMed]
152. Yates, T.A.; Tomlinson, L.A.; Bhaskaran, K.; Langan, S.; Thomas, S.; Smeeth, L.; Douglas, I.J. Lansoprazole use and tuberculosis incidence in the United Kingdom Clinical Practice Research Datalink: A population based cohort. *PLoS Med.* **2017**, *14*, e1002457. [CrossRef]
153. Chandrasekera, N.S.; Alling, T.; Bailey, M.A.; Files, M.; Early, J.V.; Ollinger, J.; Ovechkina, Y.; Masquelin, T.; Desai, P.V.; Cramer, J.W.; et al. Identification of phenoxyalkylbenzimidazoles with antitubercular activity. *J. Med. Chem.* **2015**, *58*, 7273–7285. [CrossRef]
154. Pitta, E.; Rogacki, M.K.; Balabon, O.; Huss, S.; Cunningham, F.; Lopez-Roman, E.M.; Joossens, J.; Augustyns, K.; Ballell, L.; Bates, R.H.; et al. Searching for New Leads for Tuberculosis: Design, Synthesis, and Biological Evaluation of Novel 2-Quinolin-4-yloxyacetamides. *J. Med. Chem.* **2016**, *59*, 6709–6728. [CrossRef]
155. Phummarin, N.; Boshoff, H.I.; Tsang, P.S.; Dalton, J.; Wiles, S.; Barry 3rd, C.E.; Copp, B.R. SAR and identification of 2-(quinolin-4-yloxy)acetamides as Mycobacterium tuberculosis cytochrome bc 1 inhibitors. *Medchemcomm* **2016**, *7*, 2122–2127. [CrossRef]
156. Subtil, F.T.; Villela, A.D.; Abbadi, B.L.; Rodrigues-Junior, V.S.; Bizarro, C.V.; Timmers, L.F.S.M.; de Souza, O.N.; Pissinate, K.; Machado, P.; López-Gavín, A.; et al. Activity of 2-(quinolin-4-yloxy)acetamides in Mycobacterium tuberculosis clinical isolates and identification of their molecular target by whole-genome sequencing. *Int. J. Antimicrob. Agents* **2018**, *51*, 378–384. [CrossRef]
157. Berry, E.A.; Guergova-Kuras, M.; Huang, L.S.; Crofts, A.R. Structure and function of cytochrome bc complexes. *Annu. Rev. Biochem.* **2000**, *69*, 1005–1075. [CrossRef]
158. Lamprecht, D.A.; Finin, P.M.; Rahman, M.A.; Cumming, B.M.; Russell, S.L.; Jonnala, S.R.; Adamson, J.H.; Steyn, A.J.C. Turning the respiratory flexibility of *Mycobacterium tuberculosis* against itself. *Nat. Commun.* **2016**, *7*, 12393. [CrossRef]

159. Lu, P.; Heineke, M.H.; Koul, A.; Andries, K.; Cook, G.M.; Lill, H.; van Spanning, R.; Bald, D. The cytochrome bd-type quinol oxidase is important for survival of Mycobacterium smegmatis under peroxide and antibiotic-induced stress. *Sci. Rep.* **2015**, *5*, 10333. [CrossRef] [PubMed]
160. Lu, P.; Asseri, A.H.; Kremer, M.; Maaskant, J.; Ummels, R.; Lill, H.; Bald, D. The anti-mycobacterial activity of the cytochrome bcc inhibitor Q203 can be enhanced by small-molecule inhibition of cytochrome bd. *Sci. Rep.* **2018**, *8*, 2625. [CrossRef] [PubMed]
161. Flentie, K.; Harrison, G.A.; Tükenmez, H.; Livny, J.; Good, J.A.D.; Sarkar, S.; Zhu, D.X.; Kinsella, R.L.; Weiss, L.A.; Solomon, S.D.; et al. Chemical disarming of isoniazid resistance in Mycobacterium tuberculosis. *Proc. Natl. Acad. Sci. USA* **2019**, *116*, 10510–10517. [CrossRef] [PubMed]
162. Tran, S.L.; Cook, G.M. The F1Fo-ATP synthase of Mycobacterium smegmatis is essential for growth. *J. Bacteriol.* **2005**, *187*, 5023–5028. [CrossRef]
163. Koul, A.; Dendouga, N.; Vergauwen, K.; Molenberghs, B.; Vranckx, L.; Willebrords, R.; Ristic, Z.; Lill, H.; Dorange, I.; Guillemont, J.; et al. Diarylquinolines target subunit c of mycobacterial ATP synthase. *Nat. Chem. Biol.* **2007**, *3*, 323–324. [CrossRef]
164. Haagsma, A.C.; Podasca, I.; Koul, A.; Andries, K.; Guillemont, J.; Lill, H.; Bald, D. Probing the interaction of the diarylquinoline TMC207 with its target mycobacterial ATP synthase. *PLoS ONE* **2011**, *6*, e23575. [CrossRef]
165. Nesci, S.; Trombetti, F.; Algieri, C.; Pagliarani, A. A therapeutic role for the F1FO-ATP synthase. *SLAS Discov. Adv. Life Sci. R D* **2019**, *24*, 893–903. [CrossRef]
166. Preiss, L.; Langer, J.D.; Yildiz, Ö.; Eckhardt-Strelau, L.; Guillemont, J.E.G.; Koul, A.; Meier, T. Structure of the mycobacterial ATP synthase Fo rotor ring in complex with the anti-TB drug bedaquiline. *Sci. Adv.* **2015**, *1*, e1500106. [CrossRef]
167. Hards, K.; McMillan, D.G.G.; Schurig-Briccio, L.A.; Gennis, R.B.; Lill, H.; Bald, D.; Cook, G.M. Ionophoric effects of the antitubercular drug bedaquiline. *Proc. Natl. Acad. Sci. USA* **2018**, *115*, 7326–7331. [CrossRef]
168. Hartkoorn, R.C.; Uplekar, S.; Cole, S.T. Cross-resistance between clofazimine and bedaquiline through upregulation of MmpL5 in Mycobacterium tuberculosis. *Antimicrob. Agents Chemother.* **2014**, *58*, 2979–2981. [CrossRef]
169. Almeida, D.; Ioerger, T.; Tyagi, S.; Li, S.-Y.; Mdluli, K.; Andries, K.; Grosset, J.; Sacchettini, J.; Nuermberger, E. Mutations in pepQ confer low-level resistance to bedaquiline and clofazimine in mycobacterium tuberculosis. *Antimicrob. Agents Chemother.* **2016**, *60*, 4590–4599. [CrossRef] [PubMed]
170. Xu, J.; Wang, B.; Hu, M.; Huo, F.; Guo, S.; Jing, W.; Nuermberger, E.; Lu, Y. Primary clofazimine and bedaquiline resistance among isolates from patients with multidrug-resistant tuberculosis. *Antimicrob. Agents Chemother.* **2017**, *61*, e00239-17. [CrossRef] [PubMed]
171. Ghajavand, H.; Kargarpour Kamakoli, M.; Khanipour, S.; Pourazar Dizaji, S.; Masoumi, M.; Rahimi Jamnani, F.; Fateh, A.; Siadat, S.D.; Vaziri, F. High prevalence of bedaquiline resistance in treatment-naive tuberculosis patients and verapamil effectiveness. *Antimicrob. Agents Chemother.* **2019**, *63*, e02530-18. [CrossRef] [PubMed]
172. Zimenkov, D.V.; Nosova, E.Y.; Kulagina, E.V.; Antonova, O.V.; Arslanbaeva, L.R.; Isakova, A.I.; Krylova, L.Y.; Peretokina, I.V.; Makarova, M.V.; Safonova, S.G.; et al. Examination of bedaquiline- and linezolid-resistant Mycobacterium tuberculosis isolates from the Moscow region. *J. Antimicrob. Chemother.* **2017**, *72*, 1901–1906. [CrossRef] [PubMed]
173. Villellas, C.; Coeck, N.; Meehan, C.J.; Lounis, N.; de Jong, B.; Rigouts, L.; Andries, K. Unexpected high prevalence of resistance-associated Rv0678 variants in MDR-TB patients without documented prior use of clofazimine or bedaquiline. *J. Antimicrob. Chemother.* **2017**, *72*, 684–690. [PubMed]
174. Guglielmetti, L.; Tiberi, S.; Burman, M.; Kunst, H.; Wejse, C.; Togonidze, T.; Bothamley, G.; Lange, C. QT prolongation and cardiac toxicity of new tuberculosis drugs in Europe: A tuberculosis network European trialsgroup (TBnet) study. *Eur. Respir. J.* **2018**, *52*, 1800537. [CrossRef]
175. Tong, A.S.T.; Choi, P.J.; Blaser, A.; Sutherland, H.S.; Tsang, S.K.Y.; Guillemont, J.; Motte, M.; Cooper, C.B.; Andries, K.; Van den Broeck, W.; et al. 6-Cyano analogues of bedaquiline as less lipophilic and potentially safer diarylquinolines for tuberculosis. *ACS Med. Chem. Lett.* **2017**, *8*, 1019–1024. [CrossRef]
176. Choi, P.J.; Sutherland, H.S.; Tong, A.S.T.; Blaser, A.; Franzblau, S.G.; Cooper, C.B.; Lotlikar, M.U.; Upton, A.M.; Guillemont, J.; Motte, M.; et al. Synthesis and evaluation of analogues of the tuberculosis drug bedaquiline containing heterocyclic B-ring units. *Bioorg. Med. Chem. Lett.* **2017**, *27*, 5190–5196. [CrossRef]

177. Sutherland, H.S.; Tong, A.S.T.; Choi, P.J.; Conole, D.; Blaser, A.; Franzblau, S.G.; Cooper, C.B.; Upton, A.M.; Lotlikar, M.U.; Denny, W.A.; et al. Structure-activity relationships for analogs of the tuberculosis drug bedaquiline with the naphthalene unit replaced by bicyclic heterocycles. *Bioorg. Med. Chem.* **2018**, *26*, 1797–1809. [CrossRef]
178. Blaser, A.; Sutherland, H.S.; Tong, A.S.T.; Choi, P.J.; Conole, D.; Franzblau, S.G.; Cooper, C.B.; Upton, A.M.; Lotlikar, M.; Denny, W.A.; et al. Structure-activity relationships for unit C pyridyl analogues of the tuberculosis drug bedaquiline. *Bioorg. Med. Chem.* **2019**, *27*, 1283–1291. [CrossRef]
179. Sutherland, H.S.; Tong, A.S.T.; Choi, P.J.; Blaser, A.; Conole, D.; Franzblau, S.G.; Lotlikar, M.U.; Cooper, C.B.; Upton, A.M.; Denny, W.A.; et al. 3,5-Dialkoxypyridine analogues of bedaquiline are potent antituberculosis agents with minimal inhibition of the hERG channel. *Bioorg. Med. Chem.* **2019**, *27*, 1292–1307. [CrossRef]
180. Sarathy, J.P.; Ragunathan, P.; Shin, J.; Cooper, C.B.; Upton, A.M.; Grüber, G.; Dick, T. TBAJ-876 retains bedaquiline's activity against subunits c and ε of mycobacterium tuberculosis F-ATP synthase. *Antimicrob. Agents Chemother.* **2019**, *63*, e01191-19. [CrossRef] [PubMed]
181. Kumar, S.; Mehra, R.; Sharma, S.; Bokolia, N.P.; Raina, D.; Nargotra, A.; Singh, P.P.; Khan, I.A. Screening of antitubercular compound library identifies novel ATP synthase inhibitors of mycobacterium tuberculosis. *Tuberc. Edinb. Scotl.* **2018**, *108*, 56–63. [CrossRef] [PubMed]
182. Lu, P.; Haagsma, A.C.; Pham, H.; Maaskant, J.J.; Mol, S.; Lill, H.; Bald, D. Pyrazinoic acid decreases the proton motive force, respiratory ATP synthesis activity, and cellular ATP levels. *Antimicrob. Agents Chemother.* **2011**, *55*, 5354–5357. [CrossRef] [PubMed]
183. Feng, X.; Zhu, W.; Schurig-Briccio, L.A.; Lindert, S.; Shoen, C.; Hitchings, R.; Li, J.; Wang, Y.; Baig, N.; Zhou, T.; et al. Antiinfectives targeting enzymes and the proton motive force. *Proc. Natl. Acad. Sci. USA* **2015**, *112*, E7073–E7082. [CrossRef] [PubMed]
184. Jeon, A.B.; Ackart, D.F.; Li, W.; Jackson, M.; Melander, R.J.; Melander, C.; Abramovitch, R.B.; Chicco, A.J.; Basaraba, R.J.; Obregón-Henao, A. 2-aminoimidazoles collapse mycobacterial proton motive force and block the electron transport chain. *Sci. Rep.* **2019**, *9*, 1513. [CrossRef] [PubMed]
185. Jeon, A.B.; Obregón-Henao, A.; Ackart, D.F.; Podell, B.K.; Belardinelli, J.M.; Jackson, M.; Nguyen, T.V.; Blackledge, M.S.; Melander, R.J.; Melander, C.; et al. 2-aminoimidazoles potentiate ß-lactam antimicrobial activity against Mycobacterium tuberculosis by reducing ß-lactamase secretion and increasing cell envelope permeability. *PLoS ONE* **2017**, *12*, e0180925. [CrossRef]
186. Protopopova, M.; Hanrahan, C.; Nikonenko, B.; Samala, R.; Chen, P.; Gearhart, J.; Einck, L.; Nacy, C.A. Identification of a new antitubercular drug candidate, SQ109, from a combinatorial library of 1,2-ethylenediamines. *J. Antimicrob. Chemother.* **2005**, *56*, 968–974. [CrossRef]
187. Li, K.; Schurig-Briccio, L.A.; Feng, X.; Upadhyay, A.; Pujari, V.; Lechartier, B.; Fontes, F.L.; Yang, H.; Rao, G.; Zhu, W.; et al. Multitarget drug discovery for tuberculosis and other infectious diseases. *J. Med. Chem.* **2014**, *57*, 3126–3139. [CrossRef]
188. Zhang, Y.; Wade, M.M.; Scorpio, A.; Zhang, H.; Sun, Z. Mode of action of pyrazinamide: Disruption of Mycobacterium tuberculosis membrane transport and energetics by pyrazinoic acid. *J. Antimicrob. Chemother.* **2003**, *52*, 790–795. [CrossRef]
189. Yang, C.-S.; Yuk, J.-M.; Jo, E.-K. The role of nitric oxide in mycobacterial infections. *Immune Netw.* **2009**, *9*, 46–52. [CrossRef]
190. Manjunatha, U.; Boshoff, H.I.; Barry, C.E. The mechanism of action of PA-824. *Commun. Integr. Biol.* **2009**, *2*, 215–218. [CrossRef] [PubMed]
191. Stover, C.K.; Warrener, P.; VanDevanter, D.R.; Sherman, D.R.; Arain, T.M.; Langhorne, M.H.; Anderson, S.W.; Towell, J.A.; Yuan, Y.; McMurray, D.N.; et al. A small-molecule nitroimidazopyran drug candidate for the treatment of tuberculosis. *Nature* **2000**, *405*, 962–966. [CrossRef]
192. Matsumoto, M.; Hashizume, H.; Tomishige, T.; Kawasaki, M.; Tsubouchi, H.; Sasaki, H.; Shimokawa, Y.; Komatsu, M. OPC-67683, a nitro-dihydro-imidazooxazole derivative with promising action against tuberculosis in vitro and in mice. *PLoS Med.* **2006**, *3*, e466. [CrossRef]
193. Van den Bossche, A.; Varet, H.; Sury, A.; Sismeiro, O.; Legendre, R.; Coppee, J.-Y.; Mathys, V.; Ceyssens, P.-J. Transcriptional profiling of a laboratory and clinical Mycobacterium tuberculosis strain suggests respiratory poisoning upon exposure to delamanid. *Tuberc. Edinb. Scotl.* **2019**, *117*, 18–23. [CrossRef] [PubMed]
194. Developing New TB Regimens. Available online: https://www.tballiance.org/rd/developing-new-regimens (accessed on 23 December 2019).

195. Berube, B.J.; Parish, T. Combinations of respiratory chain inhibitors have enhanced bactericidal activity against mycobacterium tuberculosis. *Antimicrob. Agents Chemother.* **2018**, *62*, e01677-17. [CrossRef] [PubMed]
196. Moraski, G.C.; Cheng, Y.; Cho, S.; Cramer, J.W.; Godfrey, A.; Masquelin, T.; Franzblau, S.G.; Miller, M.J.; Schorey, J. Imidazo[1,2-a]Pyridine-3-Carboxamides are active antimicrobial agents against mycobacterium avium infection in vivo. *Antimicrob. Agents Chemother.* **2016**, *60*, 5018–5022. [CrossRef]
197. Heinrich, N.; Dawson, R.; du Bois, J.; Narunsky, K.; Horwith, G.; Phipps, A.J.; Nacy, C.A.; Aarnoutse, R.E.; Boeree, M.J.; Gillespie, S.H.; et al. Early phase evaluation of SQ109 alone and in combination with rifampicin in pulmonary TB patients. *J. Antimicrob. Chemother.* **2015**, *70*, 1558–1566. [CrossRef]
198. Zhang, Y.; Zhu, H.; Fu, L.; Wang, B.; Guo, S.; Chen, X.; Liu, Z.; Huang, H.; Yang, T.; Lu, Y. Identifying regimens containing TBI-166, a new drug candidate against mycobacterium tuberculosis in vitro and in vivo. *Antimicrob. Agents Chemother.* **2019**, *63*, e02496-18. [CrossRef]
199. Converse, P.J.; Almeida, D.V.; Tyagi, S.; Xu, J.; Nuermberger, E.L. Shortening buruli ulcer treatment with combination therapy targeting the respiratory chain and exploiting mycobacterium ulcerans gene decay. *Antimicrob. Agents Chemother.* **2019**, *63*, e00426-19. [CrossRef]

© 2020 by the authors. Licensee MDPI, Basel, Switzerland. This article is an open access article distributed under the terms and conditions of the Creative Commons Attribution (CC BY) license (http://creativecommons.org/licenses/by/4.0/).

Review

Development of Delpazolid for the Treatment of Tuberculosis

Young Lag Cho [1] and Jichan Jang [2,*]

[1] LegoChem Biosciences, Inc., 8-26 Munoyeongseo-ro, Daedeok-gu, Daejeon 34302, Korea; young@legochembio.com

[2] Molecular Mechanisms of Antibiotics, Division of Life Science, Research Institute of Life Sciences, Gyeongsang National University, Jinju 52828, Korea

* Correspondence: jichanjang@gnu.ac.kr; Tel.: +82-(0)55-772-1368

Received: 3 March 2020; Accepted: 23 March 2020; Published: 25 March 2020

Abstract: A novel oxazolidinone with cyclic amidrazone, delpazolid (LCB01-0371), was synthesized by LegoChem BioSciences, Inc. (Daejeon, Korea). Delpazolid can improve the minimum bactericidal concentration of *Mycobacterium tuberculosis* H37Rv and significantly reduce resistance rates, especially of multi-drug-resistant tuberculosis (MDR-TB) isolates, compared with linezolid. Therefore, delpazolid can be used to treat MDR-TB. The safety, tolerability, and pharmacokinetics of delpazolid have been evaluated in a phase 1 clinical trial, which revealed that it does not cause adverse events such as myelosuppression even after three weeks of repeated dosing. Interim efficacy and safety results, particularly those from a clinical phase 2a early bactericidal activity trial including patients with drug-susceptible tuberculosis, were reported and the findings will be further analyzed to guide phase 2a studies.

Keywords: *Mycobacterium tuberculosis*; delpazolid; drug discovery; multi-drug resistance

1. Linezolid, the First Oxazolidinone Antibacterial Agent

Oxazolidinone is a heterocyclic organic compound containing both nitrogen and oxygen in a 5-membered ring and is mainly used as an antimicrobial agent. This class of antimicrobials is active against a large spectrum of Gram-positive bacteria, including methicillin-resistant *Staphylococcus aureus* (MRSA), vancomycin-resistant enterococci (VRE), vancomycin-intermediate strains, and penicillin-resistant pneumococci, and acts via inhibiting protein synthesis [1,2].

Linezolid is the first oxazolidinone antimicrobial to be developed; it exhibits a high degree of in vitro activity against various Gram-positive pathogens [3]. Linezolid exhibits bactericidal activity against *Mycobacterium tuberculosis* and has been used to treat rifampicin-resistant tuberculosis (RR-TB) or multi-drug-resistant tuberculosis (MDR-TB) [4]. Although the integration of linezolid into RR-TB or MDR-TB treatment can improve outcomes, prolonged administration is often limited by long-term side effects, including reversible myelosuppression, potentially irreversible optic neuropathy, and peripheral neuropathy [5]. Therefore, safety and tolerability are critical issues to consider when prescribing these antibiotics [6]. Less toxic alternatives are under development for diseases that require long-term therapy such as tuberculosis.

2. Development of Delpazolid (LCB01-0371)

LegoChem Biosciences (Daejeon, Korea) is a company that develops effective and safe drugs using legochemistry technology, which enables the manipulation of substances by attaching and detaching compounds around scaffold-like Lego blocks. LegoChem Biosciences searches for novel candidate substances based on the concept that a good scaffold with novel blocks, based on medicinal chemistry, can accelerate the process of improving previous scaffolds with weak activity or have side effects. Delpazolid (code No: LCB01-0371), a derivative of oxazolidinone, is the first candidate antibiotic substance identified by LegoChem Biosciences.

Delpazolid is an antibiotic that targets Gram-positive bacteria (MRSA, VRE) including *M. tuberculosis*. It is currently undergoing a phase 2 clinical trial for oral (PO) administration and a phase 1 trial for intravenous (IV) administration to treat Gram-positive (MRSA, VRE) bacteraemia. Cyclic amidrazone blocks were applied to the key scaffold of delpazolid (Figure 1). In general, after a drug is absorbed, it must be dissolved well to ensure proper secretion. Most small molecules with suboptimal pharmacokinetic (PK) profiles tend to have low solubility. In general, small-molecule ligands that bind their targets with high efficiency are more hydrophobic, and hydrophobic interactions are essential for increased ligand efficiency [7]. Hydrophobicity not only increases target binding efficacy, but also decreases the solubility of a small molecule. The cyclic amidrazone (Figure 1) on the side chain of delpazolid maintains its hydrophobicity to some extent and has a slightly basic pH similar to that of carboxylate. Therefore, it can be charged by obtaining a proton from carboxylic acid under human physiological conditions, which enhances the solubility and PK profile. Therefore, the drug is accumulated slowly and excreted well, and can be administered over the long-term with minimal side effects.

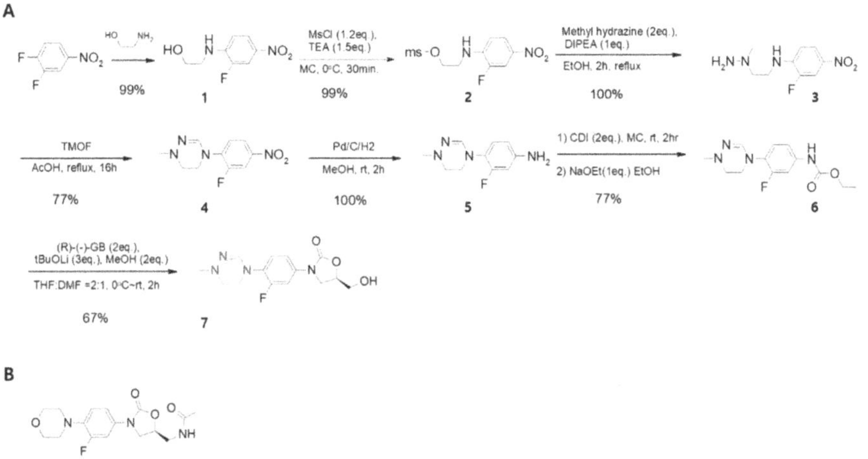

Figure 1. (**A**). Synthetic scheme showing that delpazolid can be synthesised in only seven steps with difluoro-nitrobenzene as the starting material. Each step shows a high yield and the products are easily purified without chromatography. The red color indicates cyclic amidrazone. (**B**). Chemical structure of linezolid.

3. Safety Evaluation in the Phase 1 Clinical Trial as PO

The greatest advantage associated with delpazolid is its safety. In phase 1a of a phase 1 clinical trial to evaluate its safety, as illustrated in Table 1, 64 subjects were divided into eight groups, six of whom were administered delpazolid and two who were administered the placebo. The study was the first double-blind, randomized human trial of delpazolid. To deliver single-ascending-doses (SADs), delpazolid was administered in a step-wise manner from 50 mg up to 3200 mg. Only mild adverse events were observed up to 2400 mg. At a delpazolid dose of 3200 mg, gastrointestinal (GI) tract-related adverse events were noted. In the 3200 mg dose group, volunteers had to ingest 16 tablets of 200 mg delpazolid tablets at once, resulting in GI tract-related adverse events. Therefore, the maximum tolerated dose of delpazolid was determined to be 2400 mg per day.

Table 1. Summary of phase 1a/b and 2a dose-escalation study to assess the safety, tolerability, and pharmacokinetics of delpazolid as a single agent.

Clinical Trial Phase	Experimental Design and Adverse Effects Reported
Phase 1a [9] (SAD)	Study design: Double blind, randomized, placebo control, first-in-human design
	N=64, 8 subject per group (6 active + 2 placebo)
	Doses: 50, 100, 200, 400, 800, 1,600, 2,400, and 3,200 mg
	MTD: 2,400 mg (Up to 2,400 mg, only mild adverse events were reported)
Phase 1b [8] (MAD-7 days)	Study design: Double blind, randomized, placebo control
	N=32, 8 subject per group (6 active + 2 placebo)
	Doses: 400, 800, 1,200, 1,600 mg BID for 7 days
	MTD: 1,200 mg BID (Up to 2,400 mg/day, only mild adverse events were reported)
Phase 1b [6] (MAD-21 days)	Study design: Double blind, randomized, placebo control
	N=36, 12 subject per group (10 active + 2 placebo)
	Doses: 800 mg QD and BID, 1,200 mg BID for 21 days
	MTD: 1,200 mg BID (Up to 2,400mg/day, No SAE reported)
Phase 2a [a] (EBA Trial)	Study design: Open label, randomized
	N=80, 16 subject per delpazolid group; 8 patients in active control groups, HRZE and linezolid
	Doses: Delpazolid 400 mg BID, 800 mg QD, 800 mg BID, 1,200 mg QD, HRZE and linezolid 600 mg BID for 14 days

Dose-escalation process consisted of a single-ascending-dose phase (SAD) and multiple-ascending-dose phase (MAD). MTD, maximum tolerated dose; QD, quaque die (daily); BID, bis in die (twice per day); SAE, serious adverse event; EBA, early bactericidal activity; HRZE, isoniazid (H), rifampin (R), pyrazinamide (Z), and ethambutol (E). [a] Results were not yet published.

A phase 1b study was conducted based on multiple-ascending-doses (MADs) over seven days. Thirty-two subjects were divided into eight groups, six of whom were administered delpazolid and two of whom were administered the placebo. Subjects were given delpazolid in MADs from 400 mg BID (bis in die, twice a day) up to 1600 mg BID over seven days. Doses up to 1200 mg BID for seven days were well-tolerated with no specific adverse events observed. After the 7-day MAD study, a 21-day MAD study was conducted to evaluate bone marrow toxicity, which is one of the most critical side effects of linezolid [8]. Subjects administered 800 mg once a day (QD) to 1200 mg BID delpazolid were monitored for up to three weeks to more accurately assess adverse events such as myelosuppression, as signs such as decreased platelet count may be observed even after two weeks. As illustrated in Table 1, serious adverse events were not observed under the MAD-21-day condition. In summary, no myelosuppression-related adverse events or serious adverse events were observed in phase 1a with SADs up to 2400 mg and in phase 1b with MAD up to 1200 mg BID (2400 mg per day) for 21 days. Therefore, delpazolid does not appear to exhibit adverse events associated with repeated dosing. In addition, delpazolid did not cause CYP-mediated metabolism and cardiac repolarisation issues [6,9–11].

4. Poor PK Profiles but Safe for Humans

The underlying antibacterial mechanism of delpazolid is similar to that of oxazolidinone in that it inhibits bacterial protein synthesis, which kills or inhibits the growth of bacteria [12]. However, protein synthesis also occurs in the mitochondria of eukaryotes, although mitochondria use independent protein-synthesis machinery that differs from nuclear-encoded protein synthesis in the cytoplasm. In humans, 13 genes are translated into proteins through this process, all of which participate in synthesizing membrane proteins associated with oxidative phosphorylation [13].

However, oxazolidinones uniformly inhibit human mitochondrial protein synthesis [14]. Similarly, linezolid, an oxazolidinone analogue used to treat TB, inhibits mitochondrial protein synthesis with potentially severe clinical consequences [15]. Therefore, the inhibition of protein synthesis by oxazolidinone intended to kill bacteria can impair mitochondria inside eukaryotic cells. Furthermore, myelosuppression may be a product of linezolid inhibition of mitochondrial protein synthesis [16].

As shown in Table 2, delpazolid showed a greater inhibitory effect than linezolid towards *Escherichia coli* at a 5-fold lower concentration (0.8 µg/mL).

Table 2. Antibiotic properties of oxazolidinones on bacterial and mitochondrial protein synthesis.

Compound	Bacteria	Human Mitochondria (IC$_{50}$)		Animal Mitochondria [14]
	Escherichia coli	K562 cell	AC16 cell	Rat, Rabbit (liver & heart)
Delpazolid	2.6 µM (0.8 µg/mL)	4.8 µM (1.5 µg/mL)	10.9 µM (3.4 µg/mL)	NA
Linezolid	11.6 µM (3.9 µg/mL)	3.1 µM (1.0 µg/mL)	10.0 µM (3.4 µg/mL)	12.8 µM

In addition, in a study of human cells (immortalised myelogenous leukaemia cell line K562 and human cardiomyocyte cell line AC16), delpazolid showed inhibitory effects on mitochondrial protein synthesis similar to those of linezolid. Although delpazolid exhibited activity superior to that of linezolid in prokaryotic protein synthesis inhibition, it had similar negative effects on mitochondrial protein synthesis. Therefore, delpazolid doses lower than linezolid doses would be adequate for the treatment of Gram-positive bacteria, including TB. A lower dose would effectively inhibit bacterial protein synthesis, with relatively fewer adverse effects on human mitochondrial protein synthesis.

The association between delpazolid and myelosuppression, one of the most serious side effects of linezolid, was also tested. Healthy subjects were administered delpazolid and linezolid, and the plasma area under the concentration–time curve (AUC) was determined. As shown in Figure 2, subjects were administered delpazolid at doses ranging from 400 to 1200 mg, and 600 mg linezolid as the comparator.

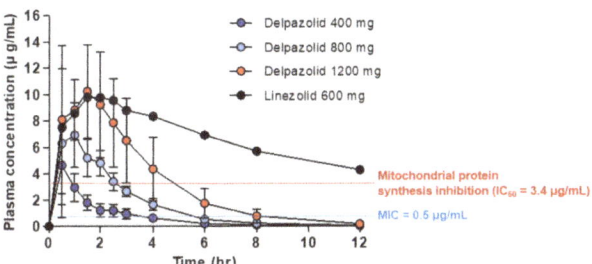

Figure 2. Mean plasma concentrations of linezolid and delpazolid in adults following oral dosing (mean ± standard deviation, $n = 6$).

As shown in Table 2, the IC$_{50}$ values of delpazolid and linezolid at which mitochondrial protein synthesis in the two human cell-lines (K562 and AC16) were similar at 3.4 µg/mL indicated that these agents killed approximately 50% of human cells at 3.4 µg/mL. Thus, 3.4 µg/mL of delpazolid and linezolid is the mitochondrial damage limit. At higher concentrations, mitochondrial protein synthesis is affected severely, leading to cell death. Therefore, considering 3.4 µg/mL as the reference value at which toxicity of the two drugs occurs, a phase 1 trial based on linezolid 600 mg BID revealed that the linezolid plasma concentration was maintained at above the IC$_{50}$ (3.4 µg/mL) for 12 h. However, delpazolid 800 mg maintained the IC$_{50}$ above the mitochondrial damage limit for only 3 h, after which it was cleared rapidly from the blood. Therefore, delpazolid provides ample time for mitochondria to recover its protein synthesis function. In addition, increasing the delpazolid dose to 1200 mg raises the IC$_{50}$

to above 3.4 µg/mL for only 5 h, after which it also clears from the blood. Therefore, the low AUC with rapid clearance in delpazolid ironically minimizes cellular toxicity. Consequently, repeated BID dosing of delpazolid results in much lower levels of myelosuppression because of the lower mitochondrial protein synthesis inhibition compared to linezolid [9,10]. Therefore, the side effects of delpazolid were much milder than those of linezolid. The difference in side effects despite the similar structure of the two drugs may be due to differences in their chemical structures. The cyclic amidrazone side chain of delpazolid facilitates more rapid clearance and prevents accumulation in the plasma compared to linezolid. Thus, rapid clearance has been demonstrated as a key advantage that reduces myelosuppression compared to linezolid. Therefore, delpazolid may replace linezolid for MDR-TB for long-term treatment [11].

5. Toxicology

In vivo animal toxicity tests on delpazolid did not reveal specific toxicity profiles for six months in rats and for nine months in dogs. Furthermore, genetic toxicity tests, including the Ames test, in vitro chromosomal aberration test, and rat micronucleus test, as well as pharmacological safety tests including the hERG safety test, cardiovascular, respiratory and neurobehavioral tests, and reproductive toxicity tests were conducted, none of which revealed a specific toxicity profile (Table 3).

Table 3. Toxicology summary (PO: per oral /IV: intravenous). The general toxicity of delpazolid in animals lasted up to six months in rats and nine months in dogs, and no unusual findings after long-term treatment [a].

General Toxicity	Status	PO	IV
Single dose acute toxicity study in rats	Completed	MTD = 2000 mpk	MTD = 1000 mpk
Single dose acute toxicity study in dogs	Completed	MTD = 1000 mpk	MTD = 500 mpk
4- week toxicity study in rats with 4 -week recovery	Completed	NOAEL = 60 mpk	NOAEL = 120 mpk
4- week toxicity study in dogs with 4 -week recovery	Completed	NOAEL (male = 20 mpk, female=10 mpk)	NOAEL = 15 mpk
26- week(6 months) toxicity study in rats with 4 -week recovery	Completed	NOAEL (male = 10 mpk, female=100→75 mpk)	-
39- week (9 months) toxicity study in dogs with 4 -week recovery	Completed	NOAEL = 10 mpk	-
Genetic Toxicity			
Ames test	Completed	Negative	
In vitro chromosomal aberration test	Completed	Negative	
Rat micronucleus test	Completed	Negative	
Safety Pharmacology			
Assessment of blockage of hERG potassium channels	Completed	Negative(IC_{50} > 100µM)	
Cardiovascular telemetry study in beagle dogs	Completed	Negative	
Respiratory (Pulmonary) study in rats	Completed	Negative	
Neurobehavioral safety evaluation in rats	Completed	Negative	
Reproductive Toxicity		PO	
Fertility and Embryonic Development to Implantation toxicity in rat	Completed	NOAEL (male = 15 mpk, female=60 mpk)	
Embryo-Fetal Development toxicity in rat	Completed	NOAEL = 15 mpk	

MTD; maximum tolerated dose, NOAEL; no-observed-adverse-effect level. [a] Results were not yet published.

In a human bioavailability study, the bioavailability of the PO form was 99–100% (800 mg) of that of the IV form. Considering that the PK profiles between the IV and PO forms are similar, conversion would be relatively easy in the future. Because delpazolid is slightly polar, it exhibits low protein binding (37% in human), rapid clearance with no accumulation, and no food-related effects (Table 4).

Table 4. Phase 1 study: summary of delpazolid pharmacokinetic parameters. IV infusion 400 mg and PO 800 mg, cross-over study IV administration of delpazolid was generally safe and well-tolerated 800 mg (PO): Bioavailability was approximately 99% switchable between the PO and IV, with no dose adjustment.

Pharmacokinetic Parameter [a]	IV Infusion; 200 mg (n=6)	IV Infusion; 400 mg (n=8)	PO; 800 mg (n=8)
C_{max} (μg/mL)	2.92 ± 0.46	5.25 ± 0.96	8.20 ± 3.47
T_{max} (hr)	0.83 ± 0.13	0.84 ± 0.13	1.22 ± 0.98
$T_{1/2}$ (hr)	1.70 ± 0.26	1.48 ± 0.16	1.64 ± 0.48
AUC_{0-24h} (μg·hr/mL)	5.59 ± 0.98	9.39 ± 1.46	18.65 ± 4.88
AUC_{inf} (μg·hr/mL)	5.63 ± 1.00	9.42 ± 1.47	18.86 ± 4.99
V_{ss}, V_z/F (L/kg)	0.90 ± 0.06	1.05 ± 0.20	1.67 ± 0.79
CL, Cl/F (L/hr/kg)	0.56 ± 0.10	0.67 ± 0.10	0.69 ± 0.18
MRT_{last} (hr)	1.55 ± 0.21	1.53 ± 0.14	2.87 ± 0.92
$C_{max, norm}$ (μg/mL)	0.95 ± 0.15	0.85 ± 0.16	0.67 ± 0.28
$AUC_{inf, norm}$ (μg·hr/mL)	1.83 ± 0.33	1.53 ± 0.24	1.53 ± 0.40
F (%)	-	-	99.8 ± 20.6

[a] Values are the means ± standard deviation (range). C_{max}, maximal drug concentration; T_{max}, time to reach C_{max}; $T_{1/2}$, half-life; AUC_{0-24}, area under the concentration-24-h curve; AUC_{inf}, AUC from time zero extrapolated to infinity; V_{ss}, steady-state volume of distribution; V_z/F, apparent volume of distribution; CL, clearance; Cl/F, apparent oral clearance; MRT_{last}, mean residence time when the drug concentration is based on values up to and including the last measured concentration; $C_{max, norm}$, C_{max} divided by dose per body weight; $AUC_{inf, norm}$, weight-normalised AUC_{inf}; F, bioavailability.

6. Activity Against TB and Combination Study of Delpazolid with Other Anti-TB Agents

Studies of the early development of delpazolid focused on Gram-positive bacteria. The efficacy of delpazolid on Gram-positive bacteria was similar or slightly better than that of linezolid. For example, in animal studies of systemic infection [17], soft tissue infection, lung infection, and thigh infection models in mice, delpazolid showed greater efficacy than linezolid (data not shown).

To evaluate the efficacy of delpazolid in TB, an in vitro susceptibility test was conducted for *M. tuberculosis* H37Rv. Compared to linezolid, the minimum inhibitory concentration (MIC) for *M. tuberculosis* H37Rv was similar to that under delpazolid; however, the minimum bactericidal concentration was more than 4-fold lower under delpazolid (Table 5).

Table 5. Drug activities and resistance rates of linezolid and delpazolid.

Drug Activities / Resistant Rate [a]	Linezolid	Delpazolid
MIC value for *M. tuberculosis* H37Rv (μg/mL)	0.5	0.5
MBC_{99} value for *M. tuberculosis* H37Rv (μg/mL)	>16	4
MDR-TB MIC_{90} (μg/mL)	1	0.5
XDR-TB MIC_{90} (μg/mL)	0.25	1
ECOFFs (epidemiological cutoff values) (μg/mL)	1.0	2.0
Resistant rate of MDR-TB (%)	6.7	0.8
Resistant rate of XDR-TB (%)	4.2	4.2

[a] A total of 240 *M. tuberculosis* isolates were tested for ECOFFS and resistant rates, including 120 MDR-TB isolates and 120 XDR-TB samples in China.

The MIC_{90} values of delpazolid for MDR/extensively drug resistant (XDR) TB isolates were 0.25 and 1 μg/mL, respectively. However, an in vitro study of MDR/XDR TB isolates from China showed that the resistance rate varied considerably. The resistance of MDR-TB to linezolid was 6.7%, whereas that to delpazolid was 0.8%, suggesting higher potential efficacy of delpazolid in the treatment of MDR-TB, although no significant difference in resistance rates was observed between linezolid and delpazolid among XDR-TB isolates [18]. Therefore, delpazolid has been considered as a targeted application for MDR-TB treatment. Considering the significantly lower resistance rate of MDR-TB against delpazolid despite its similar structure to linezolid, further studies are needed to investigate structural variations in delpazolid to evaluate the correlations between the structures of various delpazolid derivatives and their resistance rates. In addition, intracellular MICs of delpazolid that can inhibit the growth of intracellular *M. tuberculosis* H37Rv revealed efficacy levels similar to those of linezolid under low concentrations, whereas delpazolid had greater efficacy at higher concentrations (Figure 3).

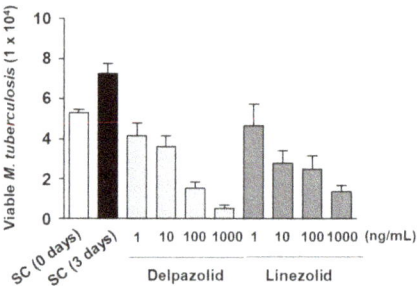

Figure 3. Intracellular activity of delpazolid. The activity of delpazolid on intracellular *M. tuberculosis* was compared to linezolid in bone marrow-derived macrophages (BMDMs) at three days after infection. The experiment was performed in triplicate, and the results are shown as the mean ± standard error of the mean (SEM). SC, solvent control.

The treatment of tuberculosis requires a combination of several antimicrobial agents and long-term therapy [19]. Therefore, evaluating synergy with other anti-TB agents is a crucial step in finding drugs that can be co-administered with delpazolid. As indicated in Table 6, a checkerboard assay was performed to identify pre-existing anti-TB medications with potential synergistic effects with delpazolid.

Table 6. MICs of selected anti-tuberculosis compounds against *M. tuberculosis* H37Rv and corresponding interaction profiles with delpazolid assessed by checkerboard.

Ref. Drug	MIC (µg/mL)	Tested TB Drugs	MIC (µg/mL)	FIC index	Activity [a]
Delpazolid	1	Isoniazid	0.13	1.13	I
		Rifampicin	0.06	0.75	Ad
		Rifapentine	0.01	0.75	Ad
		Ethambutol	0.50	1.02	I
		Cycloserine	4.0	1.02	I
		Amikacin	0.04	1.02	I
		Streptomycin	0.25	1.02	I
		Capreomycin	0.31	1.02	I
		Moxifloxacin	0.06	0.75	Ad
		Levofloxacin	0.25	0.75	Ad
		Clofazimine	0.25	0.52	pS
		Bedaquiline	0.25	0.53	pS
		Delamanid	0.02	0.75	Ad
		Ethionamide	0.5	1.03	I
		p-aminosalicylic acid	0.02	1	I
		Pyrazinamide [b]	200	0.63	pS

[a] S: synergy, pS: partial synergy, Ad: additive, I: indifference. [b] Tested in acidic condition (pH 5.2)

The assay revealed that delpazolid has partial synergism with clofazimine, bedaquiline, and pyrazinamide. Based on the results, in vitro time-kill kinetics tests were conducted by combining delpazolid with clofazimine and bedaquiline (Figure 4).

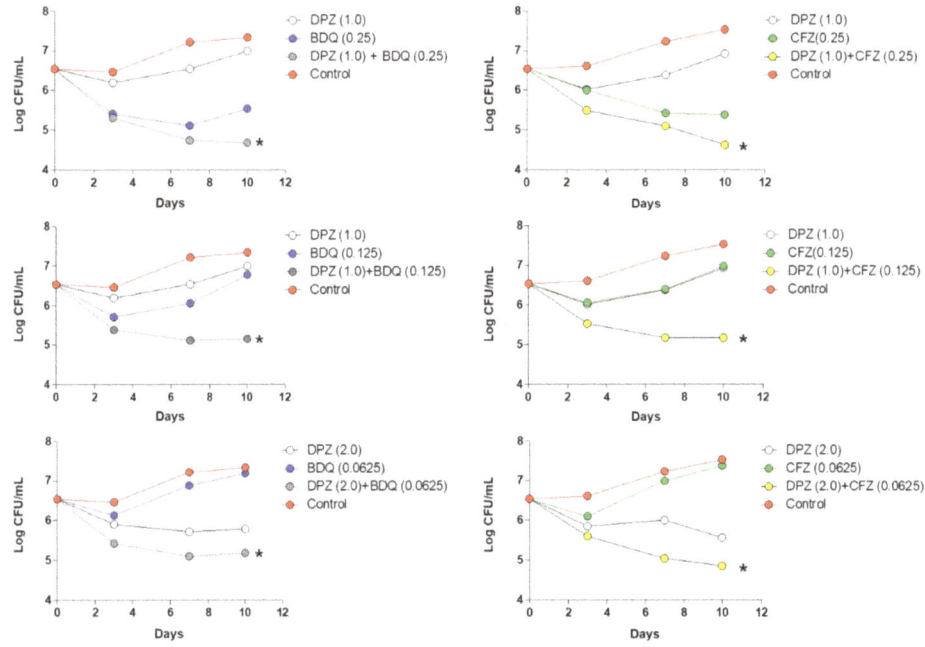

Figure 4. In vitro combination time-kill assay with anti-TB drugs. Viability of *M. tuberculosis* H37Rv was evaluated using combinations of various concentrations of delpazolid and bedaquiline or clofazimine (μg/mL).

Using the MIC against *M. tuberculosis* H37Rv for each drug, changes in colony-forming units (CFU) with monotherapy or combination therapy were evaluated. In addition, based on the MICs, synergistic effects between delpazolid plus bedaquiline and delpazolid plus clofazimine were evaluated at varying doses. Although the CFUs decreased at the MIC of a single drug, regrowth was observed over time. However, when delpazolid was combined with bedaquiline or clofazimine, using the 0.5× MIC of each drug, no regrowth was observed (Figure 4). In addition, the combination of bedaquiline and clofazimine with delpazolid consistently suppressed the growth of *M. tuberculosis* H37Rv, exhibiting high synergistic effects with 1× MIC delpazolid (1 μg/mL) and 0.5 × MIC clofazimine (0.25 μg/mL), resulting in a 2 log CFU reduction in *M. tuberculosis* H37Rv. Synergy between two new antimycobacterial compounds, such as delpazolid and bedaquiline or clofazimine, offers an attractive foundation for a new tuberculosis regimen.

7. Activity of Delpazolid on Nontuberculous Mycobacteria

Nontuberculous mycobacteria (NTM) are naturally occurring organisms found in water and soil. They are associated with biofilm formation, which enhances their disinfectant and antibiotic resistance. Particularly, *Mycobacterium avium* complex and *Mycobacterium abscessus* are the most common causes of pulmonary NTM and deadly pathogens, with high failure rates and relapse rates that may exceed 40% [20]. Although most people are not affected by such pathogens, in some individuals susceptible to conditions such as cystic fibrosis, chronic obstructive lung disease, bronchiectasis, and thoracic skeletal abnormalities, progressive and debilitating disease can occur [21].

A key concern in NTM treatment is the lack of antibiotics appropriate for long-term treatment for diverse NTM pathogens. Here, we evaluated the activity of delpazolid via in vitro susceptibility tests, as shown in the table below (Table 7) [22].

Table 7. MICs of antibiotics against clinical isolates of NTMs.

NTM Species (no. of strain tested)	Antibiotics	Range (µg/mL)	MIC$_{50}$ (µg/mL)	MIC$_{90}$ (µg/mL)
Mycobacterium avium (22)	Delpazolid	8-0.125	2	8
	Linezolid	8-0.125	2	8
	Clarithromycin	>128-≤0.125	>128	>128
Mycobacterium abscessus (20)	Delpazolid	8-0.25	2	8
	Linezolid	16-0.5	4	8
	Clarithromycin	128- ≤0.125	≤0.125	1
Mycobacterium fortuitum (21)	Delpazolid	2-0.25	1	2
	Linezolid	8-0.5	2	8
	Clarithromycin	8-≤0.125	0.25	4
Mycobacterium kansasii (22)	Delpazolid	2-0.25	1	2
	Linezolid	2-0.25	0.5	2
	Clarithromycin	0.125- ≤0.125	≤0.125	≤0.125
Mycobacterium chelonae (20)	Delpazolid	4-0.25	1	2
	Linezolid	8-0.5	2	4
	Clarithromycin	0.2- ≤0.025	0.1	0.2

Delpazolid had MICs similar to those of linezolid against *M. avium*, *M. abscessus*, *M. fortuitum*, *M. kansasii*, and *M. chelonae*, and inhibited NTM proliferation. In particular, delpazolid was effective against several *M. abscessus* strains in vitro and in a macrophage infection model. Acute infections in C57BL/6 mice, delpazolid 100 mg/kg exhibited greater in vivo efficacy than clarithromycin 200 mg/kg, a macrolide that is the main drug currently for *M. abscessus* treatment [12]. Therefore, delpazolid represents a promising novel class of oxazolidinones with improved safety for the treatment of *M. abscessus*.

8. Conclusions

As observed in clinical studies, the greatest advantage of delpazolid over linezolid is the potential for delpazolid to be used in long-term therapies. The development of delpazolid has focused on TB treatment, as this disease requires long-term treatment. In December 2016, LegoChem Biosciences entered into a license agreement with RMX Biopharma for the development, manufacture, and commercialization of delpazolid in China. In addition, delpazolid received an FDA orphan drug designation, a Qualified Infectious Disease Product Designation, and was selected as a Fast Track target drug.

On October 30, 2019, at 'The 50th Union World Conference on Lung Health,' held in Hyderabad, India, LegoChem Biosciences released the interim efficacy and safety results of a phase 2a study on delpazolid. Particularly, the results of a clinical phase 2a early bactericidal activity trial involving 79 Korean patients with drug-susceptible tuberculosis were reported. The findings will be further analyzed to determine the doses appropriate for different patient populations to guide further phase 2a studies. The phase 1 trial revealed that myelosuppression can be reduced, and phase 2a results suggested that delpazolid can replace linezolid as a therapy for TB and reduce the treatment period.

Author Contributions: Conceptualization, Y.L.C. and J.J.; writing, J.J.; review and editing, Y.L.C. and J.J.; visualization, Y.L.C. and J.J.; project administration, Y.L.C. and J.J. All authors have read and agreed to the published version of the manuscript.

Acknowledgments: We thank all those who provided important information that was part of this work. We also thank Tae ho Kim for his assistance in figure preparation.

Conflicts of Interest: The authors declare no conflict of interest.

References

1. Bozdogan, B.; Appelbaum, P.C. Oxazolidinones: Activity, mode of action, and mechanism of resistance. *Int J Antimicrob Agents.* **2004**, *23*, 113–119. [CrossRef] [PubMed]
2. Marchese, A.; Schito, G.C. The oxazolidinones as a new family of antimicrobial agent. *Clin Microbiol Infect.* **2001**, *4*, 66–74. [CrossRef] [PubMed]
3. Moellering, R.C. Linezolid: The first oxazolidinone antimicrobial. *Ann Intern Med.* **2003**, *138*, 135–142. [CrossRef] [PubMed]
4. Millard, J.; Pertinez, H.; Bonnett, L.; Hodel, E.M.; Dartois, V.; Johnson, J.L.; Caws, M.; Tiberi, S.; Bolhuis, M.; Alffenaar, J.C.; et al. Linezolid pharmacokinetics in MDR-TB: A systematic review, meta-analysis and Monte Carlo simulation. *J Antimicrob Chemother.* **2018**, *73*, 1755–1762. [CrossRef]
5. Gerson, S.L.; Kaplan, S.L.; Bruss, J.B.; Le, V.; Arellano, F.M.; Hafkin, B.; Kuter, D.J. Hematologic effects of linezolid: Summary of clinical experience. *Antimicrob Agents Chemother.* **2002**, *46*, 2723–2726. [CrossRef]
6. Choi, Y.; Lee, S.W.; Kim, A.; Jang, K.; Nam, H.; Cho, Y.L.; Yu, K.S.; Jang, I.J. Chung, J.Y. Safety, tolerability and pharmacokinetics of 21 day multiple oral administration of a new oxazolidinone antibiotic, LCB01-0371, in healthy male subjects. *J Antimicrob Chemother.* **2018**, *73*, 183–190. [CrossRef]
7. De Freitas Ferreira, R.; Schapira, M. A systematic analysis of atomic protein-ligand interactions in the PDB. *Medchemcomm.* **2017**, *8*, 1970–1981. [CrossRef]
8. Singh, B.; Cocker, D.; Ryan, H.; Sloan, D.J. Linezolid for drug-resistant pulmonary tuberculosis. *Cochrane Database Syst Rev.* **2019**, *3*, CD012836. [CrossRef]
9. Cho, Y.S.; Lim, H.S.; Cho, Y.L.; Nam, H.S.; Bae, K.S. Multiple-dose Safety, Tolerability, Pharmacokinetics, and Pharmacodynamics of Oral LCB01-0371 in Healthy Male Volunteers. *Clin Ther.* **2018**, *40*, 2050–2064. [CrossRef]
10. Cho, Y.S.; Lim, H.S.; Lee, S.H.; Cho, Y.L.; Nam, H.S.; Bae, K.S. Pharmacokinetics, Pharmacodynamics, and Tolerability of Single-Dose Oral LCB01-0371, a Novel Oxazolidinone with Broad-Spectrum Activity, in Healthy Volunteers. *Antimicrob Agents Chemother.* **2018**, *62*, e00451-18. [CrossRef]
11. Sunwoo, J.; Kim, Y.K.; Choi, Y.; Yu, K.S.; Nam, H.; Cho, Y.L.; Yoon, S.; Chung, J.Y. Effect of food on the pharmacokinetic characteristics of a single oral dose of LCB01-0371, a novel oxazolidinone antibiotic. *Drug Des Devel Ther.* **2018**, *12*, 1707–1714. [CrossRef] [PubMed]
12. Kim, T.S.; Choe, J.H.; Kim, Y.J.; Yang, C.S.; Kwon, H.J.; Jeong, J.; Kim, G.; Park, D.E.; Jo, E.K.; Cho, Y.L.; et al. Activity of LCB01-0371, a Novel Oxazolidinone, against Mycobacterium abscessus. *Antimicrob Agents Chemother.* **2017**, *61*, e02752-16. [CrossRef] [PubMed]
13. Hällberg, B.M.; Larsson, N.G. Making proteins in the powerhouse. *Cell Metab.* **2014**, *20*, 226–240. [CrossRef] [PubMed]
14. McKee, E.E.; Ferguson, M.; Bentley, A.T.; Marks, T.A. Inhibition of mammalian mitochondrial protein synthesis by oxazolidinones. *Antimicrob Agents Chemother.* **2006**, *50*, 2042–2049. [CrossRef] [PubMed]
15. De Vriese, A.S.; Coster, R.V.; Smet, J.; Seneca, S.; Lovering, A.; Van Haute, L.L.; Vanopdenbosch, L.J.; Martin, J.J.; Groote, C.C.; Vandecasteele, S.; et al. Linezolid-induced inhibition of mitochondrial protein synthesis. *Clin Infect Dis.* **2006**, *42*, 1111–1117. [CrossRef]
16. Nagiec, E.E.; Wu, L.; Swaney, S.M.; Chosay, J.G.; Ross, D.E.; Brieland, J.K.; Leach, K.L. Oxazolidinones inhibit cellular proliferation via inhibition of mitochondrial protein synthesis. *Antimicrob Agents Chemother.* **2005**, *49*, 3896–3902. [CrossRef]
17. Jeong, J.W.; Jung, S.J.; Lee, H.H.; Kim, Y.Z.; Park, T.K.; Cho, Y.L.; Chae, S.E.; Baek, S.Y.; Woo, S.H.; Lee, H.S.; et al. In vitro and in vivo activities of LCB01-0371, a new oxazolidinone. *Antimicrob Agents Chemother.* **2010**, *54*, 5359–5362. [CrossRef]
18. Zong, Z.; Jing, W.; Shi, J.; Wen, S.; Zhang, T.; Huo, F.; Shang, Y.; Liang, Q.; Huang, H.; Pang, Y. Comparison of In Vitro Activity and MIC Distributions between the Novel Oxazolidinone Delpazolid and Linezolid against Multidrug-Resistant and Extensively Drug-Resistant Mycobacterium tuberculosis in China. *Antimicrob Agents Chemother.* **2018**, *62*, e00165-18. [CrossRef]
19. Kerantzas, C.A.; Jacobs, W.R., Jr. Origins of Combination Therapy for Tuberculosis: Lessons for Future Antimicrobial Development and Application. *mBio.* **2017**, *8*, e01586-16. [CrossRef]

20. Abate, G.; Hamzabegovic, F.; Eickhoff, C.S.; Hoft, D.F. BCG Vaccination Induces M. avium and M. abscessus Cross-Protective Immunity. *Front Immunol.* **2019**, *10*, 234. [CrossRef]
21. Johnson, M.M.; Odell, J.A. Nontuberculous mycobacterial pulmonary infections. *J Thorac Dis.* **2014**, *6*, 210–220. [PubMed]
22. DeStefano, M.S.; Shoen, C.M.; Sklaney, M.R.; Cynamon, M.H. *The In Vitro Activity of Delpazolid (LCB01-0371) against Several Non-Tuberculous Mycobacteria (NTM)*; (Poster Number AAR-731, Friday, June 21, 2019); ASM Microbe: San Francisco, CA, USA, 2019.

© 2020 by the authors. Licensee MDPI, Basel, Switzerland. This article is an open access article distributed under the terms and conditions of the Creative Commons Attribution (CC BY) license (http://creativecommons.org/licenses/by/4.0/).

Review

Host-Directed Therapies and Anti-Virulence Compounds to Address Anti-Microbial Resistant Tuberculosis Infection

Raphael Gries [1,2], Claudia Sala [3] and Jan Rybniker [1,2,4,*]

1. Department I of Internal Medicine, Division of Infectious Diseases, University of Cologne, 50931 Cologne, Germany; raphael.gries@uk-koeln.de
2. Center for Molecular Medicine Cologne, University of Cologne, 50931 Cologne, Germany
3. Fondazione Toscana Life Sciences, 53100 Siena, Italy; c.sala@toscanalifesciences.org
4. German Center for Infection Research (DZIF), Partner Site Bonn-Cologne, 50931 Cologne, Germany
* Correspondence: jan.rybniker@uk-koeln.de; Tel.: +49-221-478-89611; Fax: +49-221-478-5915

Received: 9 March 2020; Accepted: 9 April 2020; Published: 13 April 2020

Abstract: Despite global efforts to contain tuberculosis (TB), the disease remains a leading cause of morbidity and mortality worldwide, further exacerbated by the increased resistance to antibiotics displayed by the tubercle bacillus *Mycobacterium tuberculosis*. In order to treat drug-resistant TB, alternative or complementary approaches to standard anti-TB regimens are being explored. An area of active research is represented by host-directed therapies which aim to modulate the host immune response by mitigating inflammation and by promoting the antimicrobial activity of immune cells. Additionally, compounds that reduce the virulence of *M. tuberculosis*, for instance by targeting the major virulence factor ESX-1, are being given increased attention by the TB research community. This review article summarizes the current state of the art in the development of these emerging therapies against TB.

Keywords: tuberculosis; *Mycobacterium tuberculosis*; host-directed therapy; anti-virulence compounds

1. Introduction

Mycobacterium tuberculosis, the etiological agent of human tuberculosis (TB), is thought to latently infect approximately one fourth of the world's population and is responsible for over one million deaths every year [1], thus representing the leading cause of mortality by an infectious disease worldwide. Immunodeficiency caused by HIV [2] and co-morbidities like diabetes [3] constitute additional risk factors for the development of active TB disease.

The current anti-TB therapy consists of a combination of four antibiotics (rifampicin, isoniazid, pyrazinamide and ethambutol) that must be administered for at least 6 months in case of drug-sensitive pulmonary TB infection [4]. However, *M. tuberculosis* displays increased resistance to first-line drugs, which has resulted in multidrug-resistant (MDR) and extensively drug-resistant (XDR) TB cases [5]. Second-line treatment regimens are therefore employed but require longer duration to be effective and are associated with severe side effects that frequently decrease patient compliance [6].

To address the increasing need for new and potent therapeutic options against TB, alternative approaches are being explored. These include host-directed therapy (HDT) and anti-virulence compounds. Within the first choice, a number of molecules that reduce inflammation, modulate autophagy and potentiate the immune response are currently in preclinical and in clinical trials. On the other hand, drugs that affect *M. tuberculosis* ability to infect and kill host cells represent a promising complement to standard antibiotic treatment.

Here we review the current state of research in the areas of HDT and anti-virulence drugs as complementary approaches to TB therapy.

2. Host Directed Therapy

2.1. Promoting Phagosome Maturation and Enhancing Autophagy

Autophagy is a natural process which protects cells against unfolded proteins and potentially dangerous aggregates, viral and bacterial infections. By means of autophagy, macrophages deliver toxic macromolecules, organelles and phagocytosed pathogens to lysosomes for degradation [7]. Modulating autophagy in order to promote bacterial killing represents one example of HDT against TB infection. This statement is supported by studies conducted by Gutierrez and co-workers who reported that stimulation of autophagy in macrophages promotes phago-lysosome maturation and impacts mycobacterial survival [8]. More recent investigations revealed increased susceptibility to TB in mice defective in autophagy pathways [9] and a positive interplay between autophagy and interferon-gamma (IFN-γ) in TB patients [10].

2.1.1. mTOR Inhibition

The best described autophagy inducer is rapamycin (sirolimus), used in patients who underwent organ transplantation [11]. Rapamycin, a macrolide produced by *Streptomyces hygroscopicus* and originally discovered on Easter Island (Rapa Nui for the inhabitants, hence the name given to the compound), inhibits TOR, the Target Of Rapamycin [12]. The mammalian Target of Rapamycin mTOR is a negative regulator of autophagy [13]. Its clinical use in infectious diseases is restricted due to its broadly immunosuppressive effects. In addition, rapamycin is metabolized by CYP3A4 [14], a hepatic enzyme induced by the first-line TB drug rifampicin, thus hampering its exploitation as HDT in TB patients. However, other molecules capable of inducing autophagy have been discovered and are now under development. Among these, vadimezan [15], Tat-beclin 1 fusion peptide [16], the calcium-channel blocker verapamil [17] and the rapamycin analogue everolimus [18]. In particular, verapamil was shown to be efficacious when combined with rifampicin and with the recently approved medication bedaquiline in mouse models of infection, where it increased the bioavailability of the antibiotic [19–21]. On the other hand, everolimus, an anti-cancer agent, may be repurposed as an anti-TB HDT therapeutic capable of inhibiting mTOR although, as with rapamycin, immunosuppression [22] and toxicity [23] may represent issues in the clinical development as an anti-TB drug.

2.1.2. Metformin

One of the most promising drugs that promotes autophagy is currently used for treatment of type 2 diabetes: metformin [24]. This compound is characterized by a good safety profile, activates 5′-adenosine monophosphate-activated protein kinase (AMPK), induces production of mitochondrial reactive oxygen species (mROS), which are deleterious to *M. tuberculosis* and was found to reduce the severity of TB disease in humans [25]. Despite these promising data, combination therapies which involved metformin in mice had contradictory results. While in one case metformin enhanced the activity of isoniazid and ethionamide [25], in another one it did not improve the efficacy of the combined first-line drugs [26]. Recent retrospective studies reported a protective effect for metformin against reactivation of latent TB in diabetic patients [27–29]. Phase II clinical trials have been initiated (Table 1).

Activity of the major virulence factor ESX-1 can be blocked by compounds BBH7 and BTP15. While BBH7 hinders the secretion mechanism by inducing zinc stress, BTP15 was shown to act by downregulating expression of the *espA-espC-espD* operon upon interaction with MprB. The secreted proteins MptpB, SapM and Zmp1, which prevent phagosomal maturation, can be directly targeted extracellularly by their respective inhibitors. The dedicated secretion systems for these three proteins are not described yet. Gene expression of PhoP-dependent virulence genes can be controlled by PhoP inhibitors which prevent binding of PhoP to specific promoter regions, thus affecting transcription.

2.1.3. Imatinib and Other Tyrosine Kinase Inhibitors

A key feature of *M. tuberculosis* is represented by its ability to inhibit phago-lysosome fusion, and thus potentially limit the efficacy of the autophagy process, thanks to the presence of specific virulence factors like lipoarabinomannan in the cell wall [30], the ESX-1 secretion system [31,32] and other key components such as the Eis protein which modulates autophagy and inflammation and suppresses host innate immune responses [33]. This inhibitory effect can be overcome by tyrosine kinase inhibitors, such as imatinib and the second-generation inhibitors nilotinib and dasatinib, which target the BCR-ABL fusion protein and are used for treating chronic myeloid leukemia [34]. Different studies explored the effect of imatinib on *M. tuberculosis*-infected macrophages and revealed that it increases acidification of lysosomes thereby halting bacterial multiplication [35]. Moreover, imatinib was shown to reduce the number of granulomatous lesions in mice and to act synergistically with first-line anti-TB drug rifampicin [36].

2.1.4. Statins

Statins, i.e., agents that lower cholesterol through inhibition of the biosynthetic pathway, also impact autophagy [37]. Given the relevance of cholesterol in *M. tuberculosis* persistence [38], statins have received considerable attention. Indeed, in addition to their cholesterol-lowering effect, statins decrease lipid body biogenesis and limit *M. tuberculosis* survival [39]. Additionally, it was discovered that atorvarstatin potentiates the effect of rifampin in *M. leprae* infection of the mouse footpad [40]. These substances are currently investigated in clinical trials (Table 1).

2.2. Vitamin D and the Induction of Anti-Microbial Peptides

Anti-microbial peptides like cathelicidins are components of the innate immune system whose synthesis is induced by mycobacterial ligands through binding to Toll-like receptors (TLRs), especially TLR2 and TLR9 [41]. Cathelicidin LL37 represents a major example of this class of molecules, is expressed by neutrophils and macrophages and participates in anti-TB defense through pore-forming capability in the bacterial membranes [42,43].

It has been shown that vitamin D promotes synthesis and release of LL37 [44], which in turn helps in autophagy [45,46]. Moreover, vitamin D enhances the ability of monocytes to respond to interferon gamma (IFN-γ) [47]. Various clinical trials which included vitamin D in addition to the standard regimen have been performed, sometimes with variable results [48–50]. It seems evident that key issues for successful use of vitamin D in TB therapy are proper dosing and possibly also genetic background and comorbidities of the patient. A recently published study by Aibana and co-workers suggested that vitamin D deficiency is associated with increased probability of developing TB in HIV-positive people [51]. However, further investigations are needed to clarify whether vitamin D supplementation might play a significant role in reducing the risk of TB.

Another vitamin whose antitubercular effects have been evaluated is vitamin A, which limits *M. tuberculosis* replication in macrophages by promoting acidification [52,53]. However, while studies in rats showed a beneficial impact of vitamin A supplementation [54], the same was not observed in humans [55,56].

Regulation of anti-microbial peptide expression is also controlled by histone deacetylase inhibitors (for instance 4-phenylbutyrate) through epigenetic mechanisms [57,58]. In the context of *M. tuberculosis* infection of human macrophages, it was demonstrated that phenylbutyrate, alone or in combination with vitamin D3, was able to counteract the suppressive effect of the bacilli on LL-37 expression, thus promoting autophagy [59].

In addition to cathelicidins, another group of anti-microbial peptides plays an important role in anti-TB mechanisms. These are defensins. Defensins are arginine-rich, cationic peptides resistant to proteolysis. They are usually stored in the granules and in the lysosomes of innate immune cells, such as neutrophils, and are released upon pathogen invasion [60]. *M. tuberculosis* stimulates production of beta defensin-2 (HBD-2), which reduces bacterial multiplication and has a chemotactic effect [61]. Despite these features, exploitation of HBD-2 in HDT against TB is far from clinical use, due to high costs and poor stability in vivo [62]. Clinical trials where these compounds are being investigated are listed in Table 1.

2.3. IFN-γ and IL-2 as Adjunct Therapy

Production of anti-microbial peptides and other antimicrobial activities exerted by macrophages are stimulated by a panel of cytokines that include TNF (Tumor Necrosis Factor), IFN-γ and interleukin 1 (IL-1). While IFN-γ plays its major role in promoting autophagy and phagosome maturation, TNF increases IFN-γ responsiveness and IL-1 counteracts the detrimental effects of Type I IFN in TB [63]. HDT against TB infection includes IFN-γ and modulators of TNF, which will be discussed later in this review. Concerning IFN-γ, it was demonstrated that its administration to TB patients via the aerosol route is well-tolerated and reduces time to sputum conversion while improving lung repair after the disease [64–66]. However, the role of IFN-γ in controlling TB is still under debate, as reported in a study by Sakai and co-workers, who showed that contribution of CD4-T cell derived IFN-γ is limited and, even worse, sometimes detrimental [67]. Another clinical study, where IL-2 was added during the first month of anti-TB treatment resulted in no benefit [68], thus questioning the relevance of adding cytokines to the existing therapy. Possible side effects and treatment costs should also be considered when exploring the administration of cytokines to TB patients.

2.4. Inhibition of M. tuberculosis Induced Inflammation and Host Cell Death

2.4.1. The Role of Corticosteroids in TB Treatment

It sounds counterintuitive to address the problem of active TB with anti-inflammatory drugs. However, for some clinical manifestations of the diseases, reduction of inflammation by using adjunctive corticosteroids has already become a well-established and lifesaving treatment approach. Addition of dexamethasone or prednisolone, two potent corticosteroids, to the antibiotic regimen for treatment of TB meningitis improves survival and is considered as a valid therapeutic approach for TB affecting the central nervous system (CNS) [69], although care should be taken since individual responses to steroid treatment might differ. Several studies have tried to improve the outcome of pulmonary TB by lowering the inflammatory response using high doses of corticosteroids in combination with antibiotics. While it was found that this therapy leads to faster resolution of symptoms and lesions in radiographic examinations and a more rapid discharge from hospitals, a statistically significant survival benefit could not be shown (Table 1) [70–72]. In addition, high dose corticosteroids may result in serious side effects such as diabetes and psychiatric symptoms. Today, corticosteroids remain the treatment of choice in specific clinical situations such as CNS TB or hyperinflammatory syndromes e.g., the immune reconstitution inflammatory syndrome (IRIS) in HIV/TB co-infected patients. Investigations at the molecular level proved that dysregulation of inflammasome signaling and of secretion of various cytokines, including IL-1γ, was associated with TB-IRIS in patients infected by HIV [73,74], thus supporting the inclusion of corticosteroids in the treatment of TB patients at risk of developing IRIS [75]. Despite these evidences, a broader application of the drugs in TB treatment is currently not justified. However, clinical studies as well as ex vivo and in vivo experiments performed with these substances indicate that a more specific or tailored modification of the TB inflammatory response may provide a suitable approach to improve patient outcomes. Understanding the exact mechanism of action of corticosteroids in TB may help overcome this hurdle. Corticosteroids are broadly immunosuppressive drugs with multiple modulatory effects on leukocytes once bound to the main target, the corticosteroid receptor. Downstream effects include repression of pro-inflammatory transcriptional regulators like NF-κB as well as impaired release of cytokines such as TNFα and IL-1 [76]. In addition, corticosteroids such as dexamethasone seem to have an *M. tuberculosis* specific inhibitory effect on necrotic host cell death in vitro [77]. This effect seems to depend on inhibition of p38 MAP kinase which impairs mitochondrial membrane stability upon infection with *M. tuberculosis*. p38 MAP kinase is activated during *M. tuberculosis* infection in vitro and in vivo and represents a possible host directed target with several clinically tested small molecule inhibitors available for repurposing.

2.4.2. Non-Steroidal Anti-Inflammatory Drugs (NSAID) and Leukotriene Inhibitors

A series of mouse studies have shown beneficial effects of non-steroidal anti-inflammatory drugs (NSAIDS) such as aspirin, diclofenac and ibuprofen when used alone or in combination with common antibiotics in *M. tuberculosis*-infected mice. The main mechanism of action seems to be inhibition of prostaglandin synthesis via inhibition of cyclooxygenase 1 and 2. Prostaglandins are known drivers of tissue damaging inflammation. It is important to note that diclofenac was shown to possess growth inhibitory effects on the bacterium itself in addition to its anti-inflammatory properties. The substances have been extensively discussed elsewhere [78]. NSAID Clinical trials initiated recently are listed in Table 1. Another category of anti-inflammatory drugs is represented by leukotriene receptor antagonists, such as zafirlukast, which was reported to have anti-mycobacterial activity in vitro and cause alterations in the transcription profile in *M. tuberculosis* [79]. The potential of these drugs in HDT against TB deserves deeper investigation given the role for leukotriene A(4) hydrolase (LTA4H) demonstrated by Tobin and colleagues in animal models of infection [80,81].

2.4.3. Necrosis

Necrotic host cell death is a highly dynamic research field increasingly linked to the release of pro-inflammatory cytokines. A better understanding of the mechanisms of *M. tuberculosis* induced cell death may provide additional starting points for HDTs. Most studies have been focusing on cell death in macrophages, however, necrosis of other cell types such as neutrophils seems to play a pivotal and additive role in *M. tuberculosis* pathogenicity. *M. tuberculosis* released by necrotic neutrophils displays improved survival and growth once phagocytosed by adjacent macrophages [82]. Neutrophil necrotic cell death is driven by reactive oxygen species (ROS) which can be abrogated by ROS inhibitors. In addition, ROS and nitric oxide (NO) have been found to show antimicrobial activity and to modulate neutrophil recruitment to the granuloma [83]. While ROS seems to increase cytokine production and to inhibit inflammasome activation, NO shows a regulatory effect on macrophages with increased expression of hypoxia-inducible factor 1 alpha (HIF-1α) and repression of nuclear factor kappa-light-chain-enhancer of activated B cells (NF-κB) [84,85]. A recent study highlights a role for ferroptotic cell death in TB. Ferroptosis is a type of regulated necrosis induced by accumulation of free iron and toxic lipid peroxides which seems to be mediated by decreased levels of glutathione peroxidase-4 (Gpx4) upon *M. tuberculosis* infection in vitro. Intraperitoneal treatment of *M. tuberculosis* infected mice with ferrostatin, a ferroptosis inhibitor resulted in reduced lung pathology and decreased bacterial load [86]. In addition to ferroptosis, efferocytosis (the physiological process of removing apoptotic cells by macrophages) is an anti-bacterial mechanism that seems to play a relevant role in TB as well [87]. Indeed, efferocytosis of apoptotic neutrophils was shown to improve control of *M. tuberculosis* in an in vitro model of HIV-*M. tuberculosis* macrophage co-infection [88,89].

2.4.4. TNF and TNF-Mediated Signaling

Further downstream of intracellular mediators or regulators of cell death, inflammation and cytokine release, there are more direct targets amenable to therapeutic interventions. These include the cytokines themselves. Biologicals targeting TNFα such as infliximab and adalimumab (monoclonal antibodies) or etanercept (TNF receptor fusion protein) may be used to limit exacerbated pathology and improve antibiotic activity. These substances are restricted for use in combination with antibiotics (adjuvant treatment) since TNFα is essential for protective immunity and granuloma integrity. Monotherapy with infliximab and other anti-TNF antibodies led to reactivation of latent TB [90]. However, when combined with anti-TB drugs, TNF neutralization enhanced *M. tuberculosis* clearance and reduced lung pathology [91]. A clinical study performed with adjuvant etanercept in patients with pulmonary TB and HIV showed a trend towards improved outcome when the TNF blocker was added to the antibiotic regimen [92].

TNF signaling is also the main target of other HDT candidates such as thalidomide, phosphodiesterase inhibitors or Janus kinase (JAK) inhibitors. Thalidomide has potent anti-inflammatory properties which led to successful application of the drug in cases where anti-TB or HIV treatment triggered hyperinflammatory syndromes such as paradoxical reactions or immune reconstitution syndrome (IRIS). A general application for adjuvant treatment approaches may be hampered due to side effects as seen in a study performed with children suffering from TB meningitis [93]. Phosphodiesterase (PDE) inhibitors seem to be more promising for broad application in TB patients. PDEs degrade cyclic AMP (cAMP), a second messenger negatively regulating TNF levels. Decreased levels of cAMP stimulate TNFα secretion, thus making PDE inhibitors interesting HDT targets. Among the five PDE subtypes, targeting PDE4 seems to be the most promising option in TB with adjuvant use of inhibitors leading to improved outcome in several animal models [94,95]. A phase II clinical trial with the PDE4 inhibitor CC-11050 is ongoing (Table 1).

2.4.5. Targeting Matrix Metalloproteinases for Improved Tissue Repair

Imbalanced inflammation eventually results in host tissue destruction, cavitation and dissemination of bacteria. A main driver of these end-stage events are matrix metalloproteinases (MMP) [96]. Once released from activated or necrotic cells, these zinc dependent proteases cleave the extracellular matrix, mostly collagen, and inhibition with small molecules should restrict tissue damage and exacerbation of the disease. Several in vitro and ex vivo studies identified elevated MMP levels in *M. tuberculosis* infected cells or tissue indicating that these enzymes are engaged [96]. In particular, Andrade and colleagues [97] evaluated the interplay between the levels of MMP and heme oxygenase-1 (HO) and discovered that the abundance of these two markers in plasma correlates with different inflammatory profiles and clinical presentations of TB. To date, the only FDA approved MMP inhibitor is doxycycline, an antibiotic with a dual mechanism of action targeting primarily MMP1 and MMP9. The drug suppressed MMP1 and 9 activities in *M. tuberculosis* infected primary human macrophages [98]. In the same study, doxycycline treatment of *M. tuberculosis*-infected guinea pigs led to reduction of the lung bacterial load compared to untreated animals. However, it is important to note that the substance shows a significant growth inhibitory effect on *M. tuberculosis* in broth (MIC 2.5 µg/mL) making it difficult to differentiate between selective host directed and antibacterial effects in these experiments. Experiments with more selective MMP inhibitors such as marimastat (BB-2516), a collagen peptidomimetic broad spectrum MMP inhibitor, showed adjuvant activity in *M. tuberculosis*-infected mice when combined with isoniazid or rifampicin [99]. In contrast to doxycycline, monotherapy with marimastat had no effect on lung bacterial burden [99].

Table 1. Candidate compounds for host-directed therapy (HDT) against tuberculosis (TB).

HDT Effect	Compound	Target or Mode of Action	Notes	Clinical Trials (ClinicalTrials.gov)	References
Promote phagosome maturation and enhance autophagy	Rapamycin (sirolimus)	Inhibition of mTOR	Metabolized by CYP3A4	—	[11–14]
	Everolimus	Inhibition of mTOR, rapamycin analogue	Anti-cancer agent	NCT02968927	[18,22,23]
	Metformin	Activates AMPK	Used to treat diabetes	Phase 2 studies planned CTRI/2018/01/011176	[25]
	Imatinib	Inhibition of BCR-ABL tyrosine kinase	Used to treat leukaemia	NCT03891901	[35,36]
	Statins	Inhibition of cholesterol biosynthetic pathway	Cholesterol is relevant in M. tuberculosis persistence	NCT03882177 NCT03456102 NCT04147286	[37,40]
	Vitamin D	Promotes synthesis of cathelicidin LL37	Variable results in clinical trials	NCT00918086 NCT01722396 NCT01130311 NCT01244204 NCT00677339 NCT01698476 NCT01137370 (all completed)	[44,47–50]
Induce anti-microbial peptides	Vitamin A	Promotes acidification of phagosome	Inconsistent results in rats and humans	NCT00057434 (completed)	[52–56]
	4-phenylbutyrate	Inhibition of histone deacetylase	Promotes autophagy	NCT01580007 NCT01698476 (all completed)	[57–59]
	Beta defensin 2 (HBD-2)	Reduces M. tuberculosis multiplication	High costs and poor stability	—	[60–62]

Table 1. Cont.

HDT Effect	Compound	Target or Mode of Action	Notes	Clinical Trials (ClinicalTrials.gov)	References
Adjunct cytokine therapy	Interferon gamma (IFN-γ)	Promotes autophagy and phagosome maturation	Reduces time to sputum conversion	NCT00201123 NCT00001407 (all completed)	[63–66]
	Interleukin 2 (IL-2)	Enhances cell-mediated response to infection	Contrasting results in clinical trials	NCT03069534	[63,68]
	Corticosteroids	Multiple anti-inflammatory effects	Standard of care for CNS TB. Other forms of TB may require high doses for beneficial effects leading to unwanted side effects	Multiple clinical trials. See meta-analysis in Critchley et al. 2013 and 2014	[69–72]
	P38 MAPK inhibitors	Protect cells from mitochondria-induced necrosis			[77]
	Ferrostatin	Decrease of glutathione peroxidase-4 (Gpx4) levels	Mouse study showing beneficial effect		[86]
Reduce inflammation/Inhibit necrotic cell death	Infliximab, adalimumab, etanercept	Inhibition of TNFα	Restricted for use in combination with antibiotics		[90–92]
	CC-11050	Phosphodiesterase (PDE) inhibition		NCT02968927	[94,95]
	Doxycycline, marimastat (BB-2516)	Inhibition of matrix metalloproteinases	Doxycycline shows growth inhibition of *M. tuberculosis*, effects probably not purely host directed	NCT02774993	[99]
	NSAID: aspirin, ibuprofen, diclofenac, etoricoxib, indomethacin	Cyclooxygenase 1 and/or 2 inhibition		NCT02781909 NCT02602509 NCT02503839	[78]

3. Targeting Bacterial Virulence

3.1. The ESX-1 Secretion System

Lately, the interest in finding novel lead compounds, which prevent infection and dissemination by inhibiting bacterial virulence factors, has increased. These anti-virulence molecules target one or more proteins in the virulence machinery with one prominent example being the ESX-1 secretion system as ESX-1 deletion mutants show strongly attenuated phenotypes in vitro and in vivo [100]. ESX-1 is a type VII secretion system essential for host cell infection, bacterial spread and macrophage escape but not for bacterial growth in axenic cultures [101]. In a whole-cell-based phenotypic screening assay selecting for compounds that abrogate ESX-1 dependent host cell death, the two ESX-1 inhibitors BTP15 and BBH7 have been found and characterized [102]. BTP15 inhibits the histidine kinase MprB that regulates ESX-1 via the *espA-espC-espD* operon. BBH7 on the other hand disturbs metal-ion homeostasis leading to zinc stress and thus hindering secretion of ESX-1 substrates such as EsxA and EsxB (Figure 1). These inhibitors can also be used to abrogate ESX-1 dependent activation of the cytosolic DNA sensor cyclic GMP-AMP synthase (cGAS), a main driver of type I interferon (IFN) secretion, thus nicely linking anti-virulence drugs to modulation of the inflammatory response [103].

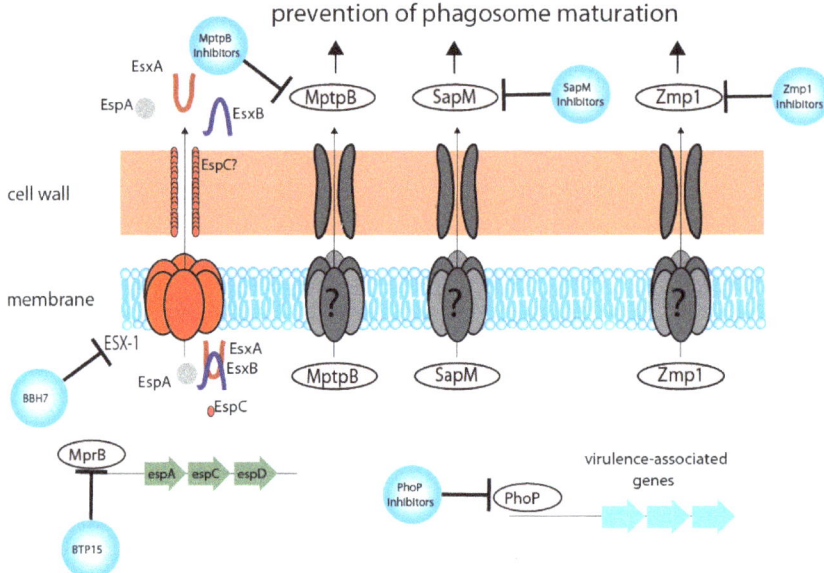

Figure 1. Schematic overview of antivirulence targets in *M. tuberculosis*.

3.2. PhoPR Inhibitors

The PhoPR two-component system plays a central role in regulating the expression of several proteins relevant for virulence of *M. tuberculosis* as mutants deficient in the effector response regulator PhoP show attenuated growth in infected THP-1 cells and in mice [104]. A microarray-based transcriptional profiling study of *M. tuberculosis* strain H37Rv revealed 110 genes that have been differently expressed in PhoP-deficient mutants [105]. This attenuated strain harbors a single nucleotide polymorphism (S219L) in the DNA-binding domain of PhoP resulting in a reduced DNA-binding capacity [105]. Further studies revealed that PhoP is involved in regulating ESX-1 and in biosynthesis of cell wall components such as sulfolipids, polyacyltrehaloses and diacyltrehaloses [106–108]. Two different approaches identified inhibitors of the PhoPR regulon.

3.2.1. Ethoxzolamide

Using a pH-inducible fluorescence reporter system Johnson et al. phenotypically screened for inhibitors of the PhoPR regulon (Figure 1). This screening discovered the carbonic anhydrase inhibitor ethoxzolamide, which inhibits PhoPR while not reducing mycobacterial growth in vitro. Chromatin immunoprecipitation followed by deep sequencing (ChIP-seq) of compound-exposed *M. tuberculosis* cultivated in medium at pH 5.7 showed downregulation of several PhoPR regulated genes involved in lipid synthesis, carbon metabolism and virulence. In addition, the presence of ethoxzolamide did not modulate the expression levels of PhoP itself indicating that the substance acts as a direct inhibitor of the core PhoPR regulon [109].

3.2.2. Inhibitors of the PhoP-DNA Complex

Three active compounds (NCGC00093547, NCGC00244580 and NCGC00161636) that directly bind to PhoP and therefore inhibit PhoP-DNA interactions were found in a screening assay based on Foster resonance energy transfer (FRET). For this screening a DNA-Protein complex of Cy3-labeled DNA and Cy5-labeled PhoP protein was exposed to compounds of interest. Inhibitors of this DNA-Protein complex led to dissociation and consequently to a reduced FRET signal [110]. Direct binding of inhibitors to PhoP was confirmed by thermal shift assays in which target-bound inhibitors stabilize the protein and increase the melting temperature, which can be quantified using the fluorescence signal of fluorophore-protein complexes. Compounds NCGC00093547 and NCGC00161636 increased PhoP melting temperature by 14 °C and 18 °C with an IC_{50} of 15.6 and 15.5 µM, respectively. Data on in vivo or ex vivo activity of these compounds is not available yet.

3.3. Phagosomal Regulation/Hindering Intracellular Survival

3.3.1. MptpB Inhibitors

The *M. tuberculosis* protein-tyrosine-phosphatase B (MptpB) is another putative target for anti-virulence compounds (Figure 1). This kinase is secreted into the cytoplasm of host macrophages allowing for inhibition outside the thick and difficult to overcome mycobacterial cell wall [111]. The function of MptpB is not fully described yet, but the protein has been reported to be necessary for bacterial survival in guinea pigs [112]. So far, it was shown that MptpB dephosphorylates host phosphotyrosine substrates, phosphoserine/threonine substrates and phosphoinositides, with the latter being essential for host macrophage maturation [113].

Several isoxazole-based molecules were created to block the primary and secondary phosphate-binding pockets of MptpB followed by phenotypic testing for activity. In these ex vivo assays, the compounds led to a reduction of mycobacterial burden in macrophages (J774 and THP-1) and in a guinea pig model, without affecting extracellular growth in broth. Additionally, attenuated growth of MDR strains of *M. tuberculosis* in the presence of compound 13 was shown in macrophages. In addition, this inhibitor caused increased sensitivity of a BCG strain to rifampicin and isoniazid in an ex vivo macrophage infection experiment [114]. A similar effect has not been published for *M. tuberculosis* yet.

3.3.2. SapM Inhibitors

Another virulence factor that affects phagocytosis and phagosome formation is the secreted acid phosphatase M (SapM) (Figure 1). SapM shows activity as a monoester alkaline phosphatase and targets two phosphoinositides ($PI(4,5)P_2$ and PI3P) important for phagosome maturation [115]. In inhibition studies, it was shown that 2-phospho-L-ascorbic acid interferes with SapM activity without attenuating extracellular bacterial growth. At 4 mM, this drug could reduce intracellular growth of *M. tuberculosis* by 39% [116].

3.3.3. Zmp1 Inhibitors

Although its role is not fully understood yet, Zmp1 is involved in mycobacterial pathogenicity as it inhibits the inflammasome and therefore prevents phagosome maturation (Figure 1) [117]. Zinc peptidases like Zmp1 are often inhibited by molecules with specific zinc binding groups (ZBG) like 8-hydroxyquinolines. 8-hydroxyquinolines are already in use as metal-interacting structures in pharmacological applications [118]. Based on this, Vickers et al. synthesized compounds consisting of an 8-hydroxyquinoline ring and a hydroxamate moiety and isosteric analogues of these. One 8-hydroxyquinoline-2-hydroxamate derivative showed a reduction in colony forming units (CFU) in infected J774 mouse macrophages while no extracellular, anti-mycobacterial activity was observed. Treatment of infected human monocyte-derived macrophages with this substance led to a decrease in bacterial burden in a dose-dependent matter. In in-vitro inhibition assays the compound inhibited Zmp1 with an IC_{50} of 0.011 µM [119].

3.4. Stress Associated Approaches

DosRST Signaling

M. tuberculosis exploits its two-component system DosRST to establish a dormant state of nonreplicating persistence (NRP) [120]. As *dosRST* mutants show attenuated growth in animal models, including nonhuman primates and guinea pigs, DosRST might be a potential target to reduce mycobacterial virulence [121]. When investigating compounds for their effect on the DosRST system, Zheng et al. identified two candidates (HC104A and HC106A) which interact with distinct members of the two-component system and decrease production of hypoxia-induced triacylglycerol by around 50% during NRP [122]. In a hypoxic shift-down model using a DosR-dependent fluorescent strain CDC1551(*hspX'*::GFP) HC106A was found to affect *M. tuberculosis* survival during NRP. HC104A on the other hand did not attenuate growth in this setting. The in vivo relevance for these interesting findings still needs to be established in a suitable animal model.

4. Conclusions

Several HDT approaches are currently being tested in a number of preclinical and clinical trials as adjuvants complementing conventional anti-TB treatment. Clinical trial results provided in the near future will present an important milestone for the implementation of HDT in routine clinical use. While representing a promising therapeutic approach in theory, most compounds targeting mycobacterial virulence factors lack in vivo proof of principle data.

Author Contributions: All of the authors participated in writing and reviewing the manuscript. All authors have read and agreed to the published version of the manuscript.

Funding: J.R. receives funding from the Thematic Translational Unit Tuberculosis (TTU TB, grant number TTU 02.806 and 02.905) of the German Centre of Infection Research (DZIF), the German Research Foundation (DFG RY 159, CRC1403), the Centre for Molecular Medicine Cologne (ZMMK–CAP8). J.R. has received funding from the Innovative Medicines Initiative 2 Joint Undertaking (JU) under grant agreement No 853989. The JU receives support from the European Union's Horizon 2020 research and innovation programme and EFPIA and Global Alliance for TB Drug Development Non Profit Organisation, Bill & Melinda Gates Foundation, University Of Dundee.

Conflicts of Interest: The authors declare no conflicts of interest.

References

1. World Health Organization (WHO). *Global Tuberculosis Report 2019*; World Health Organization (WHO): Geneva, Switzerland, 2019.
2. Du Bruyn, E.; Peton, N.; Esmail, H.; Howlett, P.J.; Coussens, A.K.; Wilkinson, R.J. Recent progress in understanding immune activation in the pathogenesis in HIV-tuberculosis co-infection. *Curr. Opin. Hiv Aids* **2018**, *13*, 455–461. [CrossRef] [PubMed]

3. Ferlita, S.; Yegiazaryan, A.; Noori, N.; Lal, G.; Nguyen, T.; To, K.; Venketaraman, V. Type 2 Diabetes Mellitus and Altered Immune System Leading to Susceptibility to Pathogens, Especially Mycobacterium tuberculosis. *J. Clin. Med.* **2019**, *8*, 2219. [CrossRef] [PubMed]
4. World Health Organization (WHO). *The End-TB Strategy*; World Health Organization (WHO): Geneva, Switzerland, 2014.
5. Mabhula, A.; Singh, V. Drug-resistance in Mycobacterium tuberculosis: Where we stand. *MedChemComm* **2019**, *10*, 1342–1360. [CrossRef] [PubMed]
6. Pontali, E.; Raviglione, M.C.; Migliori, G.B. Regimens to treat multidrug-resistant tuberculosis: Past, present and future perspectives. *Eur. Respir. Rev. Off. J. Eur. Respir. Soc.* **2019**, *28*. [CrossRef] [PubMed]
7. Levine, B.; Mizushima, N.; Virgin, H.W. Autophagy in immunity and inflammation. *Nature* **2011**, *469*, 323–335. [CrossRef] [PubMed]
8. Gutierrez, M.G.; Master, S.S.; Singh, S.B.; Taylor, G.A.; Colombo, M.I.; Deretic, V. Autophagy is a defense mechanism inhibiting BCG and Mycobacterium tuberculosis survival in infected macrophages. *Cell* **2004**, *119*, 753–766. [CrossRef]
9. Watson, R.O.; Manzanillo, P.S.; Cox, J.S. Extracellular, M. tuberculosis DNA targets bacteria for autophagy by activating the host DNA-sensing pathway. *Cell* **2012**, *150*, 803–815. [CrossRef]
10. Rovetta, A.I.; Pena, D.; Hernandez Del Pino, R.E.; Recalde, G.M.; Pellegrini, J.; Bigi, F.; Musella, R.M.; Palmero, D.J.; Gutierrez, M.; Colombo, M.I.; et al. IFNG-mediated immune responses enhance autophagy against Mycobacterium tuberculosis antigens in patients with active tuberculosis. *Autophagy* **2014**, *10*, 2109–2121. [CrossRef]
11. Nguyen, L.S.; Vautier, M.; Allenbach, Y.; Zahr, N.; Benveniste, O.; Funck-Brentano, C.; Salem, J.E. Sirolimus and mTOR Inhibitors: A Review of Side Effects and Specific Management in Solid Organ Transplantation. *Drug Saf.* **2019**, *42*, 813–825. [CrossRef]
12. Raught, B.; Gingras, A.C.; Sonenberg, N. The target of rapamycin (TOR) proteins. *Proc. Natl. Acad. Sci. USA* **2001**, *98*, 7037–7044. [CrossRef]
13. Ravikumar, B.; Vacher, C.; Berger, Z.; Davies, J.E.; Luo, S.; Oroz, L.G.; Scaravilli, F.; Easton, D.F.; Duden, R.; O'Kane, C.J.; et al. Inhibition of mTOR induces autophagy and reduces toxicity of polyglutamine expansions in fly and mouse models of Huntington disease. *Nat. Genet.* **2004**, *36*, 585–595. [CrossRef] [PubMed]
14. Kuhn, B.; Jacobsen, W.; Christians, U.; Benet, L.Z.; Kollman, P.A. Metabolism of sirolimus and its derivative everolimus by cytochrome P450 3A4: Insights from docking, molecular dynamics, and quantum chemical calculations. *J. Med. Chem.* **2001**, *44*, 2027–2034. [CrossRef] [PubMed]
15. Gao, P.; Ascano, M.; Zillinger, T.; Wang, W.; Dai, P.; Serganov, A.A.; Gaffney, B.L.; Shuman, S.; Jones, R.A.; Deng, L.; et al. Structure-function analysis of STING activation by c[G(2′,5′)pA(3′,5′)p] and targeting by antiviral DMXAA. *Cell* **2013**, *154*, 748–762. [CrossRef] [PubMed]
16. Shoji-Kawata, S.; Sumpter, R.; Leveno, M.; Campbell, G.R.; Zou, Z.; Kinch, L.; Wilkins, A.D.; Sun, Q.; Pallauf, K.; MacDuff, D.; et al. Identification of a candidate therapeutic autophagy-inducing peptide. *Nature* **2013**, *494*, 201–206. [CrossRef]
17. Gupta, S.; Tyagi, S.; Almeida, D.V.; Maiga, M.C.; Ammerman, N.C.; Bishai, W.R. Acceleration of tuberculosis treatment by adjunctive therapy with verapamil as an efflux inhibitor. *Am. J. Respir. Crit. Care Med.* **2013**, *188*, 600–607. [CrossRef]
18. Cerni, S.; Shafer, D.; To, K.; Venketaraman, V. Investigating the Role of Everolimus in mTOR Inhibition and Autophagy Promotion as a Potential Host-Directed Therapeutic Target in Mycobacterium tuberculosis Infection. *J. Clin. Med.* **2019**, *8*, 232. [CrossRef]
19. Gupta, S.; Cohen, K.A.; Winglee, K.; Maiga, M.; Diarra, B.; Bishai, W.R. Efflux inhibition with verapamil potentiates bedaquiline in Mycobacterium tuberculosis. *Antimicrob. Agents Chemother.* **2014**, *58*, 574–576. [CrossRef]
20. Gupta, S.; Tyagi, S.; Bishai, W.R. Verapamil increases the bactericidal activity of bedaquiline against Mycobacterium tuberculosis in a mouse model. *Antimicrob. Agents Chemother.* **2015**, *59*, 673–676. [CrossRef]
21. Xu, J.; Tasneen, R.; Peloquin, C.A.; Almeida, D.V.; Li, S.Y.; Barnes-Boyle, K.; Lu, Y.; Nuermberger, E. Verapamil Increases the Bioavailability and Efficacy of Bedaquiline but Not Clofazimine in a Murine Model of Tuberculosis. *Antimicrob. Agents Chemother.* **2018**, *62*. [CrossRef]

22. Hahn, D.; Hodson, E.M.; Hamiwka, L.A.; Lee, V.W.; Chapman, J.R.; Craig, J.C.; Webster, A.C. Target of rapamycin inhibitors (TOR-I; sirolimus and everolimus) for primary immunosuppression in kidney transplant recipients. *Cochrane Database Syst. Rev.* **2019**, *12*, Cd004290. [CrossRef]
23. Paluri, R.K.; Sonpavde, G.; Morgan, C.; Rojymon, J.; Mar, A.H.; Gangaraju, R. Renal toxicity with mammalian target of rapamycin inhibitors: A meta-analysis of randomized clinical trials. *Oncol. Rev.* **2019**, *13*, 455. [CrossRef] [PubMed]
24. Yew, W.W.; Chang, K.C.; Chan, D.P.; Zhang, Y. Metformin as a host-directed therapeutic in tuberculosis: Is there a promise? *Tuberculosis (Edinb. Scotl.)* **2019**, *115*, 76–80. [CrossRef] [PubMed]
25. Singhal, A.; Jie, L.; Kumar, P.; Hong, G.S.; Leow, M.K.; Paleja, B.; Tsenova, L.; Kurepina, N.; Chen, J.; Zolezzi, F.; et al. Metformin as adjunct antituberculosis therapy. *Sci. Transl. Med.* **2014**, *6*, 263ra159. [CrossRef]
26. Dutta, N.K.; Pinn, M.L.; Karakousis, P.C. Metformin Adjunctive Therapy Does Not Improve the Sterilizing Activity of the First-Line Antitubercular Regimen in Mice. *Antimicrob. Agents Chemother.* **2017**, *61*. [CrossRef] [PubMed]
27. Lin, S.Y.; Tu, H.P.; Lu, P.L.; Chen, T.C.; Wang, W.H.; Chong, I.W.; Chen, Y.H. Metformin is associated with a lower risk of active tuberculosis in patients with type 2 diabetes. *Respirology (Carltonvic)* **2018**, *23*, 1063–1073. [CrossRef] [PubMed]
28. Marupuru, S.; Senapati, P.; Pathadka, S.; Miraj, S.S.; Unnikrishnan, M.K.; Manu, M.K. Protective effect of metformin against tuberculosis infections in diabetic patients: An observational study of south Indian tertiary healthcare facility. *Braz. J. Infect. Dis. Off. Publ. Braz. Soc. Infect. Dis.* **2017**, *21*, 312–316. [CrossRef] [PubMed]
29. Pan, S.W.; Yen, Y.F.; Kou, Y.R.; Chuang, P.H.; Su, V.Y.; Feng, J.Y.; Chan, Y.J.; Su, W.J. The Risk of TB in Patients With Type 2 Diabetes Initiating Metformin vs Sulfonylurea Treatment. *Chest* **2018**, *153*, 1347–1357. [CrossRef]
30. Shui, W.; Petzold, C.J.; Redding, A.; Liu, J.; Pitcher, A.; Sheu, L.; Hsieh, T.Y.; Keasling, J.D.; Bertozzi, C.R. Organelle membrane proteomics reveals differential influence of mycobacterial lipoglycans on macrophage phagosome maturation and autophagosome accumulation. *J. Proteome Res.* **2011**, *10*, 339–348. [CrossRef]
31. Romagnoli, A.; Etna, M.P.; Giacomini, E.; Pardini, M.; Remoli, M.E.; Corazzari, M.; Falasca, L.; Goletti, D.; Gafa, V.; Simeone, R.; et al. ESX-1 dependent impairment of autophagic flux by Mycobacterium tuberculosis in human dendritic cells. *Autophagy* **2012**, *8*, 1357–1370. [CrossRef]
32. Zhang, L.; Zhang, H.; Zhao, Y.; Mao, F.; Wu, J.; Bai, B.; Xu, Z.; Jiang, Y.; Shi, C. Effects of Mycobacterium tuberculosis ESAT-6/CFP-10 fusion protein on the autophagy function of mouse macrophages. *Dna Cell Biol.* **2012**, *31*, 171–179. [CrossRef]
33. Shin, D.M.; Jeon, B.Y.; Lee, H.M.; Jin, H.S.; Yuk, J.M.; Song, C.H.; Lee, S.H.; Lee, Z.W.; Cho, S.N.; Kim, J.M.; et al. Mycobacterium tuberculosis eis regulates autophagy, inflammation, and cell death through redox-dependent signaling. *PLoS Pathog.* **2010**, *6*, e1001230. [CrossRef] [PubMed]
34. An, X.; Tiwari, A.K.; Sun, Y.; Ding, P.R.; Ashby, C.R., Jr.; Chen, Z.S. BCR-ABL tyrosine kinase inhibitors in the treatment of Philadelphia chromosome positive chronic myeloid leukemia: A review. *Leuk. Res.* **2010**, *34*, 1255–1268. [CrossRef] [PubMed]
35. Bruns, H.; Stegelmann, F.; Fabri, M.; Dohner, K.; van Zandbergen, G.; Wagner, M.; Skinner, M.; Modlin, R.L.; Stenger, S. Abelson tyrosine kinase controls phagosomal acidification required for killing of Mycobacterium tuberculosis in human macrophages. *J. Immunol. (Baltimore Md. 1950)* **2012**, *189*, 4069–4078. [CrossRef] [PubMed]
36. Napier, R.J.; Rafi, W.; Cheruvu, M.; Powell, K.R.; Zaunbrecher, M.A.; Bornmann, W.; Salgame, P.; Shinnick, T.M.; Kalman, D. Imatinib-sensitive tyrosine kinases regulate mycobacterial pathogenesis and represent therapeutic targets against tuberculosis. *Cell Host Microbe* **2011**, *10*, 475–485. [CrossRef] [PubMed]
37. Stancu, C.; Sima, A. Statins: Mechanism of action and effects. *J. Cell. Mol. Med.* **2001**, *5*, 378–387. [CrossRef] [PubMed]
38. Pandey, A.K.; Sassetti, C.M. Mycobacterial persistence requires the utilization of host cholesterol. *Proc. Natl. Acad. Sci. USA* **2008**, *105*, 4376–4380. [CrossRef] [PubMed]
39. Parihar, S.P.; Guler, R.; Khutlang, R.; Lang, D.M.; Hurdayal, R.; Mhlanga, M.M.; Suzuki, H.; Marais, A.D.; Brombacher, F. Statin therapy reduces the mycobacterium tuberculosis burden in human macrophages and in mice by enhancing autophagy and phagosome maturation. *J. Infect. Dis.* **2014**, *209*, 754–763. [CrossRef] [PubMed]

40. Lobato, L.S.; Rosa, P.S.; Ferreira Jda, S.; Neumann Ada, S.; da Silva, M.G.; do Nascimento, D.C.; Soares, C.T.; Pedrini, S.C.; Oliveira, D.S.; Monteiro, C.P.; et al. Statins increase rifampin mycobactericidal effect. *Antimicrob. Agents Chemother.* **2014**, *58*, 5766–5774. [CrossRef] [PubMed]
41. Rivas-Santiago, B.; Hernandez-Pando, R.; Carranza, C.; Juarez, E.; Contreras, J.L.; Aguilar-Leon, D.; Torres, M.; Sada, E. Expression of cathelicidin LL-37 during Mycobacterium tuberculosis infection in human alveolar macrophages, monocytes, neutrophils, and epithelial cells. *Infect. Immun.* **2008**, *76*, 935–941. [CrossRef]
42. Kahlenberg, J.M.; Kaplan, M.J. Little peptide, big effects: The role of LL-37 in inflammation and autoimmune disease. *J. Immunol. (Baltimore Md. 1950)* **2013**, *191*, 4895–4901. [CrossRef]
43. Shin, D.M.; Jo, E.K. Antimicrobial Peptides in Innate Immunity against Mycobacteria. *Immune Netw.* **2011**, *11*, 245–252. [CrossRef]
44. Liu, P.T.; Stenger, S.; Li, H.; Wenzel, L.; Tan, B.H.; Krutzik, S.R.; Ochoa, M.T.; Schauber, J.; Wu, K.; Meinken, C.; et al. Toll-like receptor triggering of a vitamin D-mediated human antimicrobial response. *Science* **2006**, *311*, 1770–1773. [CrossRef] [PubMed]
45. Liu, P.T.; Stenger, S.; Tang, D.H.; Modlin, R.L. Cutting edge: Vitamin D-mediated human antimicrobial activity against Mycobacterium tuberculosis is dependent on the induction of cathelicidin. *J. Immunol. (Baltimore Md. 1950)* **2007**, *179*, 2060–2063. [CrossRef] [PubMed]
46. Yuk, J.M.; Shin, D.M.; Lee, H.M.; Yang, C.S.; Jin, H.S.; Kim, K.K.; Lee, Z.W.; Lee, S.H.; Kim, J.M.; Jo, E.K. Vitamin D3 induces autophagy in human monocytes/macrophages via cathelicidin. *Cell Host Microbe* **2009**, *6*, 231–243. [CrossRef] [PubMed]
47. Rook, G.A.; Steele, J.; Fraher, L.; Barker, S.; Karmali, R.; O'Riordan, J.; Stanford, J. Vitamin D3, gamma interferon, and control of proliferation of Mycobacterium tuberculosis by human monocytes. *Immunology* **1986**, *57*, 159–163.
48. Mily, A.; Rekha, R.S.; Kamal, S.M.; Arifuzzaman, A.S.; Rahim, Z.; Khan, L.; Haq, M.A.; Zaman, K.; Bergman, P.; Brighenti, S.; et al. Significant Effects of Oral Phenylbutyrate and Vitamin D3 Adjunctive Therapy in Pulmonary Tuberculosis: A Randomized Controlled Trial. *PLoS ONE* **2015**, *10*, e0138340. [CrossRef]
49. Musarurwa, C.; Zijenah, L.S.; Mhandire, D.Z.; Bandason, T.; Mhandire, K.; Chipiti, M.M.; Munjoma, M.W.; Mujaji, W.B. Higher serum 25-hydroxyvitamin D concentrations are associated with active pulmonary tuberculosis in hospitalised HIV infected patients in a low income tropical setting: A cross sectional study. *BMC Pulm. Med.* **2018**, *18*, 67. [CrossRef]
50. Sudfeld, C.R.; Mugusi, F.; Aboud, S.; Nagu, T.J.; Wang, M.; Fawzi, W.W. Efficacy of vitamin D3 supplementation in reducing incidence of pulmonary tuberculosis and mortality among HIV-infected Tanzanian adults initiating antiretroviral therapy: Study protocol for a randomized controlled trial. *Trials* **2017**, *18*, 66. [CrossRef]
51. Aibana, O.; Huang, C.C.; Aboud, S.; Arnedo-Pena, A.; Becerra, M.C.; Bellido-Blasco, J.B.; Bhosale, R.; Calderon, R.; Chiang, S.; Contreras, C.; et al. Vitamin D status and risk of incident tuberculosis disease: A nested case-control study, systematic review, and individual-participant data meta-analysis. *PLoS Med.* **2019**, *16*, e1002907. [CrossRef]
52. Crowle, A.J.; Ross, E.J. Inhibition by retinoic acid of multiplication of virulent tubercle bacilli in cultured human macrophages. *Infect. Immun.* **1989**, *57*, 840–844. [CrossRef]
53. Wheelwright, M.; Kim, E.W.; Inkeles, M.S.; De Leon, A.; Pellegrini, M.; Krutzik, S.R.; Liu, P.T. All-trans retinoic acid-triggered antimicrobial activity against Mycobacterium tuberculosis is dependent on NPC2. *J. Immunol. (Baltimore Md. 1950)* **2014**, *192*, 2280–2290. [CrossRef] [PubMed]
54. Yamada, H.; Mizuno, S.; Ross, A.C.; Sugawara, I. Retinoic acid therapy attenuates the severity of tuberculosis while altering lymphocyte and macrophage numbers and cytokine expression in rats infected with Mycobacterium tuberculosis. *J. Nutr.* **2007**, *137*, 2696–2700. [CrossRef] [PubMed]
55. Lawson, L.; Thacher, T.D.; Yassin, M.A.; Onuoha, N.A.; Usman, A.; Emenyonu, N.E.; Shenkin, A.; Davies, P.D.; Cuevas, L.E. Randomized controlled trial of zinc and vitamin A as co-adjuvants for the treatment of pulmonary tuberculosis. *Trop. Med. Int. Health Tm Ih* **2010**, *15*, 1481–1490. [CrossRef] [PubMed]
56. Visser, M.E.; Grewal, H.M.; Swart, E.C.; Dhansay, M.A.; Walzl, G.; Swanevelder, S.; Lombard, C.; Maartens, G. The effect of vitamin A and zinc supplementation on treatment outcomes in pulmonary tuberculosis: A randomized controlled trial. *Am. J. Clin. Nutr.* **2011**, *93*, 93–100. [CrossRef] [PubMed]
57. Steinmann, J.; Halldorsson, S.; Agerberth, B.; Gudmundsson, G.H. Phenylbutyrate induces antimicrobial peptide expression. *Antimicrob. Agents Chemother.* **2009**, *53*, 5127–5133. [CrossRef]

58. Van der Does, A.M.; Kenne, E.; Koppelaar, E.; Agerberth, B.; Lindbom, L. Vitamin D(3) and phenylbutyrate promote development of a human dendritic cell subset displaying enhanced antimicrobial properties. *J. Leukoc. Biol.* **2014**, *95*, 883–891. [CrossRef]
59. Rekha, R.S.; Rao Muvva, S.S.; Wan, M.; Raqib, R.; Bergman, P.; Brighenti, S.; Gudmundsson, G.H.; Agerberth, B. Phenylbutyrate induces LL-37-dependent autophagy and intracellular killing of Mycobacterium tuberculosis in human macrophages. *Autophagy* **2015**, *11*, 1688–1699. [CrossRef]
60. Jarczak, J.; Kosciuczuk, E.M.; Lisowski, P.; Strzalkowska, N.; Jozwik, A.; Horbanczuk, J.; Krzyzewski, J.; Zwierzchowski, L.; Bagnicka, E. Defensins: Natural component of human innate immunity. *Hum. Immunol.* **2013**, *74*, 1069–1079. [CrossRef]
61. Rivas-Santiago, B.; Schwander, S.K.; Sarabia, C.; Diamond, G.; Klein-Patel, M.E.; Hernandez-Pando, R.; Ellner, J.J.; Sada, E. Human {beta}-defensin 2 is expressed and associated with Mycobacterium tuberculosis during infection of human alveolar epithelial cells. *Infect. Immun.* **2005**, *73*, 4505–4511. [CrossRef]
62. Miyakawa, Y.; Ratnakar, P.; Rao, A.G.; Costello, M.L.; Mathieu-Costello, O.; Lehrer, R.I.; Catanzaro, A. In vitro activity of the antimicrobial peptides human and rabbit defensins and porcine leukocyte protegrin against Mycobacterium tuberculosis. *Infect. Immun.* **1996**, *64*, 926–932. [CrossRef]
63. Moreira-Teixeira, L.; Mayer-Barber, K.; Sher, A.; O'Garra, A. Type I interferons in tuberculosis: Foe and occasionally friend. *J. Exp. Med.* **2018**, *215*, 1273–1285. [CrossRef]
64. Condos, R.; Rom, W.N.; Schluger, N.W. Treatment of multidrug-resistant pulmonary tuberculosis with interferon-gamma via aerosol. *Lancet (Lond. Engl.)* **1997**, *349*, 1513–1515. [CrossRef]
65. Koh, W.J.; Kwon, O.J.; Suh, G.Y.; Chung, M.P.; Kim, H.; Lee, N.Y.; Kim, T.S.; Lee, K.S. Six-month therapy with aerosolized interferon-gamma for refractory multidrug-resistant pulmonary tuberculosis. *J. Korean Med. Sci.* **2004**, *19*, 167–171. [CrossRef] [PubMed]
66. Suarez-Mendez, R.; Garcia-Garcia, I.; Fernandez-Olivera, N.; Valdes-Quintana, M.; Milanes-Virelles, M.T.; Carbonell, D.; Machado-Molina, D.; Valenzuela-Silva, C.M.; Lopez-Saura, P.A. Adjuvant interferon gamma in patients with drug—Resistant pulmonary tuberculosis: A pilot study. *BMC Infect. Dis.* **2004**, *4*, 44. [CrossRef] [PubMed]
67. Sakai, S.; Kauffman, K.D.; Sallin, M.A.; Sharpe, A.H.; Young, H.A.; Ganusov, V.V.; Barber, D.L. CD4 T Cell-Derived IFN-gamma Plays a Minimal Role in Control of Pulmonary Mycobacterium tuberculosis Infection and Must Be Actively Repressed by PD-1 to Prevent Lethal Disease. *PLoS Pathog.* **2016**, *12*, e1005667. [CrossRef] [PubMed]
68. Johnson, J.L.; Ssekasanvu, E.; Okwera, A.; Mayanja, H.; Hirsch, C.S.; Nakibali, J.G.; Jankus, D.D.; Eisenach, K.D.; Boom, W.H.; Ellner, J.J.; et al. Randomized trial of adjunctive interleukin-2 in adults with pulmonary tuberculosis. *Am. J. Respir. Crit. Care Med.* **2003**, *168*, 185–191. [CrossRef] [PubMed]
69. Thwaites, G.E.; Nguyen, D.B.; Nguyen, H.D.; Hoang, T.Q.; Do, T.T.; Nguyen, T.C.; Nguyen, Q.H.; Nguyen, T.T.; Nguyen, N.H.; Nguyen, T.N.; et al. Dexamethasone for the treatment of tuberculous meningitis in adolescents and adults. *N. Engl. J. Med.* **2004**, *351*, 1741–1751. [CrossRef]
70. Critchley, J.A.; Young, F.; Orton, L.; Garner, P. Corticosteroids for prevention of mortality in people with tuberculosis: A systematic review and meta-analysis. *Lancet Infect. Dis.* **2013**, *13*, 223–237. [CrossRef]
71. Dooley, D.P.; Carpenter, J.L.; Rademacher, S. Adjunctive corticosteroid therapy for tuberculosis: A critical reappraisal of the literature. *Clin. Infect. Dis.* **1997**, *25*, 872–887. [CrossRef]
72. Critchley, J.A.; Orton, L.C.; Pearson, F. Adjunctive steroid therapy for managing pulmonary tuberculosis. *Cochrane Database Syst. Rev.* **2014**, Cd011370. [CrossRef]
73. Lai, R.P.J.; Meintjes, G.; Wilkinson, K.A.; Graham, C.M.; Marais, S.; Van der Plas, H.; Deffur, A.; Schutz, C.; Bloom, C.; Munagala, I.; et al. HIV-tuberculosis-associated immune reconstitution inflammatory syndrome is characterized by Toll-like receptor and inflammasome signalling. *Nat. Commun.* **2015**, *6*, 8451. [CrossRef] [PubMed]
74. Tan, H.Y.; Yong, Y.K.; Shankar, E.M.; Paukovics, G.; Ellegard, R.; Larsson, M.; Kamarulzaman, A.; French, M.A.; Crowe, S.M. Aberrant Inflammasome Activation Characterizes Tuberculosis-Associated Immune Reconstitution Inflammatory Syndrome. *J. Immunol. (Baltimore Md. 1950)* **2016**, *196*, 4052–4063. [CrossRef] [PubMed]
75. Walker, N.F.; Stek, C.; Wasserman, S.; Wilkinson, R.J.; Meintjes, G. The tuberculosis-associated immune reconstitution inflammatory syndrome: Recent advances in clinical and pathogenesis research. *Curr. Opin. Hiv Aids* **2018**, *13*, 512–521. [CrossRef] [PubMed]

76. Coutinho, A.E.; Chapman, K.E. The anti-inflammatory and immunosuppressive effects of glucocorticoids, recent developments and mechanistic insights. *Mol. Cell. Endocrinol.* **2011**, *335*, 2–13. [CrossRef]
77. Grab, J.; Suarez, I.; van Gumpel, E.; Winter, S.; Schreiber, F.; Esser, A.; Holscher, C.; Fritsch, M.; Herb, M.; Schramm, M.; et al. Corticosteroids inhibit Mycobacterium tuberculosis-induced necrotic host cell death by abrogating mitochondrial membrane permeability transition. *Nat. Commun.* **2019**, *10*, 688. [CrossRef]
78. Kroesen, V.M.; Groschel, M.I.; Martinson, N.; Zumla, A.; Maeurer, M.; van der Werf, T.S.; Vilaplana, C. Non-Steroidal Anti-inflammatory Drugs As Host-Directed Therapy for Tuberculosis: A Systematic Review. *Front. Immunol.* **2017**, *8*, 772. [CrossRef]
79. Pinault, L.; Han, J.S.; Kang, C.M.; Franco, J.; Ronning, D.R. Zafirlukast inhibits complexation of Lsr2 with DNA and growth of Mycobacterium tuberculosis. *Antimicrob. Agents Chemother.* **2013**, *57*, 2134–2140. [CrossRef]
80. Tobin, D.M.; Roca, F.J.; Oh, S.F.; McFarland, R.; Vickery, T.W.; Ray, J.P.; Ko, D.C.; Zou, Y.; Bang, N.D.; Chau, T.T.; et al. Host genotype-specific therapies can optimize the inflammatory response to mycobacterial infections. *Cell* **2012**, *148*, 434–446. [CrossRef]
81. Tobin, D.M.; Roca, F.J.; Ray, J.P.; Ko, D.C.; Ramakrishnan, L. An enzyme that inactivates the inflammatory mediator leukotriene b4 restricts mycobacterial infection. *PLoS ONE* **2013**, *8*, e67828. [CrossRef]
82. Dallenga, T.; Repnik, U.; Corleis, B.; Eich, J.; Reimer, R.; Griffiths, G.W.; Schaible, U.E. M. tuberculosis-Induced Necrosis of Infected Neutrophils Promotes Bacterial Growth Following Phagocytosis by Macrophages. *Cell Host Microbe* **2017**, *22*, 519–530.e3. [CrossRef]
83. Remot, A.; Doz, E.; Winter, N. Neutrophils and Close Relatives in the Hypoxic Environment of the Tuberculous Granuloma: New Avenues for Host-Directed Therapies? *Front. Immunol.* **2019**, *10*, 417. [CrossRef]
84. Warnatsch, A.; Tsourouktsoglou, T.D.; Branzk, N.; Wang, Q.; Reincke, S.; Herbst, S.; Gutierrez, M.; Papayannopoulos, V. Reactive Oxygen Species Localization Programs Inflammation to Clear Microbes of Different Size. *Immunity* **2017**, *46*, 421–432. [CrossRef] [PubMed]
85. Braverman, J.; Stanley, S.A. Nitric Oxide Modulates Macrophage Responses to Mycobacterium tuberculosis Infection through Activation of HIF-1alpha and Repression of NF-kappaB. *J. Immunol. (Baltimore Md. 1950)* **2017**, *199*, 1805–1816. [CrossRef] [PubMed]
86. Amaral, E.P.; Costa, D.L.; Namasivayam, S.; Riteau, N.; Kamenyeva, O.; Mittereder, L.; Mayer-Barber, K.D.; Andrade, B.B.; Sher, A. A major role for ferroptosis in Mycobacterium tuberculosis-induced cell death and tissue necrosis. *J. Exp. Med.* **2019**, *216*, 556–570. [CrossRef] [PubMed]
87. Martin, C.J.; Booty, M.G.; Rosebrock, T.R.; Nunes-Alves, C.; Desjardins, D.M.; Keren, I.; Fortune, S.M.; Remold, H.G.; Behar, S.M. Efferocytosis is an innate antibacterial mechanism. *Cell Host Microbe* **2012**, *12*, 289–300. [CrossRef]
88. Andersson, A.M.; Larsson, M.; Stendahl, O.; Blomgran, R. Efferocytosis of Apoptotic Neutrophils Enhances Control of Mycobacterium tuberculosis in HIV-Coinfected Macrophages in a Myeloperoxidase-Dependent Manner. *J. Innate Immun.* **2019**, 1–13. [CrossRef]
89. Hosseini, R.; Lamers, G.E.; Soltani, H.M.; Meijer, A.H.; Spaink, H.P.; Schaaf, M.J. Efferocytosis and extrusion of leukocytes determine the progression of early mycobacterial pathogenesis. *J. Cell Sci.* **2016**, *129*, 3385–3395. [CrossRef]
90. Keane, J.; Gershon, S.; Wise, R.P.; Mirabile-Levens, E.; Kasznica, J.; Schwieterman, W.D.; Siegel, J.N.; Braun, M.M. Tuberculosis associated with infliximab, a tumor necrosis factor alpha-neutralizing agent. *N. Engl. J. Med.* **2001**, *345*, 1098–1104. [CrossRef]
91. Bourigault, M.L.; Vacher, R.; Rose, S.; Olleros, M.L.; Janssens, J.P.; Quesniaux, V.F.; Garcia, I. Tumor necrosis factor neutralization combined with chemotherapy enhances Mycobacterium tuberculosis clearance and reduces lung pathology. *Am. J. Clin. Exp. Immunol.* **2013**, *2*, 124–134.
92. Wallis, R.S.; Kyambadde, P.; Johnson, J.L.; Horter, L.; Kittle, R.; Pohle, M.; Ducar, C.; Millard, M.; Mayanja-Kizza, H.; Whalen, C.; et al. A study of the safety, immunology, virology, and microbiology of adjunctive etanercept in HIV-1-associated tuberculosis. *Aids (Lond. Engl.)* **2004**, *18*, 257–264. [CrossRef]
93. Schoeman, J.F.; Springer, P.; van Rensburg, A.J.; Swanevelder, S.; Hanekom, W.A.; Haslett, P.A.; Kaplan, G. Adjunctive thalidomide therapy for childhood tuberculous meningitis: Results of a randomized study. *J. Child. Neurol.* **2004**, *19*, 250–257. [CrossRef]

94. Subbian, S.; Koo, M.S.; Tsenova, L.; Khetani, V.; Zeldis, J.B.; Fallows, D.; Kaplan, G. Pharmacologic Inhibition of Host Phosphodiesterase-4 Improves Isoniazid-Mediated Clearance of Mycobacterium tuberculosis. *Front. Immunol.* **2016**, *7*, 238. [CrossRef] [PubMed]
95. Subbian, S.; Tsenova, L.; O'Brien, P.; Yang, G.; Koo, M.S.; Peixoto, B.; Fallows, D.; Dartois, V.; Muller, G.; Kaplan, G. Phosphodiesterase-4 inhibition alters gene expression and improves isoniazid-mediated clearance of Mycobacterium tuberculosis in rabbit lungs. *PLoS Pathog.* **2011**, *7*, e1002262. [CrossRef] [PubMed]
96. Sabir, N.; Hussain, T.; Mangi, M.H.; Zhao, D.; Zhou, X. Matrix metalloproteinases: Expression, regulation and role in the immunopathology of tuberculosis. *Cell Prolif.* **2019**, *52*, e12649. [CrossRef] [PubMed]
97. Andrade, B.B.; Pavan Kumar, N.; Amaral, E.P.; Riteau, N.; Mayer-Barber, K.D.; Tosh, K.W.; Maier, N.; Conceicao, E.L.; Kubler, A.; Sridhar, R.; et al. Heme Oxygenase-1 Regulation of Matrix Metalloproteinase-1 Expression Underlies Distinct Disease Profiles in Tuberculosis. *J. Immunol. (Baltimore Md. 1950)* **2015**, *195*, 2763–2773. [CrossRef]
98. Walker, N.F.; Clark, S.O.; Oni, T.; Andreu, N.; Tezera, L.; Singh, S.; Saraiva, L.; Pedersen, B.; Kelly, D.L.; Tree, J.A.; et al. Doxycycline and HIV infection suppress tuberculosis-induced matrix metalloproteinases. *Am. J. Respir. Crit. Care Med.* **2012**, *185*, 989–997. [CrossRef]
99. Xu, Y.; Wang, L.; Zimmerman, M.D.; Chen, K.Y.; Huang, L.; Fu, D.J.; Kaya, F.; Rakhilin, N.; Nazarova, E.V.; Bu, P.; et al. Matrix metalloproteinase inhibitors enhance the efficacy of frontline drugs against Mycobacterium tuberculosis. *PLoS Pathog.* **2018**, *14*, e1006974. [CrossRef]
100. Vaziri, F.; Brosch, R. ESX/Type VII Secretion Systems-An Important Way Out for Mycobacterial Proteins. *Microbiol. Spectr.* **2019**, *7*. [CrossRef]
101. Tiwari, S.; Casey, R.; Goulding, C.W.; Hingley-Wilson, S.; Jacobs, W.R., Jr. Infect and Inject: How Mycobacterium tuberculosis Exploits Its Major Virulence-Associated Type VII Secretion System, ESX-1. *Microbiol. Spectr.* **2019**, *7*. [CrossRef]
102. Rybniker, J.; Chen, J.M.; Sala, C.; Hartkoorn, R.C.; Vocat, A.; Benjak, A.; Boy-Rottger, S.; Zhang, M.; Szekely, R.; Greff, Z.; et al. Anticytolytic screen identifies inhibitors of mycobacterial virulence protein secretion. *Cell Host Microbe* **2014**, *16*, 538–548. [CrossRef]
103. Wassermann, R.; Gulen, M.F.; Sala, C.; Perin, S.G.; Lou, Y.; Rybniker, J.; Schmid-Burgk, J.L.; Schmidt, T.; Hornung, V.; Cole, S.T.; et al. Mycobacterium tuberculosis Differentially Activates cGAS- and Inflammasome-Dependent Intracellular Immune Responses through ESX-1. *Cell Host Microbe* **2015**, *17*, 799–810. [CrossRef] [PubMed]
104. Walters, S.B.; Dubnau, E.; Kolesnikova, I.; Laval, F.; Daffe, M.; Smith, I. The Mycobacterium tuberculosis PhoPR two-component system regulates genes essential for virulence and complex lipid biosynthesis. *Mol. Microbiol.* **2006**, *60*, 312–330. [CrossRef] [PubMed]
105. Lee, J.S.; Krause, R.; Schreiber, J.; Mollenkopf, H.J.; Kowall, J.; Stein, R.; Jeon, B.Y.; Kwak, J.Y.; Song, M.K.; Patron, J.P.; et al. Mutation in the transcriptional regulator PhoP contributes to avirulence of Mycobacterium tuberculosis H37Ra strain. *Cell Host Microbe* **2008**, *3*, 97–103. [CrossRef] [PubMed]
106. Frigui, W.; Bottai, D.; Majlessi, L.; Monot, M.; Josselin, E.; Brodin, P.; Garnier, T.; Gicquel, B.; Martin, C.; Leclerc, C.; et al. Control of M. tuberculosis ESAT-6 secretion and specific T cell recognition by PhoP. *PLoS Pathog.* **2008**, *4*, e33. [CrossRef]
107. Anil Kumar, V.; Goyal, R.; Bansal, R.; Singh, N.; Sevalkar, R.R.; Kumar, A.; Sarkar, D. EspR-dependent ESAT-6 Protein Secretion of Mycobacterium tuberculosis Requires the Presence of Virulence Regulator PhoP. *J. Biol. Chem.* **2016**, *291*, 19018–19030. [CrossRef]
108. Gonzalo Asensio, J.; Maia, C.; Ferrer, N.L.; Barilone, N.; Laval, F.; Soto, C.Y.; Winter, N.; Daffe, M.; Gicquel, B.; Martin, C.; et al. The virulence-associated two-component PhoP-PhoR system controls the biosynthesis of polyketide-derived lipids in Mycobacterium tuberculosis. *J. Biol. Chem.* **2006**, *281*, 1313–1316. [CrossRef]
109. Johnson, B.K.; Colvin, C.J.; Needle, D.B.; Mba Medie, F.; Champion, P.A.; Abramovitch, R.B. The Carbonic Anhydrase Inhibitor Ethoxzolamide Inhibits the Mycobacterium tuberculosis PhoPR Regulon and Esx-1 Secretion and Attenuates Virulence. *Antimicrob. Agents Chemother.* **2015**, *59*, 4436–4445. [CrossRef]
110. Wang, L.; Xu, M.; Southall, N.; Zheng, W.; Wang, S. A High-Throughput Assay for Developing Inhibitors of PhoP, a Virulence Factor of Mycobacterium tuberculosis. *Comb. Chem. High Throughput Screen* **2016**, *19*, 855–864. [CrossRef]
111. Koul, A.; Choidas, A.; Treder, M.; Tyagi, A.K.; Drlica, K.; Singh, Y.; Ullrich, A. Cloning and characterization of secretory tyrosine phosphatases of Mycobacterium tuberculosis. *J. Bacteriol.* **2000**, *182*, 5425–5432. [CrossRef]

112. Singh, R.; Rao, V.; Shakila, H.; Gupta, R.; Khera, A.; Dhar, N.; Singh, A.; Koul, A.; Singh, Y.; Naseema, M.; et al. Disruption of mptpB impairs the ability of Mycobacterium tuberculosis to survive in guinea pigs. *Mol. Microbiol.* **2003**, *50*, 751–762. [CrossRef]
113. Beresford, N.; Patel, S.; Armstrong, J.; Szoor, B.; Fordham-Skelton, A.P.; Tabernero, L. MptpB, a virulence factor from Mycobacterium tuberculosis, exhibits triple-specificity phosphatase activity. *Biochem. J.* **2007**, *406*, 13–18. [CrossRef]
114. Vickers, C.F.; Silva, A.P.G.; Chakraborty, A.; Fernandez, P.; Kurepina, N.; Saville, C.; Naranjo, Y.; Pons, M.; Schnettger, L.S.; Gutierrez, M.G.; et al. Structure-Based Design of MptpB Inhibitors That Reduce Multidrug-Resistant Mycobacterium tuberculosis Survival and Infection Burden in Vivo. *J. Med. Chem.* **2018**, *61*, 8337–8352. [CrossRef] [PubMed]
115. Vieira, O.V.; Botelho, R.J.; Rameh, L.; Brachmann, S.M.; Matsuo, T.; Davidson, H.W.; Schreiber, A.; Backer, J.M.; Cantley, L.C.; Grinstein, S. Distinct roles of class I and class III phosphatidylinositol 3-kinases in phagosome formation and maturation. *J. Cell. Biol.* **2001**, *155*, 19–25. [CrossRef] [PubMed]
116. Fernandez-Soto, P.; Bruce, A.J.E.; Fielding, A.J.; Cavet, J.S.; Tabernero, L. Mechanism of catalysis and inhibition of Mycobacterium tuberculosis SapM, implications for the development of novel antivirulence drugs. *Sci. Rep.* **2019**, *9*, 10315. [CrossRef] [PubMed]
117. Master, S.S.; Rampini, S.K.; Davis, A.S.; Keller, C.; Ehlers, S.; Springer, B.; Timmins, G.S.; Sander, P.; Deretic, V. Mycobacterium tuberculosis prevents inflammasome activation. *Cell Host Microbe* **2008**, *3*, 224–232. [CrossRef]
118. Song, Y.; Xu, H.; Chen, W.; Zhan, P.; Liu, X. 8-Hydroxyquinoline: A privileged structure with a broad-ranging pharmacological potential. *Med. Chem. Commun.* **2015**, *6*, 61–74. [CrossRef]
119. Paolino, M.; Brindisi, M.; Vallone, A.; Butini, S.; Campiani, G.; Nannicini, C.; Giuliani, G.; Anzini, M.; Lamponi, S.; Giorgi, G.; et al. Development of Potent Inhibitors of the Mycobacterium tuberculosis Virulence Factor Zmp1 and Evaluation of Their Effect on Mycobacterial Survival inside Macrophages. *ChemMedChem* **2018**, *13*, 422–430. [CrossRef]
120. Mehra, S.; Foreman, T.W.; Didier, P.J.; Ahsan, M.H.; Hudock, T.A.; Kissee, R.; Golden, N.A.; Gautam, U.S.; Johnson, A.M.; Alvarez, X.; et al. The DosR Regulon Modulates Adaptive Immunity and Is Essential for Mycobacterium tuberculosis Persistence. *Am. J. Respir. Crit. Care Med.* **2015**, *191*, 1185–1196. [CrossRef]
121. Converse, P.J.; Karakousis, P.C.; Klinkenberg, L.G.; Kesavan, A.K.; Ly, L.H.; Allen, S.S.; Grosset, J.H.; Jain, S.K.; Lamichhane, G.; Manabe, Y.C.; et al. Role of the dosR-dosS two-component regulatory system in Mycobacterium tuberculosis virulence in three animal models. *Infect. Immun.* **2009**, *77*, 1230–1237. [CrossRef]
122. Zheng, H.; Williams, J.T.; Aleiwi, B.; Ellsworth, E.; Abramovitch, R.B. Inhibiting Mycobacterium tuberculosis DosRST Signaling by Targeting Response Regulator DNA Binding and Sensor Kinase Heme. *ACS Chem. Biol.* **2019**. [CrossRef]

© 2020 by the authors. Licensee MDPI, Basel, Switzerland. This article is an open access article distributed under the terms and conditions of the Creative Commons Attribution (CC BY) license (http://creativecommons.org/licenses/by/4.0/).

Review

Model-Informed Drug Discovery and Development Strategy for the Rapid Development of Anti-Tuberculosis Drug Combinations

Rob C. van Wijk [1], Rami Ayoun Alsoud [1], Hans Lennernäs [2] and Ulrika S. H. Simonsson [1,*]

1. Department of Pharmaceutical Biosciences, Uppsala University, Uppsala 75123, Sweden; rob.vanwijk@farmbio.uu.se (R.C.v.W.); rami.alsoud@farmbio.uu.se (R.A.A.)
2. Department of Pharmacy, Uppsala University, Uppsala 75123, Sweden; hans.lennernas@farmaci.uu.se
* Correspondence: ulrika.simonsson@farmbio.uu.se; Tel.: +46-184-714-000

Received: 29 February 2020; Accepted: 25 March 2020; Published: 31 March 2020

Featured Application: Model-informed drug discovery and development (MID3) is proposed to be applied throughout the preclinical to clinical phases to provide an informative prediction of drug exposure and efficacy in humans in order to select novel anti-tuberculosis drug combinations for the treatment of tuberculosis.

Abstract: The increasing emergence of drug-resistant tuberculosis requires new effective and safe drug regimens. However, drug discovery and development are challenging, lengthy and costly. The framework of model-informed drug discovery and development (MID3) is proposed to be applied throughout the preclinical to clinical phases to provide an informative prediction of drug exposure and efficacy in humans in order to select novel anti-tuberculosis drug combinations. The MID3 includes pharmacokinetic-pharmacodynamic and quantitative systems pharmacology models, machine learning and artificial intelligence, which integrates all the available knowledge related to disease and the compounds. A translational *in vitro-in vivo* link throughout modeling and simulation is crucial to optimize the selection of regimens with the highest probability of receiving approval from regulatory authorities. *In vitro-in vivo* correlation (IVIVC) and physiologically-based pharmacokinetic modeling provide powerful tools to predict pharmacokinetic drug-drug interactions based on preclinical information. Mechanistic or semi-mechanistic pharmacokinetic-pharmacodynamic models have been successfully applied to predict the clinical exposure-response profile for anti-tuberculosis drugs using preclinical data. Potential pharmacodynamic drug-drug interactions can be predicted from *in vitro* data through IVIVC and pharmacokinetic-pharmacodynamic modeling accounting for translational factors. It is essential for academic and industrial drug developers to collaborate across disciplines to realize the huge potential of MID3.

Keywords: tuberculosis; MID3; pharmacokinetics; pharmacodynamics; drug-drug interactions; *in vitro*; *in vivo*; drug development

1. Introduction

Drug discovery and development is a challenging, lengthy, and costly process. The costs of a novel drug reaching the market can be as much as 2–3 billion dollars [1]. In the early discovery phase, libraries consisting of thousands of compounds can be synthesized chemically and tested for efficacy *in vitro* at a relatively low cost. The largest expenditures are in the late preclinical and clinical phases of drug development, where the efficacy and safety of treatment are assessed. Smart decisions need to be made regarding which compounds and regimens should progress through the preclinical phase and subsequently into clinical trials. Early characterization of each compound's exposure-response relationship, i.e., pharmacokinetic (PK)-pharmacodynamic (PD) relationship and potential interactions within regimens and with commonly co-administered drugs, can allow for informative decision making throughout preclinical development and into clinical development [2].

Tuberculosis (TB) is the leading cause of adult mortality through infectious diseases and 10 million new cases are reported globally every year [3]. Sensitive TB is currently treated with a six-month regimen of antibiotics, consisting of isoniazid, pyrazinamide, rifampicin and ethambutol, which was developed in the mid-twentieth century. This therapy is believed to be suboptimal and was not developed using modern approaches for drug development, thereby lacking important information on the PK-PD relationship. Therefore, clinical trials have recently been conducted in order to define the relationship between exposure and efficacy, as well as safety, where statistically significant exposure-response relationships for rifampicin have been identified, in order to support a higher dose of rifampicin [4–7]. Almost one in five patients will acquire multidrug-resistant tuberculosis (MDR-TB) or rifampicin-resistant tuberculosis (RR-TB) [3]. Recently, the new anti-TB drugs bedaquiline, delamanid and pretomanid were conditionally approved against MDR-TB, which led to updates to the World Health Organization (WHO) treatment guideline for MDR-TB [8]. Bedaquiline is a diarylquinoline, a new class of antibiotics. It is an inhibitor of the membrane-bound adenosine triphosphate (ATP)-synthase enzyme, therefore blocking mycobacterial ATP formation and energy metabolism. Bedaquiline is therefore bactericidal for dormant mycobacteria as well, a preferable feature for the shortening of treatment duration and prevention of relapse [9]. Delamanid is a nitroimidazole and affects the mycobacterial cell wall, thereby also improving drug penetration into the mycobacterium. It is the most potent TB drug and is active against replicating and dormant mycobacteria as well [9]. The combination of delamanid with bedaquiline is, however, not recommended, due to QT-prolongation-related cardiotoxicity [10]. Pretomanid belongs to the same class of antibiotics as delamanid [9]. Pretomanid was developed as part of a drug combination together with bedaquiline and linezolid, an oxazolidinone-class otherwise used for the treatment of pneumonia and skin infection. There is a clear need for the additional development of new effective drug combinations. The European Medical Agency (EMA) drug development guideline for TB specifies that efforts should be made to develop entirely new regimens to treat TB, rather than focusing on single drugs [11]. Due to the burden of polypharmacy for the patients and the increased risk of side effects, the focus should be on developing new regimens instead of the development of single agents as an add-on to a current regimen which was recommended in the earlier EMA TB drug development guideline [12]. Of the three new drugs against TB, only pretomanid is approved as a new combination regimen, while bedaquiline and delamanid were developed as add-ons to existing therapy [13]. The development of new combination regimens is the way forward, the acceleration of which is the objective of the new Innovative Medicines Initiative (IMI)-funded consortium European Regimen Accelerator for Tuberculosis (ERA4TB). It is important to assess drug-drug interactions (DDI), with respect to both PK and PD, to understand how the different drugs behave in certain combinations and doses in order to maximize the efficacy and potentially learn how the efficacy of the combination varies with time and concentration. The development of drug combinations is, however, challenging. It is difficult to demonstrate the contribution of an individual drug to a regimen regarding efficacy or safety [14]. The duration of treatment is lengthy, especially when considering follow-up to ensure no relapse. Moreover, the design and execution of preclinical experiments and clinical trials are complex, as the number of treatments to test grows exponentially with every added drug or dose, leading to longer development times and higher costs. Tuberculosis drug development, which focuses on regimens rather than unique drugs as an add-on treatment, thus challenges our methods to assess and identify optimal regimens. Therefore, smart experimental designs and optimized data analysis are essential. Data from larger scale *in vitro* preclinical experiments, with different drug regimens that explore the PD interaction space in order to investigate the synergism and/or antagonism of the interacting drugs, should be used to select the best regimens to determine the exposure range *in vivo*. Based on the exposure-response relationship in animals, and/or pure *in vitro* predictions, the first in-human (FIH) and early bactericidal activity (EBA) trials can be designed. These steps all require a mathematical translational approach, taking into account the PK-PD and translational factors to account for differences between preclinical species and patients [15,16].

The European Medicines Agency/European Federation of Pharmaceutical Industries and Associations (EFPIA) Modeling and Simulation joint workshop held in 2011 assembled scientists from the pharmaceutical industry, academia and regulatory authorities from across Europe, the USA and Japan to consider the future role of modeling and simulation in drug development and regulatory assessment. As a follow up to the workshop, one of the EFPIA groups' commitment to EMA was to generate a "good practice" manuscript covering aspects of planning, conduct and documentation of a variety of quantitative approaches for modeling and simulation methods where the concept of Model-Informed Drug Discovery and Development (MID3) was defined [17]. The aim of MID3 is to enable more efficient and robust research and development and regulatory decisions using an integrated model-based drug development approach [17,18]. The MID3 strategy for the development of drugs in any therapeutic area is supported by the EMA [19]. The MID3 framework has been defined as a "quantitative framework for prediction and extrapolation, centered on knowledge and inference generated from integrated models of compound, mechanism and disease level data and aimed at improving the quality, efficiency and cost effectiveness of decision making" [17]. The MID3 framework should be applied in the development of new TB drug regimens and is necessary for the reliable prediction of the optimal selection of novel TB drug combination therapies based on pre-clinical information, and subsequent decisions on which combinations to evaluate in clinical trials in order to confirm their efficacy and safety. The framework integrates all available data and information on the disease and the compounds. In addition to PK and PD models, systems biology or systems pharmacology models [18] and machine learning based on, for example, imaging data [20] or even artificial intelligence (AI) [21,22] are important tools. Figure 1 shows the proposed MID3 strategy for the rapid development of anti-TB regimens through the prediction of human-concentration-time relationships (PK), exposure-response relationships (PK-PD) and DDIs to select FIH doses, as well as the prediction of Phase II and Phase III drug regimens. Initially in a drug development program, preclinical data is mostly available. The impact of modeling and simulation increases towards the prediction of human exposure-response. With this input efficient decision can be made about the optimal combination of different drugs, and the right dose for each drug in the combination. Currently, limited modeling and simulation are required for market approval, which relies more on statistical comparison between treatment groups after phase III [23]. However, modeling and simulation can have a role in the analysis of Phase III data in order to define the relationship between exposure and clinical endpoint, evaluate PK DDI and simulate alternative potential regimens in certain subgroups, for example, patients with renal impairment [24]. A key step for successful TB drug development is to use modeling and simulation to predict the efficacy of combinations, including DDIs, for, for example, synergy. We will review the necessary steps from this perspective for the successful MID3 application to the preclinical to clinical translation of efficacious TB drug combinations, regarding the optimal doses of drugs in complex regimens.

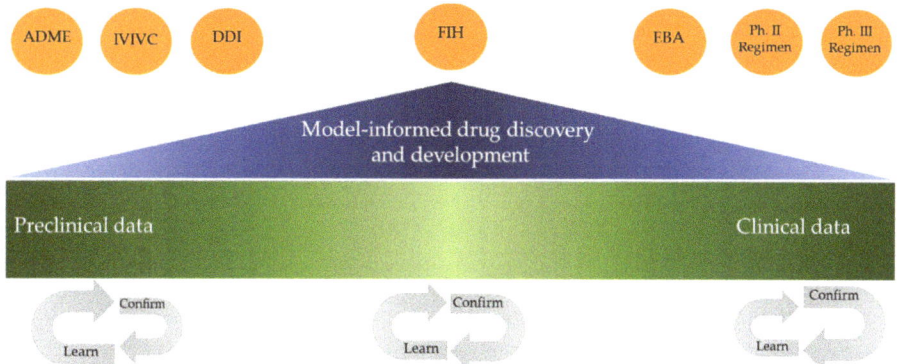

Figure 1. Illustration of the role of Model-Informed Drug Discovery and Development (MID3) and the application across preclinical to clinical drug development.

MID3, with modeling and simulation as key tools, is suggested to be applied throughout the pre-clinical to clinical drug development phases in order to optimize and inform decision making with respect to clinical trial design and the selection of drugs and doses to be carried forward from the preclinical phase and into clinical trial programs. The prediction of human-concentration-time relationships (PK), exposure-response relationships (PK-PD) in monotherapy and combination therapy, as well as drug-drug interactions, (DDI), requires the application of MID3 techniques and integration of all available data. In early preclinical drug development, preclinical data is used for predictions using, for example, *in vitro-in vivo* correlation (IVIVC), physiology-based pharmacokinetics (PBPK) and a biopharmaceutics drug disposition classification system (BDDCS) in order to define absorption, distribution, metabolism, and excretion (ADME) properties. Further down the developmental process, MID3 becomes more important in order to define exposure-response relationships and pharmacodynamic (PD) interactions using preclinical data for optimal design of first-in-human (FIH) and early bactericidal activity (EBA) trials. The need to define the optimal combination regimen using preclinical information data is evident, as the necessary number of clinical trial arms/experimental groups grow exponentially with the number of drugs within a regimen. Techniques using optimal design and simulation studies are essential and part of the MID3 framework. Throughout the process, the precision of human predictions increases. Different important drug development decision steps (circles) are subject to learn-and-confirm cycles, for example, early EBA clinical studies where the earlier defined exposure-response relationship using pre-clinical data (learning phase) is confirmed (confirming phase).

2. Model-Informed Drug Discovery and Development

Model-informed drug discovery and development is given by a quantitative framework for prediction and extrapolation, aimed at improving the quality, efficiency and cost-effectiveness of decision making in drug development [17]. It can also be utilized in early drug discovery through target identification and validation, and in describing the PK-PD and toxicological properties of the candidate drug. In addition, it increases the efficiency of trials and reduces the cost through facilitating dose and sample size selection [17]. Because of the great potential of MID3, it has been received well and implemented by drug developers [18,25]. The EMA supports MID3 and has built competence to meet the increasing modeling and simulation work in the dossiers submitted to EMA through the implementation of the modeling and Simulation Working Group (MSWG). Further, the EMA stresses that, in order to benefit from the full potential of MID3, stand-alone applications of modeling and simulation, dissociated from clinical decisions with respect to the design and objectives of clinical trials, should be avoided [19]. This is also pointed out in the MID3 white paper [17], where the implementation process is described as very important, and where the modeling and simulation work should be clearly motivated in the analysis, with clear objectives that are relevant and understandable for the entire development team. To realize the full potential of MID3, it needs to be integrated into the development plan rather than being seen as an ad-hoc activity [23]. The FDA has implemented a new Model-Informed Drug Development Paired Meeting Pilot Program which refers to the application of a wide range of quantitative approaches in drug development to facilitate the decision-making process, such as dose optimization, supportive evidence for efficacy, clinical trial design, and informing policy [26]. Despite the recent efforts within academia, EFPIA and regulatory agencies, MID3 has not been utilized to its full potential within TB drug development, where the need is great due to the complex development of new drug regimens consisting of at least three drugs.

Model-informed drug discovery and development builds upon pharmacometrics, the discipline that applies mathematical and statistical methods to understand, quantify, translate, and predict PK and PD behavior, including uncertainty in that behavior [27,28]. Pharmacometric population PK and PK-PD modeling can quantify these processes to better predict the concentration-time and exposure-response relationships of anti-TB drugs as compared to non-modeling techniques, such as non-compartmental analyses (NCA) for PK or traditional statistical analysis of, for example, the relationship between dose and baseline-reduced response at the end of treatment [29] The advantage of pharmacometric modeling is that it takes the inter-individual and inter-occasion variabilities into account. Once a population model has been developed and evaluated, various simulation techniques can be used, e.g., Monte Carlo simulations where virtual patients are drawn from the earlier quantified variance of variability in the population. Pharmacokinetic models are usually nonlinear mixed-effects models with unique parameters for fixed effects and random effects. Pharmacodynamic models can consist of a statistical method suitable for the biomarker or endpoint where time-to-positivity and relapse would be described with a time-to-event model, while colony forming unit (CFU) is a continuous variable and, as such, can be described with similar nonlinear mixed effects modeling.

Model-informed drug discovery and development is likely most impactful in the translation from preclinical to clinical, where the understanding and extrapolation of the exposure-response from preclinical to clinical is crucial. Model-informed drug discovery and development is also very important in the early clinical phases of anti-TB drug development, specifically phase II EBA trials, as it is difficult to investigate all drug combinations and associated PD interactions in clinical trials. The majority of the knowledge about the potential PD interaction space needs to come from preclinical information. Additionally, MID3 can be used to design the next preclinical or clinical study in order to optimize the likelihood of collecting informative data. A crucial step in drug development is the prediction of FIH design and associated doses. Model-informed drug discovery and development strategies and methods can be used to scale preclinical information to humans to design the FIH trial. Pharmacometric techniques have been shown to reduce the sample size needed in comparison to traditional statistical methods [29–31], while MID3 has been reported to save significant costs through its impact on decision making [17]. Preclinical experiments should be designed to be able to quantify the exposure-response relationship, including quantitative biomarkers relative to the interspecies' translation thereof [32]. An MID3 framework integrates all relevant preclinical and clinical information, and can therefore be used to back-translate results from the clinic to improve the preclinical understanding of the pathophysiology and pharmacology [33]. Even failed translations to humans are valuable in correcting the preclinical methods used. An iterative forward- and reverse-translational cycle has the potential to continuously enhance confidence in preclinical models [34]. The availability of large clinical datasets from, for example, electronic medical records accelerates reverse translation and improves the preclinical modeling of clinical manifestations [35]. Additionally, data from veterinary medicine can be utilized to guide human medicine development [36]. For this framework to really have an impact, data repositories and common languages are essential for application across different disciplines, disease areas, or stages of development [23,37]. In addition, to ensure that modeling and simulation adds value through an MID3 approach, pharmacometricians must communicate with their project teams before any data analysis starts to understand the key strategic development questions, clinical context, available data, assumptions, and decision criteria [23].

The prediction of efficacy and safety in new drug combinations with new or unknown mechanisms of action will benefit from the next paradigm in drug development and MID3, namely quantitative systems pharmacology (QSP). This is the pharmacological perspective on a systems' modeling, a body-system-wide characterization of the health and disease of an organism based on a mechanistic and molecular understanding of the individual components in the context of the holistic network [38]. QSP is the middle-out interface between systems biology and pharmacometrics, describing the pharmacological perturbation within the studied context [39]. It accounts for differences in (molecular) mechanisms of a disease [40], which is very relative for TB with its heterogeneous pathophysiology of acute, chronic, and latent infections. Because of its mathematical description of all the relevant elements of the pharmacological and pathophysiological pathways, and their differences between species, it becomes key to translational medicine [41]. Because of this quantitative understanding of the network, the prediction of the effects of drugs with new mechanisms of action improves significantly [42]. The development of QSP models in the preclinical phase is, however, uncommon, and the dedicated acquisition of experimental data like transcriptomics or metabolomics for the development of QSP models is rare [43,44]. Quantitative systems pharmacology models are intended to be applied to a wider scale than the individual questions or problems they were originally developed for [45]. For TB specifically, this could mean a systems model of the *M. tuberculosis* infection in the human context of macrophage infiltration, granuloma formation and pulmonary lesion development, with all relevant pathways and drug targets quantitatively described. The effect of new combinations, including drugs with novel mechanism of actions, can be predicted.

3. Prediction of Human Pharmacokinetics

In silico ADME-PK (absorption, distribution, metabolism, excretion, and pharmacokinetics) is the use of computer modeling to understand structure–property relationships and to predict DMPK (drug metabolism and pharmacokinetics) properties from compound structure. This is related to but distinct from physiologically based pharmacokinetic (PBPK) modeling, which strives to provide accurate predictions of the PK profile of drug candidates [46]. The focus of *in silico* ADME-PK is to guide the design of novel compounds with superior ADME properties. Most often a quantitative structure–property relationship (QSPR) approach is used to relate a compound's structure to the chemical property in question (e.g., cell permeability or metabolic clearance) measured in an *in vitro* assay. Related terms are also quantitative-structure-activity relationships (QSAR), when a set of predictor variables is related to the potency of the drug.

Orally administered products are subject to a sequence of transport and enzymatic barriers in enterohepatic systems affecting bioavailability, including extraction in the intestinal and liver tissues, which could impact the fraction of the orally administered dose that reaches the systemic circulation and thereby the site of action. Bioavailability is mainly dependent on three general and rather complex serial processes: the fraction of the oral dose that is absorbed (F_{abs}), the first-pass extraction of the drug in the gut wall (E_G), and the first-pass extraction of the drug in the liver (E_H) [47]. In general, oral products with a low F (<25%–35%) have a higher inter- and intra-individual variability in plasma exposure (coefficient of variation >60%–120%) [48]. Drugs with high degree of F_{abs} show sufficiently high solubility of the active pharmaceutical ingredient (API), no luminal degradation, and absorption along the small and/or large intestine [49,50]. The regulatory framework Biopharmaceutics Classification System (BCS) of drugs provides information relevant to understanding and predicting GI drug absorption and bioavailability in general, which is relevant to the absorption potential in the small and large intestine [51]. After the drug is absorbed, it passes to the liver, which expresses a broader range of different enzymes compared to the intestine [52], such as the family of CYP enzymes [53]. Other enzymes such as microsomal uridine 5′-diphospho-glucuronosyltransferases (UGTs), sulfotransferases, and glutathione S-transferases are found in great amounts in the human liver [54]. Humans show large inter-individual variation in the amount of different enzymes, which accounts, in part, for the large inter-individual variation reported for E_H and CL. A considerable large inter-individual variation in the expression of the different CYP isoforms has also been observed, ranging from 20-(CYP2E1 and CYP3A4) to >1000-fold (CYP2D6). The liver is also the major organ for glucuronidation in the body. Glucuronide conjugates have molecular characteristics that are associated with biliary excretion of a compound, i.e., a high molecular weight, ionized and the presence of polar groups [55].

The identification and quantification of the important PK processes described above can be investigated in relevant *in vitro* models, and predictions of PK properties like regional intestinal permeability can be made early [56]. Most importantly, an estimate of human drug clearance will determine how fast a drug is eliminated, and conversely define the dose range to study in FIH [57]. Several different models have been suggested for the prediction of oral absorption for the biopharmaceutical design of oral drug delivery systems [58,59]. The proposed BDDCS has been shown to be useful in predicting some crucial ADME parameters and especially the transport/absorption/elimination interplay [60]. Preclinical data can then be translated through *in vitro-in vivo* extrapolation, or even through PBPK modeling to generate basic PK parameters such as fraction of dose absorbed, bioavailability, clearance (CL), volume of distribution, and terminal half-life. Furthermore, the accuracy of the QSAR predictions of effective intestinal permeability (P_{eff}), is significantly improved when based on a combination of molecular physicochemical descriptors and molecular dynamics simulations from *in vitro* data [61]. Molecular simulations have been successfully used to predict the effects of cholesterol in the lipid membrane fluidity [62]. Additionally, molecular simulations have been reported to be useful as they are comparable to experimental data [63]. Among the molecular descriptors evaluated by Lipinski (e.g., polar surface area, hydrogen bond donors (HBDs)/acceptors, Log D), the number of HBDs is the most restrictive when it comes to intestinal membrane transport/absorption [64,65]. Two drugs violating this rule (i.e., >5 HBDs and low intestinal P_{eff} and F_{abs}), one of which is rifampicin, have been investigated thoroughly and offered a potential explanation for drug absorption beyond the Lipinski Rule-of-five [66]. Based on a liposomal permeation assay, it has been proposed that drug molecules with more than five HBD can be sufficiently absorbed in the intestine by passive lipoidal diffusion [66]. Some drugs are absorbed by passive lipoidal diffusion despite their unfavorable physicochemical properties. It is therefore necessary to find more complex descriptions of the molecular interaction by applying a combination of experimental data and molecular dynamic modeling and simulation to further improve the accuracy in predicting general membrane transport across the cellular membrane barrier and not only in the GI-tract [66–69].

Physiologically based pharmacokinetic modeling is considered to assist drug product development by providing quantitative predictions through a systems approach [70]. A mechanism-based model, like that of the PBPK approach, separates drug-specific from system-specific elements, which allows for the interspecies translation of the time course of the drug [41]. Physiologically based pharmacokinetic models divide the body into anatomically and physiologically meaningful compartments, including the gastrointestinal tract for absorption, the eliminating organs, and non-eliminating tissue compartments [71]. In addition, compound-specific parameters such as physicochemical and biochemical parameters (e.g., tissue/blood partitioning and metabolic CL) are incorporated into the model to predict the plasma and tissue concentration versus the time profiles of a compound in an *in vivo* system following intravenous or oral administration. Translation between species, special populations, or disease states, are the result of changing these physiological parameters accordingly. Several variations are in use, including a whole-body PBPK model describing the complete organism, and hybrid PBPK models, combining PBPK elements with empirical compartmental PK to simplify the model [72]. An important element in the physiologically based translation of PK is binding of the drug of interest to proteins in the plasma, mainly albumin, lipo- or glycoproteins, or globulins. Protein binding differs between experimental settings (*in vitro*) and species (*in vivo*) and should be taken into account because it influences tissue penetration and the free drug that can interact with its target [73]. Physiologically based pharmacokinetic modeling predictions are a valuable tool in the pharmaceutical industry due to the possibility of combining all the available and relevant information that is generated during the preclinical stage, which helps improve decision making during the selection process [74]. For instance, biorelevant dissolution-absorption PBPK modeling and simulation has been reported as applied in 88% of early drug development processes [75]. Moreover, PBPK modeling provides a powerful tool to study potential PK DDIs through incorporating the drug's physicochemical properties, PK properties, human physiological variables, and inter-individual variability estimates.

Predictions of PK parameters use information from preclinical studies in animals to transition to clinical trials, i.e., FIH studies, when no clinical information is available to guide the decision of the starting dose. Thus, the estimation of the starting dose in human subjects relies on the PK knowledge of the drug from different species. It is essential to have preclinical PK data based on blood, plasma, and/or tissue sampled longitudinally to optimally capture the complete drug profile during a dosing interval. A model-based approach to the starting dose often uses allometric scaling to predict human drug clearance and distribution volumes. Allometric scaling is based on the assumption that physiological similarities exist between different species arising from anatomical similarities, specifically similarities in body weights and body surface areas [76]. Historically, a maximum dose for FIH studies was based on the no observed adverse effect level (NOAEL) in preclinical experiments, an arbitrary safety factor, and allometric scaling. This approach is empirical by nature and therefore limited. When more mechanistic data and models are available, a minimal anticipated biological effect level (MABEL) can be estimated [42]. For example, preclinical PK-PD and interspecies differences in the target can be utilized to estimate the MABEL for a FIH trial [77]. This has the benefit of being driven by pharmacology, where FIH trials will answer pharmacological questions on PK and PD, rather than being driven by toxicology or tolerability. Taking into account MABEL and safety factors, a first study with single ascending doses (SAD study) will quantify the PK and ensure safety and tolerability. A second study with multiple administered doses (MAD) can subsequently be designed accordingly [78].

In addition to allometric scaling, the use of IVIVC has markedly increased [79–82]. It is suggested that animal PBPK models should be used as part of a stepwise approach, in which the first step uses animal data to understand the processes and verify the predictive power of *in vitro* systems, and the second step is about forecasting human PK from *in vitro* data and *in silico* methods (learn-and-confirm) [82]. The first step in predicting drug CL using IVIVC is to obtain intrinsic CL (CL_{int}) from *in vitro* data [83,84]. *In vitro* CL_{int} values determined from various systems including hepatocytes, cell transport models, liver or intestinal microsomes, or recombinant CYPs, either by substrate depletion or metabolite formation, are normalized for cell, microsomal protein or enzyme concentration. The next step consists of scaling the activity determined *in vitro* to the whole liver by the use of a scaling factor, to account for incomplete microsomal recovery from the tissue to obtain *in vivo* CL_{int}. Finally, the third step involves the use of a liver model which incorporates the effects of hepatic blood flow, plasma protein-binding and blood cell partitioning to convert the estimated *in vivo* CL_{int} into a hepatic CL (CL_H). The well-stirred liver model is most commonly used, but the dispersion model or the parallel tube model is also available [84].

Using high doses of oral anti-TB drugs may result in high plasma concentrations, leading to an increased risk of adverse effects [85] while not ensuring adequate concentrations at the site of action [85,86]. This has prompted investigation into the use of the pulmonary route to deliver anti-TB drugs directly to the site of action in the lungs. Administering anti-TB drugs as inhaled formulations ensures the delivery of the drug directly to the target organ, avoiding any unwanted systemic side effects, thereby improving patient compliance [87]. Optimal pulmonary drug delivery for locally acting drugs includes a high local lung concentration, extended lung residence time and low systemic concentration [88]. A fundamental understanding of pulmonary dissolution, residence time, and lung absorption processes is key for the successful development of inhaled products [89,90]. However, inhaled formulations have many challenges, including formulation stability, pulmonary distribution, lung toxicity, and additives safety [91]. Furthermore, dosing of inhaled drugs is more complicated than other routes of administration as their absorption in the lung is highly variable [91]. Thus, many *in vitro* and *in vivo* models have been developed to study the PK of inhaled anti-TB drugs, specifically their absorption and distribution, in order to evaluate the efficacy and safety of anti-TB drugs in the lungs. These models, while allowing a reduction in biological complexity, still face many challenges, and can be demanding to build [92]. *In vivo* animal models are the gold standard regarding the assessment of drug clearance, systemic side effects and PK after pulmonary administration [92]. However, animal models are not always able to mimic human pulmonary anatomy and physiology, or TB disease progression in humans, and they do not exhibit extrapulmonary dissemination similar to humans [86]. Translation from animal data to the clinic has been recognized as challenging [93]. Understanding the pulmonary exposure is important, and animal data can contribute specific information about the lesion to plasma ratio [94], as similar lung distribution ratios can be obtained in human. However, little is known about the factors that influence drug distribution from plasma into the range of tissues, nodules and cavities that are inhabited by the TB pathogen. Pulmonary TB lesions consist of a diversity of cell types, tissue structures and vascular architectures which suggests that the distribution of the drug is not only governed by passive equilibration between unbound drug concentration in plasma and tissue [95]. MALDI mass spectrometry imaging (MALD-MSI) is a new technique to study the distribution of small molecules in the various compartments of pulmonary lesions [96]. Information from such studies not only provides knowledge of regional differences in drug exposure, but also confirms a high exposure in regions where a high density of persistent TB bacteria is found.

4. Prediction of Human Pharmacokinetic-Pharmacodynamic Relationship

In order to translate drug effects from preclinical information to the clinical phase of drug development, defining a drug exposure-response relationship using preclinical information is of importance. The PK-PD relationship is quantitative, predictive, and reproducible and is valid in all disease models [57,58]. Thus, characterizing this relationship is of great benefit in preclinical PK-PD studies to help guide dose selection and study design in humans. Exhaustive reviews of preclinical experimental methods that quantify exposure-response relationships have previously been performed [97,98]. These methods, such as classical time-kill experiments, hollow-fiber system (HF), different murine models, rabbits and guinea pig, all mimic elements of the human pathology to a certain extent, but all have their limitations. Here, the focus is on their informativeness of the exposure-response relationship for translation to human prediction.

In vitro determination of the minimum inhibitory concentration (MIC) is informative about the sensitivity of the bacterial strain to the compound. This is especially the case when the target sizes of the *M. tuberculosis* infection, macrophages, are utilized as environmental context [97]. The MIC is a measure of the net effect of the drug on bacterial growth and survival. However, it is very crude and undynamic as it is measured at a specific concentration and after a fixed time, which might cause it to deviate from the true MIC [99]. The MIC is also limited because the resolution is determined by the chosen dilution steps, and bigger dilution steps increase the risk of under- or overestimating the MIC. In addition, the determination of MIC is based on visual inspection which makes it prone to subjective error [99]. Mouton et al. have studied the variability between MIC measurements in *Staphylococcus aureus* treated with linezolid and have concluded that over half of the variability in the MIC measurements is either due to systemic and significant inter-laboratory differences or differences between strains [100]. The other half can be explained by assay variation and different environmental conditions, such as the media used and incubation temperature [100].

Several preclinical animal models for TB are in use. The advantage of an animal model over *in vitro* systems is the holistic environment of a whole organism, including a functioning immune system, physiological feedback systems and (drug) disposition. This results in more variability in the determination of the exposure-response relationship and requires more effort to elucidate drug effect from, for example, the immune system. The most emphasis is placed on murine models of

TB [101], although there are arguments that the mouse is not a good model for TB in humans [102]. Mice can be housed in the required biosafety laboratories with ease, blood and tissue sampling is well established, and both chronic and latent infections have been successfully used [103]. However, the mice have a low susceptibility to *M. tuberculosis* and show only loosely organized granulomas, and are therefore limited when considering lesion-specific treatment. Granuloma formation in guinea pigs and rabbits is more representative of human granulomas, including caseous necrosis [102]. Guinea pigs are highly susceptible to *M. tuberculosis* which makes infection as straightforward as exposure to exhaustion from TB patients [103]. Rabbits are also utilized to study a slower response to treatment, disease relapse, and resistance development due to lung cavities, and their size makes studying drug distribution to TB lesions more feasible [98]. The experimental toolbox regarding immunologic reagents and genetic techniques is, however, more restricted in these animals, and both need more difficult and expensive husbandry. Granuloma formation can also be studied non-invasively in the zebrafish, a relatively new disease model organism in drug discovery and development [103]. Because of their transparency and easy genetic modification, fluorescence microscopy of pathogen and immune cells can be leveraged to follow infection and treatment [104]. With the small size and high fecundity of the zebrafish, high-throughput assays are available to test large numbers of compounds in short amounts of time with enough statistical power [105]. Methods to quantify internal drug exposure have also been established [106–108]. Recently, an exposure-response relationship has been developed for isoniazid in the zebrafish, which translated well to humans [109]. In general, non-invasive imaging of lesion pathology by computed tomography (CT) and positron emission tomography (PET) has the potential to improve the comparison between preclinical and clinical measurements of disease progression and treatment [102]. Ordonez et al have demonstrated this by using dynamic [^{11}C]rifampicin PET-CT imaging in patients newly diagnosed with pulmonary TB and rabbits infected with cavitary TB to noninvasively measure intralesional drug concentration-time profiles and, consequently, time to bacterial extinction [110]. They also employed integrated modeling of the PET-captured concentration-time profiles in hollow-fiber bacterial kill curve experiments to predict the rifampicin dose required to achieve a cure in 4 months, which has a huge potential in antimicrobial drug development to shorten TB treatments [110]. It is clear that no single animal model represents a heterogeneous disease such as TB. A mechanistic understanding of TB in humans will identify which elements are characterized best by which animal model [103]. Independently of which preclinical experimental method is utilized, the sampling design of both PK (e.g., drug and/or metabolite concentration) and PD (e.g., infection, bacterial burden) biomarkers is of the utmost importance. The careful selection of datapoints over the duration of the experiment and at different drug concentrations is essential for a reliable quantification of the exposure-response relationship.

Regulatory agencies suggest determining PK/PD indices based on preclinical data for antibiotics, e.g., the area under the concentration curve over MIC (AUC/MIC), the maximum concentration (C_{max}) over MIC (C_{max}/MIC), and the percent of a 24-hour time period that the drug concentration is above MIC (T > MIC), for the establishment of the PK-PD profile of antimicrobials and for deciding the most optimal dosing regimens. PK/PD indices are based on preclinical studies that describe the PK-PD relationships of antimicrobials [111]. However, PK/PD indices suffer from several clear limitations, some of which are inherent to their use of MIC, the limitations of which are discussed above. Using PK/PD indices ignores information about the time-course of individual PK and PD processes [112]. As summary endpoints, they lack the ability to track the changes in the bacterial load over time [113]. Furthermore, when using AUC/MIC as a PK/PD index, the rate of drug administration is ignored, while, when using C_{max}/MIC, bacterial killing is assumed to depend solely on the maximum drug concentration, ignoring drug half-life and infusion duration [99]. Using T > MIC assumes that the maximal drug effect has been reached when MIC is reached, regardless of whether higher concentrations were given [99]. Additionally, the colony-forming units (CFU) versus PK/PD indices profile shows great variability in the CFU observations for the same PK/PD indices value [99]. These PK/PD indices are selected and predicted as PD targets using HFS-TB to quantify a more realistic *in vitro* exposure-response relationship that is translatable to *in vivo* [98,99,114]. However, despite EMA's qualification of the preclinical HFS-TB to be used to complement existing methodologies, it still suffers from a number of limitations. The EMA advises caution when interpreting HFS-TB results, as many instances of over- and under-estimates of the drug's anti-TB activity have been reported [115]. In addition, HFS-TB cannot replace animal models or clinical studies [116], while the reproducibility of the method by other laboratories has not yet been assessed [115].

Mechanistic, or semi-mechanistic, PK-PD models in TB based on preclinical data allow for the description of the multiple mycobacterial populations present. A mechanism-based PK-PD model by Hollow-fiber systems for TB has the advantage of being able to mimic dynamic PK in comparison to more traditional static time-kill experiments. A semi-mechanistic PK-PD model can be derived using HTS data [117–119]. Khan et al. describes susceptible, resting, and non-colony-forming bacterial populations [120]. The multistate tuberculosis pharmacometric (MTP) model is a semi-mechanistic mathematical model that can describe and identify the exposure-response profile of a drug towards three bacterial subpopulations: fast-, slow-, and non-multiplying bacteria. It has been successfully applied to describe *in vitro* [121], mouse [122], and clinical data [123]. In addition, the MTP model has been successfully used in an MID3 approach, to predict observations from early clinical studies using clinical dose-response forecasting from preclinical *in vitro* studies of rifampicin and in combination with isoniazid [15,16]. This model has been selected by *The Impact and Influence Initiative* of the Quantitative Pharmacology (QP) Network of the American society of Clinical Pharmacology and Therapeutics (ASCPT) to highlight the most impactful examples of QP applications where the role of quantitative translational pharmacology has bridged science and practice to make better, faster, and more efficient decisions in drug discovery and development [25]. Another mechanism-based model is the Magombedze et al. model that mimics the disease state in TB patients by describing the mycobacterial population as logarithmic growth-phase, semi-dormant, and persister bacilli [117]. In addition, a pulmonary PK-PD model of isoniazid has been developed to better characterize the relationship between its PK and its anti-TB effects in the lungs [124].

5. Prediction of Human Drug-Drug Interactions

Tuberculosis requires a combination therapy of three different antibiotics or more, which increases the risk of DDIs. Drug-drug interactions between drugs that are intended to be used in combination should be considered as early as possible. The prediction of DDIs from preclinical data will improve the ability to predict the total efficacy of the combination in relation to the drugs in monotherapy, as well as compared to expected additivity, i.e., the sum of all effects from the drugs when given alone. DDIs that result in less efficacy in the combination than in a combination with one less drug should be avoided. However, combinations that result in an efficacy less than the expected additivity, but still result in more efficacy than when one drug is omitted, can be considered. Drug-drug interactions can relate to both PK interactions, i.e., one drug (the perpetrator) impacting the absorption, distribution, metabolism, or excretion of another drug (the victim), or PD interactions, i.e., the perpetrator impacting the potency or efficacy of the victim drug.

Regulatory guidelines on the investigation of DDIs are brief about the use of *in vitro* data, while in an MID3 context, knowledge on the relevant mechanisms of, e.g., metabolism combined with *in vitro* data can be leveraged to decide on suitable combinations of drugs without extensive experimentation [125]. Both *in vitro* studies as well as animal experiments can be utilized to assess the potential for PK DDIs [126]. *In vitro* studies make use of metabolically active hepatocytes or cells overexpressing drug transporters to determine the PK interaction potential of a new drug [127]. When studying DDIs in preclinical species, the between-species differences in transporters or enzymes should be taken into account [128]. Pharmacokinetic DDIs mostly impact drug clearance by the induction or inhibition of metabolic enzymes like those from the CYP family and, to some extent, ABC and transport proteins. Such an interaction by the perpetrator drug will greatly enhance or reduce the exposure of the victim drug. For example, rifampicin induces bedaquiline clearance 5-fold, and should therefore not be combined for therapy [129]. Because bedaquiline has a very long terminal half-life, potential DDIs are difficult to identify using traditional methods, whereas properly designed experiments and quantitative modeling are necessary to elucidate such interactions [130]. Drug distribution can also be impacted because of the induction or inhibition of drug transporters like the permeability glycoprotein (P-gp), which is present on the canalicular membrane and blood-brain barrier, among others. Physiology based pharmacokinetic modeling can be very successful to predict metabolic DDIs, and specific DDI studies can be assisted by modeling and simulations [131]. Some anti-TB drugs are reported to be substrates for different hepatic enzymes or known to be inducers or inhibitors of metabolic enzymes. Rifampicin is well known as a CYP3A4 modulator [132,133], as well as an inducer of P-gp [134]. Additionally, even though the effect of clofazimine on CYP3A4 and P-gp is still unclear, clofazimine has been shown to delay the time taken to reach C_{max} of rifampicin [135]. Horita et al. studied the effects of anti-TB and antiretroviral drugs on CYP3A4 and P-gp, and they found that clofazimine exhibits weak inductive effects on CYP3A4 [136]. Furthermore, the co-administration of bedaquiline and clofazimine has been reported to increase the risk of QT prolongation [137,138]. As described above, these potential DDIs can be predicted from *in vitro* data through, for example, *in vitro-in vivo* scaling [139] or PBPK [140]. A transcription/translation model and a PBPK model have been developed to predict rifampicin-induced DDIs with reasonable accuracy [141].

In contrast to PK interactions, due to clearly defined processes of absorption, distribution, metabolism, and excretion, PD interactions are harder to investigate and quantify. This is because, since a clinical DDI study has to study the drugs both alone and in combination, the number of arms in the study will substantially increase when studying three or more interacting drugs. The Greco model [142], which is derived from Loewe additivity, was developed to assess PD interactions. However, such a model suffers from being limited to interactions between only two drugs. On the other hand, the general pharmacodynamic interaction (GPDI) model overcomes this limitation, in addition to being flexible to different drug interaction data without requiring knowledge on the modes of action of the studied drugs [143]. The GPDI model-based approach proposes a PD interaction to be quantifiable, as multidirectional shifts in drug efficacy (E_{max}) or potency (EC_{50}) and explicates the drugs' role as victim, perpetrator or even both at the same time. The GPDI model has been utilized along with the MTP model [121] to develop a model-informed preclinical approach for the prediction of PD interactions [144]. The MTP-GPDI model has been further employed to successfully evaluate and quantify the PD interactions of anti-TB drug combinations in mice [145]. Furthermore, it has been demonstrated that the GPDI model outperforms conventional methods in the evaluation of PD interactions for TB drugs [146].

It is clear that the need for a combination therapy of TB could potentially result in DDIs in the clinic. It is therefore essential to quantitatively understand the DDIs, both PK- and PD-interactions, as early as possible in drug development. Utilizing data from *in vitro* combination experiments combined with preclinical *in vivo* data on the exposure-response relationships of the drugs in combination and early clinical data, will inform on which combinations of drugs at which doses are efficacious and safe for patients. This quantitative integration of data and translation to the clinic is possible through the MID3 model-informed framework.

6. Conclusions

The development of new combinations of anti-TB drugs is both promising and challenging. Novel drug combinations and drug delivery routes require novel and innovative techniques. Model-informed drug discovery and development is an integrated framework of preclinical and clinical data through translational models that show great promise in selecting and predicting which drug regimens to carry forward to be evaluated in clinical trials. The MID3 framework supports decision making in drug development in relation to the prediction of efficacious and safe combinations of new drugs and translates this to the clinic. It is essential for drug developers to collaborate across disciplines, and academic and industry borders and train a new type of scientist in experimental and computational innovation.

Author Contributions: All authors contributed to conceptualization of this work. All authors contributed to original draft preparation, review and editing. All authors have read and agreed to the published version of the manuscript.

Funding: This research received no external funding.

Conflicts of Interest: The authors declare no conflict of interest.

References

1. DiMasi, J.A.; Grabowski, H.G.; Hansen, R.W. Innovation in the Pharmaceutical Industry: New Estimates of R & D Costs. *J. Health Econ.* **2016**, *47*, 20–33. [CrossRef]
2. Bellanti, F.; Van Wijk, R.C.; Danhof, M.; Della Pasqua, O. Integration of PKPD Relationships into Benefit-Risk Analysis. *Br. J. Clin. Pharmacol.* **2015**, *80*, 979–991. [CrossRef] [PubMed]
3. World Health Organization. *Global Tuberculosis Report 2019*; World Health Organization: Geneva, Sweitzerland, 2019.
4. Boeree, M.J.; Diacon, A.H.; Dawson, R.; Narunsky, K.; Du Bois, J.; Venter, A.; Phillips, P.P.J.; Gillespie, S.H.; McHugh, T.D.; Hoelscher, M.; et al. A Dose-Ranging Trial to Optimize the Dose of Rifampin in the Treatment of Tuberculosis. *Am. J. Respir. Crit. Care Med.* **2015**, *191*, 1058–1065. [CrossRef] [PubMed]
5. Svensson, E.M.; Svensson, R.J.; Te Brake, L.H.M.; Boeree, M.J.; Heinrich, N.; Konsten, S.; Churchyard, G.; Dawson, R.; Diacon, A.H.; Kibiki, G.S.; et al. The Potential for Treatment Shortening with Higher Rifampicin Doses: Relating Drug Exposure to Treatment Response in Patients with Pulmonary Tuberculosis. *Clin. Infect. Dis.* **2018**, *67*, 34–41. [CrossRef] [PubMed]
6. Svensson, R.J.; Svensson, E.M.; Aarnoutse, R.E.; Diacon, A.H.; Dawson, R.; Gillespie, S.H.; Moodley, M.; Boeree, M.J.; Simonsson, U.S.H. Greater Early Bactericidal Activity at Higher Rifampicin Doses Revealed by Modeling and Clinical Trial Simulations. *J. Infect. Dis.* **2018**, *218*, 991–999. [CrossRef]
7. Susanto, B.O.; Svensson, R.J.; Svensson, E.M.; Aarnoutse, R.; Boeree, M.J.; Simonsson, U.S.H. Rifampicin Can Be given as Flat-Dosing Instead of Weight-Band Dosing. *Clin. Infect. Dis.* **2019**. [CrossRef]
8. WHO. *Consolidated Guidelines on Drug-Resistant Tuberculosis Treatment*; Licence: CC BY-NC-SA 3.0 IGO; World Health Organization: Geneva, Switzerland, 2019.
9. Bahuguna, A.; Rawat, D.S. An Overview of New Antitubercular Drugs, Drug Candidates, and Their Targets. *Med. Res. Rev.* **2020**, *40*, 263–292. [CrossRef]
10. Pontali, E.; Sotgiu, G.; Tiberi, S.; Tadolini, M.; Visca, D.; D'Ambrosio, L.; Centis, R.; Spanevello, A.; Migliori, G.B. Combined Treatment of Drug-Resistant Tuberculosis with Bedaquiline and Delamanid: A Systematic Review. *Eur. Respir. J.* **2018**, *52*. [CrossRef]
11. European Medicines Agency. *Addendum to the Guideline on the Evaluation of Medicinal Products Indicated for Treatment of Bacterial Infections to Address the Clinical Development of New Agents to Treat Pulmonary Disease Due to Mycobacterium Tuberculosis*; European Medicines Agency: London, UK, 2017.
12. European Medicines Agency. *Addendum to the Note for Guidance on Evaluation of Medicinal Products Indicated for Treatment of Bacterial Infections to Specifically Address the Clinical Development of New Agents to Treat Disease Due to Mycobacterium Tuberculosis*; European Medicines Agency: London, UK, 2010.
13. Dheda, K.; Gumbo, T.; Maartens, G.; Dooley, K.E.; Murray, M.; Furin, J.; Nardell, E.A.; Warren, R.M.; Esmail, A.; Nardell, E.; et al. The Lancet Respiratory Medicine Commission: 2019 Update: Epidemiology, Pathogenesis, Transmission, Diagnosis, and Management of Multidrug-Resistant and Incurable Tuberculosis. *Lancet Respir. Med.* **2019**, *7*, 820–826. [CrossRef]
14. European Medicines Agency. *Guideline on Strategies to Identify and Mitigate Risks for First-in-Human and Early Clinical Trials with Investigational Medicinal Products*; European Medicines Agency: London, UK, 2017.
15. Wicha, S.G.; Clewe, O.; Svensson, R.J.; Gillespie, S.H.; Hu, Y.; Coates, A.R.M.; Simonsson, U.S.H. Forecasting Clinical Dose-Response From Preclinical Studies in Tuberculosis Research: Translational Predictions With Rifampicin. *Clin. Pharmacol. Ther.* **2018**, *104*, 1208–1218. [CrossRef]

16. Susanto, B.O.; Wicha, S.G.; Hu, Y.; Coates, A.R.M.; Simonsson, U.S.H.; Biosciences, P.; Pharmacy, C.; Kingdom, U.; Simonsson, U.S.H.; Biosciences, P.; et al. Translational Model-Informed Approach for Selection of Tuberculosis Drug Combination Regimens in Early Clinical Development. *Clin. Pharmacol. Ther.* **2020**. [CrossRef] [PubMed]
17. Marshall, S.F.; Burghaus, R.; Cosson, V.; Cheung, S.; Chenel, M.; DellaPasqua, O.; Frey, N.; Hamrén, B.; Harnisch, L.; Ivanow, F.; et al. Good Practices in Model-Informed Drug Discovery and Development: Practice, Application, and Documentation. *CPT Pharmacomet. Syst. Pharmacol.* **2016**, *5*, 93–122. [CrossRef]
18. Marshall, S.; Madabushi, R.; Manolis, E.; Krudys, K.; Staab, A.; Dykstra, K.; Visser, S.A.G. Model-Informed Drug Discovery and Development: Current Industry Good Practice and Regulatory Expectations and Future Perspectives. *CPT Pharmacomet. Syst. Pharmacol.* **2019**, *8*, 87–96. [CrossRef]
19. Manolis, E.; Brogren, J.; Cole, S.; Hay, J.L.; Nordmark, A.; Karlsson, K.E.; Lentz, F.; Benda, N.; Wangorsch, G.; Pons, G.; et al. Commentary on the MID3 Good Practices Paper. *CPT Pharmacomet. Syst. Pharmacol.* **2017**, *6*, 416–417. [CrossRef]
20. Goulooze, S.C.; Zwep, L.B.; Vogt, J.E.; Krekels, E.H.J.; Hankemeier, T.; Van den Anker, J.N.; Knibbe, C.A.J. Beyond the Randomized Clinical Trial: Innovative Data Science to Close the Pediatric Evidence Gap. *Clin. Pharmacol. Ther.* **2020**. [CrossRef]
21. Ribba, B. Model-Informed Artificial Intelligence: Reinforcement Learning for Precision Dosing. *Clin. Pharmacol. Ther.* **2020**. [CrossRef]
22. Harrer, S.; Shah, P.; Antony, B.; Hu, J. Artificial Intelligence for Clinical Trial Design. *Trends Pharmacol. Sci.* **2019**, *40*, 577–591. [CrossRef]
23. Krishnaswami, S.; Austin, D.; Della Pasqua, O.; Gastonguay, M.R.; Gobburu, J.; van der Graaf, P.H.; Ouellet, D.; Tannenbaum, S.; Visser, S.A.G. MID3: Mission Impossible or Model-Informed Drug Discovery and Development? Point-Counterpoint Discussions on Key Challenges. *Clin. Pharmacol. Ther.* **2020**. [CrossRef]
24. Marshall, S.F.; Hemmings, R.; Josephson, F.; Karlsson, M.O.; Posch, M.; Steimer, J.L. Modeling and Simulation to Optimize the Design and Analysis of Confirmatory Trials, Characterize Risk-Benefit, and Support Label Claims. *CPT Pharmacomet. Syst. Pharmacol.* **2013**, *2*, 3–5. [CrossRef]
25. Gupta, N.; Bottino, D.; Simonsson, U.S.H.; Musante, C.J.; Bueters, T.; Rieger, T.R.; Macha, S.; Chenel, M.; Fancourt, C.; Kanodia, J.; et al. Transforming Translation Through Quantitative Pharmacology for High-Impact Decision Making in Drug Discovery and Development. *Clin. Pharmacol. Ther.* **2019**, 1–5. [CrossRef]
26. Wang, Y.; Zhu, H.; Madabushi, R.; Liu, Q.; Huang, S.M.; Zineh, I. Model-Informed Drug Development: Current US Regulatory Practice and Future Considerations. *Clin. Pharmacol. Ther.* **2019**, *105*, 899–911. [CrossRef] [PubMed]
27. Gieschke, R.; Steimer, J.L. Pharmacometrics: Modelling and Simulation Tools to Improve Decision Making in Clinical Drug Development. *Eur. J. Drug Metab. Pharmacokinet.* **2000**, *25*, 49–58. [CrossRef] [PubMed]
28. Ette, E.I.; Williams, P.J. Pharmacometrics: The Science of Quantitative Pharmacology. John Wiley & Sons, Inc.: Hoboken, NJ, USA, 2007; ISBN 978-0-471-67783-3.
29. Svensson, R.J.; Gillespie, S.H.; Simonsson, U.S.H. Improved Power for TB Phase IIa Trials Using a Model-Based Pharmacokinetic-Pharmacodynamic Approach Compared with Commonly Used Analysis Methods. *J. Antimicrob. Chemother.* **2017**, *72*, 2311–2319. [CrossRef] [PubMed]
30. Karlsson, K.E.; Vong, C.; Bergstrand, M.; Jonsson, E.N.; Karlsson, M.O. Comparisons of Analysis Methods for Proof-of-Concept Trials. *CPT Pharmacomet. Syst. Pharmacol.* **2013**, *2*. [CrossRef] [PubMed]
31. Chen, C.; Ortega, F.; Alameda, L.; Ferrer, S.; Simonsson, U.S.H. Population Pharmacokinetics, Optimised Design and Sample Size Determination for Rifampicin, Isoniazid, Ethambutol and Pyrazinamide in the Mouse. *Eur. J. Pharm. Sci.* **2016**, *93*, 319–333. [CrossRef]
32. Muliaditan, M.; Davies, G.R.; Simonsson, U.S.H.; Gillespie, S.H.; Della Pasqua, O. The Implications of Model-Informed Drug Discovery and Development for Tuberculosis. *Drug Discov. Today* **2017**, *22*, 481–486. [CrossRef]
33. Visser, S.A.G.; Aurell, M.; Jones, R.D.O.; Schuck, V.J.A.; Egnell, A.C.; Peters, S.A.; Brynne, L.; Yates, J.W.T.; Jansson-Löfmark, R.; Tan, B.; et al. Model-Based Drug Discovery: Implementation and Impact. *Drug Discov. Today* **2013**, *18*, 764–775. [CrossRef]
34. Kasichayanula, S.; Venkatakrishnan, K. Reverse Translation: The Art of Cyclical Learning. *Clin. Pharmacol. Ther.* **2018**, *103*, 152–159. [CrossRef]

35. Li, L. Reverse Translational Pharmacology Research Is Driven by Big Data. *CPT Pharmacomet. Syst. Pharmacol.* **2018**, *7*, 63–64. [CrossRef]
36. Schneider, B.; Balbas-Martinez, V.; Jergens, A.E.; Troconiz, I.F.; Allenspach, K.; Mochel, J.P. Model-Based Reverse Translation Between Veterinary and Human Medicine: The One Health Initiative. *CPT Pharmacometrics Syst. Pharmacol.* **2017**. [CrossRef]
37. Van der Graaf, P.H.; Benson, N. The Role of Quantitative Systems Pharmacology in the Design of First-in-Human Trials. *Clin. Pharmacol. Ther.* **2018**, *104*, 797. [CrossRef] [PubMed]
38. van der Greef, J.; McBurney, R.N. Rescuing Drug Discovery: In Vivo Systems Pathology and Systems Pharmacology. *Nat. Rev. Drug Discov.* **2005**, *4*, 961–967. [CrossRef] [PubMed]
39. Vicini, P.; Van der Graaf, P.H. Systems Pharmacology for Drug Discovery and Development: Paradigm Shift or Flash in the Pan? *Clin. Pharmacol. Ther.* **2013**, *93*, 379–381. [CrossRef] [PubMed]
40. Danhof, M.; Klein, K.; Stolk, P.; Aitken, M.; Leufkens, H. The Future of Drug Development: The Paradigm Shift towards Systems Therapeutics. *Drug Discov. Today* **2018**, *23*, 1990–1995. [CrossRef] [PubMed]
41. Sorger, P.K.; Allerheiligen, S.R.B.; Abernethy, D.R.; Altmann, R.B.; Brouwer, K.L.R.; Califano, A.; D'Argenio, D.Z.; Iyengar, R.; Jusko, W.J.; Lalonde, R.; et al. Quantitative and Systems Pharmacology in the Post-Genomic Era: New Approaches to Discovering Drugs and Understanding Therapeutic Mechanisms (White Paper). In *An NIH White Paper by the QSP Workshop Group*; Bethesda: Rockville, MD, USA, 2011; Volume 48.
42. Visser, S.A.G.; Manolis, E.; Danhof, M.; Kerbusch, T. Modeling and Simulation at the Interface of Nonclinical and Early Clinical Drug Development. *CPT Pharmacomet. Syst. Pharmacol.* **2013**, *2*, 8–10. [CrossRef]
43. Nijsen, M.J.; Wu, F.; Bansal, L.; Bradshaw-Pierce, E.; Chan, J.R.; Liederer, B.M.; Mettetal, J.T.; Schroeder, P.; Schuck, E.; Tsai, A.; et al. Preclinical QSP Modeling in the Pharmaceutical Industry: An IQ Consortium Survey Examining the Current Landscape. *CPT Pharmacomet. Syst. Pharmacol.* **2018**, *7*, 135–146. [CrossRef]
44. Schulthess, P.; Van Wijk, R.C.; Krekels, E.H.J.; Yates, J.W.T.; Spaink, H.P.; Van der Graaf, P.H. Outside-in Systems Pharmacology Combines Innovative Computational Methods with High-Throughput Whole Vertebrate Studies. *CPT Pharmacomet. Syst. Pharmacol.* **2018**, *7*, 285–287. [CrossRef]
45. Van der Graaf, P.H. Pharmacometrics and/or Systems Pharmacology. *CPT Pharmacomet. Syst. Pharmacol.* **2019**, *8*, 331–332. [CrossRef]
46. Lombardo, F.; Desai, P.V.; Arimoto, R.; Desino, K.E.; Fischer, H.; Keefer, C.E.; Petersson, C.; Winiwarter, S.; Broccatelli, F. In Silico Absorption, Distribution, Metabolism, Excretion, and Pharmacokinetics (ADME-PK): Utility and Best Practices. An Industry Perspective from the International Consortium for Innovation through Quality in Pharmaceutical Development. *J. Med. Chem.* **2017**, *60*, 9097–9113. [CrossRef]
47. Wu, C.Y.; Benet, L.Z.; Hebert, M.F.; Gupta, S.K.; Rowland, M.; Gomez, D.Y.; Wacher, V.J. Differentiation of Absorption and First-Pass Gut and Hepatic Metabolism in Humans: Studies with Cyclosporine. *Clin. Pharmacol. Ther.* **1995**, *58*, 492–497. [CrossRef]
48. Hellriegel, E.T.; Bjornsson, T.D.; Hauck, W.W. Interpatient Variability in Bioavailability Is Related to the Extent of Absorption: Implications for Bioavailability and Bioequivalence Studies. *Clin. Pharmacol. Ther.* **1996**, *60*, 601–607. [CrossRef]
49. Amidon, G.L.; Lennernäs, H.; Shah, V.P.; Crison, J.R. A Theoretical Basis for a Biopharmaceutic Drug Classification: The Correlation of in Vitro Drug Product Dissolution and in Vivo Bioavailability. *Pharm. Res.* **1995**, *12*, 413–420. [CrossRef] [PubMed]
50. Dahlgren, D.; Roos, C.; Lundqvist, A.; Abrahamsson, B.; Tannergren, C.; Hellström, P.M.; Sjögren, E.; Lennernäs, H. Regional Intestinal Permeability of Three Model Drugs in Human. *Mol. Pharm.* **2016**, *13*, 3013–3021. [CrossRef] [PubMed]
51. Tannergren, C.; Bergendal, A.; Lennernäs, H.; Abrahamsson, B. Toward an Increased Understanding of the Barriers to Colonic Drug Absorption in Humans: Implications for Early Controlled Release Candidate Assessment. *Mol. Pharm.* **2009**, *6*, 60–73. [CrossRef]
52. Shimada, T.; Yamazaki, H.; Mimura, M.; Inui, Y.; Guengerich, F.P. Interindividual Variations in Human Liver Cytochrome P-450 Enzymes Involved in the Oxidation of Drugs, Carcinogens and Toxic Chemicals: Studies with Liver Microsomes of 30 Japanese and 30 Caucasians. *J. Pharmacol. Exp. Ther.* **1994**, *270*, 414–423.
53. Rendic, S. Summary of Information on Human CYP Enzymes: Human P450 Metabolism Data. *Drug Metab. Rev.* **2002**. [CrossRef]

54. Zamek-Gliszczynski, M.J.; Hoffmaster, K.A.; Nezasa, K.; Tallman, M.N.; Brouwer, K.L.R. Integration of Hepatic Drug Transporters and Phase II Metabolizing Enzymes: Mechanisms of Hepatic Excretion of Sulfate, Glucuronide, and Glutathione Metabolites. *Eur. J. Pharm. Sci.* **2006**, *27*, 447–486. [CrossRef]
55. Luo, G.; Johnson, S.; Hsueh, M.M.; Zheng, J.; Cai, H.; Xin, B.; Chong, S.; He, K.; Harper, T.W. In Silico Prediction of Biliary Excretion of Drugs in Rats Based on Physicochemical Properties. *Drug Metab. Dispos.* **2010**, *38*, 422–430. [CrossRef]
56. Roos, C.; Dahlgren, D.; Tannergren, C.; Abrahamsson, B.; Sjögren, E.; Lennernas, H. Regional Intestinal Permeability in Rats: A Comparison of Methods. *Mol. Pharm.* **2017**, *14*, 4252–4261. [CrossRef]
57. Gumbo, T.; Lenaerts, A.J.; Hanna, D.; Romero, K.; Nuermberger, E. Nonclinical Models for Antituberculosis Drug Development: A Landscape Analysis. *J. Infect. Dis.* **2015**, *211*, S83–S95. [CrossRef]
58. Yu, L.X.; Lipka, E.; Crison, J.R.; Amidon, G.L. Transport Approaches to the Biopharmaceutical Design of Oral Drug Delivery Systems: Prediction of Intestinal Absorption. *Adv. Drug Deliv. Rev.* **1996**, *19*, 359–376. [CrossRef]
59. Heimbach, T.; Suarez-Sharp, S.; Kakhi, M.; Holmstock, N.; Olivares-Morales, A.; Pepin, X.; Sjögren, E.; Tsakalozou, E.; Seo, P.; Li, M. Dissolution and Translational Modeling Strategies Toward Establishing an In Vitro-In Vivo Link—A Workshop Summary Report. *AAPS J.* **2019**, *21*, 29. [CrossRef]
60. Wu, C.Y.; Benet, L.Z. Predicting Drug Disposition via Application of BCS: Transport/Absorption/Elimination Interplay and Development of a Biopharmaceutics Drug Disposition Classification System. *Pharm. Res.* **2005**, *22*, 11–23. [CrossRef]
61. Bennion, B.J.; Be, N.A.; McNerney, M.W.; Lao, V.; Carlson, E.M.; Valdez, C.A.; Malfatti, M.A.; Enright, H.A.; Nguyen, T.H.; Lightstone, F.C. Predicting a Drug's Membrane Permeability: A Computational Model Validated with in Vitro Permeability Assay Data. *J. Phys. Chem. B* **2017**, *121*, 5228–5237. [CrossRef]
62. Awoonor-Williams, E.; Rowley, C.N. Molecular Simulation of Nonfacilitated Membrane Permeation. *Biochim. Biophys. Acta Biomembr.* **2016**, *1858*, 1672–1687. [CrossRef]
63. Orsi, M.; Essex, J.W. Permeability of Drugs and Hormones through a Lipid Bilayer: Insights from Dual-Resolution Molecular Dynamics. *Soft Matter* **2010**, *6*, 3797–3808. [CrossRef]
64. Lipinski, C.A.; Lombardo, F.; Dominy, B.W.; Feeney, P.J. Experimental and Computational Approaches to Estimate Solubility and Permeability in Drug Discovery and Development Settings. *Adv. Drug Deliv. Rev.* **1997**, *23*, 3–25. [CrossRef]
65. Bickerton, G.R.; Paolini, G.V.; Besnard, J.; Muresan, S.; Hopkins, A.L. Quantifying the Chemical Beauty of Drugs. *Nat. Chem.* **2012**, *4*, 90. [CrossRef]
66. Krämer, S.D.; Aschmann, H.E.; Hatibovic, M.; Hermann, K.F.; Neuhaus, C.S.; Brunner, C.; Belli, S. When Barriers Ignore the "Rule-of-Five". *Adv. Drug Deliv. Rev.* **2016**, *101*, 62–74. [CrossRef]
67. Bemporad, D.; Luttmann, C.; Essex, J.W. Behaviour of Small Solutes and Large Drugs in a Lipid Bilayer from Computer Simulations. *Biochim. Biophys. Acta Biomembr.* **2005**, *1718*, 1–21. [CrossRef]
68. Kuhn, B.; Mohr, P.; Stahl, M. Intramolecular Hydrogen Bonding in Medicinal Chemistry. *J. Med. Chem.* **2010**, *53*, 2601–2611. [CrossRef] [PubMed]
69. Alex, A.; Millan, D.S.; Perez, M.; Wakenhut, F.; Whitlock, G.A. Intramolecular Hydrogen Bonding to Improve Membrane Permeability and Absorption in beyond Rule of Five Chemical Space. *MedChemComm* **2011**, *2*, 669–674. [CrossRef]
70. Agoram, B.; Woltosz, W.S.; Bolger, M.B. Predicting the Impact of Physiological and Biochemical Processes on Oral Drug Bioavailability. *Adv. Drug Deliv. Rev.* **2001**, *50*, S41–S67. [CrossRef]
71. Rowland, M.; Peck, C.; Tucker, G. Physiologically-Based Pharmacokinetics in Drug Development and Regulatory Science. *Annu. Rev. Pharmacol. Toxicol.* **2011**, *51*, 45–73. [CrossRef]
72. Zhou, Q.; Gallo, J.M. The Pharmacokinetic/Pharmacodynamic Pipeline: Translating Anticancer Drug Pharmacology to the Clinic. *AAPS J.* **2011**, *13*, 111–120. [CrossRef]
73. Woo, J.; Cheung, W.; Chan, R.; Chan, H.S.; Cheng, A.; Chan, K. In Vitro Protein Binding Characteristics of Isoniazid, Rifampicin, and Pyrazinamide to Whole Plasma, Albumin, and α-1-Acid Glycoprotein. *Clin. Biochem.* **1996**, *29*, 175–177. [CrossRef]
74. Lavé, T.; Parrott, N.; Grimm, H.P.; Fleury, A.; Reddy, M. Challenges and Opportunities with Modelling and Simulation in Drug Discovery and Drug Development. *Xenobiotica* **2007**, *37*, 1295–1310. [CrossRef]

75. Flanagan, T.; Van Peer, A.; Lindahl, A. Use of Physiologically Relevant Biopharmaceutics Tools within the Pharmaceutical Industry and in Regulatory Sciences: Where Are We Now and What Are the Gaps? *Eur. J. Pharm. Sci.* **2016**, *91*, 84–90. [CrossRef]
76. Mahmood, I. Application of Allometric Principles for the Prediction of Pharmacokinetics in Human and Veterinary Drug Development. *Adv. Drug Deliv. Rev.* **2007**, *59*, 1177–1192. [CrossRef]
77. Teitelbaum, Z.; Lave, T.; Freijer, J.; Cohen, A.F. Risk Assessment in Extrapolation of Pharmacokinetics from Preclinical Data to Humans. *Clin. Pharmacokinet.* **2010**, *49*, 619–632. [CrossRef]
78. Gumbo, T.; Angulo-Barturen, I.; Ferrer-Bazaga, S. Pharmacokinetic-Pharmacodynamic and Dose-Response Relationships of Antituberculosis Drugs: Recommendations and Standards for Industry and Academia. *J. Infect. Dis.* **2015**, *211*, S96–S106. [CrossRef] [PubMed]
79. Jones, H.M.; Parrott, N.; Jorga, K.; Lavé, T. A Novel Strategy for Physiologically Based Predictions of Human Pharmacokinetics. *Clin. Pharmacokinet.* **2006**, *45*, 511–542. [CrossRef] [PubMed]
80. Pelkonen, O.; Turpeinen, M. In Vitro-in Vivo Extrapolation of Hepatic Clearance: Biological Tools, Scaling Factors, Model Assumptions and Correct Concentrations. *Xenobiotica* **2007**, *37*, 1066–1089. [CrossRef] [PubMed]
81. Houston, J.; Galetin, A. Methods for Predicting In Vivo Pharmacokinetics Using Data from In Vitro Assays. *Curr. Drug Metab.* **2008**. [CrossRef] [PubMed]
82. Lavé, T.; Chapman, K.; Goldsmith, P.; Rowland, M. Human Clearance Prediction: Shifting the Paradigm. *Expert Opin. Drug Metab. Toxicol.* **2009**. [CrossRef]
83. Ito, K.; Houston, J.B. Prediction of Human Drug Clearance from in Vitro and Preclinical Data Using Physiologically Based and Empirical Approaches. *Pharm. Res.* **2005**, *22*, 103–112. [CrossRef]
84. Rostami-Hodjegan, A.; Tucker, G.T. Simulation and Prediction of in Vivo Drug Metabolism in Human Populations from in Vitro Data. *Nat. Rev. Drug Discov.* **2007**, *6*, 140–148. [CrossRef]
85. Aït Moussa, L.; El Bouazzi, O.; Serragui, S.; Soussi Tanani, D.; Soulaymani, A.; Soulaymani, R. Rifampicin and Isoniazid Plasma Concentrations in Relation to Adverse Reactions in Tuberculosis Patients: A Retrospective Analysis. *Ther. Adv. Drug Saf.* **2016**, *7*, 239–247. [CrossRef]
86. Muttil, P.; Wang, C.; Hickey, A.J. Inhaled Drug Delivery for Tuberculosis Therapy. *Pharm. Res.* **2009**, *26*, 2401–2416. [CrossRef]
87. Pandey, R.; Khuller, G.K. Antitubercular Inhaled Therapy: Opportunities, Progress and Challenges. *J. Antimicrob. Chemother.* **2005**, *55*, 430–435. [CrossRef]
88. Patton, J.S.; Byron, P.R. Inhaling Medicines: Delivering Drugs to the Body through the Lungs. *Nat. Rev. Drug Discov.* **2007**, *6*, 67–74. [CrossRef] [PubMed]
89. Bäckman, P.; Arora, S.; Couet, W.; Forbes, B.; de Kruijf, W.; Paudel, A. Advances in Experimental and Mechanistic Computational Models to Understand Pulmonary Exposure to Inhaled Drugs. *Eur. J. Pharm. Sci.* **2018**, *113*, 41–52. [CrossRef] [PubMed]
90. Eriksson, J.; Sjögren, E.; Thörn, H.; Rubin, K.; Bäckman, P.; Lennernäs, H. Pulmonary Absorption-Estimation of Effective Pulmonary Permeability and Tissue Retention of Ten Drugs Using an Ex Vivo Rat Model and Computational Analysis. *Eur. J. Pharm. Biopharm.* **2018**, *124*, 1–12. [CrossRef] [PubMed]
91. Scheuch, G.; Kohlhaeufl, M.J.; Brand, P.; Siekmeier, R. Clinical Perspectives on Pulmonary Systemic and Macromolecular Delivery. *Adv. Drug Deliv. Rev.* **2006**, *58*, 996–1008. [CrossRef]
92. Hittinger, M.; Juntke, J.; Kletting, S.; Schneider-Daum, N.; de Souza Carvalho, C.; Lehr, C.-M. Preclinical Safety and Efficacy Models for Pulmonary Drug Delivery of Antimicrobials with Focus on in Vitro Models. *Adv. Drug Deliv. Rev.* **2015**, *85*, 44–56. [CrossRef]
93. McGonigle, P.; Ruggeri, B. Animal Models of Human Disease: Challenges in Enabling Translation. *Biochem. Pharmacol.* **2014**, *87*, 162–171. [CrossRef]
94. Kjellsson, M.C.; Via, L.E.; Goh, A.; Weiner, D.; Low, K.M.; Kern, S.; Pillai, G.; Barry, C.E.; Dartois, V. Pharmacokinetic Evaluation of the Penetration of Antituberculosis Agents in Rabbit Pulmonary Lesions. *Antimicrob. Agents Chemother.* **2012**, *56*, 446–457. [CrossRef]
95. Dartois, V. The Path of Anti-Tuberculosis Drugs: From Blood to Lesions to Mycobacterial Cells. *Nat. Rev. Microbiol.* **2014**, *12*, 159–167. [CrossRef]
96. Prideaux, B.; Dartois, V.; Staab, D.; Weiner, D.M.; Goh, A.; Via, L.E.; Barry III, C.E.; Stoeckli, M. High-Sensitivity MALDI-MRM-MS Imaging of Moxifloxacin Distribution in Tuberculosis-Infected Rabbit Lungs and Granulomatous Lesions. *Anal. Chem.* **2011**, *83*, 2112–2118. [CrossRef]

97. Franzblau, S.G.; Degroote, M.A.; Cho, S.H.; Andries, K.; Nuermberger, E.; Orme, I.M.; Mdluli, K.; Angulo-Barturen, I.; Dick, T.; Dartois, V.; et al. Comprehensive Analysis of Methods Used for the Evaluation of Compounds against Mycobacterium Tuberculosis. *Tuberculosis* **2012**, *92*, 453–488. [CrossRef]
98. Nuermberger, E.L. Preclinical Efficacy Testing of New Drug Candidates. *Microbiol. Spectr.* **2017**, *5*. [CrossRef]
99. Nielsen, E.I.; Friberg, L.E. Pharmacokinetic-Pharmacodynamic Modeling of Antibacterial Drugs. *Pharmacol. Rev.* **2013**, *65*, 1053–1090. [CrossRef] [PubMed]
100. Mouton, J.W.; Meletiadis, J.; Voss, A.; Turnidge, J. Variation of MIC Measurements: The Contribution of Strain and Laboratory Variability to Measurement Precision. *J. Antimicrob. Chemother.* **2018**, *73*, 2374–2379. [CrossRef] [PubMed]
101. Singh, A.K.; Gupta, U.D. Animal Models of Tuberculosis: Lesson Learnt. *Indian J. Med. Res.* **2018**, *147*, 456–463. [CrossRef] [PubMed]
102. Young, D. Animal Models of Tuberculosis. *Eur. J. Immunol.* **2009**, *39*, 2011–2014. [CrossRef]
103. Myllymäki, H.; Niskanen, M.; Oksanen, K.E.; Rämet, M. Animal Models in Tuberculosis Research-Where Is the Beef? *Expert Opin. Drug Discov.* **2015**, *10*, 871–883. [CrossRef]
104. Meijer, A.H. Protection and Pathology in TB: Learning from the Zebrafish Model. *Semin. Immunopathol.* **2016**, *38*, 261–273. [CrossRef]
105. Carvalho, R.; de Sonneville, J.; Stockhammer, O.W.; Savage, N.D.L.; Veneman, W.J.; Ottenhoff, T.H.M.; Dirks, R.P.; Meijer, A.H.; Spaink, H.P. A High-Throughput Screen for Tuberculosis Progression. *PLoS ONE* **2011**, *6*, e16779. [CrossRef]
106. Kantae, V.; Krekels, E.H.; Ordas, A.; Gonzalez, O.; Van Wijk, R.C.; Harms, A.C.; Racz, P.I.; Van Der Graaf, P.H.; Spaink, H.P.; Hankemeier, T. Pharmacokinetic Modeling of Paracetamol Uptake and Clearance in Zebrafish Larvae: Expanding the Allometric Scale in Vertebrates with Five Orders of Magnitude. *Zebrafish* **2016**, *13*, 504–510. [CrossRef]
107. Van Wijk, R.C.; Krekels, E.H.J.; Kantae, V.; Harms, A.C.; Hankemeier, T.; Van der Graaf, P.H.; Spaink, H.P. Impact of Post-Hatching Maturation on the Pharmacokinetics of Exogenous Compounds in Zebrafish Larvae. *Sci. Rep.* **2018**, *9*, 2149. [CrossRef]
108. Van Wijk, R.C.; Krekels, E.H.J.; Kantae, V.; Ordas, A.; Kreling, T.; Harms, A.C.; Hankemeier, T.; Spaink, H.P.; van der Graaf, P.H. Mechanistic and Quantitative Understanding of Pharmacokinetics in Zebrafish Larvae through Nanoscale Blood Sampling and Metabolite Modelling of Paracetamol. *J. Pharmacol. Exp. Ther.* **2019**, *371*, 15–24. [CrossRef] [PubMed]
109. Van Wijk, R.C.; Krekels, E.H.J.; Hu, W.; Van der Sar, A.M.; Dijkema, S.M.; Van den Berg, D.-J.; Bahi, R.; Liu, J.; Verboom, T.; Verbeek, F.J.; et al. Translational quantitative systems pharmacology; crossing borders between experimental and computational drug development using zebrafish as model organism. *PAGE* **2021**, *29*, 9455.
110. Ordonez, A.A.; Wang, H.; Magombedze, G.; Ruiz-Bedoya, C.A.; Srivastava, S.; Chen, A.; Tucker, E.W.; Urbanowski, M.E.; Pieterse, L.; Fabian Cardozo, E.; et al. Dynamic Imaging in Patients with Tuberculosis Reveals Heterogeneous Drug Exposures in Pulmonary Lesions. *Nat. Med.* **2020**, 1–6. [CrossRef] [PubMed]
111. Committee for Human Medicinal Products (CHMP). Guideline on the Use of Pharmacokinetics and Pharmacodynamics in the Development of Antibacterial Medicinal Products. (EMA/CHMP/594085/2015). *Eur. Med. Agency* **2016**, *44*, 1–21.
112. Nielsen, E.I.; Cars, O.; Friberg, L.E. Pharmacokinetic/Pharmacodynamic (PK/PD) Indices of Antibiotics Predicted by a Semimechanistic PKPD Model: A Step toward Model-Based Dose Optimization. *Antimicrob. Agents Chemother.* **2011**, *55*, 4619–4630. [CrossRef]
113. Khan, D.D.; Friberg, L.E.; Nielsen, E.I. A Pharmacokinetic-Pharmacodynamic (PKPD) Model Based on in Vitro Time-Kill Data Predicts the in Vivo PK/PD Index of Colistin. *J. Antimicrob. Chemother.* **2016**, *71*, 1881–1884. [CrossRef]
114. Pasipanodya, J.G.; Nuermberger, E.; Romero, K.; Hanna, D.; Gumbo, T. Systematic Analysis of Hollow Fiber Model of Tuberculosis Experiments. *Clin. Infect. Dis.* **2015**, *61* (Suppl. 1), S10–S17. [CrossRef]
115. European Medicines Agency. *Qualification Opinion In-Vitro Hollow Fiber System Model of Tuberculosis (HSF-TB)*; European Medicines Agency: London, UK, 2015.
116. Chilukuri, D.; McMaster, O.; Bergman, K.; Colangelo, P.; Snow, K.; Toerner, J.G. The Hollow Fiber System Model in the Nonclinical Evaluation of Antituberculosis Drug Regimens. *Clin. Infect. Dis.* **2015**, *61*, S32–S33. [CrossRef]

117. Magombedze, G.; Pasipanodya, J.G.; Srivastava, S.; Deshpande, D.; Visser, M.E.; Chigutsa, E.; McIlleron, H.; Gumbo, T. Transformation Morphisms and Time-to-Extinction Analysis That Map Therapy Duration from Preclinical Models to Patients with Tuberculosis: Translating from Apples to Oranges. *Clin. Infect. Dis.* **2018**, *67*, S349–S358. [CrossRef]
118. Deshpande, D.; Magombedze, G.; Srivastava, S.; Bendet, P.; Lee, P.S.; Cirrincione, K.N.; Martin, K.R.; Dheda, K.; Gumbo, T. Once-a-Week Tigecycline for the Treatment of Drug-Resistant TB. *J. Antimicrob. Chemother.* **2019**, *74*, 1607–1617. [CrossRef]
119. Srivastava, S.; Deshpande, D.; Magombedze, G.; van Zyl, J.; Cirrincione, K.; Martin, K.; Bendet, P.; Berg, A.; Hanna, D.; Romero, K.; et al. Duration of Pretomanid/Moxifloxacin/Pyrazinamide Therapy Compared with Standard Therapy Based on Time-to-Extinction Mathematics. *J. Antimicrob. Chemother.* **2020**, *75*, 392–399. [CrossRef]
120. Khan, D.D.; Lagerbäck, P.; Cao, S.; Lustig, U.; Nielsen, E.I.; Cars, O.; Hughes, D.; Andersson, D.I.; Friberg, L.E. A Mechanism-Based Pharmacokinetic/Pharmacodynamic Model Allows Prediction of Antibiotic Killing from MIC Values for WT and Mutants. *J. Antimicrob. Chemother.* **2015**, *70*, 3051–3060. [CrossRef] [PubMed]
121. Clewe, O.; Aulin, L.; Hu, Y.; Coates, A.R.M.; Simonsson, U.S.H. A Multistate Tuberculosis Pharmacometric Model: A Framework for Studying Anti-Tubercular Drug Effects in Vitro. *J. Antimicrob. Chemother.* **2016**, *71*, 964–974. [CrossRef] [PubMed]
122. Chen, C.; Ortega, F.; Rullas, J.; Alameda, L.; Angulo-Barturen, I.; Ferrer, S.; Simonsson, U.S. The Multistate Tuberculosis Pharmacometric Model: A Semi-Mechanistic Pharmacokinetic-Pharmacodynamic Model for Studying Drug Effects in an Acute Tuberculosis Mouse Model. *J. Pharmacokinet. Pharmacodyn.* **2017**, *44*, 133–141. [CrossRef] [PubMed]
123. Svensson, R.J.; Simonsson, U.S.H. Application of the Multistate Tuberculosis Pharmacometric Model in Patients with Rifampicin-Treated Pulmonary Tuberculosis. *CPT Pharmacomet. Syst. Pharmacol.* **2016**, *5*, 264–273. [CrossRef]
124. Lalande, L.; Bourguignon, L.; Bihari, S.; Maire, P.; Neely, M.; Jelliffe, R.; Goutelle, S. Population Modeling and Simulation Study of the Pharmacokinetics and Antituberculosis Pharmacodynamics of Isoniazid in Lungs. *Antimicrob. Agents Chemother.* **2015**, *59*, 5181–5189. [CrossRef]
125. European Medicines Agency. *Guideline on the Investigation of Drug Interactions*; European Medicines Agency: Lonon, UK, 2012. [CrossRef]
126. Kuhlmann, J.; Mück, W. Clinical-Pharmacological Strategies to Assess Drug Interaction Potential during Drug Development. *Drug Saf.* **2001**, *24*, 715–725. [CrossRef]
127. Barton, H.A.; Lai, Y.; Goosen, T.C.; Jones, H.M.; El-Kattan, A.F.; Gosset, J.R.; Lin, J.; Varma, M.V. Model-Based Approaches to Predict Drug-Drug Interactions Associated with Hepatic Uptake Transporters: Preclinical, Clinical and Beyond. *Expert Opin. Drug Metab. Toxicol.* **2013**, *9*, 459–472. [CrossRef]
128. Chu, X.; Bleasby, K.; Evers, R. Species Differences in Drug Transporters and Implications for Translating Preclinical Findings to Humans. *Expert Opin. Drug Metab. Toxicol.* **2013**, *9*, 237–252. [CrossRef]
129. McIlleron, H.; Chirehwa, M.T. Current Research toward Optimizing Dosing of First-Line Antituberculosis Treatment. *Expert Rev. Anti. Infect. Ther.* **2019**, *17*, 27–38. [CrossRef]
130. Svensson, E.M.; Acharya, C.; Clauson, B.; Dooley, K.E.; Karlsson, M.O. Pharmacokinetic Interactions for Drugs with a Long Half-Life—Evidence for the Need of Model-Based Analysis. *AAPS J.* **2016**, *18*, 171–179. [CrossRef] [PubMed]
131. Margolskee, A.; Darwich, A.S.; Pepin, X.; Pathak, S.M.; Bolger, M.B.; Aarons, L.; Rostami-Hodjegan, A.; Angstenberger, J.; Graf, F.; Laplanche, L.; et al. IMI-Oral Biopharmaceutics Tools Project-Evaluation of Bottom-up PBPK Prediction Success Part 1: Characterisation of the OrBiTo Database of Compounds. *Eur. J. Pharm. Sci.* **2017**, *96*, 598–609. [CrossRef]
132. Chen, J.; Raymond, K. Roles of Rifampicin in Drug-Drug Interactions: Underlying Molecular Mechanisms Involving the Nuclear Pregnane X Receptor. *Ann. Clin. Microbiol. Antimicrob.* **2006**, *5*. [CrossRef] [PubMed]
133. Chattopadhyay, N.; Kanacher, T.; Kanacher, T.; Frechen, S.; Ligges, S.; Zimmermann, T.; Rottmann, A.; Ploeger, B.; Höchel, J.; Schultze-Mosgau, M.H. CYP3A4-Mediated Effects of Rifampicin on the Pharmacokinetics of Vilaprisan and Its UGT1A1-Mediated Effects on Bilirubin Glucuronidation in Humans. *Br. J. Clin. Pharmacol.* **2018**, *84*, 2857–2866. [CrossRef] [PubMed]

134. Greiner, B.; Eichelbaum, M.; Fritz, P.; Kreichgauer, H.P.; Von Richter, O.; Zundler, J.; Kroemer, H.K. The Role of Intestinal P-Glycoprotein in the Interaction of Digoxin and Rifampin. *J. Clin. Invest.* **1999**, *104*, 147–153. [CrossRef] [PubMed]
135. Mehta, J.; Gandhit, I.S.; Sanet, S.B.; Wamburkart, M.N. Effect of Clofazimine and Dapsone on Rifampicin (Lositril) Pharmacokinetics in Multibacillary and Paucibacillary Leprosy Cases. *Lepr. Rev.* **1986**, *57*, 67–76. [CrossRef]
136. Horita, Y.; Doi, N. Comparative Study of the Effects of Antituberculosis Drugs and Antiretroviral Drugs on Cytochrome P450 3a4 and P-Glycoprotein. *Antimicrob. Agents Chemother.* **2014**, *58*, 3168–3176. [CrossRef]
137. Dannemann, B.; Bakare, N.; De Marez, T.; Lounis, N.; Van Heeswijk, R.P.G.; Meyvisch, P.; Haxaire-Theeuwes, M.; Andries, K.; Everitt, D.; Upton, A. Corrected QT Interval (QTcF) Prolongation in a Phase 2 Open-Label Trial of TMC207 plus Background Regimen as Treatment for MDR-TB: Effect of Co-Administration with Clofazimine. In Proceedings of the Abstract at 52nd Interscience Conference on Antimicrobial Agents and Chemotherapy, San Francisco, CA, USA, 9–12 September 2012.
138. Pym, A.S.; Diacon, A.H.; Tang, S.J.; Conradie, F.; Danilovits, M.; Chuchottaworn, C.; Vasilyeva, I.; Andries, K.; Bakare, N.; De Marez, T.; et al. Bedaquiline in the Treatment of Multidrug- and Extensively Drugresistant Tuberculosis. *Eur. Respir. J.* **2016**, *47*, 564–574. [CrossRef]
139. Iwatsubo, T.; Suzuki, H.; Shimada, N.; Chiba, K.; Ishizaki, T.; Green, C.E.; Tyson, C.A.; Yokoi, T.; Kamataki, T.; Sugiyama, Y. Prediction of in Vivo Hepatic Metabolic Clearance of YM796 from in Vitro Data by Use of Human Liver Microsomes and Recombinant P-450 Isozymes. *J. Pharmacol. Exp. Ther.* **1997**, *282*, 909–919.
140. Zhao, P.; Zhang, L.; Grillo, J.A.; Liu, Q.; Bullock, J.M.; Moon, Y.J.; Song, P.; Brar, S.S.; Madabushi, R.; Wu, T.C.; et al. Applications of Physiologically Based Pharmacokinetic (PBPK) Modeling and Simulation during Regulatory Review. *Clin. Pharmacol. Ther.* **2011**, *89*, 259–267. [CrossRef]
141. Yamashita, F.; Sasa, Y.; Yoshida, S.; Hisaka, A.; Asai, Y.; Kitano, H.; Hashida, M.; Suzuki, H. Modeling of Rifampicin-Induced CYP3A4 Activation Dynamics for the Prediction of Clinical Drug-Drug Interactions from In Vitro Data. *PLoS ONE* **2013**, *8*. [CrossRef] [PubMed]
142. Greco, W.R.; Park, H.S.; Rustum, Y.M. Application of a New Approach for the Quantitation of Drug Synergism to the Combination of Cis-Diamminedichloroplatinum and 1-θ-d-Arabinofuranosylcytosine. *Cancer Res.* **1990**, *50*, 5318–5327. [PubMed]
143. Wicha, S.G.; Chen, C.; Clewe, O.; Simonsson, U.S.H. A General Pharmacodynamic Interaction Model Identifies Perpetrators and Victims in Drug Interactions. *Nat. Commun.* **2017**, *8*. [CrossRef] [PubMed]
144. Clewe, O.; Wicha, S.G.; de Vogel, C.P.; de Steenwinkel, J.E.M.; Simonsson, U.S.H. A Model-Informed Preclinical Approach for Prediction of Clinical Pharmacodynamic Interactions of Anti-TB Drug Combinations. *J. Antimicrob. Chemother.* **2018**, *73*, 437–447. [CrossRef]
145. Chen, C.; Wicha, S.G.; De Knegt, G.J.; Ortega, F.; Alameda, L.; Sousa, V.; De Steenwinkel, J.E.M.; Simonsson, U.S.H. Assessing Pharmacodynamic Interactions in Mice Using the Multistate Tuberculosis Pharmacometric and General Pharmacodynamic Interaction Models. *CPT Pharmacomet. Syst. Pharmacol.* **2017**, *6*, 787–797. [CrossRef]
146. Chen, C.; Wicha, S.G.; Nordgren, R.; Simonsson, U.S.H. Comparisons of Analysis Methods for Assessment of Pharmacodynamic Interactions Including Design Recommendations. *AAPS J.* **2018**, *20*, 1–12. [CrossRef]

 © 2020 by the authors. Licensee MDPI, Basel, Switzerland. This article is an open access article distributed under the terms and conditions of the Creative Commons Attribution (CC BY) license (http://creativecommons.org/licenses/by/4.0/).

Review

Structure-Based Drug Design for Tuberculosis: Challenges Still Ahead

Eduardo M. Bruch, Stéphanie Petrella and Marco Bellinzoni *

Unité de Microbiologie Structurale, Institut Pasteur, CNRS, Université de Paris, F-75015 Paris, France; eduardo.bruch@pasteur.fr (E.M.B.); stephanie.petrella@pasteur.fr (S.P.)
* Correspondence: marco.bellinzoni@pasteur.fr

Received: 30 May 2020; Accepted: 18 June 2020; Published: 20 June 2020

Abstract: Structure-based and computer-aided drug design approaches are commonly considered to have been successful in the fields of cancer and antiviral drug discovery but not as much for antibacterial drug development. The search for novel anti-tuberculosis agents is indeed an emblematic example of this trend. Although huge efforts, by consortiums and groups worldwide, dramatically increased the structural coverage of the *Mycobacterium tuberculosis* proteome, the vast majority of candidate drugs included in clinical trials during the last decade were issued from phenotypic screenings on whole mycobacterial cells. We developed here three selected case studies, i.e., the serine/threonine (Ser/Thr) kinases—protein kinase (Pkn) B and PknG, considered as very promising targets for a long time, and the DNA gyrase of *M. tuberculosis*, a well-known, pharmacologically validated target. We illustrated some of the challenges that rational, target-based drug discovery programs in tuberculosis (TB) still have to face, and, finally, discussed the perspectives opened by the recent, methodological developments in structural biology and integrative techniques.

Keywords: tuberculosis; structure-based drug design; target-based drug design; PknB; PknG; DNA gyrase; antibiotic

1. Introduction

Although the drug discovery process has historically relied on high-throughput screening (HTS) to identify hits to be developed into drug candidates, either on a given target (in most cases, an essential enzyme) or through whole-cell screenings, it still remains an experimentally laborious, expensive, and time-consuming process. Unlike traditional drug discovery, however, drug design is not necessarily based on the screening of large libraries, which is intrinsically a trial and error process, but builds on the available knowledge for a given biological target. A particular, well-known case of drug design is called structure-based drug design (SBDD), which uses the three-dimensional structural knowledge of the target to find or optimize molecules that can bind to the target with high affinity and selectivity. The potential of using structural information for discovering candidate drugs was apparent from the early days of structural biology [1], but it took several years to achieve the first successful examples, i.e., human immunodeficiency virus (HIV) protease inhibitors [2] and carbonic anhydrase inhibitors for the treatment of glaucoma [3].

A three-dimensional model of the target is, therefore, a prerequisite for SBDD. The structure can be obtained experimentally in different ways, in most cases by X-ray crystallography, although nuclear magnetic resonance (NMR) and, more recently, cryo-electron microscopy (cryo-EM) have also attracted attention [4], especially following the spectacular increase in the resolution capabilities of single-particle cryo-EM, which is commonly referred to as the resolution revolution [5,6]. For targets whose experimental structure is elusive, in silico structure prediction is also routinely performed and can be achieved, whenever suitable models are available, through homology modeling from a closely related homologous protein [7].

The identification of potential drug binding sites can be obvious when ligand-bound target structures are available. However, the capability to detect binding sites for substrates or modulators becomes highly relevant not only for targets for which no ligand-bound three-dimensional structure is available, but also to detect allosteric sites or protein–protein interaction surfaces that might be specifically targeted [8], and for these, dedicated libraries are now available [9]. Considering that binding site identification is not always straightforward, tools have been developed to infer "druggable" pockets from the identification of concave regions—that could accommodate drug-size molecules—by screening for appropriate binding properties, such as volume, hydrophobicity, hydrogen bonding, energy potential, solvent accessibility, and desolvation energy [10].

The search for potential ligands on a given target can then follow two different approaches, commonly referred to as ligand-based drug design (LBDD) or target-based drug design (TBDD). LBDD does not require, a priori, direct structural knowledge of the target, but relies on the identification of hits, usually congeneric compounds, with an established biological effect; hit-to-lead optimization can then proceed through the definition of appropriate chemical descriptors of the series, quantitative structure-activity relationship analysis (QSAR), and pharmacophore modeling [11], all followed by experimental validation. QSAR-based virtual screening approaches, alone or in combination with HTS screenings, can also be used to enlarge the panel of bioactive compounds against the target or pathway of interest, increasing the overall hit rate [12]. TBDD, in contrast, uses the physical-chemical constrains from the target three-dimensional structure, and possibly a well-defined binding pocket, to perform virtual screening of libraries, either of natural or synthetic compounds, usually applying appropriate filters like compliance to Lipinski's rules or QSAR models, or to design molecules de novo in a step-wise manner. A particular case of de novo molecule design is fragment-based drug design (FBDD), in which low-affinity target binding molecules are identified from appropriate libraries, using a variety of biophysical approaches, and then merged or linked together using the available three-dimensional information to achieve larger binders with improved properties [13,14].

It is common knowledge in the field that although SBDD approaches have proven to be successful for non-transmissible diseases and viral diseases, they have proven to be much less effective in the antibiotic discovery field [15,16]. Unfortunately, tuberculosis represents no exception to this trend [17], despite the considerable progress achieved during the last twenty years in the understanding of the pathogen molecular physiology, starting from the seminal publication of the *M. tuberculosis* genome sequence in 1998 [18], the extensive structural genomics campaigns during the following years [19–21], the advances in understanding host-pathogen interactions and the development of the disease [22,23], and notwithstanding the development of genetic and biochemical tools to allow the in vivo and in vitro validation of targets [24,25].

The purpose of this review was not to provide an exhaustive overview of the capabilities now offered by in silico approaches for antibiotic development against *M. tuberculosis*, already reviewed elsewhere [26,27], nor to make a survey of the current state of structural knowledge of the pathogen proteome and the experimental structures relevant for drug discovery, for which we point the reader to very recent, extensive work [28]. Rather, we focused here on three emblematic case studies of *M. tuberculosis* targets that attracted most efforts for anti-tuberculosis compound development by HTS campaigns, computer-aided, and structure-driven compound identification: the serine/threonine (Ser/Thr) kinases—protein kinase (Pkn)B and PknG—two amongst the most known, supposedly promising new targets offered by the post-genomic era, and the DNA gyrase, the 'old' but the well-proven target of fluoroquinolones.

2. Protein Ser/Thr Kinases as Drug Targets

Protein phosphorylation is a well-known, widespread mechanism for signal transduction and regulation of several biological functions. Protein kinases have long been known as major drug targets [29], especially for the treatment of cancer, where the deregulation of signaling mechanisms is a hallmark of the disease [30]. Protein kinases are not only pharmaceutical targets in cancer chemotherapy but also for the treatment of parasitic infections, ranging from the ones caused by Trypanosomatids or *Leishmania* [31] to malaria [32]. Since the sequence of *M. tuberculosis* H37Rv genome was first reported in 1998 [18], the pathogen has been known to possess eleven genes coding for Hanks-type Ser/Thr kinases [33], named from *pknA* to *pknL* (but no *pknC*), providing one of the first challenges to the paradigm of prokaryotic cell signaling as being entirely driven by two-component systems. One gene coding for a transmembrane Ser/Thr phosphatase, *pstP*, was also identified as lying on the same cluster as *pknA* and *pknB*, forming a putative operon [34], in addition to two genes (*ptpA* and *ptpB*) coding for Tyr phosphatases [35]. Nine out of eleven Ser/Thr kinases (all but PknG and PknK) were predicted to be integral membrane proteins, all sharing the same topology with an N-terminal catalytic domain in which the archetypal Hanks motifs could be identified [33], a single transmembrane segment and a very variable extracellular, C-terminal domain, hypothesized to be acting as a signal sensing domain. The role of such C-terminal domains is, in most cases, still puzzling, although, in some kinases like PknB, it has been suggested to be involved in the kinase activation process, possibly by controlling the kinase oligomerization state as a function of the external signal [36]. Work from Tom Alber's group in Berkeley indeed demonstrated the role of the 'back-to-back' dimerization to promote an active kinase conformation, not only for PknB [37,38] but also for PknD [39], notwithstanding the observation that mutations in the dimerization interface do not abolish kinase activity, as reported by separate groups [37,40]. PknB was also the first *M. tuberculosis* kinase for which the crystal structure of the catalytic domain was described in 2003 [41,42]. Indeed, this first crystal structure confirmed the overall conservation of the bi-lobed protein kinase fold and the initial assumptions about the presence of the known structural features of eukaryotic protein kinases, as predicted by the detection of the Hanks motifs [42], thus underlining the common origin of eukaryotic and prokaryotic kinases. Following these first milestones and the genetic proof of the in vitro essentiality of *pknB* coming both from transposon mutagenesis [43] and targeted studies [44,45], PknB then attracted the attention of tuberculosis (TB) research community as an ideal target for structure-based drug design. The community's interest rise when considering the increasing evidence, over the years, of its role in the control of peptidoglycan synthesis and cell division (recently reviewed in [46]). The first description of the potential use of PknB inhibitors as anti-mycobacterial agents reported the compound H-7 (1-(5-isoquinolinesulfonyl)-2-methylpiperazine; Table 1), well before any bacterial protein kinase structure was available [47]. Once the PknB catalytic domain structure was solved, virtual screening carried out on a library of about 40,000 compounds for hits into the PknB adenosine triphosphate-binding (ATP-binding) pocket was performed. The screening led to identifying mitoxantrone, a chemotherapeutic agent known for its DNA intercalating properties, as a sub-micromolar PknB inhibitor [48], similar to staurosporine, K-252-a, and K-252-b (Table 1), who were identified as hits after testing a few commercially available eukaryotic kinase inhibitors [44]. The crystal structure of the PknB-mitoxantrone complex was the first showing the kinase catalytic domain in complex with a non-ATP analog, kinase inhibitor. Two crucial, hydrogen bonding interactions were evidenced between the mitoxantrone hydroxyl groups and main chain atoms from the PknB hinge region that connects the two kinase lobes, i.e., the carbonyl oxygen of Glu93 and the amino group of Val95 [48], opening the way to compound optimization (Figure 1A,B). Several other PknB inhibitors were proposed in the course of the following years, starting from hits developed either from known kinase inhibitors like staurosporine analogs [37], 2-aminopurine and its derivatives, including organometallic compounds [49]), or from hits obtained from HTS on public or proprietary libraries using GarA as the substrate [50–53], or phytocompounds [54] (Table 1). Most of the published work deals with compounds inhibiting the kinase in the micro and sub-micromolar range but with limited activity on mycobacteria. The exception to this trend is IMB-YH-8 (Table 1), a compound that, despite showing an IC_{50} (half maximal inhibitory concentration) on PknB in the 20 μM range, shows good selectivity for mycobacterial PknB and PknA and a MIC (minimal inhibitory concentration) in the sub-micromolar range [53]. Compounds highly active on both PknB and PknA have also been reported recently by others in the form of substituted quinazolines, and, for a compound derived from this series (a pyrimidine analog), the crystal structures of the respective complexes with both kinases have been described [55] (Figure 1, Table 1). Both structures underline the common binding mode in the ATP pocket and the crucial interaction with the hinge region (Figure 1B). Despite the inhibition constant for both kinases falls in the nanomolar range, these dual-targeting inhibitors display promising but yet limited antibacterial effect with MIC on *M. tuberculosis* in the lower micromolar range [55]. A different PknB inhibitor for which a crystal structure in complex with the target is available is GSK690693 (Figure 1C, Table 1), also identified through a virtual screening approach on known kinase inhibitors [56]. This compound, member of the imidazopyridine aminofurazans class, also displays conserved features in its binding mode to the PknB hinge region (Figure 1B), and sub-micromolar affinity to the kinase, but no significant antimycobacterial activity on *Mycobacterium smegmatis* or *Mycobacterium bovis* BCG (bacillus Calmette-Guerin). However, the MIC is significantly lowered if the compound is associated with a sub-MIC50 concentration of meropenem, suggesting a synergistic action between PknB inhibitors and β-lactams [56].

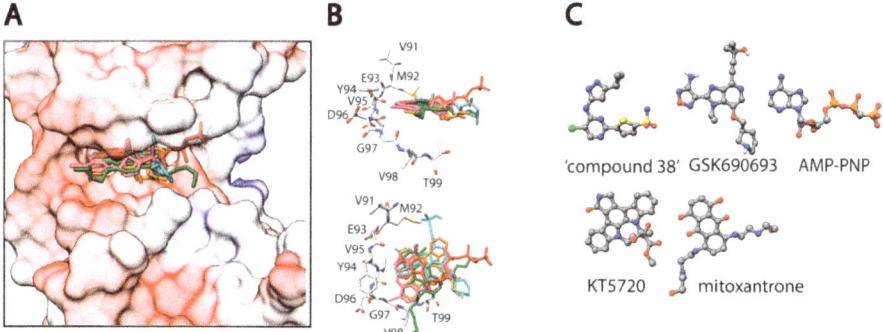

Figure 1. Experimental structures of protein kinase B (PknB)-inhibitor complexes. (**A**) Electrostatic surface from the X-ray structure of PknB in complex with AMP-PCP (β,γ-methyleneadenosine 5′-triphosphate) in red (pdb: 1O6Y), superimposed with the coordinates of four other PknB complexes with, respectively, mitoxantrone in green (pdb: 2FUM), KT5720 in orange (pdb: 3F69), GSK690693 in cyan (pdb: 5U94), and 'compound 38' in pink (pdb: 6B2P). This compound has been reported as active on both PknA and PknB [55]. (**B**) Side and top view of the superimposed ligands with the hinge region from the pdb entry 6B2P. Kinase residues are indicated. (**C**) Ball-and-stick representation of the ligands kept in the same relative binding orientation as in (**B**) and colored by element.

Table 1. *M. tuberculosis* PknB inhibitors.

Family	Name	Structure	IC$_{50}$ (µM)	Reference
Isoquinolines	H-7		ND	[47]

Table 1. Cont.

Family	Name	Structure	IC$_{50}$ (µM)	Reference
Staurosporine analogues	Staurosporine		0.6	[44]
	K-252-a		0.096	
	K-252-b		0.106	
	KT5720		~1 [a]	[37]
Anthracenediones	Mitoxantrone		0.8	[48]
Aminopurines	2-A9P		1300	[49]

Table 1. Cont.

Family	Name	Structure	IC$_{50}$ (μM)	Reference
Quinazolines	Disubstituted series		≤ 1.1	[51]
Pyrimidines	Disubstituted (-R1 also as -NHR1)		≤ 0.4	[51]
	'Compound 38'		Ki ~ 1 nM	[55]
Phytocompounds	Demethylcalabaxanthone		ND	[54]
4-oxo-crotonic acid derivatives	IMB-YH-8		20.2	[53]
Imidazopyridine aminofurazans	GSK690693		0.34	[56]

ND: not determined. [a] measured on a PknB mutant (M145L, M155V) [37].

On the other hand, another *M. tuberculosis* kinase that attracted as much attention for its potential druggability is PknG, a soluble Ser/Thr kinase that was described, by Jean Pieters and coworkers, as a virulence factor secreted into the human macrophage, where it would inhibit the phagosome-lysosome fusion [57]. Most notably, in the same work, the authors showed that chemical inhibition of PknG by the compound AX20017 (a tetrahydrobenzothiophene identified by HTS and found to inhibit PknG with IC_{50} in the sub-micromolar range; Table 2) led to the accumulation of *M. tuberculosis* inside lysosomes [57]. Noteworthy, the 2.4 Å resolution crystal structure of an N-terminal truncated form of PknG in complex with AX20017, published later [58], shows a great similarity between the AX20017 binding to PknG and the binding of ATP and mitoxantrone to PknB. Not only the compound occupies the adenine binding pocket but also makes similar interactions with the main chain atoms of residues Glu233 and Val235 (Figure 1). It was only a few years later that further structural work allowed to elucidate the binding mode of ATP to PknG and suggested a regulatory role of the rubredoxin-like domain by a reversible occlusion of the active site entrance [59]. These initial findings paved the way to further structure-based lead optimization and undoubtedly generated excitement for the potential chemical targeting of this kinase, which would allow preventing the arrest of phagosome maturation, directing *M. tuberculosis* to lysosomes. Attempts to develop more potent compounds starting from AX20017 yielded sub-micromolar inhibitors, some of which had no activity in macrophage assays [50], and no other structure of a PknG-inhibitor complex has so far been reported. It is, however, worth noting that, more than ten years later, the role of PknG in arresting the phagosome maturation is still elusive. Although progress has been made in identifying the SecA2 system as responsible for the export of PknG outside *M. tuberculosis* [60,61], and interference by PknG on the host Rab7l1 signaling pathway has been reported [62], it is largely accepted that several mechanisms, and not a single virulence factor, contribute to the *M. tuberculosis* capability to escape the phagocytic route and survive into macrophages [22,63]. Moreover, an increasing amount of evidence has since validated PknG as a key signaling element in the control of central metabolism, as first shown in *Corynebacterium glutamicum* [64,65], then in mycobacteria [66,67], where genetic and metabolomic evidence point to a role of PknG in regulating the 2-oxoglutarate node according to nutrient availability [68]. Alternative roles of PknG in biofilm formation, adaptation to oxidative stress [69], and hypoxia [70] have also been proposed. Therefore, the observed effects of the lack of PknG on the phagosome-lysosome fusion, and the decreased viability of a *pknG*-deprived *M. tuberculosis* mutant, both in vitro and in the mouse model [71], may not necessarily indicate a direct interference of the kinase on the host signaling pathways but could be ascribed to metabolic alterations caused by the depletion of PknG [68,72]. During the last years, further search for PknG inhibitors has allowed identifying sclerotiorin, an azaphilone derivative isolated from *Penicillium* sp. ZJ27 (Table 2), a known inhibitor of other unrelated enzymes like lipoxygenase, that, despite showing an IC_{50} on PknG around 76 µM and no inhibition on *M. tuberculosis* growth in vitro, displays capability to partially reduce the growth of *Mycobacterium bovis* BCG inside macrophages, and to enhance the effect of rifampicin [73]. On the other hand, the recently described PknG inhibitor NU-6027 (2,6-diamino-4-cyclohexylmethoxy-5-nitrosopyrimidine; Table 2) was identified from phenotypic screening as having an MIC99 value of 1.56 µM on *M. bovis* BCG and shown to partially inhibit PknG and PknD autophosphorylation in vitro [74]. Another series of four compounds, all inhibiting PknG in the micromolar range, was also recently reported after the screening of an 80-compound, commercially available kinase inhibitor library (Table 2); three of them were found to promote the transfer of mycobacteria to lysosomes, and two to inhibit *M. bovis* BCG growth in macrophages [75]. None of the reported compounds, however, issued from a structure-based screening approach.

Table 2. *M. tuberculosis* PknG inhibitors.

Family	Name	Structure	IC$_{50}$ (µM)	Reference
Tetrahydrobenzothiophenes	AX20017		0.39	[57]
			5.5	[75]
Azaphilones	Sclerotiorin		76	[73]
Pyrimidines	NU-6027		ND	[74]
	CYC116		35.1	[75]
Thiophenes	AZD7762		30.3	[75]
Methoxybenzenes	R406 (benzenesulfonate) and its free base R406f (also known as Tamatinib)		8.0 (R406) 16.1 (R406f)	[75]

ND: not determined.

3. DNA Gyrase: A Validated Drug Target

Topoisomerases are enzymes that participate in the overwinding or underwinding of DNA. They are commonly split into two classes, depending on the number of DNA strands cut during a single catalysis cycle: class I topoisomerases, which include eukaryotic topoisomerase I, prokaryotic topoisomerase IIIα, or reverse gyrase among others, and class II topoisomerases, whose members include (but are not restricted to) topoisomerase IV and DNA gyrase [76]. In *M. tuberculosis*, DNA gyrase is a heterotetrameric nanomachine, consisting of two GyrB and two GyrA subunits, and is essential for DNA replication, transcription, and repair in living cells [77].

M. tuberculosis is an exception in the prokaryotic world because of the presence of only one type I and one type II topoisomerase, whereas most other eubacteria have two enzymes of each type. In the canonical situation, DNA gyrase and topoisomerase IV, both type IIA topoisomerases, have distinct specific activities: DNA gyrase removes positive supercoils that accumulate ahead of replication forks [78], whereas topoisomerase IV decatenates replication intermediates [79]. Thus, in *M. tuberculosis*, DNA gyrase being the unique topoisomerase IIA, it must be able to carry out both activities in vivo [80]. Furthermore, with no other topoisomerase IIA in *M. tuberculosis*, the DNA gyrase is the unique target of antibiotics from the fluoroquinolones (FQ) family (quinolones with higher cellular penetration efficiency), which bind and stabilize the DNA–DNA gyrase complex. Stabilization of this ternary complex leads to the cytotoxic accumulation of double-strand cleaved DNA fragments within the cell, inducing bacterial death [81]. Crystallographic structures of the isolated domain and the cleavage core, either in its *apo* form or co-crystallized with several FQ, brought crucial information to elucidate the mode of binding of these drugs [82–87]. The ratio FQ/DNA–DNA gyrase in the ternary complex is two FQ molecules per complex, with one molecule binding in each cleavage site, wedging between the two ends of the cleaved DNA strands on both halves of the complex, in a drug-binding pocket whose walls are formed by DNA base stacking (Figure 2A) [82]. The emergence of strains with multidrug resistance (MDR) phenotype, or extensively-drug resistance (XDR) phenotype in which drug resistance includes this family of antibiotics, has made improving the activity of this family of drugs by using structural data to be of utmost importance. The main goals of these studies are to increase specific drug/enzyme interactions, but also to find new molecules that inhibit this validated target without cross-resistance with FQ [88].

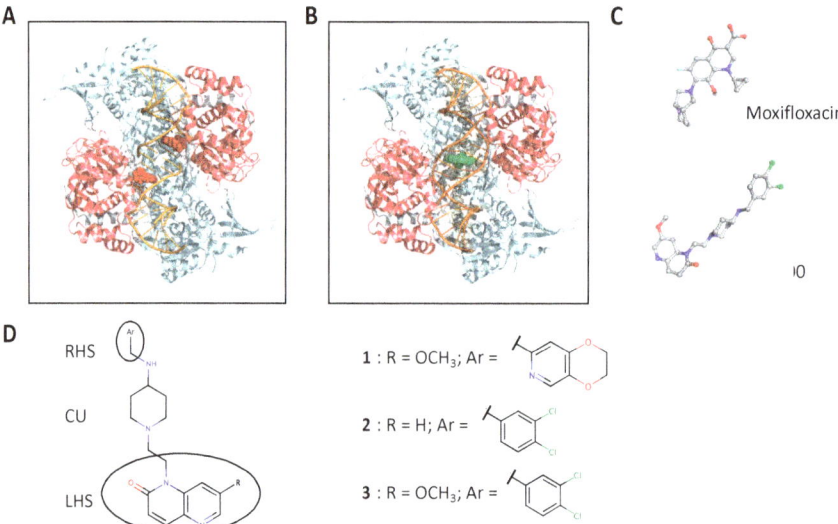

Figure 2. Inhibitors of DNA gyrase. (**A**) Top view of the X-ray structure of *M. tuberculosis* gyrase cleavage core in complex with DNA and moxifloxacin. Protein is shown in cartoon representation with transparent surfaces, with GyrB in blue and GyrA in pink. DNA is shown in orange, with moxifloxacin as red spheres (pdb: 5BS8). (**B**) Top view of the model of a complex of GSK000 (green spheres) with *M. tuberculosis* gyrase cleavage core and uncleaved DNA (in orange), based on the crystal structure of GSK299423 with *Staphylococcus aureus* gyrase (pdb: 2XCR). The same representation, as in panel A, was used for the protein. (**C**) Ball-and-stick representation of the ligands colored by element. (**D**) Generic chemical structure of the novel bacterial topoisomerase inhibitor (NBTI) series of compounds. Analog 1 is a classical NBTI, whereas analogs 2 and 3 combine the three favorable features for *M. tuberculosis* DNA gyrase inhibitor (MGI).

A significant amount of effort has thus been invested in discovering new *M. tuberculosis* gyrase inhibitors, mostly centered on the ATP- and DNA-binding sites, and particularly using in silico methods based on the emerging gyrase structural information, as extensively reviewed by Nagaraja and coworkers [89]. Although the ATP binding sites have been less successfully exploited as antibacterial targets, with the exception of the natural products, such as coumarins [90,91], it is worth noting that a candidate drug emerged from these studies. SPR720 (Table 3) is an orally bioavailable prodrug of SPR719, an aminobenzimidazole inhibitor of both gyrase and topoisomerase IV. Indeed, this novel class of antimicrobials targets the ATPase subunits of gyrase and topoisomerase IV, and it was optimized using SSBD and structure-activity relationship (SAR) studies of potency against both Gram-positive and some Gram-negative bacterial species [92,93]. Crystal structures of novobiocin (Table 3) bound to *Escherichia coli* GyrB subunit served as a starting point for ligand optimization [94], while further optimization of the metabolic profile led to the identification of SPR720 (formerly VXc-486) [95], whose suitability to be a drug candidate for the treatment of tuberculosis and non-tuberculosis mycobacterial infections was thoroughly evaluated [96,97]. The compound completed Phase I clinical trials in 2019 (study ID NCT03796910, sponsored by Spero Therapeutics, LLC), aimed at evaluating the safety, tolerability, and pharmacokinetics (PK) profile in healthy volunteers. Moreover, an in vivo combination of SPR720 with rifampicin and pyrazinamide (two first-line drugs used in TB treatment) showed comparable efficacy to the combination of three drugs, including moxifloxacin, rifampicin, and pyrazinamide [98].

Work on the DNA-binding site has also been productive, as GlaxoSmithKline (GSK) created a new class of type IIA topoisomerase inhibitors, called novel bacterial topoisomerase inhibitors (NBTIs) (Table 3) [99]. The crystallographic structures of the DNA-bound *Staphylococcus aureus* DNA gyrase with either NBTI or FQ, all obtained at high resolution, revealed that the binding sites for each drug are different [100]. Indeed, the NBTI 'bridges' the DNA and a transient non-catalytic pocket on the two-fold axis at the GyrA dimer interface, remaining close to the active sites and FQs binding sites (Figure 2B). As NBTIs display relatively poor activity against the *M. tuberculosis* DNA gyrase, the antitubercular profiles of 3000 compounds, representative of the chemical diversity of this family, were evaluated by high-throughput phenotypic screenings in vitro and in vivo [101]. MIC determination on *M. tuberculosis* H37Rv showed a high hit rate (68% of compounds with MIC values lower than 10 µM), with the most potent derivatives matching or even improving the MIC values for currently used TB drugs, including last-generation FQs. Overall, 29% of the compounds had MICs of < 1 µM, and 18% had MICs of < 0.1 µM. By using structural data provided by the work of Bax et al., the general structures of these TB-active compounds were divided into three different regions, each of them interacting with one of the three topologically important target-gyrase interacting regions [100]. While a left-hand side (LHS) is responsible for key contacts with the gyrase DNA substrate, a right-hand side (RHS) is embedded into the enzyme, potentially contributing to the protein target selectivity (Figure 2D). Last, a central linker unit (CU) establishes key interactions with the gyrase and offers the opportunity to modulate key physicochemical properties (Figure 2D). These SAR observations helped to schematize synthesis and to rationalize how to balance antimycobacterial potency with oral exposure, safety, and synthetic complexity, leading to the identification of the 7-substituted-1,5-naphthyridin-2-one core as a privileged LHS binder, the N-ethyl-4-aminopiperidines as a linker, and monocyclic aromatic rings with different substitution patterns as the best RHS binding option, in contrast to NBTIs bearing bicyclic rings as the RHS (Figure 2D, analogs 2 and 3) [101]. This work led to the identification of a subclass of naphthytidone/aminopiperidine-containing compounds that displayed activity against *M. tuberculosis*, both in vitro and in the mouse model, known as '*M. tuberculosis* DNA gyrase inhibitors' (MGIs) due to structural and activity differences with respect to NBTIs [101]. More recently, the mechanism of action of two such compounds, i.e., GSK000 and GSK325 (Figure 2C; Table 3), was assessed on the *M. tuberculosis* gyrase, showing that MGIs greatly enhanced DNA cleavage mediated by the bacterial enzyme, but they induced only single-stranded DNA breaks [102]. Their mechanism of action involves stabilizing covalent gyrase-cleaved DNA complexes and appears to suppress the ability of the enzyme to induce double-stranded breaks. Furthermore, these compounds maintained activity against mutant versions of DNA gyrase, bearing the three most commonly observed FQs resistance mutations, but displayed no activity against human topoisomerase IIα [102], suggesting good potential as candidate drugs, especially in the presence of FQ resistance. Modeling studies were carried out by the authors using the crystal structure of the *M. tuberculosis* gyrase cleavage core and the NBTI crystal structure complex of *S. aureus* gyrase [100] (Figure 2B). Most importantly, these studies allowed to confirm that FQs and MGIs do not share the same binding site and bind in a mutually exclusive manner (Figure 2), considerably reducing the risk of developing cross-resistance phenotypes.

Table 3. *M. tuberculosis* DNA gyrase inhibitors.

Family	Name	Structure	IC$_{50}$ (µM)	Reference
Fluoroquinolones	Moxifloxacin		1 2.5	[102] [101]
SPR	SPR720 (VXc-100, VXc-486)		ND	[92,93]
Coumarins	Novobiocin		0.5	[80]
NBTI	GSK000		0.5 5.4	[102] [101]
	GSK325		10.98	[101]

ND: not determined.

4. General Remarks

Several challenges need to be overcome for therapeutic molecules to be active against *M. tuberculosis*. Due to its metabolic adaptability and its known capabilities to occupy different niches inside the human host, from free aerobic bacteria to granulomas, an ideal molecule should be able to act against bacteria in very different, environmentally adapted states. The unique structure and composition of the *M. tuberculosis* cell wall are also well known to act as a barrier for potentially active compounds. To be active against latent and active infections, an ideal drug candidate should thus be able to reach bacteria in all the tubercular lesions and niches inside the human host [103], permeate the mycobacterial cell wall or be actively internalized, and deliver the desired bioactivity in bacteria under different metabolic states. To overcome these limitations and reduce the risk of further resistance development, TB drugs are delivered in combinations, starting from the standard DOTS (directly observed therapy short-course). Drug combinations, however, have to take in due account the bioavailability and pharmacokinetic properties of the single components and their associations, as extensively treated in this same journal issue [104].

In the course of the last ten years, it has become increasingly clear that some of the criteria commonly used to classify a given *M. tuberculosis* biological process, in many cases, an enzymatic step in a pathway, as a 'good' target for drug development, needed to be revised. This is particularly true for in vitro genetic essentiality. Even targets confirmed to be essential by the construction of specific conditional mutants do not necessarily show a good 'druggability', even when highly potent hits become available. The case of PknB is an emblematic example: despite the gene was early shown to be essential for *M. tuberculosis* growth, first by transposon mutagenesis [43], then through attempts to generate a knock-out mutant [44], yet several research groups, including ours, had to face the problem of the lack of correlation between hit potency in vitro and efficacy in vivo [50–52]. Similar issues were also reported for other *M. tuberculosis* kinase inhibitors, in addition to the cross-reactivity sometimes reported towards eukaryotic kinases, and the cytotoxicity shown by some compounds on human cell lines [105]. In addition, although the availability of crystal structures of both PknB and PknG in complex with hits issued from medium-throughput screens (mitoxantrone and AX20017, respectively) has looked as a promising 'proof-of-concept' [48,58], further structure-based work on these kinases has, so far, failed to produce suitable drug candidates. Many reasons might be evoked to explain the lack of *M. tuberculosis* kinase inhibitors that display significant antibacterial activity, ranging from the poor capacity to penetrate the mycobacterial cell and reach their target to the kinase redundancy in *M. tuberculosis* and their, relatively poor, substrate specificity [46], which prompted groups to seek for multiple kinase inhibition [105]. Yet, the main causes should perhaps be looked for in our, still limited, knowledge of mycobacterial molecular physiology and the regulatory networks in *M. tuberculosis*, in which Ser/Thr kinases are actors of a complex interplay [46,106]. Indeed, the activation mechanisms of both PknB and PknG are still a matter of speculation, and changes to *pknB* expression, either as depletion or overexpression, have been reported to alter the bacterial growth significantly [34]. Given the known *M. tuberculosis* metabolic plasticity, whose complexity has only recently started to be elucidated, thanks to the development of genetic and 'omics' tools [107], and the adaptability of the pathogen in the course of infection, it is now largely accepted that genetic essentiality of a putative target in laboratory conditions is not necessarily an indication of chemical vulnerability [15], and even more in clinical conditions. For these reasons, phenotypic screens have increasingly been employed in TB drug development to the detriment of target and structure-based methods, including for lead optimization [15], and the vast majority of candidate TB drugs that have been able to enter clinical trials in the last years were issued from this kind of screens [17,108].

Nevertheless, structure-based and computer-aided drug design maintain a clear potential for the future development of new anti-tubercular drug candidates. For instance, a recent, very promising success of structure-based and fragment-based approaches in TB drug discovery is the development of the so-called ethionamide boosters directed against the EthR repressor, one of which has been shown to be active both in vitro and in vivo [109]. In addition, the recent developments in X-ray crystallography [110], cryo-EM [4], and integrative structural biology methods [111] will all contribute to increasing the number of tools available to tackle the challenges that lay ahead. These opportunities are well exemplified by the case of mycobacterial DNA gyrase that, considering our capacity to obtain high-resolution structures of an almost full-length form of the *M. tuberculosis* enzyme [112], and the recent, high-resolution cryo-EM structure of *E. coli* gyrase in complex with NBTI [113], let us believe that SBDD will deliver a key contribution to developing new compounds against this 'old' but validated target. High-resolution snapshots of the complete mycobacterial gyrase machinery, especially if in complex with representative members of each known family of inhibitors, might revolutionize our knowledge of this key target and substantially increase chances of improving our therapeutic bullets. For instance, combining crystallographic and cryo-EM data could allow to perform structure-guided drug design to target these flexible complexes and identify new conformations of mycobacterial gyrase that could not, otherwise, be obtained by conventional structural methods. More generally, the integration of biophysical and structural biology data, with the notable contribution of high-resolution EM, will allow the drug discovery pipelines to work on a higher complexity level that was previously not achievable, now looking at targets in their larger biological context (e.g., complexes or cellular compartments). It is, therefore, foreseeable that, despite the technical challenges, target-based and structure-based approaches will have increasing relevance in future drug discovery and will give significant contributions in the search for new tuberculosis drugs.

Author Contributions: All authors contributed to conceptualization, the preparation of the draft, and the review of this work. All authors have read and agreed to the published version of the manuscript.

Funding: This work was partially supported by institutional grants from the Institut Pasteur and the CNRS.

Acknowledgments: We are grateful to all our past and present colleagues from the Structural Microbiology Unit, as well as collaborators from the former MM4TB Consortium, for so many fruitful exchanges and insightful discussions.

Conflicts of Interest: The authors declare no conflict of interest.

References

1. Perutz, M.F. Fundamental research in molecular biology: Relevance to medicine. *Nature* **1976**, *262*, 449–453. [CrossRef] [PubMed]
2. Wlodawer, A. Rational approach to AIDS drug design through structural biology. *Annu. Rev. Med.* **2002**, *53*, 595–614. [CrossRef] [PubMed]
3. Greer, J.; Erickson, J.W.; Baldwin, J.J.; Varney, M.D. Application of the three-dimensional structures of protein target molecules in structure-based drug design. *J. Med. Chem.* **1994**, *37*, 1035–1054. [CrossRef] [PubMed]
4. Renaud, J.-P.; Chari, A.; Ciferri, C.; Liu, W.-T.; Rémigy, H.-W.; Stark, H.; Wiesmann, C. Cryo-EM in drug discovery: Achievements, limitations and prospects. *Nat. Rev. Drug. Discov.* **2018**, *17*, 471–492. [CrossRef] [PubMed]
5. Kuehlbrandt, W. The Resolution Revolution. *Science* **2014**, *343*, 1443–1444. [CrossRef]
6. Mitra, A.K. Visualization of biological macromolecules at near-atomic resolution: Cryo-electron microscopy comes of age. *Acta Crystallogr. F Struct. Biol. Commun.* **2019**, *75*, 3–11. [CrossRef]
7. Muhammed, M.T.; Aki-Yalcin, E. Homology modeling in drug discovery: Overview, current applications, and future perspectives. *Chem. Biol. Drug. Des.* **2019**, *93*, 12–20. [CrossRef]
8. Kuenemann, M.A.; Sperandio, O.; Labbé, C.M.; Lagorce, D.; Miteva, M.A.; Villoutreix, B.O. In silico design of low molecular weight protein-protein interaction inhibitors: Overall concept and recent advances. *Prog. Biophys. Mol. Biol.* **2015**, *119*, 20–32. [CrossRef]
9. Bosc, N.; Muller, C.; Hoffer, L.; Lagorce, D.; Bourg, S.; Derviaux, C.; Gourdel, M.-E.; Rain, J.-C.; Miller, T.W.; Villoutreix, B.O.; et al. Fr-PPIChem: An academic compound library dedicated to protein–protein interactions. *ACS Chem. Biol.* **2020**. [CrossRef]
10. Yuan, Y.; Pei, J.; Lai, L. Binding site detection and druggability prediction of protein targets for structure-based drug design. *Curr. Pharm. Des.* **2013**, *19*, 2326–2333. [CrossRef]

11. Acharya, C.; Coop, A.; Polli, J.E.; Mackerell, A.D. Recent advances in ligand-based drug design: Relevance and utility of the conformationally sampled pharmacophore approach. *Curr. Comput. Aided. Drug. Des.* **2011**, *7*, 10–22. [CrossRef] [PubMed]
12. Neves, B.J.; Braga, R.C.; Melo-Filho, C.C.; Moreira-Filho, J.T.; Muratov, E.N.; Andrade, C.H. QSAR-based virtual screening: Advances and applications in drug discovery. *Front. Pharmacol.* **2018**, *9*, 1275. [CrossRef] [PubMed]
13. Davis, B.J.; Hubbard, R.E. Fragment-Based Ligand Discovery. In *Structural Biology in Drug Discovery Methods, Techniques, and Practices*; Renaud, J.-P., Ed.; Wiley: Hoboken, NJ, USA, 2020; Volume 10, pp. 79–98.
14. Mendes, V.; Blundell, T.L. Targeting tuberculosis using structure-guided fragment-based drug design. *Drug Discov. Today* **2017**, *22*, 546–554. [CrossRef] [PubMed]
15. Kana, B.D.; Karakousis, P.C.; Parish, T.; Dick, T. Future target-based drug discovery for tuberculosis? *Tuberculosis* **2014**, *94*, 551–556. [CrossRef] [PubMed]
16. Payne, D.J.; Gwynn, M.N.; Holmes, D.J.; Pompliano, D.L. Drugs for bad bugs: Confronting the challenges of antibacterial discovery. *Nat. Rev. Drug Discov.* **2007**, *6*, 29–40. [CrossRef] [PubMed]
17. Koul, A.; Arnoult, E.; Lounis, N.; Guillemont, J.; Andries, K. The challenge of new drug discovery for tuberculosis. *Nature* **2011**, *469*, 483–490. [CrossRef]
18. Cole, S.T.; Brosch, R.; Parkhill, J.; Garnier, T.; Churcher, C.; Harris, D.; Gordon, S.V.; Eiglmeier, K.; Gas, S.; Barry, C.E.; et al. Deciphering the biology of *Mycobacterium tuberculosis* from the complete genome sequence. *Nature* **1998**, *393*, 537–544. [CrossRef]
19. Holton, S.J.; Weiss, M.S.; Tucker, P.A.; Wilmanns, M. Structure-based approaches to drug discovery against tuberculosis. *Curr. Protein Pept. Sci.* **2007**, *8*, 365–375. [CrossRef]
20. Terwilliger, T.C.; Park, M.S.; Waldo, G.S.; Berendzen, J.; Hung, L.W.; Kim, C.Y.; Smith, C.V.; Sacchettini, J.C.; Bellinzoni, M.; Bossi, R.; et al. The TB structural genomics consortium: A resource for *Mycobacterium tuberculosis* biology. *Tuberculosis* **2003**, *83*, 223–249. [CrossRef]
21. Ehebauer, M.T.; Wilmanns, M. The progress made in determining the *Mycobacterium tuberculosis* structural proteome. *Proteomics* **2011**, *11*, 3128–3133. [CrossRef]
22. Huang, L.; Nazarova, E.V.; Russell, D.G. *Mycobacterium tuberculosis*: Bacterial Fitness within the Host Macrophage. *Microbiol. Spectr.* **2019**, *7*, PMC6459685. [CrossRef] [PubMed]
23. Sia, J.K.; Rengarajan, J. Immunology of *Mycobacterium tuberculosis* Infections. *Microbiol. Spectr.* **2019**, *7*, PMC6636855. [CrossRef] [PubMed]
24. Woong Park, S.; Klotzsche, M.; Wilson, D.J.; Boshoff, H.I.; Eoh, H.; Manjunatha, U.; Blumenthal, A.; Rhee, K.; Barry, C.E.; Aldrich, C.C.; et al. Evaluating the sensitivity of *Mycobacterium tuberculosis* to biotin deprivation using regulated gene expression. *PLoS Pathog.* **2011**, *7*, e1002264. [CrossRef]
25. Wei, J.-R.; Krishnamoorthy, V.; Murphy, K.; Kim, J.-H.; Schnappinger, D.; Alber, T.; Sassetti, C.M.; Rhee, K.Y.; Rubin, E.J. Depletion of antibiotic targets has widely varying effects on growth. *Proc. Natl. Acad. Sci. USA* **2011**, *108*, 4176–4181. [CrossRef]
26. Aleksandrov, A.; Myllykallio, H. Advances and challenges in drug design against tuberculosis: Application of in silico approaches. *Expert Opin. Drug Discov.*. **2019**, *14*, 35–46. [CrossRef] [PubMed]
27. Waman, V.P.; Vedithi, S.C.; Thomas, S.E.; Bannerman, B.P.; Munir, A.; Skwark, M.J.; Malhotra, S.; Blundell, T.L. Mycobacterial genomics and structural bioinformatics: Opportunities and challenges in drug discovery. *Emerg. Microbes Infect.* **2019**, *8*, 109–118. [CrossRef]
28. Pedelacq, J.D.; Nguyen, M.C.; Terwilliger, T.C.; Mourey, L. A Comprehensive Review on *Mycobacterium tuberculosis* Targets and Drug Development from a Structural Perspective. In *Structural Biology in Drug Discovery Methods, Techniques, and Practices*; Renaud, J.-P., Ed.; Wiley: Hoboken, NJ, USA, 2020; Volume 3, pp. 545–566.
29. Cohen, P. Protein kinases–the major drug targets of the twenty-first century? *Nat. Rev. Drug Discov.* **2002**, *1*, 309–315. [CrossRef]
30. Knapp, S. New opportunities for kinase drug repurposing and target discovery. *Br. J. Cancer* **2018**, *118*, 936–937. [CrossRef]
31. Schoijet, A.C.; Sternlieb, T.; Alonso, G.D. Signal Transduction Pathways as Therapeutic Target for Chagas Disease. *Curr. Med. Chem.* **2019**, *26*, 6572–6589. [CrossRef]

32. Lima, M.N.N.; Cassiano, G.C.; Tomaz, K.C.P.; Silva, A.C.; Sousa, B.K.P.; Ferreira, L.T.; Tavella, T.A.; Calit, J.; Bargieri, D.Y.; Neves, B.J.; et al. Integrative Multi-Kinase Approach for the Identification of Potent Antiplasmodial Hits. *Front. Chem.* **2019**, *7*, 773. [CrossRef]
33. Av-Gay, Y.; Everett, M. The eukaryotic-like Ser/Thr protein kinases of *Mycobacterium tuberculosis*. *Trends Microbiol.* **2000**, *8*, 238–244. [CrossRef]
34. Kang, C.-M.; Abbott, D.W.; Park, S.T.; Dascher, C.C.; Cantley, L.C.; Husson, R.N. The *Mycobacterium tuberculosis* serine/threonine kinases PknA and PknB: Substrate identification and regulation of cell shape. *Genes Dev.* **2005**, *19*, 1692–1704. [CrossRef] [PubMed]
35. Wehenkel, A.; Bellinzoni, M.; Graña, M.; Duran, R.; Villarino, A.; Fernandez, P.; Andre-Leroux, G.; England, P.; Takiff, H.; Cerveñansky, C.; et al. Mycobacterial Ser/Thr protein kinases and phosphatases: Physiological roles and therapeutic potential. *Biochim. Biophys. Acta Proteins Proteom.* **2008**, *1784*, 193–202. [CrossRef] [PubMed]
36. Alber, T. Signaling mechanisms of the *Mycobacterium tuberculosis* receptor Ser/Thr protein kinases. *Curr. Opin. Struct. Biol.* **2009**, *19*, 650–657. [CrossRef]
37. Mieczkowski, C.; Iavarone, A.T.; Alber, T. Auto-activation mechanism of the *Mycobacterium tuberculosis* PknB receptor Ser/Thr kinase. *EMBO J.* **2008**, *27*, 3186–3197. [CrossRef]
38. Lombana, T.N.; Echols, N.; Good, M.C.; Thomsen, N.D.; Ng, H.-L.; Greenstein, A.E.; Falick, A.M.; King, D.S.; Alber, T. Allosteric activation mechanism of the *Mycobacterium tuberculosis* receptor Ser/Thr protein kinase, PknB. *Structure* **2010**, *18*, 1667–1677. [CrossRef]
39. Greenstein, A.E.; Echols, N.; Lombana, T.N.; King, D.S.; Alber, T. Allosteric activation by dimerization of the PknD receptor Ser/Thr protein kinase from *Mycobacterium tuberculosis*. *J. Biol. Chem.* **2007**, *282*, 11427–11435. [CrossRef]
40. Wagner, T.; Andre-Leroux, G.; Hindie, V.; Barilone, N.; Lisa, M.-N.; Hoos, S.; Raynal, B.; Vulliez-Le Normand, B.; O'Hare, H.M.; Bellinzoni, M.; et al. Structural insights into the functional versatility of an FHA domain protein in mycobacterial signaling. *Sci. Signal.* **2019**, *12*, eaav9504. [CrossRef]
41. Young, T.A.; Delagoutte, B.; Endrizzi, J.A.; Falick, A.M.; Alber, T. Structure of *Mycobacterium tuberculosis* PknB supports a universal activation mechanism for Ser/Thr protein kinases. *Nat. Struct. Biol.* **2003**, *10*, 168–174. [CrossRef]
42. Ortiz-Lombardía, M.; Pompeo, F.; Boitel, B.; Alzari, P.M. Crystal structure of the catalytic domain of the PknB serine/threonine kinase from *Mycobacterium tuberculosis*. *J. Biol. Chem.* **2003**, *278*, 13094–13100. [CrossRef]
43. Sassetti, C.M.; Boyd, D.H. Genes required for mycobacterial growth defined by high density mutagenesis. *Mol. Microbiol.* **2003**, *48*, 77–84. [CrossRef] [PubMed]
44. Fernandez, P.; Saint-Joanis, B.; Barilone, N.; Jackson, M.; Gicquel, B.; Cole, S.T.; Alzari, P.M. The Ser/Thr protein kinase PknB is essential for sustaining mycobacterial growth. *J. Bacteriol.* **2006**, *188*, 7778–7784. [CrossRef] [PubMed]
45. Chawla, Y.; Upadhyay, S.; Khan, S.; Nagarajan, S.N.; Forti, F.; Nandicoori, V.K. Protein kinase B (PknB) of *Mycobacterium tuberculosis* is essential for growth of the pathogen in vitro as well as for survival within the host. *J. Biol. Chem.* **2014**, *289*, 13875. [CrossRef] [PubMed]
46. Bellinzoni, M.; Wehenkel, A.M.; Duran, R.; Alzari, P.M. Novel mechanistic insights into physiological signaling pathways mediated by mycobacterial Ser/Thr protein kinases. *Genes Immun.* **2019**, *20*, 383–393. [CrossRef] [PubMed]
47. Drews, S.J.; Hung, F.; Av-Gay, Y. A protein kinase inhibitor as an antimycobacterial agent. *FEMS Microbiol. Lett.* **2001**, *205*, 369–374. [CrossRef] [PubMed]
48. Wehenkel, A.; Fernandez, P.; Bellinzoni, M.; Catherinot, V.; Barilone, N.; Labesse, G.; Jackson, M.; Alzari, P.M. The structure of PknB in complex with mitoxantrone, an ATP-competitive inhibitor, suggests a mode of protein kinase regulation in mycobacteria. *FEBS Lett.* **2006**, *580*, 3018–3022. [CrossRef]
49. Bais, V.S.; Mohapatra, B.; Ahamad, N.; Boggaram, S.; Verma, S.; Prakash, B. Investigating the inhibitory potential of 2-Aminopurine metal complexes against serine/threonine protein kinases from *Mycobacterium tuberculosis*. *Tuberculosis* **2018**, *108*, 47–55. [CrossRef]
50. Székely, R.; Waczek, F.; Szabadkai, I.; Németh, G.; Hegymegi-Barakonyi, B.; Eros, D.; Szokol, B.; Pato, J.; Hafenbradl, D.; Satchell, J.; et al. A novel drug discovery concept for tuberculosis: Inhibition of bacterial and host cell signalling. *Immunol. Lett.* **2008**, *116*, 225–231. [CrossRef]

51. Chapman, T.M.; Bouloc, N.; Buxton, R.S.; Chugh, J.; Lougheed, K.E.A.; Osborne, S.A.; Saxty, B.; Smerdon, S.J.; Taylor, D.L.; Whalley, D. Substituted aminopyrimidine protein kinase B (PknB) inhibitors show activity against *Mycobacterium tuberculosis*. *Bioorg. Med. Chem. Lett.* **2012**, *22*, 3349–3353. [CrossRef]
52. Lougheed, K.E.A.; Osborne, S.A.; Saxty, B.; Whalley, D.; Chapman, T.; Bouloc, N.; Chugh, J.; Nott, T.J.; Patel, D.; Spivey, V.L.; et al. Effective inhibitors of the essential kinase PknB and their potential as anti-mycobacterial agents. *Tuberculosis* **2011**, *91*, 277–286. [CrossRef]
53. Xu, J.; Wang, J.-X.; Zhou, J.-M.; Xu, C.-L.; Huang, B.; Xing, Y.; Wang, B.; Luo, R.; Wang, Y.-C.; You, X.-F.; et al. A novel protein kinase inhibitor IMB-YH-8 with anti-tuberculosis activity. *Sci. Rep.* **2017**, *7*, 5093–5110. [CrossRef] [PubMed]
54. Appunni, S.; Rajisha, P.M.; Rubens, M.; Chandana, S.; Singh, H.N.; Swarup, V. Targeting PknB, an eukaryotic-like serine/threonine protein kinase of *Mycobacterium tuberculosis* with phytomolecules. *Comput. Biol. Chem.* **2017**, *67*, 200–204. [CrossRef] [PubMed]
55. Wang, T.; Bemis, G.; Hanzelka, B.; Zuccola, H.; Wynn, M.; Moody, C.S.; Green, J.; Locher, C.; Liu, A.; Gao, H.; et al. Mtb PKNA/PKNB Dual Inhibition Provides Selectivity Advantages for Inhibitor Design To Minimize Host Kinase Interactions. *ACS Med. Chem. Lett.* **2017**, *8*, 1224–1229. [CrossRef] [PubMed]
56. Wlodarchak, N.; Teachout, N.; Beczkiewicz, J.; Procknow, R.; Schaenzer, A.J.; Satyshur, K.; Pavelka, M.; Zuercher, W.; Drewry, D.; Sauer, J.-D.; et al. In Silico Screen and Structural Analysis Identifies Bacterial Kinase Inhibitors which Act with β-Lactams To Inhibit Mycobacterial Growth. *Mol. Pharm.* **2018**, *15*, 5410–5426. [CrossRef] [PubMed]
57. Walburger, A.; Koul, A.; Ferrari, G.; Nguyen, L.; Prescianotto-Baschong, C.; Huygen, K.; Klebl, B.; Thompson, C.; Bacher, G.; Pieters, J. Protein kinase G from pathogenic mycobacteria promotes survival within macrophages. *Science* **2004**, *304*, 1800–1804. [CrossRef] [PubMed]
58. Scherr, N.; Honnappa, S.; Kunz, G.; Mueller, P.; Jayachandran, R.; Winkler, F.; Pieters, J.; Steinmetz, M.O. Structural basis for the specific inhibition of protein kinase G, a virulence factor of *Mycobacterium tuberculosis*. *Proc. Natl. Acad. Sci. USA* **2007**, *104*, 12151–12156. [CrossRef]
59. Lisa, M.-N.; Gil, M.; Andre-Leroux, G.; Barilone, N.; Duran, R.; Biondi, R.M.; Alzari, P.M. Molecular Basis of the Activity and the Regulation of the Eukaryotic-like S/T Protein Kinase PknG from Mycobacterium tuberculosis. *Structure* **2015**, *23*, 1039–1048. [CrossRef]
60. van der Woude, A.D.; Stoop, E.J.M.; Stiess, M.; Wang, S.; Ummels, R.; van Stempvoort, G.; Piersma, S.R.; Cascioferro, A.; Jiménez, C.R.; Houben, E.N.G.; et al. Analysis of SecA2-dependent substrates in *Mycobacterium marinum* identifies protein kinase G (PknG) as a virulence effector. *Cell Microbiol.* **2014**, *16*, 280–295. [CrossRef]
61. Zulauf, K.E.; Sullivan, J.T.; Braunstein, M. The SecA2 pathway of *Mycobacterium tuberculosis* exports effectors that work in concert to arrest phagosome and autophagosome maturation. *PLoS Pathog.* **2018**, *14*, e1007011. [CrossRef]
62. Pradhan, G.; Shrivastva, R.; Mukhopadhyay, S. Mycobacterial PknG targets the Rab7l1 signaling pathway to inhibit phagosome-lysosome fusion. *J. Immunol.* **2018**, *201*, 1421–1433. [CrossRef]
63. Bussi, C.; Gutierrez, M.G. *Mycobacterium tuberculosis* infection of host cells in space and time. *FEMS Microbiol. Rev.* **2019**, *43*, 341–361. [CrossRef] [PubMed]
64. Schultz, C.; Niebisch, A.; Gebel, L.; Bott, M. Glutamate production by *Corynebacterium glutamicum*: Dependence on the oxoglutarate dehydrogenase inhibitor protein OdhI and protein kinase PknG. *Appl. Microbiol. Biotechnol.* **2007**, *76*, 691–700. [CrossRef] [PubMed]
65. Niebisch, A.; Kabus, A.; Schultz, C.; Weil, B.; Bott, M. Corynebacterial protein kinase G controls 2-oxoglutarate dehydrogenase activity via the phosphorylation status of the OdhI protein. *J. Biol. Chem.* **2006**, *281*, 12300–12307. [CrossRef] [PubMed]
66. Ventura, M.; Rieck, B.; Boldrin, F.; Degiacomi, G.; Bellinzoni, M.; Barilone, N.; Alzaidi, F.; Alzari, P.M.; Manganelli, R.; O'Hare, H.M. GarA is an essential regulator of metabolism in *Mycobacterium tuberculosis*. *Mol. Microbiol.* **2013**, *90*, 356–366. [CrossRef]
67. O'Hare, H.M.; Duran, R.; Cerveñansky, C.; Bellinzoni, M.; Wehenkel, A.M.; Pritsch, O.; Obal, G.; Baumgartner, J.; Vialaret, J.; Johnsson, K.; et al. Regulation of glutamate metabolism by protein kinases in mycobacteria. *Mol. Microbiol.* **2008**, *70*, 1408–1423. [CrossRef]
68. Rieck, B.; Degiacomi, G.; Zimmermann, M.; Cascioferro, A.; Boldrin, F.; Lazar-Adler, N.R.; Bottrill, A.R.; Le Chevalier, F.; Frigui, W.; Bellinzoni, M.; et al. PknG senses amino acid availability to control metabolism and virulence of Mycobacterium tuberculosis. *PLoS Pathog.* **2017**, *13*, e1006399. [CrossRef]

69. Wolff, K.A.; de la Peña, A.H.; Nguyen, H.T.; Pham, T.H.; Amzel, L.M.; Gabelli, S.B.; Nguyen, L. A redox regulatory system critical for mycobacterial survival in macrophages and biofilm development. *PLoS Pathog.* **2015**, *11*, e1004839. [CrossRef]
70. Khan, M.Z.; Bhaskar, A.; Upadhyay, S.; Kumari, P.; Rajmani, R.S.; Jain, P.; Singh, A.; Kumar, D.; Bhavesh, N.S.; Nandicoori, V.K. Protein kinase G confers survival advantage to *Mycobacterium tuberculosis* during latency-like conditions. *J. Biol. Chem.* **2017**, *292*, 16093–16108. [CrossRef]
71. Cowley, S.; Ko, M.; Pick, N.; Chow, R.; Downing, K.J.; Gordhan, B.G.; Betts, J.C.; Mizrahi, V.; Smith, D.A.; Stokes, R.W.; et al. The *Mycobacterium tuberculosis* protein serine/threonine kinase PknG is linked to cellular glutamate/glutamine levels and is important for growth in vivo. *Mol. Microbiol.* **2004**, *52*, 1691–1702. [CrossRef]
72. Chao, J.; Wong, D.; Zheng, X.; Poirier, V.; Bach, H.; Hmama, Z.; Av-Gay, Y. Protein kinase and phosphatase signaling in *Mycobacterium tuberculosis* physiology and pathogenesis. *Biochim. Biophys. Acta Proteins Proteom.* **2010**, *1804*, 620–627. [CrossRef]
73. Chen, D.; Ma, S.; He, L.; Yuan, P.; She, Z.; Lu, Y. Sclerotiorin inhibits protein kinase G from *Mycobacterium tuberculosis* and impairs mycobacterial growth in macrophages. *Tuberculosis* **2017**, *103*, 37–43. [CrossRef] [PubMed]
74. Kidwai, S.; Bouzeyen, R.; Chakraborti, S.; Khare, N.; Das, S.; Priya Gosain, T.; Behura, A.; Meena, C.L.; Dhiman, R.; Essafi, M.; et al. NU-6027 Inhibits Growth of *Mycobacterium tuberculosis* by Targeting Protein Kinase D and Protein Kinase G. *Antimicrob. Agents Chemother.* **2019**, *63*, 39. [CrossRef] [PubMed]
75. Kanehiro, Y.; Tomioka, H.; Pieters, J.; Tatano, Y.; Kim, H.; Iizasa, H.; Yoshiyama, H. Identification of Novel Mycobacterial Inhibitors Against Mycobacterial Protein Kinase G. *Front. Microbiol.* **2018**, *9*, 1517. [CrossRef] [PubMed]
76. Schoeffler, A.J.; Berger, J.M. DNA topoisomerases: Harnessing and constraining energy to govern chromosome topology. *Quart. Rev. Biophys.* **2008**, *41*, 41–101. [CrossRef] [PubMed]
77. Gellert, M.; Mizuuchi, K.; O'Dea, M.H.; Nash, H.A. DNA gyrase: An enzyme that introduces superhelical turns into DNA. *Proc. Natl. Acad. Sci. USA* **1976**, *73*, 3872–3876. [CrossRef]
78. Sissi, C.; Palumbo, M. In front of and behind the replication fork: Bacterial type IIA topoisomerases. *Cell. Mol. Life Sci.* **2010**, *67*, 2001–2024. [CrossRef]
79. Peng, H.; Marians, K.J. Decatenation activity of topoisomerase IV during oriC and pBR322 DNA replication in vitro. *Proc. Natl. Acad. Sci. USA* **1993**, *90*, 8571–8575. [CrossRef]
80. Aubry, A.; Fisher, L.M.; Jarlier, V.; Cambau, E. First functional characterization of a singly expressed bacterial type II topoisomerase: The enzyme from *Mycobacterium tuberculosis*. *Biochem. Biophys. Res. Commun.* **2006**, *348*, 158–165. [CrossRef]
81. Hooper, D.C.; Jacoby, G.A. Topoisomerase Inhibitors: Fluoroquinolone Mechanisms of Action and Resistance. *Cold Spring Harb. Perspect. Med.* **2016**, *6*, a025320. [CrossRef]
82. Blower, T.R.; Williamson, B.H.; Kerns, R.J.; Berger, J.M. Crystal structure and stability of gyrase–fluoroquinolone cleaved complexes from *Mycobacterium tuberculosis*. *Proc. Natl. Acad. Sci. USA* **2016**, *113*, 1706–1713. [CrossRef]
83. Fu, G.; Wu, J.; Liu, W.; Zhu, D.; Hu, Y.; Deng, J.; Zhang, X.-E.; Bi, L.; Wang, D.-C. Crystal structure of DNA gyrase B' domain sheds lights on the mechanism for T-segment navigation. *Nucleic Acids Res.* **2009**, *37*, 5908–5916. [CrossRef] [PubMed]
84. Tretter, E.M.; Schoeffler, A.J.; Weisfield, S.R.; Berger, J.M. Crystal structure of the DNA gyrase GyrA N-terminal domain from *Mycobacterium tuberculosis*. *Proteins* **2010**, *78*, 492–495. [CrossRef] [PubMed]
85. Agrawal, A.; Roue, M.; Spitzfaden, C.; Petrella, S.; Aubry, A.; Hann, M.; Bax, B.; Mayer, C. *Mycobacterium tuberculosis* DNA gyrase ATPase domain structures suggest a dissociative mechanism that explains how ATP hydrolysis is coupled to domain motion. *Biochem. J.* **2013**, *456*, 263–273. [CrossRef] [PubMed]
86. Bouige, A.; Darmon, A.; Piton, J.; Roue, M.; Petrella, S.; Capton, E.; Forterre, P.; Aubry, A.; Mayer, C. *Mycobacterium tuberculosis* DNA gyrase possesses two functional GyrA-boxes. *Biochem. J.* **2013**, *455*, 285–294. [CrossRef] [PubMed]
87. Piton, J.; Petrella, S.; Delarue, M.; Andre-Leroux, G.; Jarlier, V.; Aubry, A.; Mayer, C. Structural Insights into the Quinolone Resistance Mechanism of *Mycobacterium tuberculosis* DNA Gyrase. *PLoS ONE* **2010**, *5*, e12245. [CrossRef]

88. World Health Organization. Global Tuberculosis Report 2019. Available online: https://www.who.int/tb/publications/global_report/en/ (accessed on 19 June 2020).
89. Nagaraja, V.; Godbole, A.A.; Henderson, S.R.; Maxwell, A. DNA topoisomerase I and DNA gyrase as targets for TB therapy. *Drug Discov. Today* **2017**, *22*, 510–518. [CrossRef]
90. Vanden Broeck, A.; McEwen, A.G.; Chebaro, Y.; Potier, N.; Lamour, V. Structural Basis for DNA Gyrase Interaction with Coumermycin A1. *J. Med. Chem.* **2019**, *62*, 4225–4231. [CrossRef]
91. Mizuuchi, K.; O'Dea, M.H.; Gellert, M. DNA gyrase: Subunit structure and ATPase activity of the purified enzyme. *Proc. Natl. Acad. Sci. USA* **1978**, *75*, 5960–5963. [CrossRef]
92. Grossman, T.H.; Bartels, D.J.; Mullin, S.; Gross, C.H.; Parsons, J.D.; Liao, Y.; Grillot, A.-L.; Stamos, D.; Olson, E.R.; Charifson, P.S.; et al. Dual targeting of GyrB and ParE by a novel aminobenzimidazole class of antibacterial compounds. *Antimicrob. Agents Chemother.* **2007**, *51*, 657–666. [CrossRef]
93. Charifson, P.S.; Grillot, A.-L.; Grossman, T.H.; Parsons, J.D.; Badia, M.; Bellon, S.; Deininger, D.D.; Drumm, J.E.; Gross, C.H.; LeTiran, A.; et al. Novel dual-targeting benzimidazole urea inhibitors of DNA gyrase and topoisomerase IV possessing potent antibacterial activity: Intelligent design and evolution through the judicious use of structure-guided design and structure-activity relationships. *J. Med. Chem.* **2008**, *51*, 5243–5263. [CrossRef]
94. Holdgate, G.A.; Tunnicliffe, A.; Ward, W.H.; Weston, S.A.; Rosenbrock, G.; Barth, P.T.; Taylor, I.W.; Pauptit, R.A.; Timms, D. The entropic penalty of ordered water accounts for weaker binding of the antibiotic novobiocin to a resistant mutant of DNA gyrase: A thermodynamic and crystallographic study. *Biochemistry* **1997**, *36*, 9663–9673. [CrossRef] [PubMed]
95. Grillot, A.-L.; Le Tiran, A.; Shannon, D.; Krueger, E.; Liao, Y.; O'Dowd, H.; Tang, Q.; Ronkin, S.; Wang, T.; Waal, N.; et al. Second-generation antibacterial benzimidazole ureas: Discovery of a preclinical candidate with reduced metabolic liability. *J. Med. Chem.* **2014**, *57*, 8792–8816. [CrossRef] [PubMed]
96. Brown-Elliott, B.A.; Rubio, A.; Wallace, R.J., Jr. In Vitro Susceptibility Testing of a Novel Benzimidazole, SPR719, against Nontuberculous Mycobacteria. *Antimicrob. Agents Chemother.* **2018**, *62*, 545. [CrossRef] [PubMed]
97. Locher, C.P.; Jones, S.M.; Hanzelka, B.L.; Perola, E.; Shoen, C.M.; Cynamon, M.H.; Ngwane, A.H.; Wiid, I.J.; van Helden, P.D.; Betoudji, F.; et al. A Novel Inhibitor of Gyrase B Is a Potent Drug Candidate for Treatment of Tuberculosis and Nontuberculosis Mycobacterial Infections. *Antimicrob. Agents Chemother.* **2015**, *59*, 1455–1465. [CrossRef]
98. Shoen, C.M.; DeStefano, M.; Pucci, M.; Cynamon, M.H. Evaluating the Sterilizing Activity of SPR720 in Combination Therapy against Mycobacterium tuberculosis Infection in Mice. In Proceedings of the Conference of ASM Microbes 2019, San Francisco, CA, USA, 20–24 June 2019.
99. Coates, W.J.; Gwynn, M.N.; Hatton, I.K.; Masters, P.J.; Pearson, N.D.; Rahman, S.S.; Slocombe, B.; Warrack, J.D.; SmithKline Beecham Ltd. Quinoline Derivatives as Antibacterials. European Patent EP1051413, 4 June 2003.
100. Bax, B.D.; Chan, P.F.; Eggleston, D.S.; Fosberry, A.; Gentry, D.R.; Gorrec, F.; Giordano, I.; Hann, M.M.; Hennessy, A.; Hibbs, M.; et al. Type IIA topoisomerase inhibition by a new class of antibacterial agents. *Nature* **2010**, *466*, 935–940. [CrossRef]
101. Blanco, D.; Perez-Herran, E.; Cacho, M.; Ballell, L.; Castro, J.; González del Río, R.; Lavandera, J.L.; Remuiñán, M.J.; Richards, C.; Rullas, J.; et al. Mycobacterium tuberculosis Gyrase Inhibitors as a New Class of Antitubercular Drugs. *Antimicrob. Agents Chemother.* **2015**, *59*, 1868–1875. [CrossRef]
102. Gibson, E.G.; Blower, T.R.; Cacho, M.; Bax, B.; Berger, J.M.; Osheroff, N. Mechanism of Action of *Mycobacterium tuberculosis* Gyrase Inhibitors: A Novel Class of Gyrase Poisons. *ACS Infect. Dis.* **2018**, *4*, 1211–1222. [CrossRef]
103. Tanner, L.; Denti, P.; Wiesner, L.; Warner, D.F. Drug permeation and metabolism in *Mycobacterium tuberculosis*: Prioritising local exposure as essential criterion in new TB drug development. *IUBMB Life* **2018**, *70*, 926–937. [CrossRef]
104. Van Wijk, R.C.; Ayoun Alsoud, R.; Lennernäs, H.; Simonsson, U.S.H. Model-Informed Drug Discovery and Development Strategy for the Rapid Development of Anti-Tuberculosis Drug Combinations. *Appl. Sci.* **2020**, *10*, 2376. [CrossRef]
105. Mori, M.; Sammartino, J.C.; Costantino, L.; Gelain, A.; Meneghetti, F.; Villa, S.; Chiarelli, L.R. An Overview on the Potential Antimycobacterial Agents Targeting Serine/Threonine Protein Kinases from Mycobacterium tuberculosis. *Curr. Top. Med. Chem.* **2019**, *19*, 646–661. [CrossRef]

106. Prisic, S.; Husson, R.N. *Mycobacterium tuberculosis* Serine/Threonine Protein Kinases. *Microbiol. Spectr.* **2014**, 2. [CrossRef] [PubMed]
107. Ehrt, S.; Schnappinger, D.; Rhee, K.Y. Metabolic principles of persistence and pathogenicity in *Mycobacterium tuberculosis*. *Nat. Rev. Microbiol.* **2018**, *16*, 496–507. [CrossRef] [PubMed]
108. Sala, C.; Hartkoorn, R.C. Tuberculosis drugs: New candidates and how to find more. *Future Microbiol.* **2011**, *6*, 617–633. [CrossRef] [PubMed]
109. Villemagne, B.; Machelart, A.; Tran, N.C.; Flipo, M.; Moune, M.; Leroux, F.; Piveteau, C.; Wohlkönig, A.; Wintjens, R.; Li, X.; et al. Fragment-Based Optimized EthR Inhibitors with in Vivo Ethionamide Boosting Activity. *ACS Infect. Dis.* **2020**, *6*, 366–378. [CrossRef]
110. Maveyraud, L.; Mourey, L. Protein X-ray Crystallography and Drug Discovery. *Molecules* **2020**, *25*, 1030. [CrossRef]
111. Rout, M.P.; Sali, A. Principles for Integrative Structural Biology Studies. *Cell* **2019**, *177*, 1384–1403. [CrossRef]
112. Petrella, S.; Capton, E.; Raynal, B.; Giffard, C.; Thureau, A.; Bonneté, F.; Alzari, P.M.; Aubry, A.; Mayer, C. Overall Structures of *Mycobacterium tuberculosis* DNA Gyrase Reveal the Role of a *Corynebacteriales* GyrB-Specific Insert in ATPase Activity. *Structure* **2019**, *27*, 579–589.e5. [CrossRef]
113. Vanden Broeck, A.; Lotz, C.; Ortiz, J.; Lamour, V. Cryo-EM structure of the complete *E. coli* DNA gyrase nucleoprotein complex. *Nat. Commun.* **2019**, *10*, 4935. [CrossRef]

 © 2020 by the authors. Licensee MDPI, Basel, Switzerland. This article is an open access article distributed under the terms and conditions of the Creative Commons Attribution (CC BY) license (http://creativecommons.org/licenses/by/4.0/).

Review

Multi-Omics Technologies Applied to Tuberculosis Drug Discovery

Aaron Goff, Daire Cantillon, Leticia Muraro Wildner and Simon J Waddell *

Department of Global Health and Infection, Brighton and Sussex Medical School, University of Sussex, Brighton BN1 9PX, UK; A.Goff@bsms.ac.uk (A.G.); D.Cantillon2@bsms.ac.uk (D.C.); L.MuraroWildner@bsms.ac.uk (L.M.W.)
* Correspondence: s.waddell@bsms.ac.uk; Tel.: +44-1273-87-7572

Received: 28 May 2020; Accepted: 30 June 2020; Published: 3 July 2020

Abstract: Multi-omics strategies are indispensable tools in the search for new anti-tuberculosis drugs. Omics methodologies, where the ensemble of a class of biological molecules are measured and evaluated together, enable drug discovery programs to answer two fundamental questions. Firstly, in a discovery biology approach, to find new targets in druggable pathways for target-based investigation, advancing from target to lead compound. Secondly, in a discovery chemistry approach, to identify the mode of action of lead compounds derived from high-throughput screens, progressing from compound to target. The advantage of multi-omics methodologies in both of these settings is that omics approaches are unsupervised and unbiased to a priori hypotheses, making omics useful tools to confirm drug action, reveal new insights into compound activity, and discover new avenues for inquiry. This review summarizes the application of *Mycobacterium tuberculosis* omics technologies to the early stages of tuberculosis antimicrobial drug discovery.

Keywords: mycobacterium; tuberculosis; drug discovery; genomics; transcriptomics; proteomics; metabolomics; lipidomics; target identification; mechanism of action; antimicrobial drug resistance (AMR)

1. Introduction

Tuberculosis (TB) remains one of the top 10 causes of death worldwide, with 10 million new cases and 1.4 million deaths in 2018. The problem of antimicrobial drug resistance (AMR) is rising, with drug resistance associated with 3.4% of new TB cases globally and up to 50% of previously treated cases in some areas of the world [1]. The discovery of new drugs to treat *Mycobacterium tuberculosis* (*M.tb*) is challenging, with only pretomanid, delamanid and bedaquiline marketed for use in the last 40 years despite sustained international efforts [2]. Multiple logistical and physiological factors contribute to the difficulty of this task (reviewed eloquently elsewhere [3–5]). They include biosafety constraints of working with a slow-growing pathogenic bacterium, heterogeneity of clinical disease and bacterial phenotypes in vivo, intracellular and extracellular *M.tb* sites, drug penetration into lung pathology, the lipid-rich *M.tb* cell wall as a barrier to drug uptake and intrinsic drug resistance, limited number of validated drug targets, the requirement for combination drug therapy, and the length and cost of clinical trials. Omics technologies aim to measure and evaluate together the ensemble of molecular entities by biological class to understand the contribution of each component. Whatever the category of molecule under investigation, the key advantage of omics approaches is that they are unsupervised, and thus less biased by dogma, which is valuable for overcoming drug development bottlenecks [6,7]. Omics in a hypothesis-generating discovery biology setting, is an excellent means of identifying new targets for drug discovery. In a compound-first (discovery chemistry) approach, liberated from reductionist assays, omics technologies are useful tools to reveal or confirm drug mode of action. Mycobacterial omics are applicable throughout the drug development process from initial drug discovery to preclinical and clinical stages, at each step describing the action of compounds, derivatives, and formulations on *M.tb*. For example, identifying target drift or off-target effects during lead optimization, or characterizing drug resistance conferring mutations in clinical trials.

This review summarizes the application of *M.tb* omics strategies in the early stages of the discovery of new drugs for TB, incorporating genomics (DNA), transcriptomics (mRNA), proteomics (proteins), metabolomics (metabolites) and lipidomics (lipids). The review is not intended to be comprehensive—omics are now fully established in most drug discovery settings—but aims to highlight landmark and interesting approaches to the TB drug development problem. The review centers on omics applied directly to *M.tb*, using examples from other mycobacterial models only to illustrate groundbreaking discovery tools. We focus on (a) target identification, in this context the recognition of potentially druggable pathways worthy of drug discovery efforts in a target-based approach; (b) mode of action studies, often aimed at progressing hits from whole cell compound screening strategies on the long road to the TB clinic (Figure 1).

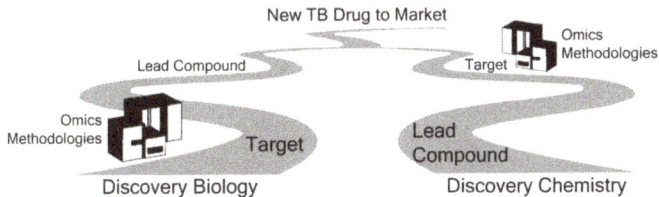

Figure 1. The application of omics tools to antimicrobial drug discovery pathways. In a discovery biology approach, omics methodologies are used early in the process to identify new targets for investigation. In a discovery chemistry setting, omics techniques are useful further down the road to identify mechanism of action of lead compounds.

2. Genomics

2.1. Target Identification

The *M.tb* H37Rv genome was sequenced in 1998, revealing ~4000 potential drug targets, of which ~50% were assigned a tentative function [8,9]. *M.tb* H37Rv was selected for sequencing as the type strain for *M.tb*. *M.tb* H37 was isolated from a patient in 1905 and serially passaged in the laboratory, a resulting strain was named H37Rv, with "R" standing for rough morphology and "*v*" for virulent [8]. It has become a commonly used laboratory strain of pathogenic *M.tb*, a genomic reference for clinical isolates, and a starting point for drug discovery. Since then, comprehensive whole genome sequencing of *M.tb* clinical isolates recovered from patients with TB has uncovered a core genome and mapped the accumulation of single nucleotide polymorphisms (SNPs) in protein coding genes [10]. Comparative genomics has also enabled drug specificity (across microbial species) and toxicity (to mammals) to be anticipated, based on the presence or absence of target protein coding sequences. The application of genomics to *M.tb* has therefore provided a framework of potential drug targets. Manipulation of gene function, through gene inactivation strategies, has aided recognition of pathways that are essential to *M.tb* in different microenvironments, thus highlighting targets for drug discovery programs [6].

Gene deletion methods designed to generate unmarked single gene knockout mutants have been employed alongside global approaches using transposons (Tn) to inactivate gene function [11]. Tn libraries are generated by disrupting genes through random insertion of a transposon throughout the genome, then applying a selective pressure to measure gene essentiality. DNA microarrays were initially used to map the changing abundance of thousands of Tn mutants, transposon site hybridization (TraSH) [12]; more recently coupled to whole genome sequencing (Tn-seq) for greater genomic resolution [13]. Tn mutant screening has been used to map the genetic pathways required for growth of *M.tb* and *Mycobacterium bovis* Bacillus Calmette-Guérin (BCG) in vitro [14]. Here, a *Himar1* based Tn delivery method using a transducing bacteriophage generated a library of ~100,000 independent clones.

A total of 614 genes essential for in vitro growth were found to be evenly distributed throughout the mycobacterial chromosome. In addition, many genes that were shown to be essential for growth appeared to be co-transcribed in operons. Essential genes were identified in amino acid, co-factor and nucleic acid biosynthesis pathways; genes of unknown function were also classified as essential. Many genes assumed to be essential that were predicted to be involved in cell wall and protein metabolism were shown to be dispensable. This is likely due to functional redundancy; for example, *purT* and *purN* both offer alternative pathways for purine de novo biosynthesis and so neither gene was essential [15], providing valuable information that allows non-essential targets to be dropped from drug discovery portfolios. A fundamental pathogenic trait of *M.tb* is the ability to survive and replicate in phagocytes, avoiding phagosome-lysosome fusion and adapting to an intracellular lifestyle [16]. Therefore, to define pathways that are essential for intra-macrophage survival, Barczak et al. mapped genes required for intracellular growth using high content imaging alongside multiplexed cytokine analysis of macrophages infected with *M.tb* Tn mutant libraries [17]. Systematic, multiparametric analysis of *M.tb* Tn mutants impaired for intracellular growth identified functional relationships between *M.tb* Tn mutants and macrophage cytokine profiles. The authors showed that production and export of the complex lipid, phthiocerol dimycocerosate (PDIM), was required for the secretion of ESX-1 substrates and permeabilization of the phagosome, revealing key virulence determinants, alongside defining pathways that are essential to the metabolism of intracellular bacilli, identifying potential targets for drug discovery. Plainly, Tn mutant libraries will only identify genes that are essential in the model under investigation, successful drug targets are likely required to be essential across several different conditions to mimic the variety of microenvironments encountered by *M.tb* during natural infection.

Gene deletion or inactivation results in the absence of gene product, this may not be reflective of drug action, where complete inhibition of protein function may not occur [18]. Conditional expression strategies that reduce rather than abrogate protein function may better represent drug action, and crucially allows essential gene targets to be investigated in the laboratory. Conditional expression using inducible promoter systems (for example, Tet ON/OFF or Pip ON/OFF) allow the expression of essential genes to be increased or reduced to understand gene function, model drug target inhibition and genetically-validate drug targets. The utility of such an inducible gene expression system was highlighted by Johnson et al. where *M.tb* mutants depleted for 474 essential genes (termed hypomorphs) were screened against a large pool of potential inhibitors, allowing for >8.5 million chemical-genetic interactions to take place [19]. An episomally-encoded *sspB* gene was introduced to control protein degradation via a carboxyl-terminal fused DAS-tag. In addition, a 20 nucleotide "barcode" was introduced to facilitate enumeration by sequencing the barcoded PCR products derived from the mutants when pooled. The expression of SspB was controlled by a TetON promoter that was induced in response to anhydrotetracycline; by using TetON promoters with varying strengths, the level of gene product targeted for degradation by the SspB gene could be titrated. Primary screening using this hypomorph methodology identified over 10-fold more hits than whole cell screening wild type *M.tb*. As expected, well known antimicrobial drug classes showed interactions with specific hypomorphs, such as fluoroquinolones with their target GyrA, as well as rifampicin with the β subunit of RNA polymerase. This screening approach identified 39 inhibitors that targeted cellular components that are already clinically-validated antimicrobial drug targets; DNA gyrase, mycolic acid synthesis, and folate biosynthesis, targeted by the fluoroquinolones, isoniazid, and para-amino-salicylic acid, respectively [20]. The 39 compounds were either novel chemical entities or known compounds with re-purposed activity, such as the plant alkaloid tryptanthrin. In addition to finding novel inhibitors for well-validated targets, hypomorph screening also identified inhibitors against novel cellular targets. Johnson et al. demonstrated that strains hypomorphic for the putative efflux pump EfpA were inhibited by the compound BRD-8000 (MIC 6 µM) but this compound showed no activity against wild type *M.tb* (MIC ≥ 50 µM). Subsequent chemical optimization improved activity of BRD-8000 against wild type *M.tb* to an MIC of 800 nM, an increase of ≥63-fold activity. *M.tb* spontaneous mutants resistant to this modified BRD-8000 compound showed a single point mutation in *efpA*, demonstrating that chemical modification of BRD-8000 had not altered target specificity.

Clustered Regularly Interspaced Short Palindromic Repeats interference (CRISPRi) has the potential to revolutionize the field, enabling precise gene silencing to identify and validate drug targets. A nuclease, dCas9 containing two mutations that eradicate its nuclease activity, is targeted to a mycobacterial gene of interest by a single guide RNA (sgRNA). Upon binding of the dCas9-sgRNA complex to the target site, the DNA duplex destabilizes and prevents gene transcription by blocking RNA polymerase promoter access [21]. Notably, the level of gene silencing can be controlled by varying the sgRNA length and sequence, this allows fine control of the expression of essential genes where traditional gene knockout approaches would be lethal. This system has been further optimized to be induced by doxycycline. This is a lipophilic drug with excellent tissue penetration properties, so that CRISPRi can be employed across a range of drug screening models including in vitro, intracellular and animal studies [22,23]. This approach was used to create libraries containing over 90,000 sgRNAs, generating pools of *M.tb* strains where the majority of genes have been targeted by CRISPRi, enabling high throughput screening approaches to be applied. The utility of CRISPRi in target-based drug discovery was demonstrated through gene silencing of folate metabolism [24]. *M.tb* and mammals require folate; however, *M.tb* must synthesize folate de novo while mammals obtain it through their diet. The variation in folate metabolism makes this biosynthetic pathway an attractive target for antimicrobial drug discovery [25]. While this pathway has been targeted in other bacteria with the antimicrobial drugs trimethoprim, inhibiting dihydrofolate reductase (FolA), and sulfamethoxazole, inhibiting dihydropteroate synthase (FolP1); the action of these drugs in mycobacteria is less clear. Generation of folate biosynthesis knockout mutants in mycobacteria has proved challenging, making target validation difficult for this pathway. *Mycobacterium smegmatis* is a non-pathogenic mycobacterium frequently used as a model organism for *M.tb* due to its low biohazard risk and tractable genomics [26]. Rock et al. utilized a panel of sgRNAs to generate hypomorphs of *folP1*, *folA* and *folC* (dihydrofolate synthase) in this model organism to show that these genes were individually essential. If weaker sgRNAs were used to decrease growth rate rather than inhibit growth completely, there was synergistic growth inhibition, demonstrating the utility of exploiting multiple targets in this pathway to maximize antimicrobial drug activity. Translation of the technology to *M.tb* will yield useful insights into target and pathway essentiality. One caveat to CRISPRi is the potential for off target effects, where dCas9 binds and silences genes that were not intended to be targeted. Bioinformatics packages exist that effectively predict this binding; however, these algorithms may not capture every off-target event [27,28].

2.2. Mode of Action

Advances in next generation sequencing technologies and the continued reduction in cost has placed whole genome sequencing (WGS) firmly into the drug discovery pipeline from mechanism of action identification through to monitoring drug resistance post approval [29]. Illumina sequencing (Illumina Inc., San Diego, CA, USA) generates short sequence reads (typically 150 bp long), and while the majority of sequencing data generated today is from Illumina platforms, alternative methodologies are also useful to antimicrobial drug discovery. Long read sequencing (~kilobases), available through platforms such as PacBio (Pacific Biosciences, Menlo Park, CA, USA) and Oxford Nanopore (Oxford Nanopore Technologies Ltd, Oxford, UK), are especially suited for sequencing repetitive regions and structural variations. WGS has been exploited to identify mutations that occur in spontaneous drug resistant colonies growing on solid agar containing a novel antimicrobial compound of interest, typically between 5-fold and 10-fold the minimum inhibitory concentration. Using drug exposure as the selective pressure, this methodology may reveal mutations that affect drug action, in the drug target itself or in gene functions that influence drug activation or efflux. Andries et al. in the discovery of bedaquiline, employed WGS of *M. smegmatis* and *M.tb* bedaquiline-resistant mutants cultured in vitro to demonstrate that this diarylquinoline targets the product of *atpE*; a subunit of the mycobacterial ATP synthase anchored in the mycobacterial membrane [30]. Kundu et al. further characterized this mode of action, showing that bedaquiline binds to the epsilon subunit [31]. Mutations can, however, be identified in other non-target genes that confer resistance, expanding understanding of drug action but complicating the interpretation of results. For example, in a murine model of TB infection, mutations in the putative Xaa-Pro aminopeptidase *pepQ* were identified that conferred low level resistance to both bedaquiline and clofazimine, neither of which target PepQ directly [32]. While the function of PepQ remains to be determined in *M.tb*, Almeida et al. postulated that this mutation prevented protein degradation of the efflux pump MmpL5, enhancing drug efflux and inducing resistance to bedaquiline and clofazimine [32].

WGS has also been used to reveal the mechanism of action of repurposed licensed pharmaceuticals in *M.tb*. Rybniker et al. identified lansoprazole from the Prestwick library of 1280 FDA approved drugs as protective to lung fibroblasts in an *M.tb* intracellular model of infection [33]. Three lansoprazole resistant mutants were sequenced; each mutant showed the same SNP, a substitution of leucine for proline, in the β-subunit of the cytochrome bc1 complex gene *qcrB*, a key element of the mycobacterial respiratory chain. To demonstrate that exposure to this commonly prescribed proton pump inhibitor did not select for lansoprazole-resistant *M.tb*, Rybniker et al. sequenced 13,559 *M.tb* Complex clinical isolates revealing only one *M. bovis* isolate with a mutation in *qcrB* [34]. Of course, this methodology to define the target of drugs with unknown mechanisms of action is only fruitful if drug-resistant colonies can be raised, which is not always the case. This was highlighted in the discovery of the novel antimicrobial drug, teixobactin from the bacterium *Eleftheria terrae*. Here, a novel iChip culture method was used to capture previously unculturable bacteria in their natural soil environment, which led to the identification of the cell wall-targeting agent teixobactin [35]. Neither *Staphylococcus aureus* nor *M.tb* drug resistant colonies could be generated even at sub-inhibitory concentrations. The authors hypothesized that this indicated general cell toxicity of the compound; however, no mammalian cell toxicity was observed. Subsequent biochemical approaches indicated that this compound exerted its antimicrobial effects through interactions with peptidoglycan pre-cursors—mostly lipid II. Therefore, the inability to generate spontaneous resistance in vitro is likely due to teixobactin inhibiting multiple non-protein targets. Like many other antimicrobial drugs, teixobactin was isolated from an environmental microorganism. It is therefore possible that bacteria in close proximity to *E. terrae* in the soil may be resistant to teixobactin, offering a scenario where teixobactin resistance mechanisms could be discovered [36]. While the inability to detect drug resistance in the laboratory does not mirror the complexity of TB clinical disease alongside host pharmacokinetic and pharmacodynamic factors, the concept of resistance-proof antimicrobial drugs is very attractive and could transform the destructive antimicrobial drug to evolution of drug resistance cycle.

WGS is also shaping our understanding of drug action and drug resistance through the large-scale sequencing of clinical isolates. Consortia mapping drug-resistance conferring mutations have revealed novel mechanisms of resistance and potential novel modes of action of existing anti-*M.tb* drugs in patients [37]. The CRyPTIC Consortium and the 100,000 Genomes Project obtained over 10,000 *M.tb* clinical isolate genomes and associated phenotypes to predict phenotypic drug susceptibility to front line agents from genome sequence. The authors successfully demonstrated that susceptibility could be correlated to the presence or absence of specific antimicrobial drug resistance mutations [37]. This approach also exposed mutations with as yet unexplained roles in drug action. For example, *M.tb* isolates have been identified with mutations in the mycothiol biosynthetic genes *mshA* and *mshC* that confer high-level resistance to ethionamide and low-level resistance to isoniazid [38]. Mycothiol is a key detoxifying and reducing molecule with similar functions to glutathione, which is absent in mycobacteria [39]. Knockout mutants of *mshA* grew normally in vitro and in immunodeficient mice but were growth defective in immunocompetent mice [40], and mycothiol-deficient mutants were less susceptible to ethionamide and isoniazid [41]. Ethionamide and isoniazid are both prodrugs that target the NADH-dependent enoyl-ACP reductase InhA; however, isoniazid is activated by KatG and ethionamide by EthA [42,43]. In order to delineate the roles mycothiol biosynthetic genes play in isoniazid and ethionamide susceptibility, Xu et al. generated null mutants of each mycothiol biosynthetic gene in *M. smegmatis* [44]. *M. smegmatis* is useful for studying the action of these two drugs, with their target InhA characterized in this bacterium to have >95% sequence identity to the *M.tb* InhA [45]. While the exact role of mycothiol in isoniazid and ethionamide resistance remains to be determined, the authors hypothesized that mycothiol is involved in the activation of isoniazid and ethionamide. *M.tb* KatG, required for isoniazid activation, likely compensates for loss of mycothiol in some settings, resulting in isoniazid susceptibility and ethionamide drug resistance in mycothiol-deficient isolates [40]. Understanding the action of current *M.tb* drugs has direct implications for the development of new inhibitors. For example, bedaquiline and clofazimine resistance can be mediated through SNPs in *Rv0678*, a transcriptional repressor of *mmpL5* and *mmpS5*, which encode an efflux pump of the resistance-nodulation-division (RND) family [46,47]. SNPs in *Rv0678* were detected in clinical isolates of *M.tb* with no prior exposure to bedaquiline or clofazimine [48]; meaning that resistance to bedaquiline was already circulating in *M.tb* isolates before the new drug was introduced to the TB clinic. While the authors could not determine the driver of these mutations, they ruled out rifampicin as a selective pressure. Studies such as these are revealing drug detoxification and efflux pathways that operate across drug class to reduce efficacy of both existing and new antimicrobial drugs.

To expand the methods available for understanding drug mechanism of action, Melief et al. constructed a library of *M.tb* strains overexpressing single genes that could be screened in a high-throughput format [49]. It was hypothesized that library mutants overexpressing the target of an antimicrobial drug should be more resistant than wild type *M.tb* to that particular drug. The library was constructed cloning each gene downstream of a tetracycline-inducible promoter. The 1733 constructs covered 40% of protein coding genes in the *M.tb* genome and contained the majority of annotated essential genes, as well as genes involved in cell wall and fatty acid biosynthesis, virulence factors, regulatory proteins and efflux. The functionality of the system was confirmed by screening the library for resistance to D-cycloserine, which identified the Alr-over-expressing mutant as the only recombinant strain that grew in the presence of the drug. Over-expression of Alr, a target of the drug, resulted in a 7-fold increase in the minimum inhibitory concentration of D-cycloserine. This library represents a new tool to discern targets and pathways that influence drug efficacy of lead compounds with unknown mechanisms of action.

3. Transcriptomics

3.1. Target Identification

Understanding the mycobacterial transcriptome is key to connecting genome information to protein target expression, highlighting potentially druggable pathways and bacterial responses to drug exposure. With the development of whole-genome technologies, such as microarrays and more recently RNAseq, gene expression studies have been able to capture a snapshot of the total abundance and differential expression of transcripts present in an organism in various conditions. Transcriptomics has become an important tool for exploring the biology of *M.tb*, providing information about adaptive responses to understand mechanisms of pathogenicity, assign gene function, discover new drug targets and explore drug action. Transcriptomics in a discovery biology setting has uncovered induction of potentially druggable pathways involved in β-oxidation of fatty acids, the glyoxylate shunt and cholesterol metabolism in *M.tb* replicating intracellularly in macrophages [50,51] and in expectorated *M.tb* in patient sputa [52], alongside expression of metal detoxification systems [53,54] amongst others. Profiling in vitro models of persistence has revealed adaptations to respiratory and metabolic networks involved in the transition of *M.tb* between different growth states [55,56] highlighting target pathways for investigation. RNA signatures from animal models of TB infection [57,58] and human tissue [59] provide important information on the expression of targets in human disease, enhancing the prospect of cidal drug action by targeting pathways active in vivo. This is valuable evidence for drug discovery decision making, since the bactericidal or bacteriostatic inhibition of an essential target in vitro does not necessarily predict in vivo drug efficacy. This was demonstrated by Pethe et al. who identified pyrimidine-imidazoles as potent antimycobacterial agents in a whole cell screen against *M.tb*; lead compounds showed activity in vitro but failed to show any inhibition in a murine model of infection [60]. Compound efficacy was linked to the accumulation of glycerol phosphate and a reduction in ATP synthesis in the presence of glycerol. Glycerol metabolism is dispensable in vivo and thus inhibition of this pathway was not cidal in animal models of TB disease. Approaches combining transcriptomic and gene essentiality datasets offer a multi-omics solution to prioritizing pathways for further investigation.

3.2. Mode of Action

The transcriptional response of *M.tb* to antimicrobial drug exposure has improved the understanding of many drugs, providing new insights for antibiotics currently in use for TB treatment with known cidal mechanisms, as well as predicting mode of action and the targets of novel compounds. This unsupervised approach is especially useful for understanding the actions of lead compounds from high-throughput screens where the mechanism of *M.tb* killing is entirely mysterious. Comparison of the *M.tb* transcriptional response to a novel compound with mRNA signatures derived from drugs of known function allows broad mode of action to be revealed (Figure 2). In the first study of *M.tb* transcriptional adaptations to drug treatment, Wilson et al. used DNA microarrays to explore changes in gene expression in response to isoniazid [61]. The authors showed that drug exposure induced several genes relevant to isoniazid's known mode of action; drug treatment caused the cluster of five genes encoding type II fatty acid synthase enzymes (*fabD-acpM-kasA-kasB-accD6*) to be upregulated. Other induced genes, such as *efpA* and *ahpC* not in the biosynthetic pathway targeted by isoniazid, were linked to the toxic effects of the drug. In subsequent years, several studies have used DNA microarrays to correlate the mRNA signatures of *M.tb* exposed to antimicrobial drugs with predicted mode of action [62–64]. Boshoff et al. generated a dataset of 430 *M.tb* gene expression profiles to measure the effect of 75 different drugs, drug combinations and growth conditions [64]. The individual RNA drug signatures were classified into groups of agents with similar modes of action (protein synthesis inhibitors, transcriptional inhibitors, cell wall synthesis inhibitors and DNA damaging agents), which have been used to predict the mechanism of action of antimycobacterial compounds of unknown function derived from whole cell screening approaches [65–68].

Figure 2. The application of transcriptomics to define drug mode of action. Principle Component Analysis (PCA) of *M.tb* responses to seven different drugs (represented by shapes) derived from RNAseq of 2–3 biological replicates per drug. *M.tb* mRNA signatures from antimicrobial drugs with similar mechanisms of action will cluster together.

Transcriptomics has been applied to investigate the mode of action of drugs now in phase I/II studies or in the clinic [2]. The *M.tb* transcriptional signature to benzothiazinone exposure most resembled that of cell wall inhibitors and ethambutol in particular, providing an early indication that the compound was targeting cell wall biosynthesis and arabinogalactan synthesis. This was confirmed to be the case with benzothiazinones inhibiting decaprenyl-phosphoribose-2′-epimerase (DprE1) in the metabolism of D-arabinose, two steps upstream of the action of ethambutol [7]. RNA profiling provided fundamental insights to elucidate the molecular mechanism of mycobacterial killing by pretomanid (formerly PA-824), a nitroimidazole active against both replicating and non-replicating *M.tb*, by inhibiting cell wall synthesis and releasing nitric oxide [69,70]. Pretomanid exposure resulted in a dual signature indicating cell wall inhibition (similar to isoniazid) and respiratory chain disruption (similar to respiratory inhibitors such as cyanide). While the upregulation of genes such as *fasI*, the *fasII* operon, *efpA* and *iniBAC* signposted the aerobic killing mechanism targeting cell wall biosynthesis, drug efficacy in anaerobic conditions was marked by the induction of the *cyd* operon encoding the non-proton-pumping cytochrome bd oxidase, the nitrate reductase *narGHIJ* and other genes involved in respiration [71]. RNAseq analysis of the *M.tb* response to the recently approved nitroimidazole delamanid (formerly OPC-67683) showed many similarities to the pretomanid signature, highlighting that respiratory poisoning plays an important role in the bactericidal effect of these compounds in anaerobic conditions [72].

Boot et al. used RNAseq responses of *M.tb* and *Mycobacterium marinum* to subinhibitory concentrations of ciprofloxacin, ethambutol, isoniazid, streptomycin and rifampicin to select genes to act as drug mode of action-specific reporters [73]. Concordance between the expression levels of *M.tb* and *M. marinum* upon drug exposure were high for the orthologous genes; however, *M. marinum* showed a more distinct stress fingerprint that facilitated simple assays for quick mode of action determination. Ten drug-specific *M. marinum* genes were selected and their promoter regions were cloned into green fluorescent protein (GFP) reporter constructs. As proof of concept that these drug reporters could accelerate TB drug discovery by identifying the mode of action of hit compounds, the *MMAR_4645*-ciprofloxacin reporter and *iniBAC*-isoniazid reporter were used to screen a library of 196 antimycobacterial compounds. The screening revealed two compounds to have a mode of action similar to isoniazid, likely targeting the mycobacterial cell wall, and one compound, similar to ciprofloxacin, that potentially inhibited DNA replication. Understanding mycobacterial responses to antimicrobial drugs may also offer new targets to enhance combination therapy. Peterson et al. used the Environment and Gene Regulatory Influence Network (EGRIN) model and Probabilistic Regulation of Metabolism (PROM) model of *M.tb* regulatory systems to demonstrate that bedaquiline pushed bacilli into a tolerant state that reduced bedaquiline killing [74]. Disruption of this network, by knocking out key transcription factors (*Rv0324* and *Rv0880*) predicted to mediate this response, significantly increased bedaquiline killing. Analysis of transcriptome data from *M.tb* exposed to antitubercular drugs identified molecules that significantly downregulated the expression of *Rv0324* or *Rv0880*, predicting synergism between bedaquiline and pretomanid through the inhibition of the *Rv0880* regulon by pretomanid. The in vitro combination of sub-inhibitory concentrations of both drugs showed an additive to mildly synergistic effect, while the effect was eliminated and a strong antagonism observed when the combinations were tested using a strain over-expressing *Rv0880*. Given the vast number of possible drug combinations, this strategy could complement other preclinical methods and accelerate the discovery of new drug regimens for TB by avoiding combinations with antagonistic interactions and prioritizing those with synergistic effects.

The interaction between drugs in combination were also explored in the computational model INDIGO-MTB, with the premise that drug synergy and antagonism occur due to coordinated, system-level molecular changes involving multiple cellular processes [75]. Using a compendium of publicly-available and in-house *M.tb* transcriptional responses to drug exposure in vitro as input, the model screened in silico more than 1 million potential combinations of 164 drugs, predicting synergistic and antagonistic regimens featuring 35 existing and potential anti-TB drugs. Combinations containing chlorpromazine, a drug used to treat psychiatric disorders, and verapamil, used to treat hypertension, were highly enriched for synergistic interactions. In contrast, combinations featuring sutezolid, an oxazolidinone anti-TB drug in phase II trials, were observed to be antagonistic. Regimens featuring combinations of bacteriostatic and bactericidal drugs, and combinations of only bactericidal drugs, had significantly more antagonistic interactions than combinations of only bacteriostatic drugs. The predictions of INDIGO-MTB were validated experimentally in vitro using checkerboard assays and the high-throughput DiaMOND method and compared to a meta-analysis of data assembled from 57 phase II clinical trials. The authors found a significant correlation between INDIGO-MTB interaction scores for drug regimen synergy and sputum culture conversion rates after 8 weeks of treatment. The model also identified *Rv1353c* as a key transcriptional regulator mediating multiple drug interactions. The upregulation of *Rv1353c* in vitro reduced drug antagonism of the bedaquiline-streptomycin combination, suggesting this transcriptional factor might be targeted to enhance drug synergies.

Most transcriptional profiling studies map the response of log phase aerobically-respiring bacilli to drug exposure in vitro, this enables direct comparison between drug signatures from axenic culture, but misses the complexity of *M.tb* in vivo phenotypes. Studies by Walter et al. and Honeyborne et al. have characterized drug responses in patient sputa during standard therapy as a measure of in-patient drug efficacy, identifying, for example, an isoniazid signature in expectorated bacilli that disappears after only 3-4 days of the start of drug therapy [76,77].

4. Proteomics

4.1. Target Identification

Due to limitations in the sensitivity of methods and in the complexity of analysis, proteomics is yet to reach the multifaceted scope of genomics and transcriptomics approaches [78]. However, the variation between protein abundance and corresponding mRNA abundance, suggests that proteomics provides a distinct and useful understanding of *M.tb* physiological responses and target expression [79]. A number of studies have revealed insights into the expression of *M.tb* proteins intracellularly and in vivo, confirming the expression of potentially druggable targets. There are broadly two proteomics strategies, firstly, a top-down approach whereby intact proteins are isolated from a biological sample and sorted by gel electrophoresis, based on their physical and chemical properties, before identification by mass spectrometry (MS) [78]. Popular in early proteomic research, this technique typically found ~100 mycobacterial proteins, accounting for approximately 3% of the total *M.tb* proteome. The second approach is the bottom-up method, where a total set of proteins are proteolytically cleaved into peptides, followed by high-performance liquid chromatography and tandem mass spectrometry (LC-MS/MS). This strategy may quantify many more proteins and track a predefined subset of proteins to greater sensitivity, however computational resolution of profiles and insufficient sensitivity currently limits proteomics discoveries [78]. Despite the lag in proteomics compared to other omics, several studies have revealed new insights into mycobacterial protein expression. Målen et al. compared expression of *M.tb* H37Rv and *M.tb* H37Ra membrane proteins using gel electrophoresis and in-gel digestion of proteins followed by MS [80]. The authors identified over 1700 proteins expressed in both strains, 29 of which were membrane-associated proteins with >5-fold difference in abundance between strains. This highlighted expression of potentially druggable proteins associated with transmembrane transport, complementing genomics and transcriptomics approaches to confirm presence of target protein, and mapped the expression of efflux systems that might impact drug efficacy. Isobaric tagging and stable isotope labelling have enhanced the identification and quantification of proteins by mass spectrometry, enabling more efficient mapping of peptides to protein and allowing comparative samples to be analyzed simultaneously [81]. Thompson et al. quantified N-terminal acetylation of cytosolic and secreted proteins in *M.tb* and *M. marinum* using an N-terminal enrichment approach coupled with stable isotope ratio mass spectrometry [82]. The authors identified 211 endogenously

N-terminally acetylated proteins in *M.tb* (11% of the total observed proteome); 31.8% of these proteins were involved in intermediary metabolism and respiration, indicating the importance and abundance of protein acetylation in mycobacteria. A difference in acylation level was detected in cytoplasmic and secreted forms of 16 mycobacterial proteins, suggesting that N-terminal acetylation might affect protein localization. Acetylation was linked to the secretion of virulence factors (ESX substrates) and antimicrobial drug detoxification systems, signifying N-terminal acetylation as a potentially druggable pathway in *M.tb*. Whilst this approach is clearly beneficial for the detection of acetylation in mycobacteria, consideration should be taken when selecting strains and growth conditions, since these factors will impact protein enrichment resulting in variation between studies, as demonstrated by the authors when comparing the proteome of H37Rv with the published literature.

Proteomics has been applied to in vitro models of infection to find new targets for drug discovery. Albrethsen et al. combined label-free LC-MS/MS and 2D difference gel electrophoresis to reveal extracellular proteins produced by *M.tb* under nutrient starvation [83]. The authors identified proteins involved in toxin-antitoxin (TA) systems, suggesting a role for these systems in the switch from active to non-replicating metabolic states and proposing that TA systems should be explored as potential therapeutic targets. Despite RNAseq often being the preferred approach for such investigations, applying proteomics revealed information regarding protein localization, binding preferences and post-transcriptional modifications that would not be generated from an mRNA analysis. *M.tb* protein expression has been mapped during intracellular infection of alveolar epithelioid cells and macrophages [84]; comparison of the *M.tb* protein expression profiles in these two cell types identified 6 mycobacterial proteins that were differentially expressed in epithelioid cells. Kruh et al. investigated *M.tb* protein expression in vivo, identifying 500 unique *M.tb* proteins that were expressed in the lungs of guinea pigs [85]. Most abundant were *M.tb* proteins implicated in cell wall function and respiration, providing a rational basis for targeted drug discovery against these pathways active during infection.

4.2. Mode of Action

Proteomics has helped to deconvolute *M.tb* responses to drug exposure providing insight into mechanisms of drug action and resistance [86]. Recently, Meneguello et al. used proteomics to explore the metabolic pathways that contribute to the activity of rifampicin [87]. A small percentage of rifampicin resistance occurs through mechanisms other than the well-characterized target, β subunit of RNA polymerase. Proteomic profiling of rifampicin-treated *M.tb* resulted in the under-expression of four proteins implicated in cell wall biosynthesis (Ino1, FabD, EsxK and PPE60), suggesting that rifampicin also affects cell wall synthesis contributing to bacterial death. The authors used a liquid chromatography-mass spectrometry (LC-MS) approach, assessing small changes to the *M.tb* proteome temporally. Consideration should be taken when conducting such experiments to minimize the introduction of false-positives as a result of weak signals from an analytical column coupled with a highly sensitive detection platform. Despite these limitations, the authors reported a protein coverage comparable to other studies that applied nano-liquid chromatography. Nano-liquid chromatography is becoming an established tool for advanced peptide separation that uses very narrow columns that are more effective for detecting low abundance compounds [88]. Similarly, proteomics has been utilized to discern the mode of action of repurposed drugs that inhibit *M.tb* in vitro. Sulfamethoxazole, a broad-spectrum antibiotic that primarily targets the folate biosynthesis pathway, exhibits a synergistic effect when combined with other anti-TB drugs. Sarkar et al. mapped the *M.tb* response to sulfamethoxazole exposure using proteomics, identifying induction of oxidative stress and electron transport chain pathways that suggested an additional mode of action for this drug [89]. Proteomics may also be exploited to determine mechanisms of drug resistance. Putim et al. employed a shotgun proteomics system to identify proteins secreted in isoniazid- and rifampicin-resistant *M.tb* compared to drug-sensitive *M.tb*. Bacterial cultures were filtered through low binding protein-cellulose acetate membranes to collect culture filtrate proteins, before sodium dodecyl sulphate-polyacrylamide gel electrophoresis (SDS-PAGE), in-gel digestion and LC-MS. Depending on the aim of the study, consideration in the liquid chromatography (LC) approach should be taken. For example, for high proteome coverage, or for detecting proteins in low abundance, a 1D gradient for 8 h or a 2D LC should be used. However, for analysis of a limited sample volume, or where characterization of low-abundance proteins is not needed, a 1D gradient for 4 h may suffice. In addition, reproducibility and sample requirements will vary greatly depending on the method used [90]. Differential abundance of proteins involved in lipid metabolism, proteasome function and ATP-binding cassette transporters (ABC transporters) between drug-resistant and drug-sensitive strains may reveal novel systems that influence drug efficacy [91].

5. Metabolomics

5.1. Target Identification

Metabolomics, the analysis of the metabolite network within a biological system, is an indispensable omics approach for drug discovery, providing information on the potentially druggable processes occurring in a cell. Metabolomics in a discovery biology setting has elucidated *M.tb* metabolic pathways in use in different microenvironments. Carvalho et al. supplemented cultures grown aerobically at 37 °C with different 13C-labelled carbon substrates followed by LC-MS to separate and identify metabolites, demonstrating that *M.tb* co-catabolizes multiple carbon sources simultaneously, through glycolytic, pentose phosphate and tricarboxylic acid pathways [92]. For example, during co-catabolism of dextrose and acetate, dextrose was preferentially metabolized into intermediates of glycolysis and the pentose phosphate pathway, whereas acetate was preferentially used for tricarboxylic acid cycle (TCA cycle) intermediates. This understanding of the *M.tb* metabolic network will help to delineate key pathways and essential metabolites that could be exploited as therapeutic targets. More recently, Serafini et al. used a similar approach to elucidate the metabolic pathways involved in the assimilation of pyruvate and lactate in *M.tb*. Although it is well-established that lipids are important carbon sources for *M.tb* during infection, the authors demonstrated a novel function for the methylcitrate cycle, highlighting that it could be reversed for the biosynthesis of propionyl-CoA and the metabolism of pyruvate and lactate, identifying new targets for drug discovery efforts [93]. This study is an excellent example of a multi-omics approach, combining transposon-directed insertion site sequencing, RNAseq transcriptomics, proteomics and metabolomics, enabling an multi-analyte functional overview of the carbon metabolism network in *M.tb*.

Agapova et al. coupled stable isotope tracing of labelled amino acids with mass spectrometry to elucidate the use of amino acids as a nitrogen source in *M.tb* [94]. The authors showed that the co-metabolism of multiple amino acids as nitrogen sources did not improve growth compared to metabolism of a single source. In addition, several amino acids were utilized as sole nitrogen sources much faster than ammonium, suggesting that *M.tb* preferentially metabolizes specific host amino acids as sources of nitrogen. As such, metabolomics provided insight into the potential for targeting specific pathways in the *M.tb* nitrogen metabolic network. The authors also suggested that greater emphasis should be placed on amino acids as sole carbon sources to better mimic physiologically relevant conditions found in the host. Borah et al. used ^{15}N-flux spectral ratio analysis to demonstrate that *M.tb* in macrophages has access to multiple amino acids for nitrogen metabolism and identified serine as an amino acid not available to intracellular bacilli [95]. The proteinogenic amino acid serine that provides the nitrogen backbone for glycine and cysteine synthesis must be synthesized by intracellular *M.tb*, highlighting this pathway, and phosphoserine transaminase in particular, as a novel target for drug discovery. In a similar target identification application, Dutta et al. used metabolomics to confirm the role of the stringent response regulator Rel in controlling transition to non-replicating states by comparing wild type *M.tb* with an *M.tb* knockout strain lacking Rel$_{Mtb}$, verifying Rel as a potential anti-TB drug target. The authors then screened a library of compounds against this target, identifying lead compounds that killed nutrient-starved *M.tb* [96].

5.2. Mode of Action

The utility of metabolomics is demonstrated in studies to elucidate the mode of action of novel compounds. In a high-throughput metabolomics approach, Zampieri et al. evaluated mass spectra of supernatants from drug-treated *M. smegmatis* cultures to profile a library of 212 antimycobacterial compounds with unknown modes of action [97]. The metabolomic signatures were first established for 62 reference compounds with 17 known targets, before assessing similarity to the test compound profiles. Over 70% of the 212 compounds could be classified with a known mechanism of action, whilst 16 compounds resulted in metabolomic profiles dissimilar to the reference compounds. Of these 16 compounds, 6 exhibited a similar metabolomic response suggestive of the inhibition of lipid and trehalose metabolism. This approach revealed new druggable pathways in *M.tb*, and importantly enables drug discovery programs to diversify target pathways, discarding molecules that likely inhibit targets of existing drugs. In this study, the compounds with unknown mechanisms of action exhibited modest inhibitory activity against *M. smegmatis* with unique metabolic patterns that likely reflect specificity in their underlying modes of action. Whilst further studies should be directed at assessing the activity and mechanism of action of these compounds in *M.tb*, this study clearly demonstrates the utility of this approach to recognize compounds that target novel pathways. In a similar approach, untargeted metabolite profiling using flow infusion electrospray ion high resolution mass spectrometry was used to explore the mode of action of pretomanid [98]. The *M. smegmatis* metabolite profile after exposure to pretomanid was distinct when compared to ampicillin, ethambutol, ethionamide, isoniazid, kanamycin, linezolid, rifampicin and streptomycin-treated cultures. Mapping of differentially abundant metabolites onto pathways highlighted the pentose phosphate pathway, suggesting that accumulation of the toxic metabolite methylglyoxal may contribute to the antibacterial activity of pretomanid. A recent LC-MS-based metabolic linkage analysis of bedaquiline-treated *M.tb* revealed that, alongside inhibition of ATP synthase, glutamine metabolism was also impacted. Since glutamine synthesis inhibitors were synergistic in combination with bedaquiline, an indirect secondary effect of bedaquiline on glutamine biosynthesis was distinguished that could be targeted therapeutically [99]. These approaches demonstrate how metabolomics may be used to elucidate the action of unknown drugs, and reveal fundamental information about the physiology of *M.tb*.

6. Lipidomics

6.1. Target Identification

Mycobacteria have unique cell envelopes, high in lipid diversity and abundance, comprising up to 40% of the bacillus dry weight [100]. Cell wall biosynthetic pathways are the target of many existing anti-TB drugs; in addition, the sequencing of the *M.tb* genome revealed many lipid biosynthesis and polyketide synthase genes that might be exploited as potential therapeutic targets. The study of this network of cellular lipids within a biological system is broadly categorized in a branch of metabolomics, known as lipidomics, which examines lipid species that are present and how they interact with other lipids, metabolites and proteins in a cell [101]. Lipidomics relies on mass spectroscopy, measuring the mass-to-charge ratio and abundance of gas-phase ions, further characterized into gas chromatography (GC)-MS, liquid chromatography (LC)-MS and direct infusion-MS [102]. The large diversity of lipids and the lack of spatial information about the distribution of these moieties within a cell complicates the inferences from lipidomic experiments [103]. Lipidomics has been employed to discover potentially druggable lipid biosynthesis pathways based on the *M.tb* response to the changing environment. Raghunandanan et al. determined the pattern of *M.tb* lipid changes in hypoxia-induced dormancy and resuscitation, showing that lipid concentration drastically decreased during dormancy and gradually increased again during re-aeration [104]. Several lipids were more abundant in non-replicating bacteria, revealing potentially targetable pathways [104]. This study demonstrates the potential of lipidomics to evaluate *M.tb* in vitro in conditions predicted to mimic in vivo microenvironments. Compared to conventional high performance-LC, the ultra-performance LC technique provided a much greater chromatographic resolution and subsequently faster analysis time [105]. Lipidomics is also a valuable tool to characterize lipid biosynthetic pathway targets. The fatty acid synthase FAS-II multi-enzyme system is essential for the biosynthesis of mycolic acids and normal cell wall function in mycobacteria. It is the target of several antimycobacterial drugs, such as isoniazid, and therefore offers additional therapeutic potential. To understand the role of HadD, a novel FAS-II enzyme, Lefebvre et al. analyzed total extractable lipids from *M. smegmatis hadD* knockout mutants by high-performance thin-layer chromatography to demonstrate that *hadD* deletion resulted in the absence of α- and epoxy-mycolic acids that disrupted the cell envelope and reduced bacterial fitness [106]. Subsequently, the authors showed that *hadD* deletion in *M.tb* resulted in a 63% reduction in keto-mycolic acids, while overexpression of *hadD* induced an 87% increase in keto-mycolic acids compared to wild type. Knockout mutants of *hadD* were attenuated in a murine model of infection, confirming *hadD* as a new target for drug discovery [107].

6.2. Mode of Action

Sharma et al. demonstrated the role for lipidomics in mechanism of action studies by mapping the impact of the natural antimycobacterial compound vanillin in *M. smegmatis*. The authors observed that vanillin changes the composition of fatty acids, glycolipids, glycerophospholipids and saccharolipids causing disruption of cell membrane homeostasis [108]. Similarly, lipidomic analysis has been instrumental in investigating the consequences of *M.tb* drug-resistance conferring mutations to well-characterized anti-TB drugs [109]. Howard et al. demonstrated that rifampicin-resistant *M.tb* isolates with mutations in *rpoB* exhibited altered lipid profiles dependent on the specific location of the SNPs [109]. Lipidomic analysis showed that different *rpoB* SNPs resulted in distinct relative abundances of short-chain and long-chain fatty acid phthiocerol dimycocerosates (PDIM), which induced alternative macrophage activation pathways and altered macrophage metabolism. Such analyses are useful not only to understand the mode of action of a compound, but to deconvolute the consequences of drug resistance-conferring mutations. Lipidomics is often used as an unbiased approach to complement findings from other methodologies. An example is the evaluation of the mycolic acid transporter, MmpL3, as a druggable target [110]. MmpL3 is an integral inner membrane transporter, with a role in the export of mycolic acids to the periplasmic space in the biosynthesis of the mycobacterial cell wall. Lipidomic analysis of *mmpL3 M.tb* knockdown mutants using thin layer chromatography of total lipids revealed a fast decline in cell wall-bound mycolic acids and trehalose dimycolates. Combined with confirmation of in vitro and in vivo essentiality, this study confirmed the mycolic acid transporter MmpL3, as a validated druggable target for *M.tb* drug discovery. Similarly, lipidomics, alongside genomics and transcriptomics, characterized the mode of action of HC2091, a novel compound that likely targets MmpL3. *M.tb* HC2091-resistant mutants were shown to have SNPs in the *mmpL3* gene, likely conferring drug resistance. Thin layer chromatography of the total extractable lipids from HC2091-treated *M.tb* cultures supplemented with radiolabelled-sodium acetate demonstrated a dose-dependent reduction in trehalose dimycolate accumulation with increasing concentration of HC2091. In combination with transcriptional profiling of HC2091-treated bacilli, the authors confirmed that HC2091 targets MmpL3 through a mechanism distinct from other MmpL3 inhibitors [66]. Lipidomics, therefore, often in combination with other omics approaches, is an effective tool, especially in the interrogation of cell wall biosynthetic pathways, a rich source of druggable *M.tb* targets.

7. Future Outlook and Conclusions

This review has focused on established mycobacterial omics technologies and their application to the early stages of *M.tb* drug discovery. New omics are emerging that will contribute to future drug development; for example, glycomics, the study of all glycans in a biological system. Glycans are vital in a broad range of processes, their direct recognition by glycan-binding proteins is important for many processes that may be essential and druggable. Recently, Kavunja et al. used a glycomics approach to identify mycolate-interacting proteins associated with synthesis and remodeling of the membrane in *M. smegmatis* that could lead to the validation of novel therapeutic targets [111]. Such techniques offer unique opportunities for biological discovery and new target identification that will expand as methodologies develop to increase sensitivity and reduce complexity.

The strength of omics technologies is multiplied when used in combination to understand bacterial metabolism and pathogenicity, leading to a true systems approach to antimicrobial drug discovery. This requires the development and maintenance of bioinformatics tools, data repositories, integration and visualization platforms. Recent initiatives in data-sharing not only showcase multi-omics studies but also aim to make omics datasets more readily accessible to the research community [112]. Increased accessibility alongside significant reductions in the cost of omics technologies and advances in data analysis platforms [113], both for data management and for improving ease-of-use, ensure that omics are ubiquitous in drug discovery, preclinical and clinical development programs across academia and pharma [114]. Where *M.tb*-focused omics technologies are combined with human omics systems in a quantitative pharmacology approach to finding and delivering new therapeutics.

The unsupervised nature of omics approaches and the separation from a priori hypotheses places omics technologies in unique and vital roles in the drug development process. For multi-omics to become fully incorporated into the drug development pipeline the challenge remains to move more quickly from initial target identification to target validation in target-based discovery, and in compound-first approaches to scale up mode of action studies to integrate with medium/high-throughput screening. Omics platforms have contributed significantly to the development of the most recent *M.tb* drugs brought to market and to multiple drug candidates now in clinical trials, alongside providing insights into *M.tb* physiology, drug action and drug resistance. Omics methodologies have become valuable tools in the search for new antimicrobial drugs that are becoming increasingly important to find.

Author Contributions: Conceptualization; Writing; Review and Editing, A.G., D.C., L.M.W. and S.J.W. All authors have read and agreed to the published version of the manuscript.

Funding: This research was funded by the National Centre for the Replacement, Refinement and Reduction of Animals in Research (NC3Rs), grant number NC/R001669/1; and the Wellcome Trust Institutional Strategic Support Fund, project G2306, grant number 192470.

Conflicts of Interest: The authors declare no conflict of interest. The funders had no role in the design of the study; in the collection, analyses, or interpretation of data; in the writing of the manuscript, or in the decision to publish the results.

References

1. World Health Organization. Global Tuberculosis Report. 2019. Available online: https://www.who.int/tb/publications/global_report/en/ (accessed on 24 May 2020).
2. The Working Group for New TB Drugs. New Drugs for TB Clinical Pipeline. Available online: https://www.newtbdrugs.org/pipeline/clinical (accessed on 26 May 2020).
3. Zumla, A.; Nahid, P.; Cole, S.T. Advances in the development of new tuberculosis drugs and treatment regimens. *Nat. Rev. Drug Discov.* **2013**, *12*, 388–404. [CrossRef]
4. Shetye, G.S.; Franzblau, S.G.; Cho, S. New tuberculosis drug targets, their inhibitors, and potential therapeutic impact. *Transl. Res.* **2020**, *220*, 68–97. [CrossRef]
5. Wellington, S.; Hung, D.T. The expanding diversity of *Mycobacterium tuberculosis* drug targets. *ACS Infect. Dis.* **2018**, *4*, 696–714. [CrossRef] [PubMed]
6. Hasin, Y.; Seldin, M.; Lusis, A. Multi-omics approaches to disease. *Genome Biol.* **2017**, *18*, 1–15. [CrossRef] [PubMed]
7. Makarov, V.; Manina, G.; Mikusova, K.; Mollmann, U.; Ryabova, O.; Saint-Joanis, B.; Dhar, N.; Pasca, M.R.; Buroni, S.; Lucarelli, A.P.; et al. Benzothiazinones kill *Mycobacterium tuberculosis* by blocking arabinan synthesis. *Science* **2009**, *324*, 801–804. [CrossRef] [PubMed]
8. Cole, S.T.; Brosch, R.; Parkhill, J.; Garnier, T.; Churcher, C.; Harris, D.; Gordon, S.V.; Eiglmeier, K.; Gas, S.; Barry, C.E., 3rd; et al. Deciphering the biology of *Mycobacterium tuberculosis* from the complete genome sequence. *Nature* **1998**, *393*, 537–544. [CrossRef] [PubMed]

9. Mazandu, G.K.; Mulder, N.J. Function prediction and analysis of *Mycobacterium tuberculosis* hypothetical proteins. *Int. J. Mol. Sci.* **2012**, *13*, 7283–7302. [CrossRef]
10. Niemann, S.; Köser, C.U.; Gagneux, S.; Plinke, C.; Homolka, S.; Bignell, H.; Carter, R.J.; Cheetham, R.K.; Cox, A.; Gormley, N.A.; et al. Genomic diversity among drug sensitive and multidrug resistant isolates of *Mycobacterium tuberculosis* with identical DNA fingerprints. *PLoS ONE* **2008**, *4*, 1–7. [CrossRef]
11. Sloan Siegrist, M.; Rubin, E.M. Phage transposon mutagenesis. In *Methods in Molecular Biology*, 2nd ed.; Humana Press: New York, NY, USA, 2009; pp. 311–325. [CrossRef]
12. Sassetti, C.M.; Boyd, D.H.; Rubin, E.J. Comprehensive identification of conditionally essential genes in mycobacteria. *Proc. Natl. Acad. Sci. USA* **2001**, *98*, 12712–12717. [CrossRef]
13. DeJesus, M.A.; Gerrick, E.R.; Xu, W.; Park, S.W.; Long, J.E.; Boutte, C.C.; Rubin, E.J.; Schnappinger, D.; Ehrt, S.; Fortune, S.M.; et al. Comprehensive essentiality analysis of the *Mycobacterium tuberculosis* genome via saturating transposon mutagenesis. *mBio* **2017**, *8*, 1–17. [CrossRef]
14. Sassetti, C.M.; Boyd, D.H.; Rubin, E.J. Genes required for mycobacterial growth defined by high density mutagenesis. *Mol. Microbiol.* **2003**, *48*, 77–84. [CrossRef] [PubMed]
15. Zhang, Y.; Morar, M.; Ealick, S.E. Structural biology of the purine biosynthetic pathway. *Cell Mol. Life Sci.* **2008**, *65*, 3699–3724. [CrossRef] [PubMed]
16. Russell, D.G.; Lee, W.; Tan, S.; Sukumar, N.; Podinovskaia, M.; Fahey, R.J.; Vanderven, B.C. The Sculpting of the *Mycobacterium tuberculosis* Genome by Host Cell-Derived Pressures. *Microbiol. Spectr.* **2014**, *2*, 1–18. [CrossRef] [PubMed]
17. Barczak, A.K.; Avraham, R.; Singh, S.; Luo, S.S.; Zhang, W.R.; Bray, M.A.; Hinman, A.E.; Thompson, M.; Nietupski, R.M.; Golas, A.; et al. Systematic, multiparametric analysis of *Mycobacterium tuberculosis* intracellular infection offers insight into coordinated virulence. *PLoS Pathog.* **2017**, *13*, 1–27. [CrossRef]
18. Schnappinger, D. Genetic Approaches to Facilitate Antibacterial Drug Development. *Cold Spring Harb. Perspect. Med.* **2015**, *5*, 1–15. [CrossRef]
19. Johnson, E.O.; LaVerriere, E.; Office, E.; Stanley, M.; Meyer, E.; Kawate, T.; Gomez, J.E.; Audette, R.E.; Bandyopadhyay, N.; Betancourt, N.; et al. Large-scale chemical-genetics yields new *M. tuberculosis* inhibitor classes. *Nature* **2019**, *571*, 72–78. [CrossRef]
20. World Health Organization. Guidelines for Treatment of Drug-Susceptible Tuberculosis and Patient Care. 2017. Update. Available online: https://www.who.int/tb/publications/2017/dstb_guidance_2017/en (accessed on 26 May 2020).
21. Qi, L.S.; Larson, M.H.; Gilbert, L.A.; Doudna, J.A.; Weissman, J.S.; Arkin, A.P.; Lim, W.A. Repurposing CRISPR as an RNA-guided platform for sequence-specific control of gene expression. *Cell* **2013**, *152*, 1173–1183. [CrossRef]
22. Rock, J. Tuberculosis drug discovery in the CRISPR era. *PLoS Pathog.* **2019**, *15*, 1–10. [CrossRef]
23. Gandotra, S.; Schnappinger, D.; Monteleone, M.; Hillen, W.; Ehrt, S. In vivo gene silencing identifies the *Mycobacterium tuberculosis* proteasome as essential for the bacteria to persist in mice. *Nat. Med.* **2007**, *13*, 1515–1520. [CrossRef]
24. Rock, J.M.; Hopkins, F.F.; Chavez, A.; Diallo, M.; Chase, M.R.; Gerrick, E.R.; Pritchard, J.R.; Church, G.M.; Rubin, E.J.; Sassetti, C.M.; et al. Programmable transcriptional repression in mycobacteria using an orthogonal CRISPR interference platform. *Nat. Microbiol.* **2017**, *2*, 1–21. [CrossRef]
25. Li, R.; Sirawaraporn, R.; Chitnumsub, P.; Sirawaraporn, W.; Wooden, J.; Athappilly, F.; Turley, S.; Hol, W.G. Three-dimensional structure of *M. tuberculosis* dihydrofolate reductase reveals opportunities for the design of novel tuberculosis drugs. *J. Mol. Biol.* **2000**, *295*, 307–323. [CrossRef] [PubMed]
26. Shiloh, M.U.; Champion, P.A. To catch a killer. What can mycobacterial models teach us about *Mycobacterium tuberculosis* pathogenesis? *Curr. Opin. Microbiol.* **2010**, *13*, 86–92. [CrossRef]
27. Haeussler, M.; Schonig, K.; Eckert, H.; Eschstruth, A.; Mianne, J.; Renaud, J.B.; Schneider-Maunoury, S.; Shkumatava, A.; Teboul, L.; Kent, J.; et al. Evaluation of off-target and on-target scoring algorithms and integration into the guide RNA selection tool CRISPOR. *Genome Biol.* **2016**, *17*, 1–12. [CrossRef] [PubMed]
28. Tsai, S.Q.; Zheng, Z.; Nguyen, N.T.; Liebers, M.; Topkar, V.V.; Thapar, V.; Wyvekens, N.; Khayter, C.; Iafrate, A.J.; Le, L.P.; et al. GUIDE-seq enables genome-wide profiling of off-target cleavage by CRISPR-Cas nucleases. *Nat. Biotechnol.* **2015**, *33*, 187–197. [CrossRef] [PubMed]
29. Galagan, J.E. Genomic insights into tuberculosis. *Nat. Rev. Genet.* **2014**, *15*, 307–320. [CrossRef] [PubMed]

30. Andries, K.; Verhasselt, P.; Guillemont, J.; Gohlmann, H.W.; Neefs, J.M.; Winkler, H.; Van Gestel, J.; Timmerman, P.; Zhu, M.; Lee, E.; et al. A diarylquinoline drug active on the ATP synthase of *Mycobacterium tuberculosis*. *Science* **2005**, *307*, 223–227. [CrossRef]
31. Kundu, S.; Biukovic, G.; Gruber, G.; Dick, T. Bedaquiline Targets the epsilon Subunit of Mycobacterial F-ATP Synthase. *Antimicrob. Agents Chemother.* **2016**, *60*, 6977–6979. [CrossRef]
32. Almeida, D.; Ioerger, T.; Tyagi, S.; Li, S.Y.; Mdluli, K.; Andries, K.; Grosset, J.; Sacchettini, J.; Nuermberger, E. Mutations in pepQ Confer Low-Level Resistance to Bedaquiline and Clofazimine in *Mycobacterium tuberculosis*. *Antimicrob. Agents Chemother.* **2016**, *60*, 4590–4599. [CrossRef]
33. Rybniker, J.; Vocat, A.; Sala, C.; Busso, P.; Pojer, F.; Benjak, A.; Cole, S.T. Lansoprazole is an antituberculous prodrug targeting cytochrome bc1. *Nat. Commun.* **2015**, *6*, 1–8. [CrossRef]
34. Rybniker, J.; Kohl, T.A.; Barilar, I.; Niemann, S. No Evidence for Acquired Mutations Associated with Cytochrome bc 1 Inhibitor Resistance in 13,559 Clinical *Mycobacterium tuberculosis* Complex Isolates. *Antimicrob. Agents Chemother.* **2019**, *63*, 1–3. [CrossRef]
35. Ling, L.L.; Schneider, T.; Peoples, A.J.; Spoering, A.L.; Engels, I.; Conlon, B.P.; Mueller, A.; Schaberle, T.F.; Hughes, D.E.; Epstein, S.; et al. A new antibiotic kills pathogens without detectable resistance. *Nature* **2015**, *517*, 455–459. [CrossRef] [PubMed]
36. Piddock, L.J. Teixobactin, the first of a new class of antibiotics discovered by iChip technology? *J. Antimicrob. Chemother.* **2015**, *70*, 2679–2680. [CrossRef] [PubMed]
37. Allix-Beguec, C.; Arandjelovic, I.; Bi, L.; Beckert, P.; Bonnet, M.; Bradley, P.; Cabibbe, A.M.; Cancino-Munoz, I.; et al.; CRyPTIC Consortium; The 100000 Genomes Project Prediction of Susceptibility to First-Line Tuberculosis Drugs by DNA Sequencing. *N. Engl. J. Med.* **2018**, *379*, 1403–1415. [CrossRef]
38. Vilcheze, C.; Jacobs, W.R., Jr. Resistance to Isoniazid and Ethionamide in *Mycobacterium tuberculosis*: Genes, Mutations, and Causalities. *Microbiol. Spectr.* **2014**, *2*, 1–21. [CrossRef] [PubMed]
39. Newton, G.L.; Buchmeier, N.; Fahey, R.C. Biosynthesis and functions of mycothiol, the unique protective thiol of Actinobacteria. *Microbiol. Mol. Biol. Rev.* **2008**, *72*, 471–494. [CrossRef]
40. Vilcheze, C.; Av-Gay, Y.; Attarian, R.; Liu, Z.; Hazbon, M.H.; Colangeli, R.; Chen, B.; Liu, W.; Alland, D.; Sacchettini, J.C.; et al. Mycothiol biosynthesis is essential for ethionamide susceptibility in *Mycobacterium tuberculosis*. *Mol. Microbiol.* **2008**, *69*, 1316–1329. [CrossRef]
41. Vilcheze, C.; Av-Gay, Y.; Barnes, S.W.; Larsen, M.H.; Walker, J.R.; Glynne, R.J.; Jacobs, W.R., Jr. Coresistance to isoniazid and ethionamide maps to mycothiol biosynthetic genes in *Mycobacterium bovis*. *Antimicrob. Agents Chemother.* **2011**, *55*, 4422–4423. [CrossRef]
42. Zhang, Y.; Heym, B.; Allen, B.; Young, D.; Cole, S. The catalase-peroxidase gene and isoniazid resistance of *Mycobacterium tuberculosis*. *Nature* **1992**, *358*, 591–593. [CrossRef]
43. DeBarber, A.E.; Mdluli, K.; Bosman, M.; Bekker, L.G.; Barry, C.E., 3rd. Ethionamide activation and sensitivity in multidrug-resistant *Mycobacterium tuberculosis*. *Proc. Natl. Acad. Sci. USA* **2000**, *97*, 9677–9682. [CrossRef]
44. Xu, X.; Vilcheze, C.; Av-Gay, Y.; Gomez-Velasco, A.; Jacobs, W.R., Jr. Precise null deletion mutations of the mycothiol synthesis genes reveal their role in isoniazid and ethionamide resistance in *Mycobacterium smegmatis*. *Antimicrob. Agents Chemother.* **2011**, *55*, 3133–3139. [CrossRef]
45. Banerjee, A.; Dubnau, E.; Quemard, A.; Balasubramanian, V.; Um, K.S.; Wilson, T.; Collins, D.; de Lisle, G.; Jacobs, W.R., Jr. inhA, a gene encoding a target for isoniazid and ethionamide in *Mycobacterium tuberculosis*. *Science* **1994**, *263*, 227–230. [CrossRef] [PubMed]
46. Hartkoorn, R.C.; Uplekar, S.; Cole, S.T. Cross-resistance between clofazimine and bedaquiline through upregulation of MmpL5 in *Mycobacterium tuberculosis*. *Antimicrob. Agents Chemother.* **2014**, *58*, 2979–2981. [CrossRef] [PubMed]
47. Radhakrishnan, A.; Kumar, N.; Wright, C.C.; Chou, T.H.; Tringides, M.L.; Bolla, J.R.; Lei, H.T.; Rajashankar, K.R.; Su, C.C.; Purdy, G.E.; et al. Crystal structure of the transcriptional regulator Rv0678 of *Mycobacterium tuberculosis*. *J. Biol. Chem.* **2014**, *289*, 16526–16540. [CrossRef] [PubMed]
48. Villellas, C.; Coeck, N.; Meehan, C.J.; Lounis, N.; de Jong, B.; Rigouts, L.; Andries, K. Unexpected high prevalence of resistance-associated Rv0678 variants in MDR-TB patients without documented prior use of clofazimine or bedaquiline. *J. Antimicrob. Chemother.* **2017**, *72*, 684–690. [CrossRef]
49. Melief, E.; Kokoczka, R.; Files, M.; Bailey, M.A.; Alling, T.; Li, H.; Ahn, J.; Misquith, A.; Korkegian, A.; Roberts, D.; et al. Construction of an overexpression library for *Mycobacterium tuberculosis*. *Biol. Methods Protoc.* **2018**, *3*, 1–9. [CrossRef]

50. Schnappinger, D.; Ehrt, S.; Voskuil, M.I.; Liu, Y.; Mangan, J.A.; Monahan, I.M.; Dolganov, G.; Efron, B.; Butcher, P.D.; Nathan, C.; et al. Transcriptional Adaptation of *Mycobacterium tuberculosis* within Macrophages: Insights into the Phagosomal Environment. *J. Exp. Med.* **2003**, *198*, 693–704. [CrossRef]
51. Rienksma, R.A.; Suarez-Diez, M.; Mollenkopf, H.J.; Dolganov, G.M.; Dorhoi, A.; Schoolnik, G.K.; Martins Dos Santos, V.A.; Kaufmann, S.H.; Schaap, P.J.; Gengenbacher, M. Comprehensive insights into transcriptional adaptation of intracellular mycobacteria by microbe-enriched dual RNA sequencing. *BMC Genom.* **2015**, *16*, 1–15. [CrossRef]
52. Garton, N.J.; Waddell, S.J.; Sherratt, A.L.; Lee, S.M.; Smith, R.J.; Senner, C.; Hinds, J.; Rajakumar, K.; Adegbola, R.A.; Besra, G.S.; et al. Cytological and transcript analyses reveal fat and lazy persister-like bacilli in tuberculous sputum. *PLoS Med.* **2008**, *5*, 634–645. [CrossRef]
53. Tailleux, L.; Waddell, S.J.; Pelizzola, M.; Mortellaro, A.; Withers, M.; Tanne, A.; Castagnoli, P.R.; Gicquel, B.; Stoker, N.G.; Butcher, P.D.; et al. Probing host pathogen cross-talk by transcriptional profiling of both *Mycobacterium tuberculosis* and infected human dendritic cells and macrophages. *PLoS ONE* **2008**, *3*, 1–14. [CrossRef]
54. Botella, H.; Peyron, P.; Levillain, F.; Poincloux, R.; Poquet, Y.; Brandli, I.; Wang, C.; Tailleux, L.; Tilleul, S.; Charriere, G.M.; et al. Mycobacterial p(1)-type ATPases mediate resistance to zinc poisoning in human macrophages. *Cell Host Microbe.* **2011**, *10*, 248–259. [CrossRef]
55. Rustad, T.R.; Harrell, M.I.; Liao, R.; Sherman, D.R. The enduring hypoxic response of *Mycobacterium tuberculosis*. *PLoS ONE* **2008**, *3*, 1–8. [CrossRef] [PubMed]
56. Salina, E.G.; Waddell, S.J.; Hoffmann, N.; Rosenkrands, I.; Butcher, P.D.; Kaprelyants, A.S. Potassium availability triggers *Mycobacterium tuberculosis* transition to, and resuscitation from, non-culturable (dormant) states. *Open Biol.* **2014**, *4*, 1–15. [CrossRef] [PubMed]
57. Talaat, A.M.; Lyons, R.; Howard, S.T.; Johnston, S.A. The temporal expression profile of *Mycobacterium tuberculosis* infection in mice. *Proc. Natl. Acad. Sci. USA* **2004**, *101*, 4602–4607. [CrossRef] [PubMed]
58. Talaat, A.M.; Ward, S.K.; Wu, C.W.; Rondon, E.; Tavano, C.; Bannantine, J.P.; Lyons, R.; Johnston, S.A. Mycobacterial bacilli are metabolically active during chronic tuberculosis in murine lungs: Insights from genome-wide transcriptional profiling. *J. Bacteriol.* **2007**, *189*, 4265–4274. [CrossRef]
59. Rachman, H.; Strong, M.; Ulrichs, T.; Grode, L.; Schuchhardt, J.; Mollenkopf, H.; Kosmiadi, G.A.; Eisenberg, D.; Kaufmann, S.H. Unique transcriptome signature of *Mycobacterium tuberculosis* in pulmonary tuberculosis. *Infect. Immun.* **2006**, *74*, 1233–1242. [CrossRef] [PubMed]
60. Pethe, K.; Sequeira, P.C.; Agarwalla, S.; Rhee, K.; Kuhen, K.; Phong, W.Y.; Patel, V.; Beer, D.; Walker, J.R.; Duraiswamy, J.; et al. A chemical genetic screen in *Mycobacterium tuberculosis* identifies carbon-source-dependent growth inhibitors devoid of in vivo efficacy. *Nat. Commun.* **2010**, *1*, 1–8. [CrossRef] [PubMed]
61. Wilson, M.; DeRisi, J.; Kristensen, H.H.; Imboden, P.; Rane, S.; Brown, P.O.; Schoolnik, G.K. Exploring drug-induced alterations in gene expression in *Mycobacterium tuberculosis* by microarray hybridization. *Proc. Natl. Acad. Sci. USA* **1999**, *96*, 12833–12838. [CrossRef] [PubMed]
62. Betts, J.C.; McLaren, A.; Lennon, M.G.; Kelly, F.M.; Lukey, P.T.; Blakemore, S.J.; Duncan, K. Signature gene expression profiles discriminate between isoniazid-, thiolactomycin-, and triclosan-treated *Mycobacterium tuberculosis*. *Antimicrob. Agents Chemother.* **2003**, *47*, 2903–2913. [CrossRef]
63. Waddell, S.J.; Stabler, R.A.; Laing, K.; Kremer, L.; Reynolds, R.C.; Besra, G.S. The use of microarray analysis to determine the gene expression profiles of *Mycobacterium tuberculosis* in response to anti-bacterial compounds. *Tuberculosis* **2004**, *84*, 263–274. [CrossRef] [PubMed]
64. Boshoff, H.I.; Myers, T.G.; Copp, B.R.; McNeil, M.R.; Wilson, M.A.; Barry, C.E., 3rd. The transcriptional responses of *Mycobacterium tuberculosis* to inhibitors of metabolism: Novel insights into drug mechanisms of action. *J. Biol. Chem.* **2004**, *279*, 40174–40184. [CrossRef]
65. Liang, J.; Zeng, F.; Guo, A.; Liu, L.; Guo, N.; Li, L.; Jin, J.; Wu, X.; Liu, M.; Zhao, D.; et al. Microarray analysis of the chelerythrine-induced transcriptome of *Mycobacterium tuberculosis*. *Curr. Microbiol.* **2011**, *62*, 1200–1208. [CrossRef]
66. Zheng, H.; Williams, J.T.; Coulson, G.B.; Haiderer, E.R.; Abramovitch, R.B. HC2091 Kills *Mycobacterium tuberculosis* by Targeting the MmpL3 Mycolic Acid Transporter. *Antimicrob. Agents Chemother.* **2018**, *62*, 1–12. [CrossRef]

67. Foo, C.S.; Lupien, A.; Kienle, M.; Vocat, A.; Benjak, A.; Sommer, R.; Lamprecht, D.A.; Steyn, A.J.C.; Pethe, K.; Piton, J.; et al. Arylvinylpiperazine Amides, a New Class of Potent Inhibitors Targeting QcrB of *Mycobacterium tuberculosis*. *mBio* **2018**, *9*, 1–13. [CrossRef] [PubMed]
68. Lupien, A.; Foo, C.S.; Savina, S.; Vocat, A.; Piton, J.; Monakhova, N.; Benjak, A.; Lamprecht, D.A.; Steyn, A.J.C.; Pethe, K.; et al. New 2-Ethylthio-4-methylaminoquinazoline derivatives inhibiting two subunits of cytochrome bc1 in *Mycobacterium tuberculosis*. *PLoS Pathog.* **2020**, *16*, 1–19. [CrossRef] [PubMed]
69. Stover, C.K.; Warrener, P.; VanDevanter, D.R.; Sherman, D.R.; Arain, T.M.; Langhorne, M.H.; Anderson, S.W.; Towell, J.A.; Yuan, Y.; McMurray, D.N.; et al. A small-molecule nitroimidazopyran drug candidate for the treatment of tuberculosis. *Nature* **2000**, *405*, 962–966. [CrossRef] [PubMed]
70. Singh, R.; Manjunatha, U.; Boshoff, H.I.; Ha, Y.H.; Niyomrattanakit, P.; Ledwidge, R.; Dowd, C.S.; Lee, I.Y.; Kim, P.; Zhang, L.; et al. PA-824 kills nonreplicating *Mycobacterium tuberculosis* by intracellular NO release. *Science* **2008**, *322*, 1392–1395. [CrossRef]
71. Manjunatha, U.; Boshoff, H.I.; Barry, C.E. The mechanism of action of PA-824: Novel insights from transcriptional profiling. *Commun. Integr. Biol.* **2009**, *2*, 215–218. [CrossRef]
72. Van den Bossche, A.; Varet, H.; Sury, A.; Sismeiro, O.; Legendre, R.; Coppee, J.Y.; Mathys, V.; Ceyssens, P.J. Transcriptional profiling of a laboratory and clinical *Mycobacterium tuberculosis* strain suggests respiratory poisoning upon exposure to delamanid. *Tuberculosis* **2019**, *117*, 18–23. [CrossRef]
73. Boot, M.; Commandeur, S.; Subudhi, A.K.; Bahira, M.; Smith, T.C., 2nd; Abdallah, A.M.; van Gemert, M.; Lelievre, J.; Ballell, L.; Aldridge, B.B.; et al. Accelerating Early Antituberculosis Drug Discovery by Creating Mycobacterial Indicator Strains That Predict Mode of Action. *Antimicrob. Agents Chemother.* **2018**, *62*, 1–16. [CrossRef]
74. Peterson, E.J.R.; Ma, S.; Sherman, D.R.; Baliga, N.S. Network analysis identifies Rv0324 and Rv0880 as regulators of bedaquiline tolerance in *Mycobacterium tuberculosis*. *Nat. Microbiol.* **2016**, *1*, 16078. [CrossRef]
75. Ma, S.; Jaipalli, S.; Larkins-Ford, J.; Lohmiller, J.; Aldridge, B.B.; Sherman, D.R.; Chandrasekaran, S. Transcriptomic Signatures Predict Regulators of Drug Synergy and Clinical Regimen Efficacy against Tuberculosis. *mBio* **2019**, *10*, 1–16. [CrossRef] [PubMed]
76. Honeyborne, I.; McHugh, T.D.; Kuittinen, I.; Cichonska, A.; Evangelopoulos, D.; Ronacher, K.; van Helden, P.D.; Gillespie, S.H.; Fernandez-Reyes, D.; Walzl, G.; et al. Profiling persistent tubercule bacilli from patient sputa during therapy predicts early drug efficacy. *BMC Med.* **2016**, *14*, 1–13. [CrossRef] [PubMed]
77. Walter, N.D.; Dolganov, G.M.; Garcia, B.J.; Worodria, W.; Andama, A.; Musisi, E.; Ayakaka, I.; Van, T.T.; Voskuil, M.I.; de Jong, B.C.; et al. Transcriptional Adaptation of Drug-tolerant *Mycobacterium tuberculosis* During Treatment of Human Tuberculosis. *J. Infect. Dis.* **2015**, *212*, 990–998. [CrossRef] [PubMed]
78. Bespyatykh, J.A.; Shitikov, E.A.; Ilina, E.N. Proteomics for the Investigation of Mycobacteria. *Acta Nat.* **2017**, *9*, 15–25. [CrossRef]
79. Vogel, C.; Marcotte, E.M. Insights into the regulation of protein abundance from proteomic and transcriptomic analyses. *Nat. Rev. Genet.* **2012**, *13*, 227–232. [CrossRef] [PubMed]
80. Malen, H.; De Souza, G.A.; Pathak, S.; Softeland, T.; Wiker, H.G. Comparison of membrane proteins of *Mycobacterium tuberculosis* H37Rv and H37Ra strains. *BMC Microbiol.* **2011**, *11*, 18. [CrossRef] [PubMed]
81. Rauniyar, N.; Yates, J.R., 3rd. Isobaric labeling-based relative quantification in shotgun proteomics. *J. Proteome. Res.* **2014**, *13*, 5293–5309. [CrossRef]
82. Thompson, C.R.; Champion, M.M.; Champion, P.A. Quantitative N-Terminal Footprinting of Pathogenic Mycobacteria Reveals Differential Protein Acetylation. *J. Proteome. Res.* **2018**, *17*, 3246–3258. [CrossRef]
83. Albrethsen, J.; Agner, J.; Piersma, S.R.; Hojrup, P.; Pham, T.V.; Weldingh, K.; Jimenez, C.R.; Andersen, P.; Rosenkrands, I. Proteomic profiling of *Mycobacterium tuberculosis* identifies nutrient-starvation-responsive toxin-antitoxin systems. *Mol. Cell Proteom.* **2013**, *12*, 1180–1191. [CrossRef]
84. Agarwal, S.; Ghosh, S.; Sharma, S.; Kaur, K.; Verma, I. *Mycobacterium tuberculosis* H37Rv expresses differential proteome during intracellular survival within alveolar epithelial cells compared with macrophages. *Pathog. Dis.* **2018**, *76*, 1–9. [CrossRef]
85. Kruh, N.A.; Troudt, J.; Izzo, A.; Prenni, J.; Dobos, K.M. Portrait of a pathogen: The *Mycobacterium tuberculosis* proteome in vivo. *PLoS ONE* **2010**, *5*, 1–13. [CrossRef] [PubMed]
86. Sharma, D.; Bisht, D.; Khan, A.U. Potential Alternative Strategy against Drug Resistant Tuberculosis: A Proteomics Prospect. *Proteomes* **2018**, *6*, 1–9. [CrossRef] [PubMed]

87. Meneguello, J.E.; Arita, G.S.; Silva, J.V.O.; Ghiraldi-Lopes, L.D.; Caleffi-Ferracioli, K.R.; Siqueira, V.L.D.; Scodro, R.B.L.; Pilau, E.J.; Campanerut-Sa, P.A.Z.; Cardoso, R.F. Insight about cell wall remodulation triggered by rifampicin in *Mycobacterium tuberculosis*. *Tuberculosis* **2020**, *120*, 1–5. [CrossRef] [PubMed]
88. Wilson, S.R.; Vehus, T.; Berg, H.S.; Lundanes, E. Nano-LC in proteomics: Recent advances and approaches. *Bioanalysis* **2015**, *7*, 1799–1815. [CrossRef] [PubMed]
89. Sarkar, R.; Mdladla, C.; Macingwana, L.; Pietersen, R.D.; Ngwane, A.H.; Tabb, D.L.; van Helden, P.D.; Wiid, I.; Baker, B. Proteomic analysis reveals that sulfamethoxazole induces oxidative stress in *M. tuberculosis*. *Tuberculosis* **2018**, *111*, 78–85. [CrossRef]
90. Hinzke, T.; Kouris, A.; Hughes, R.A.; Strous, M.; Kleiner, M. More Is Not Always Better: Evaluation of 1D and 2D-LC-MS/MS Methods for Metaproteomics. *Front. Microbiol.* **2019**, *10*, 1–13. [CrossRef]
91. Putim, C.; Phaonakrop, N.; Jaresitthikunchai, J.; Gamngoen, R.; Tragoolpua, K.; Intorasoot, S.; Anukool, U.; Tharincharoen, C.S.; Phunpae, P.; Tayapiwatana, C.; et al. Secretome profile analysis of multidrug-resistant, monodrug-resistant and drug-susceptible *Mycobacterium tuberculosis*. *Arch. Microbiol.* **2018**, *200*, 299–309. [CrossRef]
92. de Carvalho, L.P.; Fischer, S.M.; Marrero, J.; Nathan, C.; Ehrt, S.; Rhee, K.Y. Metabolomics of *Mycobacterium tuberculosis* reveals compartmentalized co-catabolism of carbon substrates. *Chem. Biol.* **2010**, *17*, 1122–1131. [CrossRef]
93. Serafini, A.; Tan, L.; Horswell, S.; Howell, S.; Greenwood, D.J.; Hunt, D.M.; Phan, M.D.; Schembri, M.; Monteleone, M.; Montague, C.R.; et al. *Mycobacterium tuberculosis* requires glyoxylate shunt and reverse methylcitrate cycle for lactate and pyruvate metabolism. *Mol. Microbiol.* **2019**, *112*, 1284–1307. [CrossRef]
94. Agapova, A.; Serafini, A.; Petridis, M.; Hunt, D.M.; Garza-Garcia, A.; Sohaskey, C.D.; de Carvalho, L.P.S. Flexible nitrogen utilisation by the metabolic generalist pathogen *Mycobacterium tuberculosis*. *Elife* **2019**, *8*, 1–22. [CrossRef]
95. Borah, K.; Beyss, M.; Theorell, A.; Wu, H.; Basu, P.; Mendum, T.A.; Nh, K.; Beste, D.J.V.; McFadden, J. Intracellular *Mycobacterium tuberculosis* Exploits Multiple Host Nitrogen Sources during Growth in Human Macrophages. *Cell Rep.* **2019**, *29*, 3580–3591. [CrossRef]
96. Dutta, N.K.; Klinkenberg, L.G.; Vazquez, M.J.; Segura-Carro, D.; Colmenarejo, G.; Ramon, F.; Rodriguez-Miquel, B.; Mata-Cantero, L.; Porras-De Francisco, E.; Chuang, Y.M.; et al. Inhibiting the stringent response blocks *Mycobacterium tuberculosis* entry into quiescence and reduces persistence. *Sci. Adv.* **2019**, *5*, 1–13. [CrossRef] [PubMed]
97. Zampieri, M.; Szappanos, B.; Buchieri, M.V.; Trauner, A.; Piazza, I.; Picotti, P.; Gagneux, S.; Borrell, S.; Gicquel, B.; Lelievre, J.; et al. High-throughput metabolomic analysis predicts mode of action of uncharacterized antimicrobial compounds. *Sci. Transl. Med.* **2018**, *10*, 1–13. [CrossRef] [PubMed]
98. Baptista, R.; Fazakerley, D.M.; Beckmann, M.; Baillie, L.; Mur, L.A.J. Untargeted metabolomics reveals a new mode of action of pretomanid (PA-824). *Sci. Rep.* **2018**, *8*, 1–7. [CrossRef] [PubMed]
99. Wang, Z.; Soni, V.; Marriner, G.; Kaneko, T.; Boshoff, H.I.M.; Barry, C.E., 3rd; Rhee, K.Y. Mode-of-action profiling reveals glutamine synthetase as a collateral metabolic vulnerability of *M. tuberculosis* to bedaquiline. *Proc. Natl. Acad. Sci. USA* **2019**, *116*, 19646–19651. [CrossRef] [PubMed]
100. Chiaradia, L.; Lefebvre, C.; Parra, J.; Marcoux, J.; Burlet-Schiltz, O.; Etienne, G.; Tropis, M.; Daffe, M. Dissecting the mycobacterial cell envelope and defining the composition of the native mycomembrane. *Sci. Rep.* **2017**, *7*, 1–12. [CrossRef]
101. Wu, Z.; Shon, J.C.; Liu, K.H. Mass Spectrometry-based Lipidomics and Its Application to Biomedical Research. *J. Lifestyle Med.* **2014**, *4*, 17–33. [CrossRef]
102. Griffiths, W.J.; Wang, Y. Mass spectrometry: From proteomics to metabolomics and lipidomics. *Chem. Soc. Rev.* **2009**, *38*, 1882–1896. [CrossRef]
103. Wenk, M.R. Lipidomics: New tools and applications. *Cell* **2010**, *143*, 888–895. [CrossRef]
104. Raghunandanan, S.; Jose, L.; Gopinath, V.; Kumar, R.A. Comparative label-free lipidomic analysis of *Mycobacterium tuberculosis* during dormancy and reactivation. *Sci. Rep.* **2019**, *9*, 1–12. [CrossRef]
105. Zhao, Y.Y.; Wu, S.P.; Liu, S.; Zhang, Y.; Lin, R.C. Ultra-performance liquid chromatography-mass spectrometry as a sensitive and powerful technology in lipidomic applications. *Chem. Biol. Interact.* **2014**, *220*, 181–192. [CrossRef] [PubMed]

106. Lefebvre, C.; Boulon, R.; Ducoux, M.; Gavalda, S.; Laval, F.; Jamet, S.; Eynard, N.; Lemassu, A.; Cam, K.; Bousquet, M.P.; et al. HadD, a novel fatty acid synthase type II protein, is essential for alpha- and epoxy-mycolic acid biosynthesis and mycobacterial fitness. *Sci. Rep.* **2018**, *8*, 1–15. [CrossRef] [PubMed]

107. Lefebvre, C.; Frigui, W.; Slama, N.; Lauzeral-Vizcaino, F.; Constant, P.; Lemassu, A.; Parish, T.; Eynard, N.; Daffe, M.; Brosch, R.; et al. Discovery of a novel dehydratase of the fatty acid synthase type II critical for ketomycolic acid biosynthesis and virulence of *Mycobacterium tuberculosis*. *Sci. Rep.* **2020**, *10*, 1–12. [CrossRef] [PubMed]

108. Sharma, S.; Hameed, S.; Fatima, Z. Lipidomic insights to understand membrane dynamics in response to vanillin in Mycobacterium smegmatis. *Int. Microbiol.* **2020**, *23*, 263–276. [CrossRef] [PubMed]

109. Howard, N.C.; Marin, N.D.; Ahmed, M.; Rosa, B.A.; Martin, J.; Bambouskova, M.; Sergushichev, A.; Loginicheva, E.; Kurepina, N.; Rangel-Moreno, J.; et al. *Mycobacterium tuberculosis* carrying a rifampicin drug resistance mutation reprograms macrophage metabolism through cell wall lipid changes. *Nat. Microbiol.* **2018**, *3*, 1099–1108. [CrossRef]

110. Li, W.; Obregon-Henao, A.; Wallach, J.B.; North, E.J.; Lee, R.E.; Gonzalez-Juarrero, M.; Schnappinger, D.; Jackson, M. Therapeutic Potential of the *Mycobacterium tuberculosis* Mycolic Acid Transporter, MmpL3. *Antimicrob. Agents Chemother.* **2016**, *60*, 5198–5207. [CrossRef]

111. Kavunja, H.W.; Biegas, K.J.; Banahene, N.; Stewart, J.A.; Piligian, B.F.; Groenevelt, J.M.; Sein, C.E.; Morita, Y.S.; Niederweis, M.; Siegrist, M.S.; et al. Photoactivatable Glycolipid Probes for Identifying Mycolate-Protein Interactions in Live Mycobacteria. *J. Am. Chem. Soc.* **2020**, *142*, 7725–7731. [CrossRef]

112. Conesa, A.; Beck, S. Making multi-omics data accessible to researchers. *Sci. Data* **2019**, *6*, 1–4. [CrossRef]

113. Ulfenborg, B. Vertical and horizontal integration of multi-omics data with miodin. *BMC Bioinform.* **2019**, *20*, 1–10. [CrossRef]

114. Pinu, F.R.; Beale, D.J.; Paten, A.M.; Kouremenos, K.; Swarup, S.; Schirra, H.J.; Wishart, D. Systems Biology and Multi-Omics Integration: Viewpoints from the Metabolomics Research Community. *Metabolites* **2019**, *9*, 1–31. [CrossRef]

© 2020 by the authors. Licensee MDPI, Basel, Switzerland. This article is an open access article distributed under the terms and conditions of the Creative Commons Attribution (CC BY) license (http://creativecommons.org/licenses/by/4.0/).

Review

Post-Tuberculosis (TB) Treatment: The Role of Surgery and Rehabilitation

Dina Visca [1,2,†], Simon Tiberi [3,4,†], Rosella Centis [5], Lia D'Ambrosio [6], Emanuele Pontali [7], Alessandro Wasum Mariani [8,9], Elisabetta Zampogna [1], Martin van den Boom [10], Antonio Spanevello [1,2] and Giovanni Battista Migliori [5,*]

1. Division of Pulmonary Rehabilitation, Istituti Clinici Scientifici Maugeri IRCCS, 21049 Tradate, Italy; dina.visca@icsmaugeri.it (D.V.); elisabetta.zampogna@icsmaugeri.it (E.Z.); antonio.spanevello@icsmaugeri.it (A.S.)
2. Department of Medicine and Surgery, Respiratory Diseases, University of Insubria, 21100 Varese, Italy
3. Blizard Institute, Barts and The London School of Medicine and Dentistry, Queen Mary University of London, London E1 4NS, UK; simon.tiberi@nhs.net
4. Division of Infection, Royal London Hospital, Barts Health NHS Trust, London E1 1BB, UK
5. Servizio di Epidemiologia Clinica delle Malattie Respiratorie, Istituti Clinici Scientifici Maugeri IRCCS, 21049 Tradate, Italy; rosella.centis@icsmaugeri.it
6. Public Health Consulting Group, 6900 Lugano, Switzerland; lia.dambrosio59@gmail.com
7. Department of Infectious Diseases, Galliera Hospital, 16128 Genova, Italy; pontals@yahoo.com
8. Departamento de Cirurgia Toracica, Instituto do Coracao, Hospital das Clinicas HC-FMUSP, Sao Paulo 05403-000, Brazil; awmariani@gmail.com
9. Disciplina de Cirurgia Torácica, Faculdade de Medicina, Universidade de São Paulo, São Paulo 05508-220, Brazil
10. Joint TB, HIV and Viral Hepatitis Programme, Division of Health Emergencies and Communicable Diseases, WHO Regional Office for Europe, 2100 Copenhagen, Denmark; vandenboomm@who.int

* Correspondence: giovannibattista.migliori@icsmaugeri.it
† equally contributed.

Received: 3 March 2020; Accepted: 10 April 2020; Published: 15 April 2020

Abstract: Even though the majority of tuberculosis (TB) programmes consider their work completed when a patient is 'successfully' cured, patients often continue to suffer with post-treatment or surgical sequelae. This review focuses on describing the available evidence with regard to the diagnosis and management of post-treatment and surgical sequelae (pulmonary rehabilitation). We carried out a non-systematic literature review based on a PubMed search using specific key-words, including various combinations of 'TB', 'MDR-TB', 'XDR-TB', 'surgery', 'functional evaluation', 'sequelae' and 'pulmonary rehabilitation'. References of the most important papers were retrieved to improve the search accuracy. We identified the main areas of interest to describe the topic as follows: 1) 'Surgery', described through observational studies and reviews, systematic reviews and meta-analyses, IPD (individual data meta-analyses), and official guidelines (GRADE (Grading of Recommendations Assessment, Development and Evaluation) or not GRADE-based); 2) Post-TB treatment functional evaluation; and 3) Pulmonary rehabilitation interventions. We also highlighted the priority areas for research for the three main areas of interest. The collection of high-quality standardized variables would allow advances in the understanding of the need for, and effectiveness of, pulmonary rehabilitation at both the individual and the programmatic level. The initial evidence supports the importance of the adequate functional evaluation of these patients, which is necessary to identify those who will benefit from pulmonary rehabilitation.

Keywords: TB; post-treatment sequelae; surgery; pulmonary rehabilitation

1. Introduction

Pulmonary and pleural tuberculosis (TB) may be severe and challenging even with drug susceptible strains of *Mycobacterium tuberculosis* and may require a multidisciplinary approach for best management. Moreover, drug-resistant tuberculosis (TB) and, in particular, multidrug-resistant (MDR) and extensively drug-resistant (XDR) TB frequently occur in patients who have had prior TB episodes and may worsen previously damaged lungs [1–3]. Managing these cases is difficult, requiring a multidisciplinary team approach [4] and expensive treatment (which is toxic and with treatment success still below expectations) [2,5].

The availability of new drugs (bedaquiline, delamanid, and pretomanid) after many years of neglect provides new perspectives, improved success rates and a reduced prevalence of adverse events [6–8]. The rapid detection of TB is also key in order to catch the disease process early and preserve lung function.

As new evidence is made available and more is known about drugs and regimens, more patients are surviving [9–12], and it is emerging that other aspects require attention: the importance of preventing transmission [13], ensuring adequate nutrition, considering adjuvant surgery, and post-treatment sequelae [2,3,14]. These were emphasised in a comprehensive review of the Global Tuberculosis Network (GTN) based on the consensus of about 100 global experts [2]. This review is focused on describing the available evidence on adjuvant surgery and diagnosis and management of post-treatment sequelae (pulmonary rehabilitation).

2. Materials and Methods

We carried out a non-systematic literature review based on a PubMed search using specific key-words, including various combinations of 'TB', 'MDR-TB', 'XDR-TB', 'surgery', 'functional evaluation', 'sequelae', and 'pulmonary rehabilitation'. References of the existing reviews were retrieved to improve accuracy.

Manuscripts written in English, Spanish, and Russian were selected, including full articles and relevant abstracts.

The main areas of interest we identified to describe the topic are as follows:

1) 'Surgery', described through observational studies and reviews, systematic reviews, and meta-analyses, IPD (individual data meta-analyses) and official guidelines (GRADE or not GRADE-based). Due to the scant evidence on thoracoplasty and other less frequent surgical procedures, we concentrated on lung resection.
2) Post-TB treatment lung functional evaluations.
3) Pulmonary rehabilitation interventions.

The priorities for research have been identified for each main area of interest.

3. Surgery and TB

3.1. Observational Studies and Reviews

A limited number of observational studies and reviews are available on the topic; the majority suggests that adjuvant surgery in selected patients may be useful to improve treatment outcomes [2,15–27]. However, the strength of the conclusions from these studies is somewhat limited by the risk of bias related to the variability of the centres' procedures, patients' profiles, treatment regimens, timing and types of surgical procedures, and it is difficult, if not impossible, to identify homogenous patients to compare among studies.

3.2. Traditional Meta-analyses

A few meta-analyses are available on the topic [28–32]. A traditional meta-analysis of 15 reports of surgical resection found that treatment success was achieved in 84% (95% confidence interval (CI), 78%–89%) of patients, noting substantial heterogeneity among the studies [29].

Two other meta-analyses of MDR-TB patients who had either resection or non-resection surgery found that surgical patients had better outcomes than those who did not [28,30]; however, there was no distinction between the different forms of resection surgery.

In the Marrone's meta-analysis [28], 24 studies identified a significant association between surgery and successful treatment compared to non-surgical interventions (OR 2.24, 95%CI 1.68–2.97). The meta-analysis from 23 single-arm studies demonstrated that, respectively, 92% (95%CI 88.1–95) and 87% (95% CI 83–91) of surgical patients achieved successful short and long-term outcomes. In the sub-group analysis (studies reporting both surgical and non-surgical treatment outcomes) favourable surgical outcomes (treatment success) were associated with increased drug-resistance, i.e., better results for XDR-TB patients than for MDR-TB ones.

Confounding by indication (a form of bias that occurs when the patients most likely to benefit are selected for therapy) was a major limitation in each meta-analysis.

Furthermore, antibiotic regimens were not standardized across studies, meaning that the studies could not account for factors such as the individual drug regimens or the timing of surgery in relation to culture conversion.

3.3. Evaluating the Role of Surgery Through IPD

A recent IPD based on the large MDR-TB cohort coordinated by McGill University [33–35] utilized a sophisticated analysis (propensity score matching) to evaluate the benefits offered by surgery. Individual patient data from 26 cohort studies were analysed, including clinical features and information on both medical and surgical therapy. Primary analyses compared treatment success (cure and completion) to a combined outcome of failure, relapse, or death. The effects of all forms of resection surgery, pneumonectomy, and partial lung resection were evaluated [35].

The final analysis was conducted on 4,238 patients from 18 surgical studies and 2,193 from 8 non-surgical ones. Pulmonary resection surgery (478 patients) was associated with improved treatment success (adjusted odds ratio (aOR), 3.0; 95% confidence interval (CI), 1.5–5.9), but pneumonectomy was not (aOR, 1.1; 95% CI, 0.6–2.3). Treatment success was achieved in 95.2% of patients undergoing surgery after culture conversion compared with 91.2% of those who had surgery before it (aOR, 2.6; 95% CI, 0.9–7.1).

Patients undergoing partial lung resection achieved better treatment success and lower failure/death rates than patients who had either pneumonectomy or no surgery. The median duration of medical therapy was 20 months (interquartile range [IQR], 13.7–24.0 months) for those who had surgery after culture conversion versus 29 months (IQR, 22–45 months) for those undergoing surgery before conversion. The loss to follow-up was lower among patients who had surgery (11%; 95% CI, 4–17%) than among those who had not (22%; 95% CI, 14–31%).

The authors concluded that, among MDR-TB patients, partial lung resection (but not pneumonectomy) was associated with improved treatment success, although selection bias cannot be excluded [35]. This finding can be explained with the lower rate of mortality among surgical versus non-surgical TB patients. Furthermore, patients undergoing surgery had, overall, more severe drug-resistance profiles and more extensive diseases [35]. Importantly, both surgical and non-surgical patients were rather young with a low probability of confounding co-morbidities [35].

A summary of the available evidence is reported in Table 1

Table 1. Main reports on the indications, type, and outcomes of surgery performed in patients affected by pulmonary tuberculosis complications.

First Author, Year, Country	Reference	Type of Paper; Patients Number and MDR/XDR-TB Proportion (%)	Indications for Surgery; Type of Surgery Assessed	Timing of Surgery	Favorable Outcome (Treatment Success)	Post-operative Complications/ Mortality (within 30 Days from Surgery)	Favouring Surgery
Sayir, 2019, Turkey	[24]	Case series; 9 pts, unspecified MDR/XDR %	Emergency: hemorrhage Elective: empyema Unspecified type of surgery	- After sputum smear negativity was achieved - After negative culture for empyema	Unspecified	Complications: unspecified Mortality: 11.1%	N/A
Yablonskii, 2019, Russia	[25]	Review; N/A pts and MDR/XDR %	Emergency: hemorrhage, spontaneous pneumothorax Urgent: recurrent haemoptysis Elective: localized cavitary disease with persistent sputum positivity Complications of TB diseases Assessment of all types of surgery	Unspecified	Treatment success: 67%–100%	Complications: 9%–30.8% Mortality: 0–5.5%	N/A
Borisov, 2019, Several Countries	[20]	Case series; 55 pts, 43.6% at least MDR, 56.4% XDR	Localised disease allowing for resection, failed bacteriological conversion, disease worsening Mostly lobectomy, segmentectomy, pneumonectomy	After 8 months of therapy (median); range: 5-13	Treatment success 69.1%	Complications: 23.7% (out of 38 evaluable patients) Mortality: 0	N/A
Chen, 2018, China	[26]	Case series; 32 pts, 12.5% MDR	Haemoptysis (100%), continuously positive smear (28.1%), pulmonary aspergillosis (40.6%) Regional arterial embolization followed by pulmonary resection	After at least 6 months of standard therapy	Unspecified	Complications 18.75% Mortality: unspecified	N/A
Giller, 2018, Russia	[27]	Case series; 5,599 pts, unspecified MDR/XDR %	Fibro-cavitary and cavitary pulmonary TB (58.5%), tuberculoma with destruction (18.8%), tuberculous pleural empyema (18.8%), caseous pneumonia (3.4%), intrathoracic lymph nodes (0.5%) Unspecified type of surgery	After 1-3 years of treatment (for 84% of patients)	Treatment success: 92.1%–98%	Complications 1.9% Mortality: 0.1%	N/A
Fox, 2016, N/A	[35]	Meta-analysis; 478 pts (18 studies) 100% MDR (of whom 8.6% XDR)	Unknown indication for surgery Pneumonectomy (118 pts), partial lung resection (227 pts), unspecified (132 pts)	Unknown	Treatment success* 81% (pneumonectomy 69%; partial lung resection 90%)	Complications: unspecified Mortality: 8.4% (70% of deaths > 30 days from surgery)	Yes

Table 1. Cont.

First Author, Year, Country	Reference	Type of Paper; Patients Number and MDR/XDR-TB Proportion (%)	Indications for Surgery; Type of Surgery Assessed	Timing of Surgery	Favorable Outcome (Treatment Success)	Post-operative Complications/ Mortality (within 30 Days from Surgery)	Favouring Surgery
Subotic, 2016, N/A	[33]	Review; N/A pts and MDR/XDR %	Persistently positive sputum smear and/or culture despite appropriate chemotherapy, relapse, high risk of relapse based on drug resistance profile; Lobectomy, pneumonectomy, resection of tuberculoma	For infectious TB patients after at least 6–8 months of appropriate anti-TB therapy	Treatment success after surgery between 75% and 98%	Complications: 9%–26% Mortality: <5% after lung resection for TB	N/A
Marrone, 2013, N/A	[28]	Systematic review and meta-analysis; 706 pts, 100% MDR/XDR	Standardized or non-standardized indication for surgery Unspecified type of surgery	Unspecified	Treatment success: 87% at twelve months post-surgery; more favorable outcome for XDR than for MDR	Complications: 3% short-term; 8% long-term Mortality: unspecified	Yes
Xu, 2011, N/A	[29]	Systematic review and meta-analysis; 949 pts, 100% MDR	Unspecified indication for surgery Mainly pneumonectomy, lobectomy, segmentectomy	Unspecified	Treatment success: 84%	Complications: unspecified; Mortality: 3%	Unspecified

Legend: MDR: multidrug –resistant; XDR: extensively drug-resistant; * excluding lost to follow up; TB: tuberculosis; N/A: not applicable, Pts: patients.

3.4. World Health Organization (WHO), International Union Against Tuberculosis and Lung Disease (The UNION), and ATS/CDC/ERS/IDSA (American Thoracic Society/Centers for Disease Control and prevention/European Respiratory Society/Infectious Diseases Society of America) guidelines

In the consolidated WHO 2019 MDR-TB guidelines (and in the preceding 2016 and 2011 ones) the following recommendation were given (based on GRADE): in patients with rifampicin-resistant (RR)-TB or MDR-TB, elective partial lung resection (lobectomy or wedge resection) may be used alongside a recommended MDR-TB regimen [36].

In a regional WHO European guidance the indications and contra-indications for surgery were clearly defined [19,37]. Surgical interventions may have emergency (life threatening conditions), urgent (irreversible TB and haemoptysis), and elective natures.

Elective surgery indications include localised unilateral forms of bacteriologically-confirmed cavitary disease, MDR-/XDR-TB failing medical treatment, and complications/sequelae (spontaneous pneumothorax /pyopneumothorax; pleural empyema with or without bronchopleural fistula; aspergilloma; nodular-bronchial fistula; broncholith; and pachypleuritis/pericarditis with respiratory and blood circulation insufficiency; trachea/large bronchi stenosis; and post-TB bronchiectasis).

The following contra-indications have been identified [19,37]:

- Bilateral, extensive cavities;
- Impaired pulmonary function (forced expiratory volume in one second FEV1 (forced expiratory volume in 1 s) <1.5 L for lobectomy and < 2.0 L for pneumonectomy);
- Pulmonary-heart failure III–IV (New York Hart Association functional classification);
- Body mass index (BMI) up to 40%–50% of normality;
- Severe co-morbid conditions (uncontrolled diabetes, ulcer exacerbation, and liver/renal insufficiency);
- Active bronchial TB.

The UNION guidelines (which are not designed with the GRADE approach) suggest that 'surgery should be considered for treating drug-resistant (DR)-TB only in patients meeting the three following conditions: 1) a fairly localised lesion, 2) an adequate respiratory reserve, and 3) a lack of sufficient available drugs to design a regimen potent enough to ensure a cure. Ideally, surgery needs to be performed at the moment chemotherapy has achieved the lowest possible bacillary load (sputum smear and culture converted to negative) within a complete cycle of chemotherapy [38].

In the recently published ATS/CDC/ERS/IDSA guidelines [39] the PICO (population, intervention, comparator, outcomes) question 19 was on 'Surgery for MDR-TB' as follows: 'Should elective lung resection surgery (i.e., a lobectomy or pneumonectomy) be used as an adjunctive therapeutic option in combination with antimicrobial therapy, versus medical therapy alone for adults with MDR-TB?'

The following recommendations were issued:

'Recommendation 19A: We suggest elective partial lung resection (e.g., a lobectomy or wedge resection), rather than medical therapy alone, for adults with MDR-TB receiving antimicrobial-based therapy (conditional recommendation, very low certainty in the evidence). The writing committee believes this option would be beneficial for patients for whom clinical judgement, supported by bacteriological and radiographic data, suggest a strong risk of treatment failure or relapse with medical therapy alone.

Recommendation 19B: We suggest medical therapy alone, rather than including elective total lung resection (pneumonectomy), for adults with MDR-TB receiving antimicrobial therapy (conditional recommendation, very low certainty of evidence)' [39].

In summary, all major guidelines are consistent in recommending surgery in selected cases, following chemotherapy and favouring elective partial lung resection when possible, based on specific indications: failure of drug therapy, relapse, localized (e.g., cavity) or extensive pulmonary TB, clinical complications (e.g., haemoptysis or empyema) [39]. However, recent evidence suggests that bilateral surgery can also be safe and effective [40].

The patients undergoing surgery are candidates for pulmonary rehabilitation [20,39].

3.5. Priorities for Research

The ATS/CDC/ERS/IDSA guidelines proposed the following priorities for TB research on TB and surgery [39]: ideal timing for surgery; optimal drug regimens and duration before and after surgery; the role of surgery in special populations and patients with co-morbidities (e.g., HIV co-infection), optimal surgical approaches, optimal infection control measures to be implemented peri-operatively, and the role of pulmonary rehabilitation.

4. Post-TB Treatment Sequelae and Rehabilitation

There is evidence that patients with pulmonary TB have up to a five to six times higher probability of abnormal pulmonary function when compared with LTBI (latent TB infection) individuals [41]. TB sequelae are likely to follow delayed diagnosis, extensive disease, and long and/or repeated treatments [42]. TB sequelae are risk factors for bronchiectasis and COPD (chronic obstructive pulmonary disease), both conditions are more common in smokers and in the presence of in-door or out-door drug pollution [43]. The most common alterations are represented by obstructions with or without restriction. Airflow obstruction is usually without response to the bronchodilator, and often coupled with bronchiectasis and/or tracheobronchial stenosis, alterations of the lung parenchyma (cavities and pulmonary fibrosis) or of the pleura (empyema, fibrothorax, bronchopleural fistula, and pneumothorax). Restriction can affect gas exchange, as well as other vascular complications including pulmonary or bronchial arteritis, thrombosis, artery dilatation, Rasmussen aneurysm, or 'cor pulmonale' [43]. Both mechanical and gas exchange alterations can limit daily activities, exercise capacity, and impair quality of life (QoL) [43].

4.1. Post-TB Treatment Functional Evaluation

A baseline examination with functional evaluation can be performed safely when the patient is smear and culture negative (on at least two samples two weeks apart) and is undergoing effective treatment; otherwise, infection control measures are necessary [13]. As the patient might need a different approach when resting and when making exercise (e.g., walking), a careful evaluation should be ideally performed both at rest and under exercise conditions [43,44].

At rest, spirometry with response to the bronchodilator, diffusing capacity of the lung for carbon monoxide (DLCO), arterial blood gases analysis are recommended to study lung mechanics, complemented by plethysmography at the initial evaluation (if feasible) (Figure 1) [43].

Spirometry is the most widely accepted test to assess lung function impairment. It can be conducted with a simple spirometer, which costs a minimum of 150$ and can be used at point-of-care or with a sophisticated apparatus which includes plethysmography (which is able to diagnose lung restriction and 'air trapping' by measuring the Residual Volume (RV)). The core parameters evaluated by spirometry are forced expiratory volume (FEV)$_1$ (low FEV$_1$ indicates airflow obstruction), FVC (Forced Vital Capacity) and their ratio (FEV$_1$/FVC) [43]. *DLCO* describes the status of gas exchanges at the pulmonary level, which can be hampered even in the presence of normal spirometry and plethysmography.

Under exercise conditions it is useful to have the patient undergo the 6-min walking test (6MWT) or the cardiopulmonary exercise test (CPET) which provides additional information on the physiological reserve (and, indirectly on QoL) [43,44]. The *6MWT* measures the distance covered (in metres) in 6 min. It can be done in any setting, is cheap and easy to interpret: it correlates with QoL and improves after rehabilitation [43–45]. *CPET* is a more sophisticated, expensive, and technology-dependent tool which cannot be performed in all centres. It provides information on the exercise capacity-limiting determinants (respiratory: mechanical or as exchange-related; muscular; and cardio-vascular).

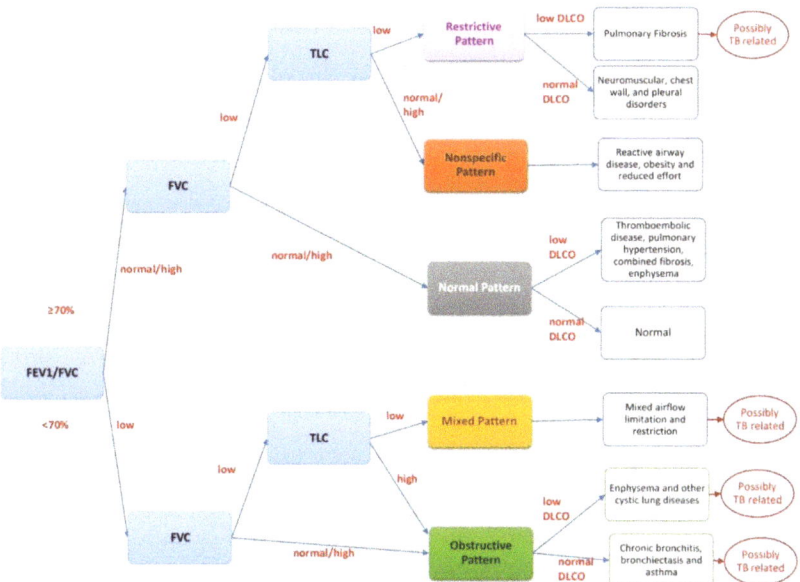

Figure 1. Interpreting spirometry. **Legend**: FEV1: forced expiratory volume in 1 s; FVC: forced vital capacity; FEV1/FVC ratio: the percentage of the FVC expired in one second; DLCO: diffusing capacity of the lung for carbon monoxide; TLC: total lung capacity.

Different tools exist to evaluate QoL, including generic questionnaires (e.g., 36-item Short Form (SF) health survey or SF-36 and its shortened version with 12 questions, the SF-12) or specific tools as the SGRQ (St. Georges's Respiratory Questionnaire) specifically investigating QoL in chronic respiratory diseases (Table 2).

Table 2. Available tests to evaluate health-related quality of life (HRQoL).

Questionnaire and Items	Domains	Mode and Time of Administration	Score
NON-DISEASE SPECIFIC			
The Short Form (36) Health Survey SF-36v2 36 items	1. Vitality 2. Physical functioning 3. Bodily pain 4. General health perceptions 5. Physical role functioning 6. Emotional role functioning 7. Social role functioning 8. Mental health	Self-administered 10 ± 8 min	Higher scores indicate better HRQoL. The correct calculation of SF-36 requires the use of special algorithms, which are strictly controlled by a private company *
Euroqol 5 dimensions EQ-5D//	1. Mobility 2. Self-care 3. Usual activities 4. Pain/discomfort 5. Anxiety/depression and additional perceived health status measured through a visual-analogue scale (VAS)	Self-administered 5/10 min.	Two scores, one for the 5 domains and another for the VAS. 5 domains score: 1–scores indicate worse HRQoL VAS score: 0-100 Higher scores indicate better HRQoL
World Health Organization Quality of Life questionnaire WHOQOL-100 100 items	1. Physical Health 2. Psychological 3. Level of Independence 4. Social Relations 5. Environment 6. Spirituality/Religion/Personal beliefs	Self-administered if respondents have sufficient ability: otherwise, interviewer assisted/administered 30 min.	Produces scores relating to particular facets of QoL, scores relating to larger domains and a score relating to overall QoL and general health. Higher scores indicate better HRQoL *
World Health Organization Quality of Life questionnaire WHOQOL-BREF 26 items	1. Physical health 2. Psychological 3. Social relationships 4. Environment	Self-administered if respondents have sufficient ability: otherwise, interviewer assisted/administered 10/15 min.	Produces a quality of life profile. It is possible to derive four domain scores. Higher scores indicate better HRQoL #

Table 2. Cont.

	NON-DISEASE SPECIFIC		
Questionnaire and Items	Domains	Mode and Time of Administration	Score
	DISEASE SPECIFIC		
	TUBERCULOSIS		
Functional Assessment of Chronic Illness Therapy FACIT-TB 47 items	1. Physical Well-Being 2. Psychological Well-Being 3. Function Well-Being 4. Social Well-Being 5. Spiritual Well-Being 6. Environment 7. Perception	Self-administered 16.3 ± 3.1 min.	0–180 Higher scores indicate better HRQoL
Pulmonary Tuberculosis Scale of the System of Quality of Life Instruments for Chronic Diseases QLICD-PT 40 items	1. Physical domain 2. Psychological domain 3. Social domain 4. TB Specific domain	Self-administered Approximately 10 min.	0–100
	OTHER PULMONARY DISEASES		
Saint George Respiratory Questionnaire SGRQ (*Asthma and COPD*) 76 items	1. Symptoms 2. Activity 3. Impacts	Self-administered 15/20 min.	0–100 Higher scores indicate worse HRQoL
Maugeri Respiratory Failure Questionnaire MRF-28 (*Chronic respiratory failure*) 28 items	1. Daily activities 2. Cognition 3. Invalidity, and additional items related to fatigue, depression and problems with treatment	Self-administered 15 ± 6 min	0–100 Higher scores indicate worse HRQoL

Table 2. *Cont.*

Questionnaire and Items	Domains	Mode and Time of Administration	Score
NON-DISEASE SPECIFIC			
Quality of Life-Bronchiectasis QOL-B (*Non CF-Bronchiectasis*) 37 items	1. Respiratory Symptoms 2. Physical 3. Role 4. Emotional 5. Social Functioning 6. Vitality 7. Health Perceptions 8. Treatment Burden	Self-administered 4–5 min.	0–100 Higher scores indicate better HRQoL
Asthma Quality of Life Questionnaire AQLQ (*Asthma*) 32 items	1. Symptoms 2. Activity Limitation 3. Emotional Function 4. Environmental Exposure	Self-administered 4–5 min.	1–7 Higher scores indicate better HRQoL

HRQoL: Health-Related Quality of Life; QoL: Quality of Life; COPD: Chronic Obstructive Pulmonary Disease; TB: tuberculosis. * https://www.optum.com/solutions/life-sciences/answer-research/patient-insights/sf-health-surveys.html; # Details on scoring are included in manuals available from The WHOQOL Group: https://www.who.int/healthinfo/survey/whoqol-qualityoflife/en/index2.html.

4.2. Pulmonary Rehabilitation

Pulmonary rehabilitation is a non-pharmacological intervention aimed at improving the physical and psychological conditions of individuals affected by chronic lung diseases [46]. It includes different interventions including, among others, the integration of an optimised medical treatment (drugs, Long-Term Oxygen Therapy-LTOT, ventilation) with physiotherapy, exercise training, education, and behavioural changes [39,43,44].

4.2.1. LTOT and Ventilation

The importance of LTOT and mechanical ventilation in supporting the management of post-TB treatment sequelae is well known [44]. Intermittent positive pressure ventilation through a nasal mask (NIPPV) applied during exercise in patients with pulmonary TB sequelae improved arterial blood gas measurements, reduced breathlessness, and increased exercise tolerance [47]. The use of a poncho (wraparound) ventilator and mouth intermittent positive pressure ventilation (MIPPV) was studied, showing beneficial results [48].

4.2.2. Physiotherapy

The role of physiotherapy in expectorating secretions is well known and largely utilised [44]. Mechanical methods of vibration massage have been proposed to prevent early post-resectional complications (atelectasis, non-specific pneumonia, residual post-resection pleural cavity, and bronchial fistulas) after surgical interventions for TB and to improve the functional status [49].

4.2.3. Exercise Training

Post TB sequelae may cause obstructive or restrictive damage and decrease the effort tolerance [50]. Patients undergoing long and/or multiple rounds of treatment may suffer from cachexia, asthenia, and muscle fatigue [44]. Before initiating a specific rehabilitation programme, patients should undergo a complete lung functional assessment, including spirometry and exercise capacity testing to enable an appropriate exercise training regime. Exercise capacity is usually based on cardiopulmonary exercise testing or walking tests (6MWT, incremental shuttle walking test (ISWT)) in order to set physical training sessions that exceed the physical loads of daily life activities [45,51–58]. Some studies documented the positive role of aerobic training on symptoms, anxiety, depression, and QoL [45,51–58].

4.2.4. Education and Psychological Counselling

Education about lung disease and its management is an important aspect of pulmonary rehabilitation [43–45]. It implies that specialists teach patients about respiratory diseases and support them through self-management training. Knowledge about the disease helps patients to understand, recognize, and treat their symptoms in order to achieve a better control of the disease in daily life. It may consist of one or more interventions, including smoking cessation, oxygen therapy, nutrition, physical activity, and the proper use of medications [43–45].

Psychological support is extremely important because depression and anxiety are often associated with TB and may contribute to fatigue and reduce physical activity. Psychological counselling should be offered either individually or in small groups as discussion will enable patients to feel more comfortable with their disease and favour their participation in social activities [43–45]

4.3. Effectiveness of Pulmonary Rehabilitation in TB

The available information on the effectiveness of rehabilitation is summarised in Table 3.

Table 3. Summary of studies reporting pre/post pulmonary rehabilitation (PR) interventions and their effects on outcome measures in post-TB sequelae patients.

Author, Year; [Ref] Country	Reference	Study Population Design Lung Function and Exercise Tests PR Duration and Setting	Outcome Measures Significantly Improved		Outcome Measures Not Significantly Improved		Gain in 6MWT or ISWT (meters/m)
			LFT	Exercise Capacity and QoL	LFT	Exercise Capacity and QoL	
Visca, 2019, Italy	[45]	43 patients; Retrospective study Spirometry, BGA, 6MWT Spontaneous walking SpO$_2$ 3 weeks, inpatient	FEV1 LFEV1% FVC% PaO$_2$ SaO$_2$	6MWD m 6MWD % HR average SpO$_2$ average (6MWT) SpO$_2$ min (6MWT) Modified Borg dyspnoea final Modified Borg fatigue final SpO$_2$ baseline (SW) SpO$_2$ min (SW)	FVC L * FEV1/FVC * RV ˆ PaCO$_2$ * pH ˆ FiO$_2$ ˆ (BGA)	HR baseline ˆ HR max * SpO$_2$ baseline (6MWT) ˆ Modified Borg dyspnoea baseline ˆ Modified Borg fatigue baseline ˆ SpO$_2$ baseline (SW) * FiO$_2$ ˆ (6MWT, SW)	+35 m
Tada, 2002, Japan	[51]	37 patients Study design # Spirometry, BGA, 6MWT, Pimax, QoL 3.9 weeks, inpatient	VC L FEV1 L PaO$_2$ Pimax	6MWD m Dyspnoea # QoL #	_#	_#	+36 m
Jones, 2017, Uganda	[52]	29 patients Prospective study Spirometry (baseline only), ISWT Sit-to-Stand, BMI, Mid upper arm circumference, QoL 6 weeks, outpatient and homebased	BMI (kg/m^2) Mid upper arm circumference (cm)	ISWT (m) Borg score after ISWT Sit-to-stand time (seconds) CCQ total score CCQ symptom score CCQ mental state score CCQ functional state scoreP HQ-9 total score Karnofsky score	Not specified	Not specified	+90 m
Ando, 2003, Japan	[53]	32 patients Prospective non randomized open trial Spirometry, BGA, 6MWT, HRQL 9 weeks, outpatient	VC L	6MWD m TDI MRC ADL	FEV1 L * PaO$_2$ * PaCO$_2$ *	Not specified	+42 m
Singh, 2018, India	[54]	29 patients Prospective cohort study Spirometry, 6MWT, HRQL 8 weeks, inpatient	Not specified	6MWD m CRQ	FEV1 L * FVC L * FEV1/FVC ˆ	MMRC *	+38 m

Table 3. Cont.

Author, Year, [Ref] Country	Reference	Study Population Design Lung Function and Exercise Tests PR Duration and Setting	Outcome Measures Significantly Improved		Outcome Measures Not Significantly Improved		Gain in 6MWT or ISWT (meters/m)
			LFT	Exercise Capacity and QoL	LFT	Exercise Capacity and QoL	
Yoshida, 2006, Japan	[55]	10 patients Observational study Spirometry, BGA, treadmill test 6MWT 3 weeks, inpatient	Not specified	VO$_2$ peak (Treadmill) 6MWD m	VC L ^ VC % ^ FEV1 L ^ FEV1/FVC% * MVV L * PaO$_2$ * PaCO$_2$ ^ pH ^	Treadmill: VEmax L/min ^ VEmax/MVV * HRmax ^ 6MWT: Modified Borg dyspnoea final * Modified Borg fatigue final * 6MWT: SpO$_2$% ^ Pulse rate ^ Modified Borg dyspnoea final * Modified Borg fatigue final *	+68 m
Wilches, 2009, Colombia (Case report)	[56]	1 patient § Case report Spirometry, BGA,6MWT, HRQL 32 weeks, inpatient	Not specified	Not specified	Not specified	6MWD * HADS * SF36 * MRC * Borg dyspnoea final *	+110 m
Betancourt-Peña, 2015, Colombia	[57]	11 patients Quasi-experimental study Spirometry (baseline only) 6MWT BMI, HRQL 8 weeks, outpatient	Not specified	6MWD VO$_2$ peak (Treadmill) SGRQ HADS	BMI (kg/m^2) *	SpO$_2$ baseline ^ SpO$_2$ final ^ 6MWT desaturation ^ MRC *	+110.2 m
Rivera Motta, 2016, Colombia (conference abstract)	[58]	8 patients Spirometry, 6MWT, Treadmill test QoL 8 weeks, inpatient	Not specified	VO$_2$ peak (Treadmill) 6MWD m SF36 SGRQ	Not specified	Not specified	+63.6

* with positive trend; ^ stable or worsened; # article in Japanese, details not reported in abstract; § post MDR-TB patient; ADL: activities of daily living score; BGA: blood gas analysis; BMI: body mass index; CCQ: clinical chronic obstructive pulmonary disease (COPD) questionnaire; CRQ: chronic respiratory disease questionnaire; FEV1: forced expiratory volume in 1 s; FVC: forced vital capacity; FEV1/FVC ratio: the percentage of the FVC expired in one second; FiO$_2$: fraction of inspired oxygen; HADS: Hospital Anxiety and Depression Scale; HR: heart rate; HRQL: health-related quality of life; 6MWD: 6-min walk distance; 6MWT: 6-min walk test; ISWT: incremental shuttle walking test; LFT: lung function tests; MMRC: Modified Medical Research Council dyspnoea scale; MRC: Medical Research Council dyspnoea scale; MVV: maximum voluntary ventilation; PaCO$_2$: partial pressure of arterial carbon dioxide; PaO$_2$: partial pressure of arterial oxygen; pH: potential of hydrogen; PHQ-9: patient health questionnaire-9; PImax: maximal inspiratory pressure; PR: pulmonary rehabilitation; QoL: quality of life; RV: residual volume; SaO$_2$: oxygen saturation in arterial blood; SF36: 36-Item Short Form Health Survey; SGRQ: St. George Respiratory Questionnaire; SpO$_2$: peripheral capillary oxygen saturation; SW: spontaneous walking; TB: tuberculosis; TDI: transition dyspnoea index; VC: vital capacity; VEmax: maximum minute expired ventilation; VO$_2$: oxygen consumption.

In the vast majority of the studies, the spirometry parameters, oxygen saturation, and exercise capacity tests (6MWT or ISWT, range 35–110 m) improved significantly. When QoL tests were performed, they also improved significantly [45,51–58]. Unfortunately the different studies reported different parameters, making a meta-analytic evaluation difficult.

4.4. Priorities for Research

A comprehensive review on TB and rehabilitation [44] recommended that future studies investigating pulmonary rehabilitation include the following information to ensure comparative analyses:

a) Patients' characteristics (age, sex, ethnicity, etc);
b) A description of the TB disease, (history of previous treatment, bacteriological status, drug-resistance profile, treatment history -drugs and regimens; and adverse events observed) [7,8];
c) The physiopathological status, spirometry with response to bronchodilator, assessment of lung volumes through plethysmography, DLCO, arterial blood gas analysis, 6MWT, radiological evaluation—ideally a computerized tomography (CT) scan, a QoL evaluation with both general and a specific tools (St. George's questionnaire);
d) Rationale and design of the pulmonary rehabilitation plan, with a pre-/post-test comparison;
e) Cost-assessment and evaluation of programmatic feasibility [59].

5. Conclusions

This review describes the evidence available on adjuvant surgery (as described in the most important recent guidelines), as well as on the diagnosis and management of patients with post-treatment sequelae. The initial evidence supports the importance of adequate functional evaluations of these patients, which is necessary to identify those who will benefit from pulmonary rehabilitation.

A collection of high-quality standardised variables would allow the research to advance in the understanding of the need for, and the effectiveness of, pulmonary rehabilitation both at the individual and at the programmatic level.

Author Contributions: Conceptualization, D.V., S.T. and G.B.M.; methodology, R.C., L.D., G.B.M.; data curation, E.P., R.C., D.V.; Tables and Figure conceptualization: E.P., D.V., R.C., E.Z., G.B.M.; writing—original draft preparation, all authors; surgery: A.W.M.; Functional evaluation and rehabilitation: D.V., E.Z., A.S.; Public health aspects: M.v.s.B.; overall writing—review and editing, all authors; supervision, G.B.M. All authors have read and agreed to the published version of the manuscript.

Funding: This research received no external funding.

Acknowledgments: The article is part of the activities of the Global Tuberculosis Network (GTN; Committees on TB Treatment, Working Group on Pulmonary Rehabilitation and Global TB Consilium) and of the WHO Collaborating Centre for Tuberculosis and Lung Diseases, Tradate, ITA-80, 2017-2020- GBM/RC/LDA).

Conflicts of Interest: The authors declare no conflict of interest.

References

1. Lönnroth, K.; Migliori, G.B.; Abubakar, I.; D'Ambrosio, L.; de Vries, G.; Diel, R.; Douglas, P.; Falzon, D.; Gaudreau, M.A.; Goletti, D.; et al. Towards tuberculosis elimination: An action framework for low-incidence countries. *Eur. Respir. J.* **2015**, *45*, 928–952. [CrossRef]
2. Migliori, G.B.; Tiber, I.S.; Zumla, A.; Petersen, E.; Chakaya, J.M.; Wejse, C.; Muñoz Torrico, M.; Duarte, R.; Alffenaar, J.W.; Members of the Global Tuberculosis Network; et al. MDR/XDR-TB management of patients and contacts: Challenges facing the new decade. The 2020 clinical update by the Global Tuberculosis Network. *Int. J. Infect. Dis.* **2020**, S1201-9712(20)30045-X. [CrossRef]

3. Lange, C.; Aarnoutse, R.E.; Alffenaar, J.W.C.; Bothamley, G.; Brinkmann, F.; Costa, J.; Chesov, D.; van Crevel, R.; Dedicoat, M.; Dominguez, J.; et al. Management of patients with multidrug-resistant tuberculosis. *Int. J. Tuberc. Lung. Dis.* **2019**, *23*, 645–662. [CrossRef] [PubMed]
4. Pontali, E.; Tadolini, M.; Migliori, G.B. TB consilia and quality of tuberculosis management. *Int. J. Tuberc. Lung. Dis.* **2019**, *23*, 1048–1049. [CrossRef] [PubMed]
5. World Health Organization. *Global Tuberculosis Report 2019*; WHO/CDS/TB/2019.15; World Health Organization: Geneva, Switzerland, 2019; Available online: https://apps.who.int/iris/bitstream/handle/10665/329368/9789241565714-eng.pdf?ua=1 (accessed on 26 February 2020).
6. Borisov, S.E.; Dheda, K.; Enwerem, M.; Romero Leyet, R.; D'Ambrosio, L.; Centis, R.; Sotgiu, G.; Tiberi, S.; Alffenaar, J.W.; Maryandyshev, A.; et al. Effectiveness and safety of bedaquiline-containing regimens in the treatment of MDR- and XDR-TB: A multicentre study. *Eur. Respir. J.* **2017**, *49*, 1700387. [CrossRef]
7. Akkerman, O.; Aleksa, A.; Alffenaar, J.W.; Al-Marzouqi, N.H.; Arias-Guillén, M.; Belilovski, E.; Bernal, E.; Boeree, M.J.; Borisov, S.E.; Bruchfeld, J.; et al. Members of the International Study Group on new anti-tuberculosis drugs and adverse events monitoring. Surveillance of adverse events in the treatment of drug-resistant tuberculosis: A global feasibility study. *Int. J. Infect. Dis.* **2019**, *83*, 72–76. [CrossRef]
8. Borisov, S.; Danila, E.; Maryandyshev, A.; Dalcolmo, M.; Miliauskas, S.; Kuksa, L.; Manga, S.; Skrahina, A.; Diktanas, S.; Codecasa, L.R.; et al. Surveillance of adverse events in the treatment of drug-resistant tuberculosis: First global report. *Eur. Respir. J.* **2019**, *54*, 1901522. [CrossRef]
9. Pontali, E.; Raviglione, M.C.; Migliori, G.B.; writing group members of the Global TB Network Clinical Trials Committee. Regimens to treat multidrug-resistant tuberculosis: Past, present and future perspectives. *Eur. Respir. Rev.* **2019**, *28*, 190035. [CrossRef]
10. Pontali, E.; Sotgiu, G.; Tiberi, S.; Tadolini, M.; Visca, D.; D'Ambrosio, L.; Centis, R.; Spanevello, A.; Migliori, G.B. Combined treatment of drug-resistant tuberculosis with bedaquiline and delamanid: A systematic review. *Eur. Respir. J.* **2018**, *52*, 1800934. [CrossRef]
11. Pontali, E.; D'Ambrosio, L.; Centis, R.; Sotgiu, G.; Migliori, G.B. Multidrug-resistant tuberculosis and beyond: An updated analysis of the current evidence on bedaquiline. *Eur. Respir. J.* **2017**, *49*, 1700146. [CrossRef]
12. Pontali, E.; Sotgiu, G.; D'Ambrosio, L.; Centis, R.; Migliori, G.B. Bedaquiline and MDR-TB: A systematic and critical analysis of the evidence. *Eur. Respir. J.* **2016**, *47*, 394–402. [CrossRef] [PubMed]
13. Migliori, G.B.; Nardell, E.; Yedilbayev, A.; D'Ambrosio, L.; Centis, R.; Tadolini, M.; van den Boom, M.; Ehsani, S.; Sotgiu, G.; Dara, M. Reducing tuberculosis transmission: A consensus document from the World Health Organization Regional Office for Europe. *Eur. Respir. J.* **2019**, *53*, 1900391. [CrossRef] [PubMed]
14. Akkerman, O.W.; Ter Beek, L.; Centis, R.; Maeurer, M.; Visca, D.; Muñoz-Torrico, M.; Tiberi, S.; Migliori, G.B. Rehabilitation, optimized nutritional care, and boosting host internal milieu to improve long-term treatment outcomes in tuberculosis patients. *Int. J. Infect. Dis.* **2020**, S1201-9712(20)30031-X. [CrossRef] [PubMed]
15. Chan, E.D.; Iseman, M.D. Surgery for MDR-TB? *Int. J. Tuberc. Lung. Dis.* **2013**, *17*, 710. [CrossRef] [PubMed]
16. Chan, E.D.; Laurel, V.; Strand, M.J.; Chan, J.F.; Huynh, M.L.; Goble, M.; Iseman, M.D. Treatment and outcome analysis of 205 patients with multidrug-resistant tuberculosis. *Am. J. Respir. Crit. Care Med.* **2004**, *169*, 1103–1109. [CrossRef]
17. Iseman, M. Treatment of multidrug-resistant tuberculosis. *N. Engl. J. Med.* **1993**, *329*, 784.
18. Iseman, M.D.; Madsen, L.; Goble, M.; Pomerantz, M. Surgical intervention in the treatment of pulmonary disease caused by drug resistant Mycobacterium tuberculosis. *Am. Rev. Respir. Dis.* **1990**, *141*, 623. [CrossRef]
19. Dara, M.; Sotgiu, G.; Zaleskis, R.; Migliori, G.B. Untreatable tuberculosis: Is surgery the answer? *Eur. Respir. J.* **2015**, *45*, 577–582. [CrossRef]
20. Borisov, S.E.; D'Ambrosio, L.; Centis, R.; Tiberi, S.; Dheda, K.; Alffenaar, J.W.; Amale, R.; Belilowski, E.; Bruchfeld, J.; Canneto, B.; et al. Outcomes of patients with drug-resistant-tuberculosis treated with bedaquiline-containing regimens and undergoing adjunctive surgery. *J. Infect.* **2019**, *78*, 35–39. [CrossRef]
21. Tiberi, S.; Torrico, M.M.; Rahman, A.; Krutikov, M.; Visca, D.; Silva, D.R.; Kunst, H.; Migliori, G.B. Managing severe tuberculosis and its sequelae: From intensive care to surgery and rehabilitation. *J. Bras. Pneumol.* **2019**, *45*, e20180324. [CrossRef]
22. Calligaro, G.L.; Moodley, L.; Symons, G.; Dheda, K. The medical and surgical treatment of drug-resistant tuberculosis. *J. Thorac. Dis.* **2014**, *6*, 186–195. [PubMed]

23. Subotic, D.; Yablonskiy, P.; Sulis, G.; Cordos, I.; Petrov, D.; Centis, R.; D'Ambrosio, L.; Sotgiu, G.; Migliori, G.B. Surgery and pleuro-pulmonary tuberculosis: A scientific literature review. *J. Thorac. Dis.* **2016**, *8*, E474–E485. [CrossRef] [PubMed]
24. Sayir, F.; Ocakcioglu, I.; Şehitoğulları, A.; Çobanoğlu, U. Clinical analysis of pneumonectomy for destroyed lung: A retrospective study of 32 patients. *Gen. Thorac. Cardiovasc. Surg.* **2019**, *67*, 530–536. [CrossRef] [PubMed]
25. Yablonskii, P.K.; Kudriashov, G.G.; Avetisyan, A.O. Surgical Resection in the Treatment of Pulmonary Tuberculosis. *Thorac. Surg. Clin.* **2019**, *29*, 37–46. [CrossRef] [PubMed]
26. Chen, G.; Zhong, F.M.; Xu, X.D.; Yu, G.C.; Zhu, P.F. Efficacy of regional arterial embolization before pleuropulmonary resection in 32 patients with tuberculosis-destroyed lung. *Bmc. Pulm. Med.* **2018**, *18*, 156. [CrossRef]
27. Giller, D.B.; Giller, B.D.; Giller, G.V.; Shcherbakova, G.V.; Bizhanov, A.B.; Enilenis, I.I.; Glotov, A.A. Treatment of pulmonary tuberculosis: Past and present. *Eur. J. Cardiothorac. Surg.* **2018**, *53*, 967–972. [CrossRef]
28. Marrone, M.T.; Venkataramanan, V.; Goodman, M.; Hill, A.C.; Jereb, J.A.; Mase, S.R. Surgical interventions for drug-resistant tuberculosis: A systematic review and metaanalysis. *Int. J. Tuberc. Lung. Dis.* **2013**, *17*, 6–16. [CrossRef]
29. Xu, H.B.; Jiang, R.H.; Li, L. Pulmonary resection for patients with multidrug-resistant tuberculosis: Systematic review and meta-analysis. *J. Antimicrob. Chemother.* **2011**, *66*, 1687–1695. [CrossRef]
30. Johnston, J.C.; Shahidi, N.C.; Sadatsafavi, M.; Fitzgerald, J.M. Treatment outcomes of multidrug-resistant tuberculosis: A systematic review and meta-analysis. *PLoS ONE* **2009**, *4*, e6914. [CrossRef]
31. Orenstein, E.W.; Basu, S.; Shah, N.S.; Andrews, J.R.; Friedland, G.H.; Moll, A.P.; Gandhi, N.R.; Galvani, A.P. Treatment outcomes among patients with multidrug-resistant tuberculosis: Systematic review and meta-analysis. *Lancet. Infect. Dis.* **2009**, *9*, 153–161. [CrossRef]
32. Hannink, G.; Gooszen, H.G.; van Laarhoven, C.J.; Rovers, M.M. A systematic review of individual patient data meta-analyses on surgical interventions. *Syst. Rev.* **2013**, *2*, 52. [CrossRef] [PubMed]
33. Collaborative Group for the Meta-Analysis of Individual Patient Data in MDR-TB treatment–2017; Ahmad, N.; Ahuja, S.D.; Akkerman, O.W.; Alffenaar, J.C.; Anderson, L.F.; Baghaei, P.; Bang, D.; Barry, P.M.; Bastos, M.L.; et al. Treatment correlates of successful outcomes in pulmonary multidrug-resistant tuberculosis: An individual patient data meta-analysis. *Lancet* **2018**, *392*, 821–834. [CrossRef]
34. Campbell, J.R.; Falzon, D.; Mirzayev, F.; Jaramillo, E.; Migliori, G.B.; Mitnick, C.D.; Ndjeka, N.; Menzies, D. Improving Quality of Patient Data for Treatment of Multidrug- or Rifampin-Resistant Tuberculosis. *Emerg. Infect. Dis.* **2020**, *26*, 2020. [CrossRef] [PubMed]
35. Fox, G.J.; Mitnick, C.D.; Benedetti, A.; Chan, E.D.; Becerra, M.; Chiang, C.Y.; Keshavjee, S.; Koh, W.J.; Shiraishi, Y.; Viiklepp, P.; et al. Surgery as an Adjunctive Treatment for Multidrug-Resistant Tuberculosis: An Individual Patient Data Metaanalysis. *Clin. Infect. Dis.* **2016**, *62*, 887–895. [CrossRef]
36. World Health Organization. *WHO Consolidated Guidelines on Drug-Resistant Tuberculosis Treatment*; WHO/CDS/TB/2019.7; World Health Organization: Geneva, Switzerland, 2019; Available online: https://apps.who.int/iris/bitstream/handle/10665/311389/9789241550529-eng.pdf?ua=1 (accessed on 26 February 2020).
37. World Health Organization Regional Office for Europe. *The Role of Surgery in the Treatment of Pulmonary TB and Multidrug and Extensively Drug-Resistant TB*; WHO Regional Office for Europe: Copenhagen, Denmark, 2014; Available online: http://www.euro.who.int/__data/assets/pdf_file/0005/259691/The-role-of-surgery-in-the-treatment-of-pulmonary-TB-and-multidrug-and-extensively-drug-resistant-TB.pdf?ua=1 (accessed on 26 February 2020).
38. Caminero, J.A. (Ed.) *Guidelines for Clinical and Operational Management of Drug-Resistant Tuberculosis*; International Union Against Tuberculosis and Lung Disease: Paris, France, 2013; pp. 1–232.
39. Nahid, P.; Mase, S.R.; Migliori, G.B.; Sotgiu, G.; Bothamley, G.H.; Brozek, J.L.; Cattamanchi, A.; Cegielski, J.P.; Chen, L.; Daley, C.L.; et al. Treatment of Drug-Resistant Tuberculosis. An Official ATS/CDC/ERS/IDSA Clinical Practice Guideline. *Am. J. Respir. Crit. Care. Med.* **2019**, *200*, e93–e142. [CrossRef]
40. Marfina, G.Y.; Vladimirov, K.B.; Avetisian, A.O.; Starshinova, A.A.; Kudriashov, G.G.; Sokolovich, E.G.; Yablonskii, P.K. Bilateral cavitary multidrug- or extensively drug-resistant tuberculosis: Role of surgery. *Eur. J. Cardiothorac. Surg.* **2018**, *53*, 618–624. [CrossRef]
41. Pasipanodya, J.G.; Miller, T.L.; Vecino, M.; Munguia, G.; Garmon, R.; Bae, S.; Drewyer, G.; Weis, S.E. Pulmonary impairment after tuberculosis. *Chest* **2007**, *131*, 1817–1824. [CrossRef]

42. Hnizdo, E.; Singh, T.; Churchyard, G. Chronic pulmonary function impairment caused by initial and recurrent pulmonary tuberculosis following treatment. *Thorax* **2000**, *55*, 32–38. [CrossRef]
43. Muñoz-Torrico, M.; Cid-Juárez, S.; Galicia-Amor, S.; Troosters, T.; Spanevello, A. Sequelae Assessment and Rehabilitation. In *Tuberculosis (ERSMonograph)*; Migliori, G.B., Bothamley, G., Duarte, R., Rendon, A., Eds.; European Respiratory Society: Sheffield, UK, 2018; pp. 326–342. [CrossRef]
44. Muñoz-Torrico, M.; Rendon, A.; Centis, R.; D'Ambrosio, L.; Fuentes, Z.; Torres-Duque, C.; Mello, F.; Dalcolmo, M.; Pérez-Padilla, R.; Spanevello, A.; et al. Is there a rationale for pulmonary rehabilitation following successful chemotherapy for tuberculosis? *J. Bras. Pneumol.* **2016**, *42*, 374–385. [CrossRef]
45. Visca, D.; Zampogna, E.; Sotgiu, G.; Centis, R.; Saderi, L.; D'Ambrosio, L.; Pegoraro, V.; Pignatti, P.; Muñoz-Torrico, M.; Migliori, G.B.; et al. Pulmonary rehabilitation is effective in patients with tuberculosis pulmonary sequelae. *Eur. Respir. J.* **2019**, *53*, 1802184. [CrossRef]
46. Spruit, M.A.; Singh, S.J.; Garvey, C.; ZuWallack, R.; Nici, L.; Rochester, C.; Hill, K.; Holland, A.E.; Lareau, S.C.; Man, W.D.; et al. ATS/ERS Task Force on Pulmonary Rehabilitation. An official American Thoracic Society/European Respiratory Society statement: Key concepts and advances in pulmonary rehabilitation. *Am. J. Respir. Crit. Care. Med.* **2013**, *188*, e13–e64. [CrossRef] [PubMed]
47. Tsuboi, T.; Ohi, M.; Chin, K.; Hirata, H.; Otsuka, N.; Kita, H.; Kuno, K. Ventilatory support during exercise in patients with pulmonary tuberculosis sequelae. *Chest* **1997**, *112*, 1000–1007. [CrossRef] [PubMed]
48. Yang, G.F.; Alba, A.; Lee, M. Respiratory rehabilitation in severe restrictive lung disease secondary to tuberculosis. *Arch. Phys. Med. Rehabil.* **1984**, *65*, 556–558. [PubMed]
49. Strelis, A.A.; Strelis, A.K.; Roskoshnykh, V.K. Vibration massage in the prevention of postresection complications and in the clinical rehabilitation of patients with pulmonary tuberculosis after surgical interventions. *Probl. Tuberk. Bolezn. Legk.* **2004**, *11*, 29–34.
50. Daniels, K.J.; Irusen, E.; Pharaoh, H.; Hanekom, S. Post-tuberculosis health-related quality of life, lung function and exercise capacity in a cured pulmonary tuberculosis population in the Breede Valley District, South Africa. *S. Afr. J. Physiother.* **2019**, *75*, 1319. [CrossRef]
51. Tada, A.; Matsumoto, H.; Soda, R.; Endo, S.; Kawai, H.; Kimura, G.; Yamashita, M.; Okada, C.; Takahashi, K. Effects of pulmonary rehabilitation in patients with pulmonary tuberculosis sequelae. *Nihon. Kokyuki. Gakkai. Zasshi.* **2002**, *40*, 275–281.
52. Jones, R.; Kirenga, B.J.; Katagira, W.; Singh, S.J.; Pooler, J.; Okwera, A.; Kasiita, R.; Enki, D.G.; Creanor, S.; Barton, A. A pre-post intervention study of pulmonary rehabilitation for adults with post-tuberculosis lung disease in Uganda. *Int. J. Chron. Obstruct. Pulmon. Dis.* **2017**, *12*, 3533–3539. [CrossRef]
53. Ando, M.; Mori, A.; Esaki, H.; Shiraki, T.; Uemura, H.; Okazawa, M.; Sakakibara, H. The effect of pulmonary rehabilitation in patients with post-tuberculosis lung disorder. *Chest* **2003**, *123*, 1988–1995. [CrossRef]
54. Singh, S.K.; Naaraayan, A.; Acharya, P.; Menon, B.; Bansal, V.; Jesmajian, S. Pulmonary Rehabilitation in Patients with Chronic Lung Impairment from Pulmonary Tuberculosis. *Cureus* **2018**, *10*, e3664. [CrossRef]
55. Yoshida, N.; Yoshiyama, T.; Asai, E.; Komatsu, Y.; Sugiyama, Y.; Mineta, Y. Exercise training for the improvement of exercise performance of patients with pulmonary tuberculosis sequelae. *Intern. Med.* **2006**, *45*, 399–403. [CrossRef]
56. Wilches, E.C.; Rivera, J.A.; Mosquera, R.; Loaiza, L.; Obando, L. Pulmonary rehabilitation in multi-drug resistant tuberculosis (TB MDR): A case report. *Colomb. Med.* **2009**, *40*, 436–441.
57. Betancourt-Peña, J.; Muñoz-Erazo, B.E.; Hurtado-Gutiérrez, H. Efecto de la rehabilitación pulmonar en la calidad de vida y la capacidad funcional en pacientes con secuelas de tuberculosis [Effect of pulmonary rehabilitation in quality of life and functional capacity in patients with tuberculosis sequela]. *Nova* **2015**, *13*, 47–54. [CrossRef]
58. Rivera Motta, J.A.; Wilches, E.C.; Mosquera, R.P. Pulmonary rehabilitation on aerobic capacity and health-related quality of life in patients with sequelae of pulmonary TB. *Am. J. Respir. Crit. Care. Med.* **2016**, *193*, A2321. [CrossRef]
59. Visca, D.; Centis, R.; D'Ambrosio, L.; Muñoz-Torrico, M.; Chakaya, J.M.; Tiberi, S.; Spanevello, A.; Sotgiu, G.; Migliori, G.B. Post-TB treatment pulmonary rehabilitation: Do we need more? *Int. J. Tuberc. Lung. Dis.* **2020**, in press.

© 2020 by the authors. Licensee MDPI, Basel, Switzerland. This article is an open access article distributed under the terms and conditions of the Creative Commons Attribution (CC BY) license (http://creativecommons.org/licenses/by/4.0/).

Editorial

Drugs and Vaccines Will Be Necessary to Control Tuberculosis

Rino Rappuoli [1,2,3]

1. GSK, 53100 Siena, Italy; rino.r.rappuoli@gsk.com
2. Monoclonal Antibody Discovery Lab, Fondazione Toscana Life Sciences, 53100 Siena, Italy
3. Imperial College London, London SW7 2AZ, UK

Received: 5 June 2020; Accepted: 9 June 2020; Published: 10 June 2020

For most infectious diseases, vaccines are used to prevent infection and drugs are used for acute therapy and eradication of established infections. This is not the case for tuberculosis, where BCG, the vaccine against tuberculosis, which is given to most children in low- and middle-income countries, does not prevent infection, but it is only effective in decreasing infant mortality. The consequence is that infection occurs in one third of the global population and, following infection, the bacterium establishes a lifelong chronic presence. More than 2 billion people today are chronically infected. The immune system of the chronically infected individuals is effective to keep the bacterium at bay for most of the time, however there are circumstances where the immune system becomes temporarily or permanently weaker and the bacterium escapes the immunity and causes severe disease. The consequence are 1.5 million deaths every year. In addition to the only partially effective BCG vaccine against tuberculosis, also the available therapies for tuberculosis are suboptimal. They require months to eradicate the infection and they are largely ineffective due to the increasing rate of bacteria resistant to antibiotics. The question we have is why in 2020 we still rely on a vaccine developed a century ago (in 1922) and why we do not have better drugs. There are two main reasons for this unsatisfactory situation. The first one is the very low investment in TB research. The second one is the science gap. Indeed, over the last few decades we have been trying to use the emerging modern vaccine technologies to get better vaccines than BCG. The efforts to make better TB vaccines are summarized in the chapter by Carlos Martin in this book [1]. Remarkably, most of the new vaccines failed in animal models and the few promising ones failed in clinical trials, showing that the improvements in science were not able to deliver vaccines better than BCG and that the science gap in understanding the mechanism of immune protection remains. Very recently, however, some encouraging results have suggested that we may be at the breaking point and make significant progress after a century of stagnation. The three encouraging features are a new, molecularly designed live attenuated *Mycobacterium tuberculosis* entering Phase III clinical trials ([2,3] and VPM1002 in Carlos Martin's chapter), new data showing that a BCG boost can reduce new infections in adolescents ([4] and BCG revaccination in Carlos Martin's chapter), and finally, the remarkable observation that a protein-based adjuvanted vaccine given to chronically infected young people is able to reduce significantly severe disease ([5,6] and M72/AS01E in Carlos Martin's chapter). I find this latter study not only interesting for tuberculosis but also for the entire vaccinology because it is the first time that a vaccine is able to control a chronic infection that has been already established. In the case of viruses, the protein based, adjuvanted vaccine against Herpes Zoster has also shown to be able to control an already established chronic infection. We need to capture the breakthrough innovation of these trials to quickly develop vaccines to prevent infection in the naïve population and to prevent recurrences in chronically infected people. Combination of vaccines and drugs to improve the efficacy and reduce the time required for therapy should also be encouraged. At the same time, we need to boost basic science to understand the molecular mechanisms behind pathogenesis, protection and immunity. The only way to control and possibly eliminate tuberculosis from our planet is to develop effective vaccines and drugs.

Funding: This research received no external funding.

Conflicts of Interest: Dr. Rappuoli is an employee of GSK group of companies.

References

1. Martin, C.; Aguilo, N.; Marinova, D.; Gonzalo-Asensio, J. Update on TB vaccine pipeline. *Appl. Sci.* **2020**, *10*, 2632. [CrossRef]
2. Grode, L.; Seiler, P.; Baumann, S.; Hess, J.; Brinkmann, V.; Nasser Eddine, A.; Mann, P.; Goosmann, C.; Bandermann, S.; Smith, D.; et al. Increased vaccine efficacy against tuberculosis of recombinant Mycobacterium bovis bacille Calmette-Guérin mutants that secrete listeriolysin. *J. Clin. Investig.* **2005**, *115*, 2472–2479. [CrossRef] [PubMed]
3. Loxton, A.G.; Knaul, J.K.; Grode, L.; Gutschmidt, A.; Meller, C.; Eisele, B.; Johnstone, H.; van der Spuy, G.; Maertzdorf, J.; Kaufmann, S.H.E.; et al. Safety and Immunogenicity of the Recombinant Mycobacterium bovis BCG Vaccine VPM1002 in HIV-Unexposed Newborn Infants in South Africa. *Clin. Vaccine Immunol. CVI* **2017**, *24*. [CrossRef] [PubMed]
4. Nemes, E.; Geldenhuys, H.; Rozot, V.; Rutkowski, K.T.; Ratangee, F.; Bilek, N.; Mabwe, S.; Makhethe, L.; Erasmus, M.; Toefy, A. Prevention of M. tuberculosis Infection with H4:IC31 Vaccine or BCG Revaccination. *N. Engl. J. Med.* **2018**, *379*, 138–149. [CrossRef]
5. Tait, D.R.; Hatherill, M.; Van Der Meeren, O.; Ginsberg, A.M.; Van Brakel, E.; Salaun, B.; Scriba, T.J.; Akite, E.J.; Ayles, H.M.; Bollaerts, A.; et al. Final Analysis of a Trial of M72/AS01(E) Vaccine to Prevent Tuberculosis. *N. Engl. J. Med.* **2019**, *381*, 2429–2439. [CrossRef] [PubMed]
6. Van Der Meeren, O.; Hatherill, M.; Nduba, V.; Wilkinson, R.J.; Muyoyeta, M.; Van Brakel, E.; Ayles, H.M.; Henostroza, G.; Thienemann, F.; Scriba, T.J.; et al. Phase 2b Controlled Trial of M72/AS01(E) Vaccine to Prevent Tuberculosis. *N. Engl. J. Med.* **2018**, *379*, 1621–1634. [CrossRef] [PubMed]

© 2020 by the author. Licensee MDPI, Basel, Switzerland. This article is an open access article distributed under the terms and conditions of the Creative Commons Attribution (CC BY) license (http://creativecommons.org/licenses/by/4.0/).

Review

Update on TB Vaccine Pipeline

Carlos Martin [1,2,3,*], Nacho Aguilo [1,2], Dessislava Marinova [1,2] and Jesus Gonzalo-Asensio [1,2]

1. Grupo de Genética de Micobacterias, Microbiología, Facultad de Medicina Universidad de Zaragoza, 50009 Zaragoza, Spain; naguilo@unizar.es (N.A.); tbvacman@unizar.es (D.M.); jagonzal@unizar.es (J.G.-A.)
2. CIBERES Enfermedades Respiratorias, Instituto de Salud Carlos III, 28029 Madrid, Spain
3. Servicio de Microbiología, Hospital Universitario Miguel Servet, 3ISS Aragón, 50009 Zaragoza, Spain
* Correspondence: carlos@unizar.es

Received: 10 March 2020; Accepted: 31 March 2020; Published: 10 April 2020

Abstract: In addition to antibiotics, vaccination is considered among the most efficacious methods in the control and the potential eradication of infectious diseases. New safe and effective vaccines against tuberculosis (TB) could be a very important tool and are called to play a significant role in the fight against TB resistant to antimicrobials. Despite the extended use of the current TB vaccine Bacillus Calmette-Guérin (BCG), TB continues to be transmitted actively and continues to be one of the 10 most important causes of death in the world. In the last 20 years, different TB vaccines have entered clinical trials. In this paper, we review the current use of BCG and the diversity of vaccines in clinical trials and their possible indications. New TB vaccines capable of protecting against respiratory forms of the disease caused by sensitive or resistant *Mycobacterium tuberculosis* strains would be extremely useful tools helping to prevent the emergence of multi-drug resistance.

Keywords: BCG; tuberculosis vaccines; TBVI; EDCTP; IAVI; CTVD

1. Introduction

To date, vaccines have been able to overcome the evolution of antibiotic-resistant strains, which makes vaccination one of the most cost-effective measures for fighting infectious diseases [1]. *Mycobacterium tuberculosis* (*Mtb*) is included as 'critical' in the WHO priority list of research and development for new antibiotics effective against current resistant strains of tuberculosis (TB) [2]. Concurrently, new efficacious vaccines will be a very important tool to fight antimicrobial resistant TB (AMR TB) and are called to play an important role against this serious health issue [1].

The potential of using TB vaccines to combat AMR TB has generally been undervalued. This could be partly due to the lack of integral efficacy of the present vaccine Bacillus Calmette-Guérin (BCG) failing to reduce the numbers of TB cases, which makes new efficacious vaccines against respiratory forms of TB a critical necessity to help combat AMR TB. Currently, one of the biggest threats in TB is the emergence of multidrug-resistant (MDR) *Mtb* strains, resistant to isoniazide and rifampicin, and extensively drug-resistant (XDR) strains, resistant to at least four of the core anti-TB drugs. In 2018, half a million people were diagnosed with MDR-TB and is estimated that fifty million people were infected with MDR *Mtb* strains, creating a reservoir for future cases of active TB making treatment extremely difficult [2]. AMR poses a threat in TB; both the World Health Organization (WHO) and International Centres for Disease Control and Prevention (CDC) have expressed concern about antibiotic treatments for TB. New TB vaccines are necessary to complement existing and in-development pipeline TB treatment and diagnostic strategies. Considering there has been no evidence to suggest that molecular mechanisms of drug resistance in *Mtb* could affect immune control susceptibility, it is likely that vaccine efficacy against MDR-TB and drug-sensitive TB will be equivalent [3].

In the present work we review the current use of BCG and the non-specific effect against other pathogens and we summarized the diversity of new TB vaccine candidates in clinical trials and their indications. We discuss the need to keep the pipeline of new TB vaccine candidates and the current clinical trial designs employed in the field for efficacy determination of new TB vaccines which include designs for prevention of infection (POI), prevention of disease (POD), and prevention of recurrence of TB (POR), or trials for evaluating the therapeutic effect of TB vaccine candidates when applied in combination with current TB drug regimens with the aim to shorten duration of treatment. The pipeline includes new prime TB vaccines for administration at birth, which are expected to protect better than BCG, and for use in adolescents and adults, as revaccination strategies in individuals previously vaccinated with BCG at birth, as per WHO Preferred Product Characteristics (PPC) for new TB vaccines [3,4]. Novel TB vaccine strategies, which are safe and effective, are imperative in the global efforts to halt dissemination of drug-sensitive and AMR TB [3].

2. BCG the Current TB Vaccine in Use

Despite its variable efficacy against respiratory forms of TB, BCG remains the only marketed vaccine in use against TB, with more than 90% coverage in countries with high TB incidence [5–7] (Figure 1). BCG is an attenuated vaccine derived from *Mycobacterium bovis*, the etiologic agent of TB in cattle [8]. BCG was first introduced in clinic almost a hundred years ago, when in 1921 it was administered orally to a baby whose mother had died of TB the day after her birth. The baby showed no adverse effects to BCG vaccination and, more importantly, did not develop TB. Between 1921 and 1926 more than 50,000 children were vaccinated. Mortality among vaccinated children was 1.8%, compared to a mortality greater than 25% in unvaccinated children, showing its effectiveness in reducing infant mortality, not only due to TB if not due to other respiratory diseases [9]. Today we know that the main cause of attenuation of BCG is due to the loss of Region of Difference 1 (RD1) associated with the loss of the virulence factor of the secreted immunodominant antigen of 6 kDa (ESAT-6) [8].

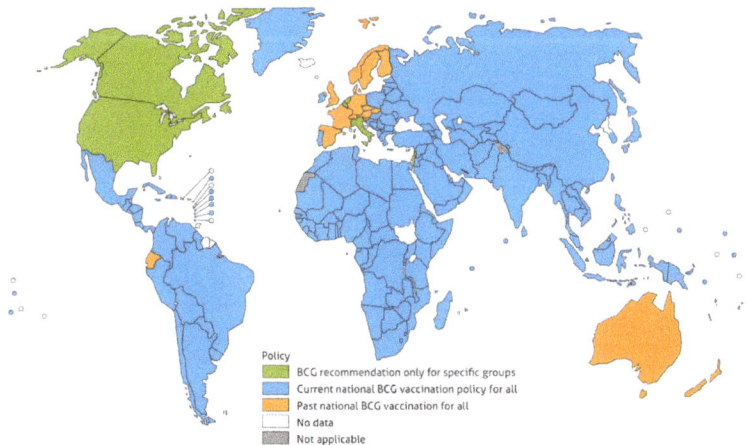

Figure 1. Bacillus Calmette-Guérin (BCG) vaccination coverage by country. Data from the World Health Organization about the BCG coverage in each country [10,11]. A total of 113 countries reported coverage of at least 90% [2].

In 1976, WHO established the Expanded Program on Immunization (EPI) to ensure universal access of mothers and infants/children to routinely recommended infant/childhood vaccines. Initially, there were six vaccine-preventable diseases included in the EPI: TB, poliomyelitis, diphtheria, tetanus, pertussis and measles. Intradermal vaccination with BCG at birth has been included in the WHO EPI, resulting in more than four billion vaccines administered worldwide to date and approximately 200 million doses given each year. Depending on strain and manufacturer of BCG, concentration of live bacteria in vaccine vials can range between 50,000 to three million per dose [12,13]. BCG vaccination is recommended as part of national childhood immunization programmes according to a country's TB epidemiology. In 2018, BCG vaccination was reported for 153 countries and 113 of these countries reported at least 90% BCG coverage [2,10,11] (Figure 1).

Today, BCG vaccination is recommended by the WHO in all newborn infants in countries where TB incidence is high. In countries where the incidence of TB is not high, BCG is recommended if the child is continually exposed to TB patient who does not respond to treatment and patient's separation is not possible, or when the child is continually exposed to a patient who has infectious pulmonary TB caused by MDR or XDR strains [14]. The recommendation of BCG vaccination for adults in endemic areas with high exposure to resistant TB remains controversial. Considering the potential risks of anti-TB treatment failure and complications related to BCG vaccination of immunocompromised individuals, administration of BCG could be recommended in unvaccinated individuals, tuberculin-negative or interferon gamma release assays (IGRA)-negative individuals exposed to MDR TB. Thus, trials evaluating protective efficacy of BCG in the context of exposure to MDR TB in adults are needed [14].

Commercial BCG is not a unique product as different formulations exist in terms of BCG strain, composition and/or dosage. Currently, six BCG strains are used worldwide in international immunization programs: BCG Pasteur 1173 P2, BCG Danish1331, BCG Glaxo 107, BCG Tokyo 172-1, BCG Russia-I and BCG Brazil [15]. These different BCG strains in use [16] exhibit different characteristics of attenuation and protection in animal models [17]. The genomic analysis of BCG sub-strains shows multiple differences, including other deletions other than RD1 that contribute to phenotypic variations between them [17], with clear attenuation differences but without being shown to contribute to differences in their efficacy [5]. A recent systematic head-to-head comparison study of BCG vaccine formulations demonstrated marked variations in content of viable mycobacteria which correlated with age-specific induction of cytokines in vitro [13]. According to the same study, these differences in viability possibly contribute to an observed formulation-dependent activation of innate and adaptive immune responses that could account for the variable protection observed in clinic [13].

3. Nonspecific Effects of BCG

One of the reasons why BCG is still used universally in middle and low income countries is because numerous studies indicate that similar to other live attenuated vaccines in use today, BCG have additional beneficial effects on the initially intended protection against TB [18–20]. Neonatal BCG is able to induce a strong Th1 cytokine response shown to enhance immune responses to other infant vaccines of the EPI [21]. In countries of low TB endemicity, BCG administration at birth has been related to reduction of childhood hospitalizations due to unrelated respiratory infections and sepsis [22,23]. In addition, there is emerging evidence that BCG may induce nonspecific resistance (T- and B-cell dependent) to other pathogens [20], which should be taken into consideration for AMR TB.

The non-specific beneficial effects ascribed to BCG have been attributed to the vaccine's ability to functionally and epigenetically reprogram innate immune cells, such as monocytes, macrophages, and NK (natural killer) cells, a process termed 'trained immunity' [24]. In human monocytes, BCG induced trained immunity has been attributed to the induction of metabolic pathways, which are regulated by epigenetic mechanisms at the level of chromatin organization [25]. In this context, future clinical trials could provide insight on the potential therapeutic role that modulation of these pathways may have during vaccination [26]. A recent large, multinational study conducted in sub-Saharan Africa suggests association of BCG vaccination with a reduced risk of malaria in children under the age of 5 years [27,28]. If these results are corroborated, they would denote that timely BCG vaccination could aid the global efforts to decrease malaria burden, including resistant forms of the disease.

To date, the preclinical and clinical down-selection process of new TB vaccine candidates has employed BCG as the reference gold standard comparator, because of its well-established safety profile [29,30]. Today, WHO also encourages incorporation of the nonspecific beneficial effects ascribed to BCG's ability to induce trained immunity in the design and development of novel TB vaccine candidates, especially those intended for BCG-replacement.

4. New TB Vaccine Candidates from Discovery to Clinical Trials

The most effective licensed vaccines against different infectious diseases confer protective immunity by /inducing neutralizing antibodies. Whereas, while for other diseases, such as HIV, malaria or TB, a strong response of cellular immunity is necessary, and we don't have correlated protection that could anticipate the efficacy of a new vaccine candidate [31]. The last 20 years have seen important breakthroughs in TB vaccinology, ranging from novel adjuvant systems or viral vectors for intracellular antigen presentation to advanced genetic engineering techniques for rational attenuation of *Mtb*. All these advances in vaccinology have led to the development of new TB vaccine candidates including novel prime and prime–boost regimes that could reach all age groups and TB population spectrums. The biology behind the host immune responses to *Mtb* are not yet understood [32,33]. Developing a protective vaccine requires not only finding the right antigens, but also activating the right ratio of protective and suppressive immune cells against these pathogens. In Douglas Young's words, "we need efficacious vaccines to understand immunology of TB" and we need "immunology studies to design a good TB vaccine".

Given the lack of BCG protection against respiratory forms of TB, in the last 20 years an enormous effort has been made in the research and development of new vaccines against TB. In the discovery phase, thousands of potential candidates were identified, of which hundreds have passed to preclinical evaluation in animal models and only a little more than a dozen have happened to be tested in early clinical studies in humans to date. There are different stages that each vaccine candidate needs to perform, including early first-in-human Phase I, then Phase II, and finally Phase III clinical trials, to reach marketing authorization. In the United States, for nearly two decades former Aeras (now International AIDS Vaccine Initiative, IAVI) [34], supported by the Bill and Melinda Gates Foundation, were dedicated to the discovery of new TB vaccine candidates. At the end of 2018, Aeras transferred its preclinical assets and clinical programs, biorepository, clinical staff, funding and other assets to International AIDS Vaccine Initiative (IAVI) [34]. In Europe, the research promoted by the different European Commission (EC) Framework Programs has made possible for hundreds of candidates to pass to preclinical evaluation of which several are currently in clinical trials [35]. In 2008, thanks to the EC funding programmes, the European TB Vaccine Initiative (TBVI) [36] was founded integrating at least 50 R&D consortium partners from the public (academia) and private (industry) sectors. TBVI is a non-profit product development partnership that facilitates the discovery and development of new, safe and effective TB vaccines and biomarkers that are accessible and affordable for global use. Clinical trials of a small number of selected TB vaccine candidates are supported by IAVI with funding sources from US NIH funding programmes and by the European and Developing Countries Clinical Trials Partnership (EDCTP) [37].

Given that the protection of BCG administered intradermally shows such limited results of efficacy, new studies changing the routes of administration of BCG have demonstrated encouraging results in non-human primates (NHP) applying BCG by the intravenous (IV) route against virulent *M. tuberculosis* challenge [38]. Although the difficulties of using the IV route for mass vaccination campaigns, the efficacy results demonstrating ability of IV BCG to substantially limit *Mtb* infection in highly susceptible rhesus macaques could have important implications in the preclinical evaluation of new candidates, as it could provide a prototype for identifying immune biomarkers and mechanisms of vaccine-induced protection against TB. The respiratory administration of BCG has demonstrate to be very promising by conferring very good immunity and protection in NHP [39]. If these results are confirmed in clinical studies, the aerosol route could be considered a possible universal vaccination route for BCG and new TB vaccine strategies.

Other live attenuated vaccines in preclinical testing fuelling the TB pipeline include the search for new candidates based on recombinant BCG has shown very promising results in preclinical animal models. On one hand, deletion of *zmp1* gene improves BCG-mediated protection in guinea pigs against TB [40] and on the other, inclusion of virulence genes, such as the RD1 region, genetically modified so as to not increase the virulence of BCG [41,42].

5. TB Vaccine Candidates in Clinical Trials

As we previously mentioned, the lack of immune marker(s) for prediction of vaccine-elicited protection makes finding effective vaccines against TB extremely challenging, as it requires long and expensive efficacy trials with thousands of volunteers (Phase IIb proof-of-concept trials and Phase III efficacy trials) in endemic countries with a high incidence of TB after obtaining robust safety and immunogenicity data in previous trials with tens (Phase I) and then with hundreds (Phase II) of volunteers [32] (Figure 2).

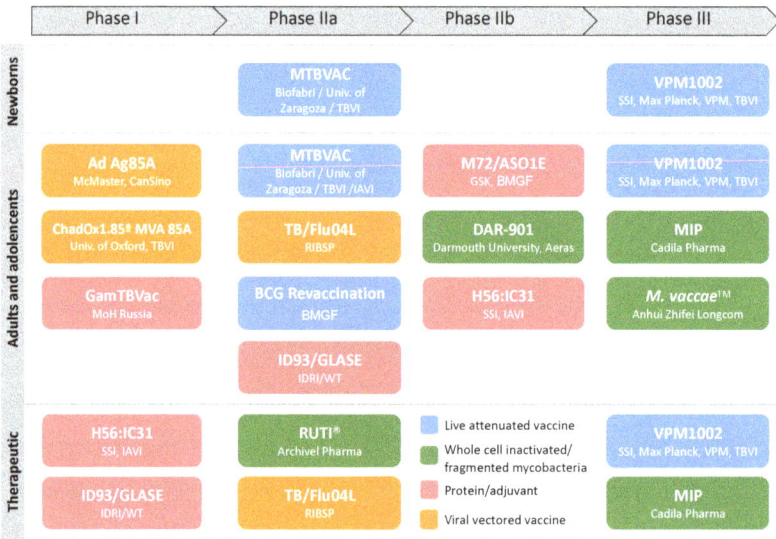

Figure 2. Tuberculosis (TB) vaccine candidates in the pipeline of clinical trials. The diagram shows the advance in clinical trials of the 14 vaccine candidates coloured according to each vaccine strategy.

Unlike HIV and malaria, twenty years ago there was no new candidate for a vaccine against TB in clinical trials. After more than 10 years of previous clinical trials, in 2012 MVA85A was the first candidate since BCG in 1921 to enter efficacy evaluation as preventive TB vaccine in infants [43,44]. MVA85A vaccine was tested in a double-blind, placebo-control Phase IIb efficacy study for its ability to increase (or boost) the immunity in healthy infants HIV-uninfected (aged four to six months) recently vaccinated with BCG at birth, living in a highly endemic region for TB (Worcester, South Africa). A total of 2797 children were vaccinated (1399 with MVA85A and 1398 with a placebo) and followed up every three months for more than three years. Results showed that 32 children (2%) of the 1399 in the MVA85A arm, were diagnosed with TB and 39 children (3%) of the 1398 vaccinated with BCG + placebo. The difference between the two groups was not significant and the interpretation was absence of efficacy of MVA85A. For the TB vaccine scientific community, this study resulted highly informative, and it was a great step forward in the research of new vaccines paving the way for new TB vaccine efficacy studies. The Worcester study was coordinated by the South African TB Vaccine Initiative (SATVI) and the site is highly prepared for testing new promising TB vaccines. Years after this study, the scientific community continues learning and drawing conclusions about the immunology of the disease. After three years of follow-up of the children in the study, the QuantiFERON (QFT) and the risk of disease were studied by SATVI and Oxford University teams [45]. The extended immunogenicity analysis of the trial data showed that both children that remained QFT- (less than 0.35 IU/mL) or those who became QFT + (with less than 4 UI/mL) had a lower risk of developing TB as compared to QFT + children with more than 4 IU/mL, whose risk of TB disease is high. This result may accelerate studies with other vaccines [45].

Current efficacy trials of new prophylactic TB vaccines attempt to measure either the prevention of infection against *Mtb* (POI), or the prevention of acquiring TB disease (POD) and the prevention of recurrent TB disease (POR) [46]. POR trials evaluate therapeutic vaccines administered as an adjunct to drug treatment to increase the effectiveness and shorten the duration of TB treatment in patients undergoing or completing TB treatment for active disease. The efficacy endpoints of POR trials are prevention of reactivation of existing infections and/or prevention of disease due to new infections. Considering lack of correlates of protection, POD trials provide the most reliable endpoint of vaccine efficacy and acceptance for Regulatory authorities.

In newborns, new TB vaccines should provide evidence of significant superiority over BCG. Global vaccination strategies targeting adolescents and adults would preferably include individuals with and without pre-existing *Mtb* infection, thus avoiding use of IGRAs which are expensive and can interfere with vaccine-induced immune results. In addition, testing candidates in both uninfected and infected individuals in future trials could avoid risks of excluding potentially efficacious candidates in the pipeline against infection but which lack efficacy in IGRA-positive individuals [47].

New TB vaccine candidates today in clinical trials could be divided into whole cell vaccines and subunit vaccines. Whole cell vaccines include live mycobacterial vaccines derived from live attenuated *Mtb* strains, *M. bovis* BCG or recombinant BCG and killed mycobacterial vaccines that could be formulated from other saprophytic mycobacterial species or *Mtb* [48] (Figure 3). Subunit vaccines contain *Mtb* antigens expressed as recombinant proteins that are formulated with different adjuvants or expressed by recombinant viral vectors that are used as vehicles for the administration of antigens [47]. Most of the current subunit vaccine candidates are vaccines with limited antigen diversity which are designed to enhance prior immunity mediated by T cells [49] (Figure 3).

There is a total of 14 TB vaccine candidates in clinical trials today, seven are based on subunits and seven consist on whole-cell mycobacteria [47,50] (Figure 3). Of the subunit candidates, four are mycobacterial fusion protein(s) in new adjuvant formulations (ID93: GLA-SE, H56.IC31, M72:ASO1E, GamTBVac) and three are based on recombinant live-attenuated or replication-deficient virus-vectored expressing one or more *Mtb* proteins (Ad5Ag85, ChadOx1.85/MVA85A, TB/FLU-04L). Of the whole-cell mycobacterial candidates, there are four candidates based on inactivated/extracts of mycobacteria (*M. vaccae*, MIP, DAR-901, RUTI) and three vaccine candidates are based on live attenuated mycobacteria, one is BCG revaccination, since a positive signal has been seen in prevention studies of *Mtb* infection [51], other is based on the use of recombinant BCG (VPM1002) and the third is derived from rationally attenuated *Mtb* (MTBVAC).

5.1. Whole-Cell Vaccine Candidates

5.1.1. Live Attenuated Vaccines

Attenuated live vaccines based in *Mtb* in humans is expected to stimulate specific host-immune responses mimicking natural TB infection without causing disease (similar to with latent TB infection (LTBI)). In support of this rationale, prospective cohort studies with individuals exposed to patients with active TB indicate that those persons LTBI could be close to 80% more protected against secondary *Mtb* infection than individuals naïve to *Mtb* infection [52]. MTBVAC is based on the Pasteurian approach of vaccinology, to attenuate the pathogen from a human clinical isolate. The rational attenuation of MTBVAC is due to deletions in the major virulence genes *phoP* and *fadD26*. Following the Geneva consensus for attenuated TB vaccines [30], antibiotic markers were eliminated in MTBVAC [53]. As an attenuated derivative from the human pathogen, MTBVAC contains all antigens present in *Mtb* [54], including those contained in the RD1 region, which is deleted in BCG and responsible for increasing the protection in animal models [55] (Figure 3). The primary target population of MTBVAC are neonates; the secondary target population includes adolescents and adults (as a booster vaccine). A Phase Ia trial in adults showed that MTBVAC is safe and immunogenic [56] and a second Phase Ib in neonates in an endemic country showed that MTBVAC is as safe as BCG and more immunogenic [57]. Dose-defining Phase IIa trials in both target populations started in 2019 with funding support from US NIH in collaboration with IAVI in adults (ClinicalTrials.gov NCT02933281) and the EDCTP in collaboration with TBVI in neonates (ClinicalTrials.gov NCT03536117) (Figure 2). MTBVAC development is included in TBVI pipeline. MTBVAC was discovered and constructed at the University of Zaragoza (Spain) in collaboration with Pasteur Institute (France). Recently it was demonstrated that similarly to BCG, MTBVAC is able to induce trained immunity through the induction of glycolysis and glutaminolysis, accumulation of histone methylation marks at the promoters of pro-inflammatory genes, facilitating an enhanced response after secondary challenge. Importantly, these findings in human primary myeloid cells are complemented by a strong MTBVAC induced heterologous protection against a lethal challenge *with Streptococcus pneumoniae* in an experimental murine model of pneumonia (Tarancon et al. in press PLOS Pathogens 2020). The Spanish Biopharmaceutical company Biofabri is vaccine manufacturer of MTBVAC responsible for industrial and clinical development of the vaccine under the umbrella of TBVI.

BCG/BCG revaccination strategy showed a positive signal in prevention studies of *Mtb* infection [51] and has been included for application in adolescents and adults vaccinated with BCG at birth (Figure 2). BCG revaccination in adults reduced the rate of upper respiratory tract infections as compared to a subunit vaccine or placebo groups (2.1%, 9.4% and 7.9%, respectively; $P<0.001$ for both comparisons) [51] suggesting that BCG revaccination could prevent respiratory diseases including AMR forms of these diseases. The POI trial tested the ability of BCG revaccination to prevent *Mtb* infection using IGRA conversion QuantiFERON test in healthy South African adolescents [51]. However, considering the difficulty of understanding what IGRA conversion and reversion means in terms of developing TB disease, POD trials should be performed in order to obtain the authorization by the regulatory agencies for licensure of new TB vaccines. BCG/BCG revaccination strategy is included in the pipe line of the Bill and Melinda Gates Foundation.

VPM1002 is a recombinant *M. bovis* BCG, which is developed by the Max Planck Institute in Berlin to express listeriolysin from *Listeria monocytogenes* and with a deletion of the gene coding for urease C [58] (Figure 3). VPM1002 is in clinical development led by Vakzine Projekt Management and in collaboration with Serum Institute of India and is included in TBVI pipeline. The rationale is to improve the effectiveness of BCG by inserting additional genes. Two Phase I trials for safety and immunogenicity in adults and newborns have been published [59,60] and a Phase IIb efficacy trial is currently being carried out in South Africa to assess the safety and immunogenicity and protective efficacy of the vaccine including uninfected HIV-exposed newborns, with support by EDCTP. Other clinical trials undergoing with VPM1002 include a phase II/III POR and a Phase III POD trial in India [61] (Figure 2).

5.1.2. Inactivated Whole-Cell Mycobacteria

Most of inactivated whole cell mycobacterial TB vaccine candidates have been designed as "therapeutic vaccines" seeking to reduce treatment length in people infected with latent TB, or reduce the likelihood of recurrence after the end of a treatment [48].

M. vaccae™ vaccine is a lysate comprised of inactivated *Mycobacterium vaccae* (non-tuberculous mycobacteria) developed as an immunotherapeutic agent to help shorten TB treatment for patients with drug-susceptible TB and licensed by the China Food and Drug Administration (FDA). A Phase III study of efficacy has been recently published [2,62] (Figures 2 and 3).

MIP Immuvac is a heat-killed *Mycobacterium indicus pranii* vaccine, approved by the drug controller general of India and FDA as an immune-therapeutic and immunoprophylactic strategy for use in multibacillary leprosy patients (as an adjunct to standard multidrug therapy), and for preventing the development of leprosy among close contacts of leprosy patients. A Phase III efficacy and safety trial for preventing pulmonary TB among healthy house-hold contacts of sputum smear-positive TB patients is underway in India [2] (Figure 2).

DAR-901 is based on a heat-inactivated non-tuberculous bacterium *M. vaccae* renamed as *Mycobacterium obuense* which was produced by former Aeras (Rockville, MD, USA) [63] for use as booster vaccine in BCG vaccines (Figure 3). It is the scalable manufacturing version of candidate vaccine SRL172, which showed efficacy in a Phase III trial among HIV-infected adults in Tanzania. DAR-901 is in a Phase IIb prevention of infection trial among BCG-primed adolescents, also in the United Republic of Tanzania. The trial is scheduled for completion in 2020 [2] (Figure 2).

RUTI®is a liposomal formulation containing fragmented, detoxified *Mtb* grown under stress conceived conditions, as a potential therapeutic vaccine (Figure 3). RUTI is sponsored by the Archivel Farma (Spain), and is currently in a Phase IIa of clinical trials [64,65] (Figure 2). When administered one month after isoniazid treatment, RUTI showed safety and immunogenicity in individuals with latent TB infection (LTBI). Plans for evaluation in a POD trial among HIV-infected and uninfected patients with LTBI are underway [47].

5.2. Subunit Vaccines

5.2.1. Subunit Viral Vectored Vaccines

There are three subunit vaccines that use attenuated viral vectors by different routes of administration.

Ad5HuAg85A vaccine, developed by the University of McMaster in Canada, consists of a human adenovirus serotype 5 vector that expresses Ag85A and is administered intramuscularly (Figure 3). AdHu5Ag85A has been evaluated in a Phase 1 safety and immunogenicity study in BCG-naïve and BCG-immunized healthy adults by the intramuscular route showing adequate safety and tolerability [47,66] (Figure 2).

ChAdOx85A/MVA85A vaccine strategy, developed by the University of Oxford, like the recombinant pox vaccine MVA85A (mentioned above), ChAdOx85A is a simian adenovirus which also expresses Ag85A (Figure 3). Both vaccine candidates are tested together in a joint heterologous prime-boost regimen delivered through both systemic and mucosal routes in BCG-vaccinated individuals. A Phase I trial of intramuscular administration of ChAdOx85A in BCG vaccinated adults in the United Kingdom, tested alone and as part of a prime-boost strategy with MVA85A, has been completed. Aerosol administration of ChAdOx185A is currently evaluated in a Phase I trial among BCG-vaccinated adults in Switzerland (Figure 2). Safety and immunogenicity of aerosol administration of MVA85A in BCG-vaccinated individuals and in people with a latent TB infection have been evaluated. In 2019, plans for a Phase IIa safety and immunogenicity trial among adults and adolescents in Uganda were underway with the aim of evaluating intramuscular administration of ChAdOx185A and MVA85A [2].

TB/FLU-04L is based on an attenuated replication-deficient influenza virus vector expressing antigens Ag85A and ESAT-6 (Figure 3). It was designed as a preventive booster vaccine in BCG-vaccinated infants, adolescents and adults. It is about to start a Phase IIa study in individuals with latent TB [2] (Figure 2).

5.2.2. Adjuvanted Subunit Vaccines

D93 + GLA-SE vaccine was developed by the Infectious Disease Research Institute (IDRI) in the United States. ID93+GLA-SE comprises four *Mtb* antigens, of which three are associated with virulence (Rv2608, Rv3619 and Rv3620) and one (Rv1813), with latency) formulated with the adjuvant GLA-SE for delivery [67] (Figure 3). It has been evaluated Phase IIa trial in South Africa among HIV-negative TB patients who have recently completed TB treatment for active pulmonary disease [68] (Figure 2). This is in preparation for Phase II safety and immunogenicity trials among adults undergoing active TB therapy. A POI Phase IIa trial in BCG-vaccinated healthy health care workers is also underway [2].

H56:IC31 is a fusion protein of three *Mtb* antigens (Ag85B, ESAT-6 and Rv2660c) and delivered in the adjuvant IC31© (Valneva Austria GmBH, Vienna, Austria) (Figure 3). H56:IC31 was discovered and developed by the Statem Serum Institut (SSI) of Copenhagen. It has been tested in three Phase I or I/IIa trials for safety and immunogenicity in BCG-vaccinated adults (Figure 2) showing acceptable safety and immunogenicity. It has completed a Phase Ib safety and immunogenicity trial in adolescents. A POR Phase IIb trial funded by EDCTP and coordinated by IAVI is underway in South Africa and the United Republic of Tanzania [2].

M72/AS01E is also a subunit candidate vaccine comprising two *Mtb* antigens (32A and 39A) formulated in the AS01 adjuvant for delivery (Figure 3), also used in the formulation of the malaria vaccine (RTS, S/AS01, GlaxoSmithKline) and the recombinant zoster vaccine Shingrix, GlaxoSmithKline [69]. It was evaluated in a IIb efficacy trial in in Kenya, South Africa and Zambia among *Mtb*-infected HIV-negative adults whose data showed 54.0% protective efficacy in *Mtb*–infected young adult women [69] (Figure 2). The immunogenicity analysis after end of the three-year follow-up showed that M72/AS01 elicited an immune response and provided protection against progression to pulmonary TB disease for at least three years [70]. This is the first time a proof-of-principle trial demonstrates vaccine-induced protection against clinical TB disease. However, whether M72/AS01 could provide protection against TB among *Mtb*-uninfected and HIV-negative individuals and in people from other geographical areas remain key questions to be answered. M72/AS01E has been exclusively licensed to the Medical Research Institute of Bill and Melinda Gates Foundation for further development.

GamTBvac is a fusion protein comprising *Mtb* antigens Ag85A and ESAT6–CFP-10 with the dextran-binding domain immobilized on dextran. It is formulated with an adjuvant consisting of a DEAE-dextran core and CpG oligodeoxynucleotides (TLR9 agonist) (Figure 3). GamTBvac is undergoing a Phase IIa safety and immunogenicity evaluation in healthy BCG-vaccinated adults, following a successful Phase 1 safety and immunogenicity trial in Russia [47] (Figure 2).

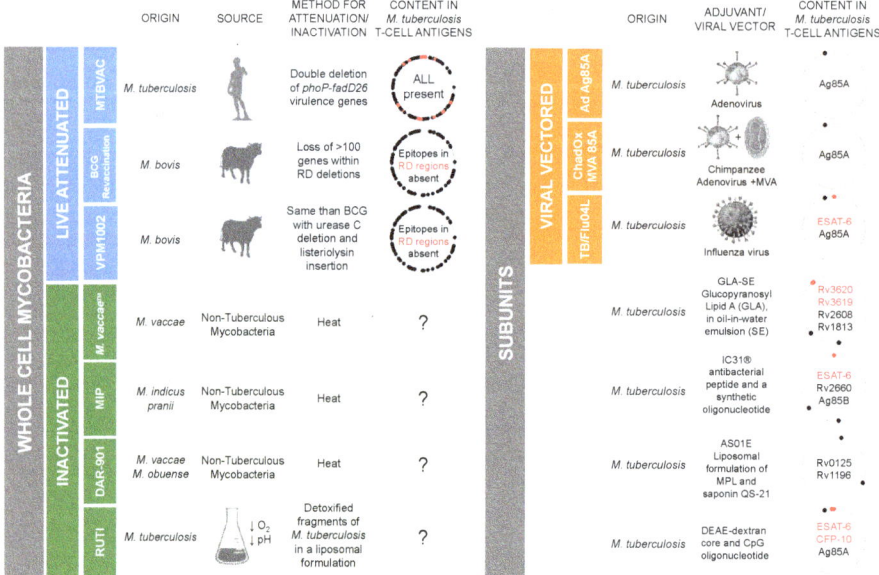

Figure 3. The diversity of TB vaccine candidates in clinical trials. Schematic table showing the main characteristics of the vaccine candidates. The table is coloured according to vaccine strategies indicated in Figure 2 and contains representative information for each candidate, including the mycobacterial origin from each vaccine and their antigenic content.

6. Target Population for a New TB Vaccine

In its Preferred Product Characteristics (PPC) document for TB vaccines, WHO defines two target populations for which a new TB vaccine could play a very important role in the fight against TB including resistant forms of the disease [3,4]. The WHO prioritizes two types of strategies one, safe, effective and affordable TB vaccines for adolescents and adults and the other, a TB vaccine for neonates and infants with improved safety and efficacy as compared to BCG.

Following the negative results of the first efficacy study which was conducted with MVA5A in infants, the priority population was changed to mainly adults. Modelling studies suggest that new effective vaccines for adolescents and adults, who are responsible for disease transmission, would have the greatest impact in halting TB incidence [71,72]. These studies suggest that although disease incidence in children under five years of age is considerable, TB transmission among this age-group is not common [73].

Prophylactic prime vaccines should be compared to BCG, since whole-cell vaccines seek to protect better than BCG. The immune responses elicited by live mycobacterial vaccines are considered to be specific and long-lasting, and these responses are not obtained with subunit vaccines. Vaccine persistence or restricted replication in vivo could account for the differences of immune responses, as observed for other live human vaccines, such as polio, measles, and yellow fever [48]. Considering the variability in BCG formulations, new TB vaccine trials which use BCG as reference comparator, should be interpreted cautiously with reference to a specific BCG formulation avoiding generalization of data to all BCGs [13].

Since the majority of the adolescent and adult population in countries endemic of TB has been previously vaccinated with BCG at birth, what is sought with subunit candidates comprising specific *Mtb* antigens is to potentiate the pre-existing immunity induced by BCG [2,46,62]. Recent studies indicate that new TB vaccines that are compared to BCG should be interpreted cautiously with reference to a specific BCG formulation and not presumed to generalize to all BCGs [13]. In addition, testing the new concepts of vaccines in relevant animal models such as NHP, should be key before advancing into expensive clinical trials of efficacy [74,75]. Something that has been questioned in the MVA85A efficacy study [38] was that the efficacy experiments in NHP, which showed lack of efficacy by the tested clinical route and dose of administration [76,77]. We should be very careful not to repeat mistakes with new candidates. Recently, it was published in NHP, that boosting BCG with M72/ASO1E or H56/CAF01or rAd5 failed to enhance BCG-induced protection against TB [78]. Thus, care should be taken with the advanced clinical development of such candidates.

Healthy newborns represent the most sensitive population without pre-existing immunity to BCG or environmental mycobacteria, which in older groups can lead to possible effects of masking and blocking vaccination [79,80]. We, therefore, think that the efficacy should be studied in newborns as first step, and once the efficacy of a new vaccine has been established, then such new candidates should be tested for efficacy in adolescents and adults where the impact on TB will be greater given that the pulmonary forms are responsible for this transmission [80].

Author Contributions: Writing—review and editing, C.M., N.A., D.M. and J.G.-A. All authors have read and agreed to the published version of the manuscript.

Funding: This work has been funded by the Ministry of Science (RTI2018-097625-B-I00) and the European and Developing Countries Clinical Trials Partnership (EDCTP) RIA2016V-1637.

Acknowledgments: In this section, you can acknowledge any support given which is not covered by the author contribution or funding sections. This may include administrative and technical support, or donations in kind (e.g., materials used for experiments).

Conflicts of Interest: C.M., D.M., N.A. and J.G.-A. are inventors of patents related to the tuberculosis vaccine, the owner of which is the University of Zaragoza and of which the Spanish biotechnology company Biofabri is responsible for industrial and clinical development of MTBVAC.

References

1. Bloom, D.E.; Black, S.; Salisbury, D.; Rappuoli, R. Antimicrobial resistance and the role of vaccines. *Proc. Natl. Acad. Sci. USA* **2018**, *115*, 12868–12871. [CrossRef] [PubMed]
2. WHO. *Global Tuberculosis Report 2019*; WHO, Ed.; WHO: Geneva, Switzerland, 2019; pp. 1–297.
3. WHO. *Preferred Product Characteristics for New Tuberculosis Vaccines*; WHO: Geneva, Switzerland, 2018; pp. 1–28.
4. Schrager, L.K.; Chandrasekaran, P.; Fritzell, B.H.; Hatherill, M.; Lambert, P.-H.; Mcshane, H.; Tornieporth, N.; Vekemans, J. WHO preferred product characteristics for new vaccines against tuberculosis. *Lancet Infect. Dis.* **2018**, *18*, 828–829. [CrossRef]
5. Mangtani, P.; Abubakar, I.; Ariti, C.; Beynon, R.; Pimpin, L.; Fine, P.E.M.; Rodrigues, L.C.; Smith, P.G.; Lipman, M.; Whiting, P.F.; et al. Protection by BCG vaccine against tuberculosis: A systematic review of randomized controlled trials. *Clin. Infect. Dis.* **2014**, *58*, 470–480. [CrossRef] [PubMed]
6. Zwerling, A.; Behr, M.A.; Verma, A.; Brewer, T.F.; Menzies, D.; Pai, M. The BCG World Atlas: A Database of Global BCG Vaccination Policies and Practices. *PLoS Med.* **2011**, *8*, e1001012. [CrossRef] [PubMed]
7. Fine, P.E. ScienceDirect.com—The Lancet—Variation in protection by BCG: Implications of and for heterologous immunity. *Lancet* **1995**, *346*, 1339–1345. [CrossRef]
8. Brosch, R.; Gordon, S.V.; Garnier, T.; Eiglmeier, K.; Frigui, W.; Valenti, P.; Santos, S.D.; Duthoy, S.; Lacroix, C.; Garcia-Pelayo, C.; et al. Genome plasticity of BCG and impact on vaccine efficacy. *Proc. Natl. Acad. Sci. USA* **2007**, *104*, 5596–5601. [CrossRef]
9. Gheorghiu, M.; Lagranderie, M.; Balazuc, A.-M. *Vaccines: A Biography*; Artenstein, A.W., Ed.; Springer New York: New York, NY, USA, 2009; pp. 125–140.
10. The BCG World Atlas Is a Database of Global BCG Vaccination Policies and Practices. Available online: http://www.bcgatlas.org (accessed on 1 April 2020).
11. The European Centre for Disease Prevention and Control Is an Agency of the European Union and Provides Access to the Vaccine Schedules in All Countries of the European Union. Available online: http://vaccine-schedule.ecdc.europa.eu/Pages/Scheduler.aspx (accessed on 1 April 2020).
12. Lambach, D.P. *Who Informative Sheet BCG*; WHO: Geneva, Switzerland, 2012; pp. 1–5.
13. Angelidou, A.; Conti, M.-G.; Diray-Arce, J.; Benn, C.S.; Shann, F.; Netea, M.G.; Liu, M.; Potluri, L.P.; Sanchez-Schmitz, G.; Husson, R.; et al. Licensed Bacille Calmette-Guérin (BCG) formulations differ markedly in bacterial viability, RNA content and innate immune activation. *Vaccine* **2020**, *38*, 2229–2240. [CrossRef]
14. Martinón-Torres, F.; Martin, C. Tuberculosis Vaccines. In *Pediatric Vaccines and Vaccinations*; Springer International Publishing: Cham, Switzerland, 2017; Volume 359, pp. 149–160.
15. Ho, M.M.; Southern, J.; Kang, H.-N.; Knezevic, I. WHO Informal Consultation on standardization and evaluation of BCG vaccines Geneva, Switzerland 22–23 September 2009. *Vaccine* **2010**, *28*, 6945–6950. [CrossRef]
16. Behr, M.A. BCG–different strains, different vaccines? *Lancet Infect. Dis.* **2002**, *2*, 86–92. [CrossRef]
17. Zhang, L.; Ru, H.-W.; Chen, F.-Z.; Jin, C.-Y.; Sun, R.-F.; Fan, X.-Y.; Guo, M.; Mai, J.-T.; Xu, W.-X.; Lin, Q.-X.; et al. Variable Virulence and Efficacy of BCG Vaccine Strains in Mice and Correlation with Genome Polymorphisms. *Mol. Ther.* **2016**, *24*, 398–405. [CrossRef] [PubMed]
18. Benn, C.S.; Netea, M.G.; Selin, L.K.; Aaby, P. A small jab—A big effect: Nonspecific immunomodulation by vaccines. *Trends Immunol.* **2013**, *34*, 431–439. [CrossRef] [PubMed]
19. Cauchi, S.; Locht, C. Non-specific Effects of Live Attenuated Pertussis Vaccine Against Heterologous Infectious and Inflammatory Diseases. *Front. Immunol.* **2018**, *9*, 2872. [CrossRef] [PubMed]
20. Kleinnijenhuis, J.J.; Quintin, J.J.; Preijers, F.F.; Joosten, L.A.B.L.; Ifrim, D.C.D.; Saeed, S.S.; Jacobs, C.C.; van Loenhout, J.J.; de Jong, D.D.; Stunnenberg, H.G.H.; et al. Bacille Calmette-Guerin induces NOD2-dependent nonspecific protection from reinfection via epigenetic reprogramming of monocytes. *Proc. Natl. Acad. Sci. USA* **2012**, *109*, 17537–17542. [CrossRef] [PubMed]
21. Ota, M.O.C.; Vekemans, J.; Schlegel-Haueter, S.E.; Fielding, K.; Sanneh, M.; Kidd, M.; Newport, M.J.; Aaby, P.; Whittle, H.; Lambert, P.-H.; et al. Influence of *Mycobacterium bovis* bacillus Calmette-Guérin on antibody and cytokine responses to human neonatal vaccination. *J. Immunol.* **2002**, *168*, 919–925. [CrossRef]

22. de Castro, M.J.; Pardo-Seco, J.; Martinón-Torres, F. Nonspecific (Heterologous) Protection of Neonatal BCG Vaccination Against Hospitalization Due to Respiratory Infection and Sepsis. *Clin. Infect. Dis.* **2015**, *60*, 1611–1619. [CrossRef]
23. Iglesias, M.J.; Martin, C. Editorial Commentary: Nonspecific Beneficial Effects of BCG Vaccination in High-income Countries, Should We Extend Recommendation of BCG Vaccination? *Clin. Infect. Dis.* **2015**, *60*, 1620–1621. [CrossRef]
24. Netea, M.G.; Joosten, L.A.B.; Latz, E.; Mills, K.H.G.; Natoli, G.; Stunnenberg, H.G.; O'Neill, L.A.J.; Xavier, R.J. Trained immunity: A program of innate immune memory in health and disease. *Science* **2016**, *352*, aaf1098. [CrossRef]
25. Arts, R.J.W.; Moorlag, S.J.C.F.M.; Novakovic, B.; Li, Y.; Wang, S.-Y.; Oosting, M.; Kumar, V.; Xavier, R.J.; Wijmenga, C.; Joosten, L.A.B.; et al. BCG Vaccination Protects against Experimental Viral Infection in Humans through the Induction of Cytokines Associated with Trained Immunity. *Cell Host Microbe* **2018**, *23*, 89–100.e5. [CrossRef]
26. Arts, R.J.W.; Carvalho, A.; La Rocca, C.; Palma, C.; Rodrigues, F.; Silvestre, R.; Kleinnijenhuis, J.; Lachmandas, E.; Gonçalves, L.G.; Belinha, A.; et al. Immunometabolic Pathways in BCG-Induced Trained Immunity. *Cell Rep.* **2016**, *17*, 2562–2571. [CrossRef]
27. Walk, J.; de Bree, L.C.J.; Graumans, W.; Stoter, R.; van Gemert, G.-J.; van de Vegte-Bolmer, M.; Teelen, K.; Hermsen, C.C.; Arts, R.J.W.; Behet, M.C.; et al. Outcomes of controlled human malaria infection after BCG vaccination. *Nat. Commun.* **2019**, *10*, 874–878. [CrossRef]
28. Berendsen, M.L.; van Gijzel, S.W.; Smits, J.; de Mast, Q.; Aaby, P.; Benn, C.S.; Netea, M.G.; van der Ven, A.J. BCG vaccination is associated with reduced malaria prevalence in children under the age of 5 years in sub-Saharan Africa. *BMJ Glob. Health* **2019**, *4*, e001862-11. [CrossRef] [PubMed]
29. Walker, K.B.; Brennan, M.J.; Ho, M.M.; Eskola, J.; Thiry, G.; Sadoff, J.; Dobbelaer, R.; Grode, L.; Liu, M.A.; Fruth, U.; et al. The second Geneva Consensus: Recommendations for novel live TB vaccines. *Vaccine* **2010**, *28*, 2259–2270. [CrossRef] [PubMed]
30. Kamath, A.T.; Fruth, U.; Brennan, M.J.; Dobbelaer, R.; Hubrechts, P.; Ho, M.M.; Mayner, R.E.; Thole, J.; Walker, K.B.; Liu, M.; et al. AERAS Global TB Vaccine Foundation; World Health Organization New live mycobacterial vaccines: The Geneva consensus on essential steps towards clinical development. *Vaccine* **2005**, *23*, 3753–3761. [CrossRef] [PubMed]
31. Rappuoli, R.; Aderem, A. A 2020 vision for vaccines against HIV, tuberculosis and malaria. *Nature* **2011**, *473*, 463–469. [CrossRef]
32. Marinova, D.; Gonzalo-Asensio, J.; Aguilo, N.; Martin, C. Recent developments in tuberculosis vaccines. *Expert Rev. Vaccines* **2013**, *12*, 1431–1448. [CrossRef] [PubMed]
33. FMedSci, H.M. Tuberculosis 2019 3 Insights and challenges in tuberculosis vaccine development. *Lancet Respir. Med.* **2019**, *7*, 810–819.
34. On 1st Oct 2018, The International Vaccine Initiative (IAVI) https://www.iavi.org/ and Aeras, a Nonprofit Organization Dedicated to Developing Tuberculosis (TB) Vaccines, Today Announced the Transfer to IAVI of Aeras' TB Vaccine Clinical Research Programs and Assets, Consisting of Certain Clinical Staff, Clinical Programs, Biorepository, Funding Commitments, and Other Assets. The Transaction, Effective Today, Will Enable the Continuity of Aeras' Core TB Vaccine Clinical Programs and Will Expand IAVI's Clinical Development Capabilities and Network, Incorporating an Experienced Clinical Team and South African Clinical Partner Network with a Strong Track Record in Later-Stage Clinical Trials and Work with Adolescent and Adult Populations. Available online: https://www.iavi.org/newsroom/press-releases/2018/iavi-acquires-aeras-tb-vaccine-clinical-programs-and-assets (accessed on 1 April 2020).
35. Kaufmann, S.H.E.; Dockrell, H.M.; Drager, N.; Ho, M.M.; Mcshane, H.; Neyrolles, O.; Ottenhoff, T.H.M.; Patel, B.; Roordink, D.; Spertini, F.; et al. TBVAC2020 Consortium TBVAC2020: Advancing Tuberculosis Vaccines from Discovery to Clinical Development. *Front. Immunol.* **2017**, *8*, 1203. [CrossRef]
36. The TuBerculosis Vaccine Initiative (TBVI) Is a Non-Profit Foundation that Facilitates the Discovery and Development of New, Safe and Effective TB Vaccines that Are Accessible and Affordable for All People. Available online: https://www.tbvi.eu/ (accessed on 1 April 2020).

37. The European & Developing Countries Clinical Trials Partnership (EDCTP) Funds Clinical Research to Accelerate the Development of New or Improved Drugs, Vaccines, Microbicides and Diagnostics against HIV/AIDS, Tuberculosis and Malaria as well as Other Poverty-Related Infectious Diseases in sub-Saharan Africa, with a Focus on Phase II and III Clinical Trials. Available online: https://www.edctp.org/ (accessed on 1 April 2020).
38. Darrah, P.A.; Zeppa, J.J.; Maiello, P.; Hackney, J.A.; Wadsworth, M.H.; Hughes, T.K.; Pokkali, S.; Swanson, P.A.; Grant, N.L.; Rodgers, M.A.; et al. Prevention of tuberculosis in macaques after intravenous BCG immunization. *Nature* **2019**, *577*, 95–102. [CrossRef]
39. Dijkman, K.; Sombroek, C.C.; Vervenne, R.A.W.; Hofman, S.O.; Boot, C.; Remarque, E.J.; Kocken, C.H.M.; Ottenhoff, T.H.M.; Kondova, I.; Khayum, M.A.; et al. Prevention of tuberculosis infection and disease by local BCG in repeatedly exposed rhesus macaques. *Nat. Med.* **2019**, *25*, 255–262. [CrossRef]
40. Sander, P.; Clark, S.; Petrera, A.; Vilaplana, C.; Meuli, M.; Selchow, P.; Zelmer, A.; Mohanan, D.; Andreu, N.; Rayner, E.; et al. Deletion of zmp1 improves *Mycobacterium bovis* BCG-mediated protection in a guinea pig model of tuberculosis. *Vaccine* **2015**, *33*, 1353–1359. [CrossRef]
41. Gröschel, M.I.; Sayes, F.; Shin, S.J.; Frigui, W.; Pawlik, A.; Orgeur, M.; Canetti, R.; Honoré, N.; Simeone, R.; van der Werf, T.S.; et al. Recombinant BCG Expressing ESX-1 of *Mycobacterium marinum* Combines Low Virulence with Cytosolic Immune Signaling and Improved TB Protection. *Cell Rep.* **2017**, *18*, 2752–2765. [CrossRef] [PubMed]
42. Kroesen, V.M.; Madacki, J.; Frigui, W.; Sayes, F.; Brosch, R. Mycobacterial virulence: Impact on immunogenicity and vaccine research. *F1000Res* **2019**, *8*, 2025. [CrossRef]
43. Tameris, M.D.; Hatherill, M.; Landry, B.S.; Scriba, T.J.; Snowden, M.A.; Lockhart, S.; Shea, J.E.; McClain, J.B.; Hussey, G.D.; Hanekom, W.A.; et al. Safety and efficacy of MVA85A, a new tuberculosis vaccine, in infants previously vaccinated with BCG: A randomised, placebo-controlled phase 2b trial. *Lancet* **2013**, *381*, 1021–1028. [CrossRef]
44. Tameris, M.; Mcshane, H.; McClain, J.B.; Landry, B.; Lockhart, S.; Luabeya, A.K.K.; Geldenhuys, H.; Shea, J.; Hussey, G.; van der Merwe, L.; et al. Lessons learnt from the first efficacy trial of a new infant tuberculosis vaccine since BCG. *Tuberculosis* **2013**, *93*, 143–149. [CrossRef]
45. Andrews, J.R.; Nemes, E.; Tameris, M.; Landry, B.S.; Mahomed, H.; McClain, J.B.; Fletcher, H.A.; Hanekom, W.A.; Wood, R.; Mcshane, H.; et al. Serial QuantiFERON testing and tuberculosis disease risk among young children: An observational cohort study. *Lancet Respir. Med.* **2017**, *5*, 282–290. [CrossRef]
46. Verver, S.; Warren, R.M.; Beyers, N.; Richardson, M.; van der Spuy, G.D.; Borgdorff, M.W.; Enarson, D.A.; Behr, M.A.; Van Helden, P.D. Rate of reinfection tuberculosis after successful treatment is higher than rate of new tuberculosis. *Am. J. Respir. Crit. Care Med.* **2005**, *171*, 1430–1435. [CrossRef]
47. Sable, S.B.; Posey, J.E.; Scriba, T.J. Tuberculosis Vaccine Development: Progress in Clinical Evaluation. *Clin. Microbiol. Rev.* **2019**, *33*, 16076. [CrossRef]
48. Scriba, T.J.; Kaufmann, S.H.E.; Henri Lambert, P.; Sanicas, M.; Martin, C.; Neyrolles, O. Vaccination Against Tuberculosis with Whole-Cell Mycobacterial Vaccines. *J. Infect. Dis.* **2016**, *214*, 659–664. [CrossRef]
49. Marinova, D.; Gonzalo-Asensio, J.; Aguilo, N.; Martin, C. MTBVAC from discovery to clinical trials in tuberculosis-endemic countries. *Expert Rev. Vaccines* **2017**, *16*, 565–576. [CrossRef]
50. TBVI is Continuously Working on the Development of New Vaccine Candidates. The Current TB Vaccine Pipeline (Last Update October 2019), Is as Follows. Available online: https://www.tbvi.eu/what-we-do/pipeline-of-vaccines/ (accessed on 1 April 2020).
51. Nemes, E.; Geldenhuys, H.; Rozot, V.; Rutkowski, K.T.; Ratangee, F.; Bilek, N.; Mabwe, S.; Makhethe, L.; Erasmus, M.; Toefy, A.; et al. C-040-404 Study Team Prevention of *M. tuberculosis* Infection with H4:IC31 Vaccine or BCG Revaccination. *N. Engl. J. Med.* **2018**, *379*, 138–149. [CrossRef]
52. Andrews, J.R.; Noubary, F.; Walensky, R.P.; Cerda, R.; Losina, E.; Horsburgh, C.R. Risk of progression to active tuberculosis following reinfection with *Mycobacterium tuberculosis*. *Clin. Infect. Dis.* **2012**, *54*, 784–791. [CrossRef]
53. Arbués, A.; Aguilo, J.I.; Gonzalo-Asensio, J.; Marinova, D.; Uranga, S.; Puentes, E.; Fernandez, C.; Parra, A.; Cardona, P.-J.; Vilaplana, C.; et al. Construction, characterization and preclinical evaluation of MTBVAC, the first live-attenuated *M. tuberculosis*-based vaccine to enter clinical trials. *Vaccine* **2013**, *31*, 4867–4873.
54. Aguilo, N. MTBVAC: Attenuating the Human Pathogen of Tuberculosis (TB) Toward a Promising Vaccine against the TB Epidemic. *Front. Immunol.* **2017**, *8*, 1803.

55. Aguilo, N.; Gonzalo-Asensio, J.; Alvarez-Arguedas, S.; Marinova, D.; Gomez, A.B.; Uranga, S.; Spallek, R.; Singh, M.; Audran, R.; Spertini, F.; et al. Reactogenicity to major tuberculosis antigens absent in BCG is linked to improved protection against *Mycobacterium tuberculosis*. *Nat. Commun.* **2017**, *8*, 16085. [CrossRef] [PubMed]
56. Spertini, F.; Audran, R.; Chakour, R.; Karoui, O.; Steiner-Monard, V.; Thierry, A.-C.; Mayor, C.E.; Rettby, N.; Jaton, K.; Vallotton, L.; et al. Safety of human immunisation with a live-attenuated *Mycobacterium tuberculosis* vaccine: A randomised, double-blind, controlled phase I trial. *Lancet Respir. Med.* **2015**, *3*, 953–962. [CrossRef]
57. Tameris, M.; Mearns, H.; Penn-Nicholson, A.; Gregg, Y.; Bilek, N.; Mabwe, S.; Geldenhuys, H.; Shenje, J.; Luabeya, A.K.-K.; Murillo, I.; et al. Live-attenuated *Mycobacterium tuberculosis* vaccine MTBVAC versus BCG in adults and neonates: A randomised controlled, double-blind dose-escalation trial. *Lancet Respir. Med.* **2019**, *7*, 757–770. [CrossRef]
58. Grode, L. Increased vaccine efficacy against tuberculosis of recombinant *Mycobacterium bovis* bacille Calmette-Guerin mutants that secrete listeriolysin. *J. Clin. Investig.* **2005**, *115*, 2472–2479. [CrossRef]
59. Grode, L.; Ganoza, C.A.; Brohm, C.; Weiner, J., 3rd; Eisele, B.; Kaufmann, S.H.E. Safety and immunogenicity of the recombinant BCG vaccine VPM1002 in a phase 1 open-label randomized clinical trial. *Vaccine* **2013**, *31*, 1340–1348. [CrossRef]
60. Loxton, A.G.; Knaul, J.K.; Grode, L.; Gutschmidt, A.; Meller, C.; Eisele, B.; Johnstone, H.; van der Spuy, G.; Maertzdorf, J.; Kaufmann, S.H.E.; et al. Safety and Immunogenicity of the Recombinant *Mycobacterium bovis* BCG Vaccine VPM1002 in HIV-Unexposed Newborn Infants in South Africa. *Clin. Vaccine Immunol.* **2017**, *24*, e00439-16. [CrossRef]
61. Frick, M. *TB VACCINES Pipeline Report 2019*; Treatment Action Group, TAG: New York, NY, USA, 2019; pp. 1–25.
62. Bourinbaiar, A.S.; Batbold, U.; Efremenko, Y.; Sanjagdorj, M.; Butov, D.; Damdinpurev, N.; Grinishina, E.; Mijiddorj, O.; Kovolev, M.; Baasanjav, K.; et al. Phase III, placebo-controlled, randomized, double-blind trial of tableted, therapeutic TB vaccine (V7) containing heat-killed *M. vaccae* administered daily for one month. *J. Clin. Tuberc. Other Mycobact. Dis.* **2020**, *18*, 100141. [CrossRef]
63. Ginsberg, A.M.; Ruhwald, M.; Mearns, H.; Mcshane, H. TB vaccines in clinical development. *Tuberculosis* **2016**, *99*, S16–S20. [CrossRef] [PubMed]
64. Vilaplana, C.; Montané, E.; Pinto, S.; Barriocanal, A.M.; Domenech, G.; Torres, F.; Cardona, P.-J.; Costa, J. Double-blind, randomized, placebo-controlled Phase I Clinical Trial of the therapeutical antituberculous vaccine RUTI®. *Vaccine* **2010**, *28*, 1106–1116. [CrossRef] [PubMed]
65. Nell, A.S.; D'lom, E.; Bouic, P.; Sabaté, M.; Bosser, R.; Picas, J.; Amat, M.; Churchyard, G.; Cardona, P.-J. Safety, Tolerability, and Immunogenicity of the Novel Antituberculous Vaccine RUTI: Randomized, Placebo-Controlled Phase II Clinical Trial in Patients with Latent Tuberculosis Infection. *PLoS ONE* **2014**, *9*, e89612. [CrossRef] [PubMed]
66. Smaill, F.; Jeyanathan, M.; Smieja, M.; Medina, M.F.; Thanthrige-Don, N.; Zganiacz, A.; Yin, C.; Heriazon, A.; Damjanovic, D.; Puri, L.; et al. A Human Type 5 Adenovirus-Based Tuberculosis Vaccine Induces Robust T Cell Responses in Humans Despite Preexisting Anti-Adenovirus Immunity. *Sci. Transl. Med.* **2013**, *5*, 205ra134. [CrossRef] [PubMed]
67. Coler, R.N.; Day, T.A.; Ellis, R.; Piazza, F.M.; Beckmann, A.M.; Vergara, J.; Rolf, T.; Lu, L.; Alter, G.; Hokey, D.; et al. TBVPX-113 Study Team The TLR-4 agonist adjuvant, GLA-SE, improves magnitude and quality of immune responses elicited by the ID93 tuberculosis vaccine: First-in-human trial. *NPJ Vaccines* **2018**, *3*, 34. [CrossRef]
68. Penn-Nicholson, A.; Tameris, M.; Smit, E.; Day, T.A.; Musvosvi, M.; Jayashankar, L.; Vergara, J.; Mabwe, S.; Bilek, N.; Geldenhuys, H.; et al. TBVPX-114 study team Safety and immunogenicity of the novel tuberculosis vaccine ID93 + GLA-SE in BCG-vaccinated healthy adults in South Africa: A randomised, double-blind, placebo-controlled phase 1 trial. *Lancet Respir. Med.* **2018**, *6*, 287–298. [CrossRef]
69. Van Der Meeren, O.; Hatherill, M.; Nduba, V.; Wilkinson, R.J.; Muyoyeta, M.; Van Brakel, E.; Ayles, H.M.; Henostroza, G.; Thienemann, F.; Scriba, T.J.; et al. Phase 2b Controlled Trial of M72/AS01E Vaccine to Prevent Tuberculosis. *N. Engl. J. Med.* **2018**, *379*, 1621–1634. [CrossRef]
70. Tait, D.R.; Hatherill, M.; Van Der Meeren, O.; Ginsberg, A.M.; Van Brakel, E.; Salaun, B.; Scriba, T.J.; Akite, E.J.; Ayles, H.M.; Bollaerts, A.; et al. Final Analysis of a Trial of M72/AS01E Vaccine to Prevent Tuberculosis. *N. Engl. J. Med.* **2019**, *381*, 2429–2439. [CrossRef]

71. Yates, T.A.; Khan, P.Y.; Knight, G.M.; Taylor, J.G.; McHugh, T.D.; Lipman, M.; White, R.G.; Cohen, T.; Cobelens, F.G.; Wood, R.; et al. The transmission of *Mycobacterium tuberculosis* in high burden settings. *Lancet Infect. Dis.* **2016**, *16*, 227–238. [CrossRef]
72. Tovar, M.; Arregui, S.; Marinova, D.; Martin, C.; Sanz, J.; Moreno, Y. Bridging the gap between efficacy trials and model-based impact evaluation for new tuberculosis vaccines. *Nat. Commun.* **2019**, *10*, 5457. [CrossRef]
73. Blaser, N.; Zahnd, C.; Hermans, S.; Salazar-Vizcaya, L.; Estill, J.; Morrow, C.; Egger, M.; Keiser, O.; Wood, R. Tuberculosis in Cape Town: An age-structured transmission model. *Epidemics* **2016**, *14*, 54–61. [CrossRef] [PubMed]
74. Kashangura, R.; Sena, E.S.; Young, T.; Garner, P. Effects of MVA85A vaccine on tuberculosis challenge in animals: Systematic review. *Int. J. Epidemiol.* **2015**, *44*, 1970–1981. [CrossRef] [PubMed]
75. Godlee, F. We need better animal research, better reported. *BMJ* **2018**, k124. [CrossRef]
76. Cohen, D. Oxford vaccine study highlights pick and mix approach to preclinical research. *BMJ* **2018**. [CrossRef]
77. Macleod, M. Learning lessons from MVA85A, a failed booster vaccine for BCG. *BMJ* **2018**. [CrossRef] [PubMed]
78. Darrah, P.A.; DiFazio, R.M.; Maiello, P.; Gideon, H.P.; Myers, A.J.; Rodgers, M.A.; Hackney, J.A.; Lindenstrøm, T.; Evans, T.; Scanga, C.A.; et al. Boosting BCG with proteins or rAd5 does not enhance protection against tuberculosis in rhesus macaques. *NPJ Vaccines* **2019**, *4*, 21. [CrossRef]
79. Barreto, M.L.; Pilger, D.; Pereira, S.M.; Genser, B.; Cruz, A.A.; Cunha, S.S.; Sant'Anna, C.; Hijjar, M.A.; Ichihara, M.Y.; Rodrigues, L.C. Causes of variation in BCG vaccine efficacy: Examining evidence from the BCG REVAC cluster randomized trial to explore the masking and the blocking hypotheses. *Vaccine* **2014**, *32*, 3759–3764. [CrossRef]
80. Arregui, S.; Sanz, J.; Marinova, D.; Martin, C.; Moreno, Y. On the impact of masking and blocking hypotheses for measuring the efficacy of new tuberculosis vaccines. *PeerJ* **2016**, *4*, e1513. [CrossRef]

© 2020 by the authors. Licensee MDPI, Basel, Switzerland. This article is an open access article distributed under the terms and conditions of the Creative Commons Attribution (CC BY) license (http://creativecommons.org/licenses/by/4.0/).

MDPI
St. Alban-Anlage 66
4052 Basel
Switzerland
Tel. +41 61 683 77 34
Fax +41 61 302 89 18
www.mdpi.com

Applied Sciences Editorial Office
E-mail: applsci@mdpi.com
www.mdpi.com/journal/applsci

Lightning Source UK Ltd.
Milton Keynes UK
UKHW050801011220
374381UK00003B/330

9 783039 432363